A text book for Engineering & A.M.I.E Students
(With over 150 Solved Examples)

Engineering Heat Transfer

C. P. Gupta
M.Sc., B.E. (Mech.), Ph.D.(Wis.)
Professor (Emeritus)

Rajendra Prakash
B.Sc., B.E. (Mech.), M.S. (Mich St.), Ph.D.
Professor (Emeritus)

Mechanical & Industrial Engineering Department
Indian Institute of Technology
Roorkee

Eighth edition 2015

Nem Chand & Bros, Roorkee, U.A., India

First	Edition	1968
Second	Edition	1969
Third	Edition	1971
Fourth	Edition	1976
Reprinted		1979
Fifth	Edition	1988
Sixth	Edition	1994
Seventh	Edition	1999
Reprint		2005
Reprint		2009
Eighth	Edition	2015

ISBN 978- 81-85240-81-7

Published by
Nem Chand & Bros., Civil Lines, Roorkee 247 667, India
Website: www.nemchandbros.com, e-mail : ncb_rke@rediffmail.com
Photo set and Printed at the
Roorkee Press, Mahavir Marg, Roorkee 247 667

Preface

To the First Edition of 'Engineering Thermodynamics & Heat Transfer'

The book is intended to serve as a text book for courses in *Thermodynamics and Heat Transfer* taught to engineering degree students. Although meant primarily for the use of Mechanical Engineering students, it will be equally useful as a text book for students of Civil, Electrical, Metallurgical, Chemical and Electronics Engineering.

In our country, until about ten years ago, Thermodynamics and Heat Transfer were not taught as such. In place of Thermodynamics, Heat Engines or Prime Movers were included in the syllabus, thereby placing emphasis on the use of the laws and equations of Thermodynamics rather than the understanding of the meaning of these laws and the derivation of these equations. Heat Transfer was not taught at all. In recent years, the syllabi of a majority of Technical Institutions have been revised to incorporate the understanding of basic principles of Thermodynamics and Heat Transfer by including these subjects as separate courses and thus giving due importance to them. However, no text book written by an Indian author is available on these subjects which could cater for the needs arising out of such revision of curriculum. The present volume is meant to fill this gap.

Throughout the text, attempt has been made to present the subject matter in a simple, lucid and precise manner. Stress has been laid in making the student understand the physical picture. For example equations based on energy or mass balances have been developed after writing out these balances in words, so that the exact meaning of every symbol is clearly understood. Numerous solved examples are given to illustrate the use of the equations developed in the text. A large amount of reference material and properly data is included in the appendices to make the test self-sufficient.

It is regretted, however, that despite careful scrutiny of the proofs, several printing mistakes have nevertheless remained in the book. If such mistakes are pointed out by readers, the same will be gratefully received.

Authors have drawn heavily from the existing literature on the subjects, for which they gratefully acknowledge. They also thank Sri S.C. Sharma of Earthquake Engineering School for his help in preparing the drawings for the figures and the Publishers for their kind cooperation during the printing of the book.

Roorkee
June 20, 1968

C.P.Gupta
R. Prakash

To the Eighth Edition

The book has been revised and updated with additional information introduced within the text. The additional 35 new solved examples have been incorporated in the corresponding Chapters. Further, the pages have been composed with the aid of a Computer. It has made the final setting of pages look more beautiful and overall getup of the book immensely improved.

It is hoped that this attractive edition of the book will be liked by the readers.

Roorkee
June 20, 2007

C.P.Gupta
R. Prakash

Contents

3. Heat Transfer by Conduction (Un-Steady-State)

4. Heat Transfer by Convection

5. Heat Transfer with Change of Phase

6. Thermal Radiation

7. Heat Exchangers

8. Solar Thermal Collectors

Appendices

Index

CHAPTER
1

Fundamentals of Heat Transfer

1.1 INTRODUCTION

During the study of thermodynamics it is found necessary to distinguish between the thermal energy in storage and that in transit from one system to another. The former is referred to as *internal energy* while the latter as *heat*. Thermal energy can originate from any kind of energy according to the first law of thermodynamics which governs all energy transformations quantitatively, without any restriction on the direction. Transfer of heat is due to unique property of matter, temperature, and is governed by the second law of thermodynamics which dictates that free flow of heat is possible only from a body of higher temperature to that at a lower temperature.

All heat transfer processes, therefore, involve the transfer of energy, and obey the first as well as the second law of thermodynamics. It may initially appear that the principles of heat transfer are obtainable from the basic laws of thermodynamics. It is however not true. The science of thermodynamics deals with equilibrium. Since the heat flow will occur due to temperature non-equilibrium, its quantitative analysis could be derived on the basis of some other branch of science designated as *Heat Transfer*. The subject of heat transfer is, in reality, firmly grounded in the sciences of *thermodynamics* and *fluid mechanics*.

1.2 ENGINEERING HEAT TRANSFER

Whenever a temperature gradient exists within a system, or when two systems at different temperatures are brought into contact or interaction, thermal energy is transferred. This energy in transit can not be measured or observed directly, but the effects it produces can be observed and measured. Engineering heat transfer therefore consists of the study of thermal energy movement by the measurement of such effects.

It has become exceedingly important for engineers to possess a clear understanding of the principles of heat transfer and its application to a number of problems. With the growth of human population and its standard of living, the need of energy has increased manifold. Strangely enough, energy is abundantly available in nature. We have, however, to discover its source, release it and then control its release. After the release the heat energy is required to be transported to the medium where it will be put to meaningful use. From an engineering viewpoint, the determination of the *rate of heat transfer* is the item which needs special consideration.

The design of boilers, refrigerators or other types of heat exchangers say, for the storage, transfer and use of cryogenic substances and for a number of similar applications involve consideration of heat transfer rates. These are only a few examples of the applications of engineering heat transfer in day to day practice.

Heat exchanger design will not only depend upon the amount of heat to be transmitted, but rather on the rate of transfer under the specified conditions. An engineer will have to make a detailed heat transfer analysis in order to finalize the design or size of the equipment necessary to transfer a specified amount of heat in a given time.

It is not only in the field of utilization of energy where we are involved in heat transfer studies. There may be other situations where heat is being generated during a certain operation, and may be detrimental to the equipment. In such cases we are often faced with the task of removing the undesirable amount of heat as far as possible. For instance, gas turbine blades, walls of an I.C. engine combustion chamber, outer surface of a space vehicle, all depend for their durability on rapid heat removal from their surface.

1.3 ANALOGIES BETWEEN VARIOUS TRANSPORT PROCESSES

Lately, it has been found interesting to make an analogy between various *Transfer Processes* — the transport of *energy, mass and momentum*. The transfer of any of these entities is due to the existence of a certain driving force or *potential gradient*. Further, the transfer occurs according to a certain fundamental law. It is observed that the fundamental relations for these transfer processes are similar in nature; and under specified conditions the results for one type of transfer are adaptable to the other types of phenomena. The transfer analogy shown in Table 1.1 illustrates the similarity of the processes.

Table 1.1 Analogy Between Transport Process

Transferred Entity	Flux	Driving Force or Potential Gradient	Law	Equation	Diffusivity (cm^2/sec)	Transport Property
Energy (Heat)	$q = \frac{Q}{A}$ (J/cm^2 sec)	Temperature gradient (°C/cm)	Fourier's Law (ρ, C_p const)	$\frac{Q}{A} = -k\frac{\partial T}{\partial y} = -\alpha\frac{\partial}{\partial y}(\rho C_p T)$	$\alpha = \frac{k}{\rho C_p}$	Conductivity k (J/cm sec C)
Mass	$i = \frac{m}{A}$ (g/cm^2 sec)	Concentration gradient $\left(\frac{g/cm^3}{cm}\right)$	Fick's Law	$\frac{m}{A} = -D_v\frac{\partial(\rho)}{\partial y}$	D_v	Diffusion constant D (cm^2/sec)
Momentum	$\tau_{xy} = \frac{F}{A}$ (N/cm^2)	Velocity gradient $\left(\frac{cm/sec}{cm}\right)$	Newton's Law (ρ Const.)	$\tau_{xy} = -\mu\frac{\partial(V)}{\partial y} = -\nu\frac{\partial(\rho V)}{\partial y}$	$\nu = \frac{\mu}{\rho}$	Viscosity μ (g/cm sec)

These relationships respectively state that (a) heat transfer occurs due to a *temperature gradient,* (b) mass transfer occurs due to gradient in *mass concentration* and (c) momentum transfer occurs due to a *velocity gradient*. It may, however, be pointed out that these analogies do not apply in two and three- dimensional cases as shear stress, τ_{xy} is a tensor quantity with nine components while heat and mass flow, q and j are vectors with three components only.

1.4 MODES OF HEAT TRANSFER

An instructor, giving a test to the students in a class room, may cause the question paper to reach the farther-most student in a row by any of the three possible methods, explained below :

(a) He could deliver the paper to the student nearest to him who delivers the same to the next and so on till it finally reaches the last student,

(b) He could also carry the paper himself to the other end of the row.

(c) As a last alternative, he could throw the question paper so that it reaches the last student.

The transport of heat energy from one region to another occurs by any (or a combination) of similar methods. In literature such three methods of heat transmission are recognized by the terms *conduction, convection* and *radiation* respectively. In the above example, the question paper corresponds to heat energy while the students correspond to the molecules of the substance.

If the flow of heat is a result of transfer of internal energy from one molecule to other, the process is called *conduction.* Through solids, this is the only possible mode of heat transmission.

In liquid and gases, however, the molecules are no longer confined to a certain point but constantly change their positions even if the substance is at rest. The heat energy is transported along with the motion of these molecules from one region to another. This process is called *convection.*

All solid bodies as well as liquids and gases have a tendency of radiating thermal energy in the form of electromagnetic waves and of absorbing similar energy emerging from the neighboring bodies. This type of heat transport is known as *thermal radiation.*

Judging from the nature of these it may appear that heat *convection* is different in behaviour as compared to *conduction* and *radiation* as far as the strict compliance of definition of heat transfer is concerned. The processes of conduction and radiation depend for their operation on the fact of temperatures being different while the process of convection needs the assistance of mechanical mass transport also and is, therefore, governed by fluid motion as well.

On the contrary, it will be observed later that the basic heat transfer equations are similar for the cases of conduction and convection in as much as the flux is *linearly* related to temperature difference while that for radiation it is different from the other two. Here the heat flux is related to the difference in the *fourth power* of the absolute temperatures.

In engineering or industrial processes, heat transfer may occur due to one or due to a combination of more than one of the three modes of transport. An analysis of the problem has therefore to be made carefully, before applying the relevant basic laws and equations.

1.5 VARIATION IN TRANSFER OF HEAT DUE TO TIME

The knowledge of only the mode of heat transfer is not enough for the solution of heat transfer problems but in addition the designer should know whether there are variations in heat transfer rates or in temperatures with *time*. The following discussion describes the various types of the heat flow processes.

When the rate of heat flow is constant i.e. it does not change with time, resulting in a steady state condition, the process is called a *steady* heat transfer process. The internal energy of the system does not change in such a process. Heat transfer to water in boiler tubes from the hot gases, and from a hot fluid to a cold fluid in a heat exchanger are typical examples of such processes.

When the temperature at any point in a system changes with time, the process is called *unsteady* or *transient* heat transfer process. The internal energy of the system changes in such a process. Typical examples of unsteady processes are the cooling of castings in a foundry or that of a hot steel job on a blacksmith's anvil.

In case, however, the variation in temperature of an unsteady state system undergoes a particular cycle, the process is called a *periodic* or *quasi-steady* state heat transfer process. In such problems the temperatures at each point return periodically to the same value. Similarly the rate of heat flow and that of energy storage undergoes periodic variation. Examples of such a system are the heating or cooling of a building wall during the 24 hour cycle of the day or that of the walls of an I. C. engine.

In the succeeding sections of this chapter we will review the basic laws governing the modes of heat transfer, with a view to obtaining a bird's eye view of the field without getting involved into details.

1.6 FOURIER'S LAW

Although the fundamental law of heat conduction was proposed by Biot in 1804, it could not become popular until 1822, when Fourier confirmed the same by his outstanding contribution in the field.

He proposed that the fundamental conditions for heat transfer by *conduction* within a solid are (a) a temperature gradient exists and (b) heat flows in the direction of decreasing temperature.

If one-dimensional heat conduction is considered across a solid wall as shown in Fig.1.1, Fourier's law states that the rate of heat transfer q is the product of :

(a) Area A, normal to heat flow path,

(b) $\partial t/\partial x$, the *temperature gradient* at the section i.e. the rate of change of temperature with reference to the distance in the direction of heat flow.

(c) k, a property of the conducting material known as *thermal conductivity*.

The above statement expressed mathematically results in the basic *heat conduction equation* :

$$q = \frac{dQ}{d\tau} = -kA\frac{\partial t}{\partial x} \tag{1.1}$$

where dQ = quantity of heat conducted in time $d\tau$.

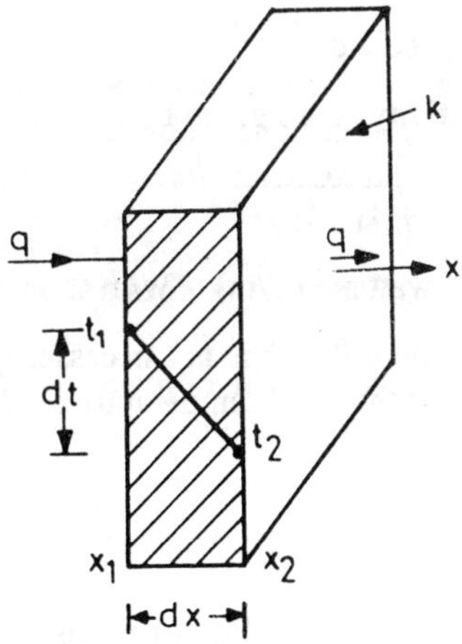

Fig. 1.1 Conduction of Heat

It may be noted that since the heat flow always occurs in the direction of decreasing temperature, the temperature gradient ($\partial t/\partial x$) will be negative, hence the negative sign appears in Eq. (1.1). Also note that the following assumptions are implied in this equation :

(a) The heat flow in y or z direction is zero.

(b) The temperature in any plane perpendicular to the x-axis is uniform throughout the plane.

Eq. (1.1) may be integrated for the case of the steady state heat conduction through a uniform slab as shown in Fig. 1.1 by method of separation of variables :

$$q\,dx = -kA\,dt$$

or
$$q\int_{x_1}^{x_2} dx = -kA\int_{t_1}^{t_2} dt$$

or
$$q(x_2 - x_1) = -kA(t_2 - t_1)$$

or
$$q = kA\frac{t_1 - t_2}{x_2 - x_1}$$

$$q = (k/l)A(t_1 - t_2)$$
$$= -(k/l)A\,\Delta t \qquad (1.2)$$

where $l = (x_2 - x_1)$ = thickness of the slab

Adopting a consistent set of engineering units as follows,

$$A = (\text{meter})^2$$

$$l = \text{meter}$$

$$t = {}^\circ\text{C}$$

$$q = \text{W or J/s}$$

$$\text{and } \tau = \text{second}$$

The units of thermal conductivity, *k* are given by :

$$k = \frac{q}{A(t_1 - t_2)/l}$$

$$= \text{watt/m}^2 \text{ (C/m)} = \text{W/m C}$$

It should also be mentioned here that *k* is not necessarily uniform (a constant) for a material but may be a function of the local temperature *t*. We shall discuss this in detail in Chapter 2.

1.7 NEWTON'S LAW

In 1701, Newton while studying the cooling of fluids proposed that the heat transfer between a surface and a fluid takes place by some phenomenon other than pure conduction and the *convection* process is more prominent than the conduction process. He proposed the fundamental *heat convection equation* :

$$q = h A \Delta t \tag{1.3}$$

where q = average rate of heat transfer by convection, (watt) or (J/s)

A = heat transfer area, (m^2)

Δt = difference between the temperature of the surface and that of the fluid at some specified location, (°C).

h = an average value of the proportionality constant called *convective heat transfer coefficient*, (watt/m^{2}C)

It is pointed out that this relationship gives only the *definition* of *h* rather than any phenomenological interpretation of the process of convection. The study of convection deals primarily with the evaluation of the convective heat transfer coefficient, *h*.

The determination of h also involves the knowledge of *fluid mechanics* and its value depends on many factors including the geometry of surface, the velocity and the physical properties of the fluid and also the temperature difference Δt.

1.8 STEFAN BOLTZMANN LAW

The heat exchange by radiation was studied experimentally by Stefan, who in 1879 proposed that the energy radiated by a body is proportional to the fourth power of its absolute temperature. It was confirmed by Boltzmann in 1884 by theoretical studies resulting in the *Stefan Boltzmann law* which may be stated as :

$$q = \sigma A T^4 \tag{1.4}$$

where q = maximum rate of heat emission by a body at the given temperature (black body emission), (watt)

A = area of the body, (m^2)

T = absolute temperature, (K)

σ = a constant known as *Stefan-Boltzmann constant*

= 17.14×10^{-10} btu/hr ft^2 R^4

= 487.6×10^{-10} kcal/hr m^2 K^4

= 5.6697×10^{-8} W/m^2 K^4

Net rate of heat radiation from a black body of area A_1 at absolute temperature T_1 to another black body at a lower temperature T_2 will therefore be given by :

$$q = \sigma A_1 (T_1^4 - T_2^4)$$

For the case of real bodies this equation becomes,

$$q = \sigma A_1 F_{12} (T_1^4 - T_2^4) \tag{1.5}$$

where q = net rate of heat exchange by radiation

F_{12} = a shape factor, depending upon the geometry, emissivities and relative location of the two surfaces.

1.9 THERMAL CONDUCTANCE AND RESISTANCE

A review of Eq. (1.2) will indicate that if k is independent of temperature, the rate of heat flow is given by ;

$$q = \left(\frac{k}{l}\right) A\,(t_1 - t_2) = K\,(t_1 - t_2) \tag{1.6}$$

$$\text{where} \quad K = \frac{kA}{l}, \text{(W/C)} \tag{1.7}$$

K is known as *internal thermal conductance* and is equal to the rate of heat conducted through a particular solid of area A and thickness *l*, per degree temperature difference. The term k/*l* is known as the *internal conductance* per unit area or *unit thermal conductance.*

The inverse of conductance is known as the *internal thermal resistance* R, and is given by ;

$$q = \left(\frac{k}{l}\right) A\,(t_1 - t_2) = \frac{t_1 - t_2}{R} \tag{1.8}$$

where $R = l/kA$, (C/W)

Example 1.1

The outer surface of a 0.2 m thick concrete wall (10 m × 3 m) is kept at a temperature of 5°C while the inner surface is kept at 30 °C. Thermal conductivity of the concrete is 1.2 W/mK. Determine the thermal resistance of the wall and rate of heat loss through it.

Solution :

$$R = l/kA \tag{1.8}$$

$$= \frac{0.2}{1.2 \times (10 \times 3)} = 0.00556 \text{ C/W} \qquad \text{Ans.}$$

$$q = -kA\frac{dt}{dx} = kA\frac{t_i - t_o}{l} \qquad (1.2)$$

$$= \frac{t_i - t_o}{R} = \frac{30-5}{0.00556} = 4500 \text{ W} \qquad \text{Ans.}$$

Example 1.2

The outer surface temperature of a roof is 37°C and that of the ambient air is 7°C. Calculate the rate of heat exchange between ambient air and 200 m^2 of outer roof area if the value of convective heat transfer coefficient is 10 W/m^{2}K.

Solution :

$$q = hA\Delta t \qquad (1.3)$$

$$= 10 \times (200) \times (37 - 7)$$

$$= 60{,}000 \text{ W}$$

$$= 60 \text{ kW} \qquad \text{Ans.}$$

Example 1.3

A 2 cm dia rod with a surface emissivity (ε) of 0.9 is maintained at 727°C. It is placed in a vacuum chamber whose walls are maintained at 527°C. Calculate the rate of heat loss from the rod per meter length.

Solution :

Since the chamber is totally enclosed and is maintained at vacuum, heat transfer is assumed to be purely due to radiation.

Further, let $F_{12} = \varepsilon_1$

$$q = \sigma A_1 \varepsilon_1 (T_1^4 - T_2^4) \qquad (1.5)$$

$$= (5.6697 \times 10^{-8}) \times \left(\frac{\pi \times 2 \times 1}{100}\right) \times (0.9) \times (1000^4 - 800^4)$$

$$= 1895 \text{ W} \qquad \text{Ans.}$$

1.10 CONVECTIVE AND RADIATIVE CONDUCTANCE

The term *K*, the thermal conductance does not possess any physical significance for the case of convective and radiative heat transfer. It is, however, observed that in many engineering problems involving these modes of heat transfer, the usage of this term simplifies their solution. The values of *K* the conductance and *R* the resistance for convective (K_c and R_c) and radiative (K_r and R_r) heat transfer processes respectively, are obtained as below :

(i) Convection Process

$$q = hA(t_1 - t_2) \qquad (1.3)$$

$$= K_c (t_1 - t_2)$$

where $K_c = h\,A$, (W/C) (1.9)

and therefore $R_c = \dfrac{1}{hA}$, (C/W) (1.10)

(ii) Radiation Process

$$q = \sigma A_1 F_{12} (T_1^4 - T_2^4) \quad (1.5)$$
$$= K_r (t_1 - t_2)$$

where $K_r = \dfrac{\sigma A_1 F_{12} (T_1^4 - T_2^4)}{t_1 - t_2}$, (W/C) (1.11)

and therefore $R_r = \dfrac{t_1 - t_2}{\sigma A_1 F_{12} (T_1^4 - T_2^4)}$, (C/W) (1.12)

1.11 COMBINED HEAT TRANSFER PROCESS

In the preceeding sections, the three mechanisms of heat transfer have been discussed separately. In most of the engineering applications, however, heat is transferred in successive steps by similar or different mechanisms. In a particular location the transfer may occur by one mechanism or by more than one mechanism working in parallel.

For instance, let us consider the case of the heating of water in a tube laid in the combustion chamber of a boiler. The water will receive heat from the products of combustion which emit and absorb radiation. The heat will flow by combination of different modes through successive steps as indicated below (Fig.1.2).

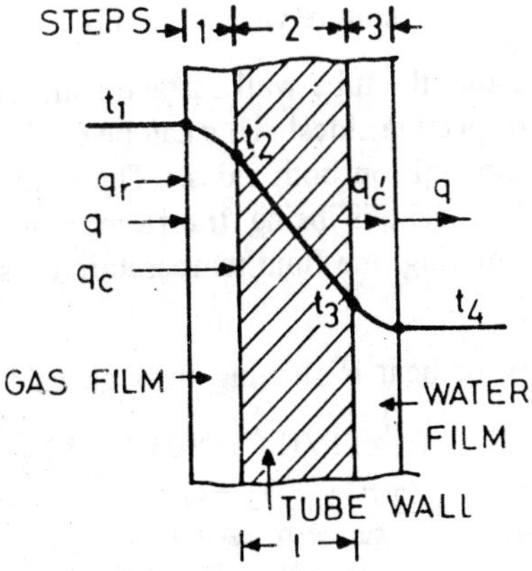

Fig. 1.2 Combined Heat Transfer Process

Step 1 : Transfer of heat from hot gas to the outer surface of area A of the tube by convection and radiation working in parallel.

$$q = q_c + q_r$$
$$= h\,A\,(t_1 - t_2) + h_r\,A\,(t_1 - t_2) \quad (1.13)$$
$$= (K_c + K_r)(t_1 - t_2)$$

$$= \frac{t_1 - t_2}{R_1}$$

where R_1 = resistance to heat flow through step 1.

$$= \frac{1}{K_c + K_r}$$

and t_1, t_2 = temperature of hot gas and that of outside wall respectively.

Step 2 : Conduction of heat through the metallic wall of tube

$$q = \frac{kA}{l}(t_2 - t_3) \tag{1.14}$$

$$= K(t_2 - t_3)$$

$$= \frac{t_2 - t_3}{R_2}$$

where R_2 = resistance to heat flow through step 2.

t_3 = inner wall temperature

Step 3 : Finally, the heat will pass to the water from the inner wall by convection.

$$q = q_c' = h' A (t_3 - t_4) \tag{1.15}$$

$$= \frac{t_3 - t_4}{R_3}$$

where R_3 = resistance to heat flow through step 3.

t_4 = water temperature within the tube.

It may be clarified here that the tube wall surface temperatures are different than the fluid temperatures on the respective sides. This can be explained by assuming that a *thin layer* of fluid adheres to the wall on both sides. The temperature gradient exists only within this thin layer. The heat is being transferred through this layer mainly by convection. As a result of mixing, the fluid temperature outside this layer is assumed to be uniform.

Since the same quantity of heat is flowing through each step we obtain from Eqs. (1.13), (1.14) and (1.15) ;

$$q = \frac{t_1 - t_2}{R_1} = \frac{t_2 - t_3}{R_2} = \frac{t_3 - t_4}{R_3} \tag{1.16}$$

$$= \frac{(t_1 - t_2) + (t_2 - t_3) + (t_3 - t_4)}{R_1 + R_2 + R_3}$$

$$q = \frac{t_1 - t_4}{R_1 + R_2 + R_3} = \frac{t_1 - t_4}{\Sigma R} \tag{1.17}$$

This indicates that the total resistance to heat flow from temperature t_1 to temperature t_4 is the summation of all the individual *resistances* of different layers.

If we designate the reciprocal of the resistance by *UA*, then;

$$UA = \frac{1}{R_1 + R_2 + R_3}$$

or, $$q = UA\,(t_1 - t_4) \tag{1.18}$$

This equation represents the heat flow from the hot gas to water. Here U is known as the *overall unit conductance* or the *overall coefficient of heat transfer*.

It may be re-emphasized here that the concept of resistance or conductance will be found to be of great use in evaluating the heat transfer in cases similar to the one just discussed.

Example 1.4

The combustion chamber of a boiler has its wall construction as shown in Fig. 1.3. The temperature of hot gas is 760°C, and that of the outside air is 25°C. The inside and outside film conductances are 25 and 5 W/m^{2}C respectively. The values of ε for inside and outside walls are 0.90 and 0.93 respectively.

Calculate the overall heat transfer coefficient, assuming the value of k (W/m C) for fire brick as 0.12, for insulating brick as 0.15 and for steel plate as 50.

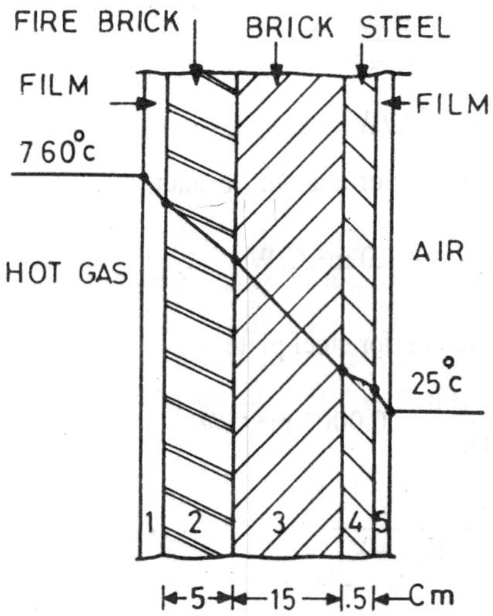

Fig. 1.3 Example 1.4

Solution :

The hot gas at 760°C will heat the inner surface of the fire brick by convection and radiation. Considering unit area of wall, the overall heat transfer coefficient will be given by :

$$UA = \frac{1}{R_1 + R_2 + R_3 + R_4 + R_5}$$

(i) $$R_1 = \frac{1}{K_{c1} + K_{r1}}$$

Now $$K_{c1} = 25 \text{ W/m}^2 \text{ K.}$$

$$K_{r1} = \frac{\sigma F_{12} A (T_1^4 - T_2^4)}{t_1 - t_2} \quad (1.11)$$

Let us assume, in the first instance, that the inner wall surface temperature is 758°C (1031 K).

so $$K_{r1} = \frac{(5.6697 \times 10^{-8}) \times 0.90 \times 1 \times (1033^4 - 1031^4)}{(1033 - 1031)}$$

(assume $F_{12} = \varepsilon_1$)

$$= 213 \text{ W/m}^2\text{C.}$$

Hence, $$R_1 = \frac{1}{25 + 213}$$

$$= 0.0042 \text{ m}^2 \text{ C/W}$$

so $$q = (1033 - 1031)/0.0042 = 476 \text{ W/m}^2$$

(ii) R_2 = Resistance for fire bricks

$$= \frac{l}{kA} = \frac{5/100}{0.12 \times 1}$$

$$= 0.416 \text{ m}^2 \text{ C/W}$$

(iii) R_3 = Resistance for masonry bricks

$$= \frac{15/100}{0.15 \times 1} = 1.0 \text{ m}^2 \text{ C/W}$$

(iv) R_4 = Resistance for steel plate

$$= \frac{0.5/100}{50 \times 1} = 0.0001 \text{ m}^2 \text{ C/W}$$

(v) $$R_5 = \frac{1}{K_5} = \frac{1}{K_{c_o} + K_{r_o}}$$

Now $$K_{co} = 5 \text{ W/m}^2 \text{ C}$$

and $$K_{ro} = \frac{\sigma F_{12} A (T_5^4 - T_6^4)}{t_5 - t_6}$$

In order to find value of t_5, we use Eq. (1.17) ;

$$t_1 - t_5 = q (R_1 + R_2 + R_3 + R_4)$$

$$= 476 (0.0042 + 0.416 + 1.0 + 0.0001)$$

$$= 676$$

or $t_5 = 760 - 676$

$= 84$ C

thus $$K_{ro} = \frac{(5.6697 \times 10^{-8}) \times 0.93 \times 1(357^4 - 298^4)}{84 - 25}$$

$= 7.45$ W/m^{2}C

Hence $$R_5 = \frac{1}{5 + 7.45} = 0.0804$$

Therefore the total resistance of the wall system ;

$$R = 0.0042 + 0.416 + 1.0 + 0.0001 + 0.0804$$

$$= 1.5007 = 1/UA$$

Hence $$U = \frac{1}{1.5007 \times 1}$$

$= 0.66$ W/m^{2}C Ans.

As a check for the assumption $t_2 = 758$°C, the heat flow is now calculated as

$$q = UA\,\Delta t = 0.66 \times 1 \times (760 - 25)$$

$$= 485.1 \text{ W/m}^2$$

This value is reasonable as compared to the value of 476 W/m^2 calculated in para (i) above. It is interesting to note that out of total heat flow of 476 W, only 25 × 1 (760 – 758) = 50 W are flowing by convection at the inner surface ,while 213 × 1 (760 – 758) = 426 W are being transmitted due to radiation i.e. 89 % of the energy is being transmitted by radiation at the inner surface. Similarly it will be observed that about 45 % of heat is dissipated to the surrounding air by radiation at the outer surface.

1.12 ANALOGY BETWEEN FLOW OF HEAT AND ELECTRICITY

It may be noted that the name thermal resistance has been given in much the same way as that of electrical resistance offered to the flow of current across potential difference.

Basic equation for the flow of current in a conductor is given by *Ohm's law*,

$$i = -\frac{\Delta V}{R_e} \tag{1.19}$$

where i = current, (amp)

ΔV = difference in voltage, (volts)

R_e = electrical resistance, (ohm)

The equation for the flow of heat is ;

$$q = \frac{dQ}{d\tau} = -(k/l)\,A\,\Delta t = -\frac{\Delta t}{l/Ak} \tag{1.2}$$

A comparison of equations (1.2) and (1.19) will indicate that they are similar. Therefore a heat flow system can be considered to be analogous to an electrical flow system. Further, the equation describing the behaviour of one system can be transformed to describe the behaviour of the other system by changing the variables suitably.

The electrical and thermal quantities which are analogous to each other may be listed as in Table 1.2.

Table 1.2 : Analogous Electrical and Thermal Quantities

Electrical			Thermal		
Voltage	:	V (volts)	Temperature	:	(°C)
Resistance	:	R_e (ohm)	Resistance	:	l/Ak (°C/W)
Current	:	i (amp)	Heat flow rate	:	q (W)
Charge	:	Q (Coulomb)	Heat flow	:	Q (J)

The analogy just discussed has been found useful in the solution of many problems of particular engineering nature involving steady and unsteady heat flow. For instance a *thermal circuit* of an actual problem can be transformed into a corresponding *electrical circuit*. The purpose of doing so is to make an experimental setup which could measure those quantities which are easily measurable. Evidently the electrical quantities are easier to measure as compared to the heat flow quantities. Thus any heat flow problem can be analysed experimentally by making an analogous electrical circuit, measuring the electrical quantities involved, and then transferring the data back to heat quantities for obtaining the final result. This is done with the help of an *analogue computer*.

Example 1.5

Draw an analogous electrical circuit for the case given in Example 1.4 and determine the value of electrical resistances in ohms if a 12 volt D.C. battery is to be used.

Solution

Recalling the data of Example 1.4, we have :

heat flow rate = 476 W

Let heat flow rate of 100 W correspond to 1 amp, then the current flow in the circuit

i = 4.76 amp

Voltage V = 12 Volts

hence resistance of the circuit,

$$R_e = \frac{12}{4.76} = 2.54 \text{ ohm}$$

For the two circuits to be analogous, the ratio of each electrical resistance to the overall resistance should be same as corresponding ratio in the thermal circuit.

Now, $R_2 = 0.416 \text{ m}^2 \text{ C/W}$

so
$$R_{e2} = \frac{0.416}{1.5007} \times 2.54$$
$$= 0.70 \text{ ohm.}$$

Similarly all the values of electrical resistances of the circuit shown in Fig. 1.4 may be calculated.

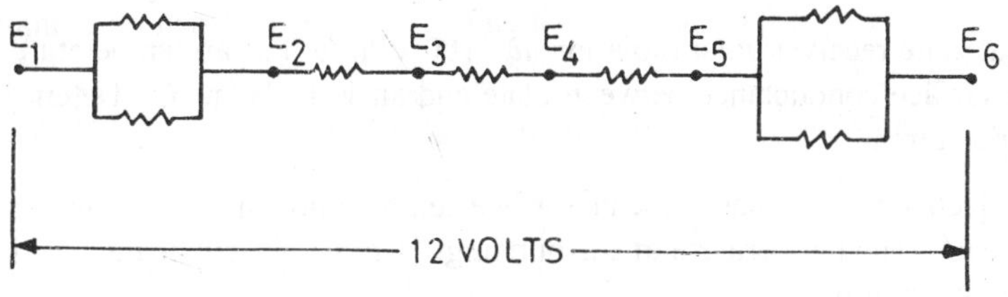

Fig. 1.4 Example 1.5

The values obtained by this method will have the following equivalence :

		Electrical	*Thermal*
(i)	Potential difference	12 V	735°C
(ii)	Quantity flow rate	4.76 amp	476 W
(iii)	Resistance	2.54 ohm	1.5007 m^2 C/W

PROBLEMS

(Note : 1 Nm = 1 W/s = 1 J)

1.1 Differentiate between conductivity and conductance. What are their units ?

1.2 Fill up the blanks ;

(a) 1 Btu/hr ft F = — W/m°C

(b) 1 Btu/hr ft F = — h.p./ft^2F/in

(c) 1 kcal/hr m C = — W/m K

(d) 1 kcal/hr m^2 C = — W/m^2K

1.3 Determine the conductivity of a test panel 15 cm by 16 cm and 1.25 cm thick, if during a two hour test 8.35 kJ of heat is conducted through the panel having temperatures of 35 C and 25 C at the two faces. [Ans : 0.645 W/m C]

1.4 Find the rate of heat transfer of a homogeneous plane wall, given the following : Temperature of outer and inner surface 100 C and 20 C respectively ; thermal conductivity 1.75 W/m K and wall thickness 20 cm. [Ans : 700 W/m^2]

1.5 Determine the rate of heat flow per unit area of a furnace wall and the exterior surface temperature for the following conditions :

(i) h_i = 55 W/m^2 C

(ii) Heat flow by radiation to interior wall from hot gas at 2000 C, Q_r = 58,250 W/m^2

(iii) Interior surface temperature 850 C.

(v) Thermal conductance of wall 232 W/m^2 C [Ans : 120, 200 W/m^2, 324 C]

1.6 Determine the thermal radiation exchanged between black bodies at 150 C and 1500 C. [Ans : 527,500 W/m^2]

1.7 A flat plate receives solar radiation @ 1100 W/m^2. The air temperature is 27 C and unit surface conductance between plate and air is 11 W/m^2 C. Determine the plate temperature. [Ans : 127 C]

1.8 An electronic tube has a mean surface temperature of 55 C. Its surface area is 150 cm^2. The temperature of surroundings is 25 C. Calculate the heat lost from the tube by radiation.

1.9 Compute the coefficient of heat transfer due to convection if 75,000 kJ are transferred per hr per m^2 from a surface at 200 C to a fluid at 100 C.

[Ans : 210 W/m^2C]

1.10 The walls of a refrigerated van consist of 1.5 mm of steel sheet at the outer surface, 10 mm plywood at the inner surface and 2 cm of glasswool in between. Calculate the rate of heat flow if the temperatures at the inside and outside surfaces are – 15 C and 25 C. The value of k (W/m C) for steel, glasswool and plywood is 23.2, 0.014 and 0.052 respectively. [Ans : 6 W/m^2]

1.11 A laminated block is made of *m* sheets of material **A** and (*m* + 1) sheets of material **B**. The thermal conductivities are k_1 and k_2 and the thickness of sheets are b_1 and b_2 respectively. The block has the face dimensions of $l \times w$, in the direction normal to heat flow.

Obtain a relationship between b_1 and b_2 so that heat flow rate in the direction of dimension l is 200 times that in the normal direction to the face.

1.12 Determine the quantity of heat passing per hr per m^2 of a boiler wall if it is 20 mm thick, and made of material whose k = 58 W/m C. The inner surface of the wall is covered by a layer of scale (k = 1.16 W/m C) 2 mm thick. The two surface temperatures are 250 C and 200 C. Also determine the inner surface temperature of steel plate. [Ans : 24,000 W/m^2, 241.7 C]

1.13 Determine the equivalent thermal conductivity of a package of transformer laminations numbering n, if each is 0.5 mm thick and a sheet of leatheroid paper 0.05 mm thick is used in between each of them. $k_{steel} = 63$; $k_{paper} = 0.16$ W/mK.

1.14 Compute the heat loss per day through 100 m^2 of brick wall (k = 84 W/m K), 30 cm thick, if the inner face is at 20°C and the outer face at 10°C. How much oil must be burned to compensate for this loss if the heat of combustion is 41,900 kJ/kg and the efficiency of the furnace is 60 %. [Ans : 20 × 10^4 kJ ; 9.59 kg.]

1.15 Calculate the heat loss in 5 hr through a glass pane 8 mm thick and 1 m^2 area if the two surfaces differ by 1.5°C in temperature. k = 0.82 W/m K. [Ans. 2780 kJ]

1.16 An electric furnace is of 15 cm × 15 cm × 20 cm inside dimensions. Its walls are made of 12 cm thick brick (k = 0.0021 cal/sec cm°C). Calculate the rate of heat loss in watt. The inner surface of the walls is at 1600°C and the outer surface at 200°C. [Ans : 914 W]

1.17 Calculate the temperature drop through a boiler shell 1.25 cm thick if the rate of evaporation of water is 15 kg/hr per cm^2 of area of the shell. k = 35 W/m K and L = 2250 kJ/kg.

1.18 Calculate the rate of heat loss through a 15 cm thick wall composed of 5 cm pine wood (k = 1.05), 1.25 cm of asbestos (k = 1.55) and 8.75 cm of mineral wool (k = 0.042). The inside and outside surfaces are at 20 °C and 5 °C, respectively. Also calculate the temperature drop through each of the layers. The units of k are W/m K.

1.19 Calculate the heat loss per m^2 of furnace wall made up of 27 cm of fire brick, 12 cm of insulating brick and 10 cm of red brick. The inside and outside surface temperatures of the wall are 1000°C and 40°C. The thermal conductivities of fire brick, insulating brick and red brick are 1.2, 0.14 and 1.7 W/m K, respectively.

1.20 A large composite wall is built up of the following materials : 7.5 cm marble (k = 1.15); 7.5 cm brick (k = 0.6), 2.5 cm pine (conductance = 0.3) and 2 cm inside plaster (k = 0.06). All values of k are in W/m K.

The outside and inside unit surface conductances are respectively 80 and 8 W/m^2 K, and temperatures are 200 C and 20 C. Calculate (a) the overall heat transfer coefficient (b) the temperature of the brick-pine interface.

1.21 A thermocouple (0.8 mm OD wire) is used to measure temperature of quiescent gas in a furnace. The thermocouple reading is 165°C. It is known, however, that the rate of radiant heat flow per meter length from hotter furnace walls to the thermocouple wire is 1.1 W/m and the unit conductance between the wire and the gas is 6.8 W/m^2 K. With this information, estimate the true gas temperature. State your assumption and indicate the equations used. [Ans : 100.6°C]

1.22 Water at a temperature of 76.6°C is to be evaporated slowly in a vessel. The water is in a low-pressure container which is surrounded by steam. The steam is condensing at 107°C. The overall heat transfer coefficient between the water and the steam is 1100 W/m^2 K. Calculate the surface area of the container which would be required to evaporate water at a rate of 0.01 kg/s. [Ans : 0.69 m^2]

1.23 Heat is transferred through a plane wall from the inside of a room at 22°C to the outside air at - 2°C. The unit-surface conductances at the inside and outside surface are 12 and 28 W/m^2 K, respectively. The thermal resistance of the wall per unit area is 0.5 m^2 K/W. Determine the temperature at the outer surface of the wall and the rate of heat flow through the wall per unit area. [Ans : 0.1°C, 37.6 W/m^2]

1.24 A flat plate placed in the sunlight receives 600 W/m^2 of radiant heat from the sun and the atmosphere. If the air temperature is 27°C and the unit-surface conductance between the plate and the air is 12 W/m^2 K, determine the plate temperature. Neglect heat losses from the bottom of the plate.

[Ans : 77°C for $\varepsilon = 0$; 35 °C for $\varepsilon = 1$]

1.25 How much fibreglass insulation (k = 0.035 W/m K) is needed to enable a guarantee that the outside temperature of a kitchen oven will not exceed 43°C ? The maximum oven temperature to be maintained by the conventional type of thermostatic control is 290°C, the kitchen temperature may vary from 15°C to 33°C and the average heat transfer coefficient between the oven surface and the kitchen is 12 W/m^2 K. [Ans. 7.2 cm]

1.26 In a manufacturing process, a large sheet of plastic 20 mm thick is to be glued to a 20 mm thick corkboard. To effect a good bond the glue should be maintained at 50°C for a long time. This is achieved by heating the outer plastic surface by radiation. The convective heat transfer coefficient is 10 W/m C on both sides. The conductivities of plastic and cork are 1 and 0.2 W/m C, respectively, and the ambient temperature is 20 °C. Find the temperature of outer plastic surface and radiant heat flux needed. Neglect losses due to radiation and the thermal resistance of the glue.

1.27 List the significant modes of heat transfer in the following cases :

(a) Heat loss from a vacuum flask.

(b) Burning of a finger tip when placed on a hot surface.

(c) Generation of steam in tubes inside a boiler combustion chamber.

(d) A central heating radiator.

(e) Quenching of a white hot steel ingot in water.

CHAPTER
2

Heat Transfer by Conduction (Steady State)

2.1 CONCEPT OF HEAT CONDUCTION

The process of conduction of heat was briefly discussed in Section 1.4, and the basic law governing this process was given in Section 1.6. We shall now study the physical phenomenon and the governing mathematical equations for heat flow in solids.

In solids, as already mentioned, the heat flow takes place exclusively by the process of conduction, while in gases and liquids it may occur by the processes of conduction, convection and radiation simultaneously. The main interest, therefore, will now be with the flow of heat in solids. In certain specific cases, however, the principles of heat conduction will be applicable to liquids and gases also when convection and radiation heat exchange is prevented or is negligible.

The phenomenon of heat conduction is believed to be a simple exchange of energy by lattice vibration. The increased temperature is associated with increased vibration energy of atoms. Thus, in the hotter portion of a solid, atoms which have larger vibration energy, communicate a part of it through the neighbouring low energy molecules and so on throughout the whole length of the body.

The conduction of heat is also partly achieved by the quanta of radiation between the molecules through electromagnetic waves. An electron at a certain quantum level, as soon as it emits heat to a neighbouring atom, will revolve in an orbit at a lower quantum level, or closer to the nucleus. This is also called *intermolecular radiation.* In the cases of porous solids, however, ordinary radiation and convection also take place inside the pores in addition to the intermolecular radiation.

In *metals* the conduction of heat is greatly assisted by the flow of the so-called free electrons which are similar to other electrons but are more easily separable from the remainder of the atom. *Conduction* of heat is caused by the migration of high energy free electrons to regions of lower energy.

Though the actual mechanism of heat conduction is complicated and has not been understood well, the basic laws which govern it have been established and verified experimentally.

2.2 GENERAL HEAT CONDUCTION EQUATION

The equation (1.1) represented the fundamental heat conduction law for unidirectional flow of heat :

$$q = \frac{dQ}{d\tau} = -kA\frac{\partial t}{\partial x} \tag{1.1}$$

In general, however, temperature variations will exist throughout the body. Heat will also flow in other directions of the body, because there may be nothing present in the body to prevent the same. It is, therefore, necessary that a general mathematical formulation be obtained with the help of this equation, which would be applicable to heat flow in all the three directions. Such a mathematical treatment can lead to an expression of *temperature distribution* as a function of space coordinates and time. Thus the heat flow at any particular location in any direction, at any instant, can be evaluated. Further, a general three dimensional heat conduction equation can be used for one or two dimensional cases with suitable modifications.

Consider an infinitesimal rectangular parallelepiped of a solid as shown in Fig. 2.1. Let the edges parallel to the three axes be respectively *dx*, *dy* and *dz*. The volume is given by :

$$dV = dx\,dy\,dz$$

Let dQ_x represent the total quantity of heat entering the face area *dy dz* in time $d\tau$ as shown in the figure.

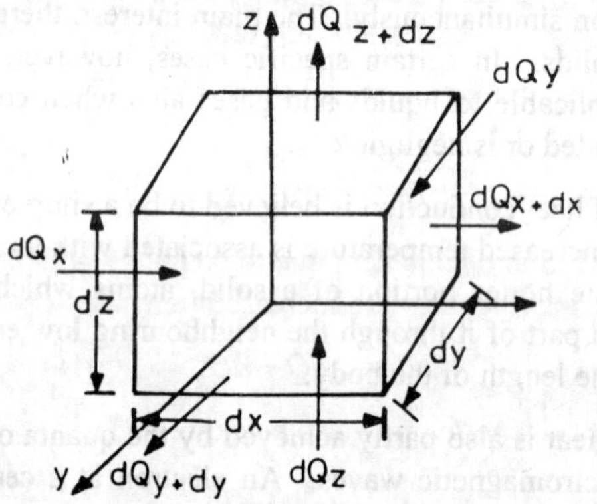

Fig. 2.1 Conduction of Heat in a Solid

So on the basis if Eqs. (1.1) we have

$$\frac{dQ_x}{d\tau} = -\mathrm{k}\,(dy\,dz)\,\frac{\partial t}{\partial x} \tag{2.1}$$

The gradient is expressed as a partial derivative of the temperature *t* as it is a function of *x*, *y*, *z* and τ.

Now, a corresponding quantity of heat will be leaving the element. Let it be dQ_{x+dx}, value of which may be obtained by Taylor expansion of dQ_x ;

so $$dQ_{x+dx} = dQ_x + \frac{\partial}{\partial x}(dQ_x)\,dx + ...$$

The net heat flow due to conduction in x-direction will therefore be :

$$dQ_x - dQ_{x+dx} = -\frac{\partial}{\partial x}(dQ_x)\,dx \qquad \text{[Ignoring all other terms]}$$

$$= \frac{\partial}{\partial x}\left(k\, dy\, dz \frac{\partial t}{\partial x} dx\right) d\tau$$

[by substituting Eq. (2.1)]

$$= \frac{\partial}{\partial x}\left(k \frac{\partial t}{\partial x}\right)(dx\, dy\, dz)\, d\tau$$

$$= \frac{\partial}{\partial x}(k\, \partial t/\partial x)\, dV\, d\tau \qquad (2.2)$$

Similarly the net heat conduction in y and z directions may be obtained as follows :

$$dQ_y - dQ_{y+dy} = \frac{\partial}{\partial y}\left(k \frac{\partial t}{\partial y}\right) dV\, d\tau \qquad (2.3)$$

$$dQ_z - dQ_{z+dz} = \frac{\partial}{\partial z}\left(k \frac{\partial t}{\partial z}\right) dV\, d\tau \qquad (2.4)$$

So net amount of heat, dQ stored in the element $dx\ dy\ dz$ will be obtained by the addition of Eqs. (2.2), (2.3) and (2.4).

Hence
$$dQ = dV\left\{\frac{\partial}{\partial x}\left(k \frac{\partial t}{\partial x}\right) + \frac{\partial}{\partial y}\left(k \frac{\partial t}{\partial y}\right) + \frac{\partial}{\partial z}\left(k \frac{\partial t}{\partial z}\right) d\tau\right\} \qquad (2.5)$$

Further, considering a general case, let there be internal heat generated in the solid element at the rate of Q' J/m^{3}s.

so heat generated in time $d\tau$, $$dQ_g = Q'\, dV\, d\tau \qquad (2.6)$$

Now, the heat is stored in the element due to conduction and also generated within the element. Both reasons together serve to increase internal energy of the element by dE,

Thus
$$dE = dQ + dQ_g \qquad (2.7)$$

The increase in internal energy may also be expressed as

$$dE = c\rho\, dV \frac{\partial t}{\partial \tau}\, d\tau \qquad (2.8)$$

where c = specific heat

ρ = density of the solid element

Hence from Eqs. (2.5) to (2.8) we get ;

$$c\,\rho\, dV \frac{\partial t}{\partial \tau}\, d\tau = dV\left\{\frac{\partial}{\partial x}\left(k \frac{\partial t}{\partial x}\right) + \frac{\partial}{\partial y}\left(k \frac{\partial t}{\partial y}\right) + \frac{\partial}{\partial z}\left(k \frac{\partial t}{\partial z}\right)\right\} d\tau + (Q'\, dV\, d\tau)$$

or
$$\frac{\partial t}{\partial \tau} = \frac{1}{c\rho}\left\{\frac{\partial}{\partial x}\left(k \frac{\partial t}{\partial x}\right) + \frac{\partial}{\partial y}\left(k \frac{\partial t}{\partial y}\right) + \frac{\partial}{\partial z}\left(k \frac{\partial t}{\partial z}\right)\right\} + \frac{Q'}{c\rho} \qquad (2.9)$$

The Eq. (2.9) is the *general heat conduction equation* in rectangular coordinates.

If the thermal conductivity k is constant, i.e., independent of x, y and z, Eq. (2.9) becomes ;

$$\frac{\partial t}{\partial \tau} = \frac{k}{c\rho}\left(\frac{\partial^2 t}{\partial x^2} + \frac{\partial^2 t}{\partial y^2} + \frac{\partial^2 t}{\partial z^2}\right) + \frac{Q'}{c\rho} \quad (2.10)$$

A new property of the conducting solid, α, referred to as *thermal diffusivity* (m^2/sec) is defined by ;

$$\alpha = \frac{k}{c\rho} \quad (2.11)$$

Incorporating α we have the *general heat conduction equation* for constant value of k :

$$\frac{1}{\alpha}\frac{\partial t}{\partial \tau} = \left(\frac{\partial^2 t}{\partial x^2} + \frac{\partial^2 t}{\partial y^2} + \frac{\partial^2 t}{\partial z^2}\right) + \frac{Q'}{k} \quad (2.12)$$

Now, three particular cases of Eq. (2.12) may arise :

Case (i) No Heat Generation ($Q' = 0$)

Eq. (2.12) takes the form ;

$$\frac{\partial^2 t}{\partial x^2} + \frac{\partial^2 t}{\partial y^2} + \frac{\partial^2 t}{\partial z^2} = \frac{1}{\alpha}\frac{\partial t}{\partial \tau}$$

or $$\nabla^2 t = \frac{1}{\alpha}\frac{\partial t}{\partial \tau} \quad \text{(Fourier's Equation)} \quad (2.13)$$

Case (ii) Steady State Heat Conduction ($\partial t/\partial \tau = 0$)

Eq. (2.12) takes the form ;

or $$\frac{\partial^2 t}{\partial x^2} + \frac{\partial^2 t}{\partial y^2} + \frac{\partial^2 t}{\partial z^2} + \frac{Q'}{k} = 0$$

or $$\nabla^2 t + \frac{Q'}{k} = 0 \quad \text{(Poisson's Equation)} \quad (2.14)$$

Case (iii) Steady State Heat Conduction with no Heat generation ($\partial t/\partial \tau = 0$, $Q' = 0$)

Eq. (2.12) takes the form :

$$\frac{\partial^2 t}{\partial x^2} + \frac{\partial^2 t}{\partial y^2} + \frac{\partial^2 t}{\partial z^2} = 0$$

or $$\nabla^2 t = 0 \quad \text{(Laplace Equation)} \quad (2.15)$$

Eq. (2.12) represents in general the condition which governs the flow of heat in a solid having k as constant. Solution of this equation for any specified case will yield the temperature distribution in the body and the quantity of heat flow.

2.3 THERMAL DIFFUSIVITY

It can be seen from Eqs. (2.14) and (2.15) that during steady state heat conduction thermal conductivity is the only property of a substance which determines the temperature distribution. However, in the unsteady state, the temperature distribution is influenced by the thermal conductivity as well as the thermal storage capacity of the substance. The *thermal storage capacity* is simply the product of the density and specific heat (ρc) of a substance. In this case Eq. (2.12) shows that the behaviour of different substances will vary with the ratio of the thermal conductivity to the thermal storage capacity ($k/c\rho$). It is for this reason that this combination of properties is defined as a new single property and is given the name *thermal diffusivity*, α.

thus
$$\alpha = k/c\rho \tag{2.11}$$

2.4 HEAT CONDUCTION EQUATION IN OTHER COORDINATE SYSTEMS

It will be observed that in a particular situation only one definite coordinate system is simpler to use. Hence the general heat conduction equation will now be derived in other coordinate systems as well.

(a) Cylindrical Coordinates

The problem of heat conduction in systems which have cylindrical geometry like pipes, rods etc. can be most conveniently handled by suitably modifying Eq. (2.12) into cylindrical coordinates.

The rectangular and cylindrical coordinates are inter-related by the relation :

$$x = r\cos\theta,\ y = r\sin\theta,\ z = z$$

Using these relations, Eq. (2.12) can be transformed into the cylindrical coordinates system as follows :

$$\frac{1}{\alpha}\frac{\partial t}{\partial \tau} = \left(\frac{\partial^2 t}{\partial x^2} + \frac{\partial^2 t}{\partial y^2} + \frac{\partial^2 t}{\partial z^2}\right) + \frac{Q'}{k} \tag{2.12}$$

Since x and y are functions of r and θ while z is independent of r and θ,

$$\frac{\partial t}{\partial r} = \frac{\partial t}{\partial x}\frac{\partial x}{\partial r} + \frac{\partial t}{\partial y}\frac{\partial y}{\partial r}$$

$$= \cos\theta\frac{\partial t}{\partial x} + \sin\theta\frac{\partial t}{\partial y}$$

$$\frac{\partial t}{\partial \theta} = \frac{\partial t}{\partial x}\frac{\partial x}{\partial \theta} + \frac{\partial t}{\partial y}\frac{\partial y}{\partial \theta}$$

$$= -\mathrm{r}\sin\theta\frac{\partial t}{\partial x} + \mathrm{r}\cos\theta\frac{\partial t}{\partial y}$$

so
$$\cos\theta\frac{\partial t}{\partial r} = \cos^2\theta\frac{\partial t}{\partial x} + \cos\theta\sin\theta\frac{\partial t}{\partial y}$$

$$\frac{\sin\theta}{r}\frac{\partial t}{\partial \theta} = -\sin^2\theta\frac{\partial t}{\partial x} + \cos\theta\sin\theta\frac{\partial t}{\partial y}$$

or
$$\frac{\partial t}{\partial x} = \cos\theta\frac{\partial t}{\partial r} - \frac{\sin\theta}{\mathrm{r}}\frac{\partial t}{\partial \theta} \tag{i}$$

Similarly

$$\frac{\partial t}{\partial y} = \sin\theta\frac{\partial t}{\partial r} + \frac{\cos\theta}{r}\frac{\partial t}{\partial \theta} \tag{ii}$$

If we put $\frac{\partial t}{\partial x}$ for t, Eq. (i) becomes :

$$\frac{\partial}{\partial x}\left(\frac{\partial t}{\partial x}\right) = \cos\theta\,\frac{\partial}{\partial r}\left(\frac{\partial t}{\partial x}\right) - \frac{\sin\theta}{r}\,\frac{\partial}{\partial\theta}\left(\frac{\partial t}{\partial x}\right)$$

$$= \cos\theta\,\frac{\partial}{\partial r}\left[\cos\theta\,\frac{\partial t}{\partial r} - \frac{\sin\theta}{r}\,\frac{\partial t}{\partial\theta}\right] - \frac{\sin\theta}{r}\,\frac{\partial}{\partial\theta}\left[\cos\theta\,\frac{\partial t}{\partial r} - \frac{\sin\theta}{r}\,\frac{\partial t}{\partial\theta}\right]$$

or $$\frac{\partial^2 t}{\partial x^2} = \left[\cos^2\theta\,\frac{\partial^2 t}{\partial r^2} + \frac{\cos\theta\sin\theta}{r^2}\,\frac{\partial t}{\partial\theta} + \frac{\sin^2\theta}{r}\,\frac{\partial t}{\partial r} + \frac{\sin^2\theta}{r^2}\,\frac{\partial^2 t}{\partial\theta^2} + \frac{\sin\theta\cos\theta}{r^2}\,\frac{\partial t}{\partial\theta}\right] \quad \text{(a)}$$

Similarly, Eq. (ii) will become,

$$\frac{\partial^2 t}{\partial y^2} = \left\{\sin\theta\,\frac{\partial}{\partial r} + \frac{\cos\theta}{r}\,\frac{\partial}{\partial\theta}\right\}\left\{\sin\theta\,\frac{\partial t}{\partial r} + \frac{\cos\theta}{r}\,\frac{\partial t}{\partial\theta}\right\}$$

$$= \left[\sin^2\theta\,\frac{\partial^2 t}{\partial r^2} + \frac{\cos^2\theta}{r}\,\frac{\partial t}{\partial r} - \frac{\cos\theta\sin\theta}{r^2}\,\frac{\partial t}{\partial\theta} + \frac{\cos^2\theta}{r^2}\,\frac{\partial^2 t}{\partial\theta^2} + \left(-\frac{\cos\theta\sin\theta}{r^2}\,\frac{\partial t}{\partial\theta}\right)\right] \quad \text{(b)}$$

Adding Eqs (a) and (b) ;

$$\frac{\partial^2 t}{\partial x^2} + \frac{\partial^2 t}{\partial y^2} = \left[(\cos^2\theta+\sin^2\theta)\,\frac{\partial^2 t}{\partial r^2} + \left(\frac{\cos^2\theta+\sin^2\theta}{r}\right)\frac{\partial t}{\partial r} + \left(\frac{\cos^2\theta+\sin^2\theta}{r^2}\right)\frac{\partial^2 t}{\partial\theta^2}\right]$$

$$= \left\{\frac{\partial^2 t}{\partial r^2} + \frac{1}{r}\,\frac{\partial t}{\partial r} + \frac{1}{r^2}\,\frac{\partial^2 t}{\partial\theta^2}\right\} \quad \text{(c)}$$

Substituting Eq. (c) in Eq. (2.12) we get the *general heat conduction equation* in cylindrical coordinates as given below :

$$\frac{1}{\alpha}\,\frac{\partial t}{\partial\tau} = \left(\frac{\partial^2 t}{\partial r^2} + \frac{1}{r}\,\frac{\partial t}{\partial r} + \frac{1}{r^2}\,\frac{\partial^2 t}{\partial\theta^2} + \frac{\partial^2 t}{\partial z^2}\right) + \frac{Q'}{k} \quad (2.16)$$

(b) Spherical Coordinates

Similarly the general heat conduction equation can be obtained for spherical coordinates. Such a coordinate system is interrelated with rectangular coordinate system by the relations (Fig. 2.2) :

$$x = r\sin\theta\cos\phi,\ y = r\cos\theta,\ z = r\sin\theta\sin\phi$$

It is, however, more convenient to derive the conduction equation in spherical coordinates directly rather than modify Eq. (2.12) into spherical coordinates.

With reference to Fig. 2.2, consider the flow of heat through an elemental volume of a sphere. The sides of this volume are dr, $r\,d\theta$, and $r\sin\theta\,d\phi$. We assume that k, ρ and c do not vary with position. We then obtain :

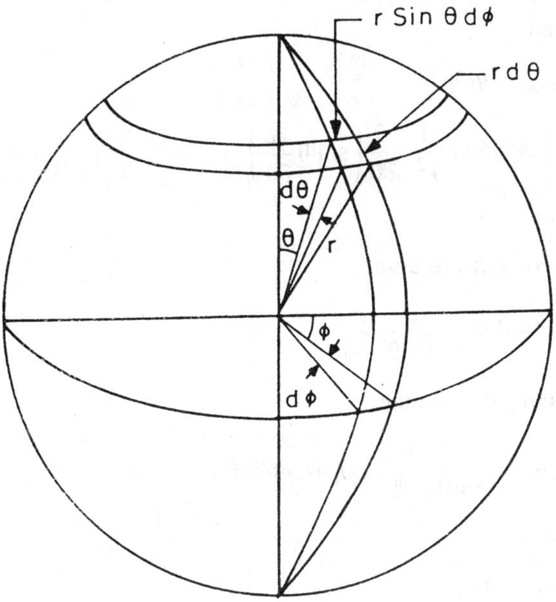

Fig. 2.2 Spherical Coordinates

(a) For direction r :

(i) Rate of heat entering the volume,

$$(q_r)_E = -k\,(r\,d\theta\; r\sin\theta\; d\phi)\frac{\partial t}{\partial r} = -k\,(r^2 \sin\theta\; d\theta\; d\phi)\frac{\partial t}{\partial r}$$

(ii) Rate of heat leaving the volume,

$$(q_r)_L = (q_r)_E + \frac{\partial}{\partial r}(q_r)_E\, dr$$

$$= (q_r)_E - k\sin\theta\; d\theta\; d\phi\; dr\;\frac{\partial}{\partial r}\left(r^2\frac{\partial t}{\partial r}\right)$$

(iii) Heat storage,

$$= (q_r)_E - (q_r)_L$$

$$= k\sin\theta\; d\theta\; d\phi\; dr\;\frac{\partial}{\partial r}\left(r^2\frac{\partial t}{\partial r}\right) \qquad \text{(A)}$$

(b) For direction θ :

(i) Rate of heat entering the volume,

$$(q_\theta)_E = -k\,(r\sin\theta\; d\phi\; dr)\,\frac{1}{r}\,\frac{\partial t}{\partial \theta}$$

(ii) Rate of heat leaving the volume,

$$(q_\theta)_L = (q_\theta)_E + \frac{1}{r}\,\frac{\partial}{\partial\theta}(q_\theta)_E\,(r\,d\theta)$$

$$= (q_\theta)_E - k\,r^2\,d\theta\,d\phi\,dr\,\frac{1}{r^2}\,\frac{\partial}{\partial\theta}\left(\sin\theta\,\frac{\partial t}{\partial\theta}\right)$$

(iii) Heat storage rate,

$$= (q_\theta)_E - (q_\theta)_L$$

$$= kr^2\,d\theta\,d\phi\,dr\,\frac{1}{r^2}\,\frac{\partial}{\partial\theta}\left(\sin\theta\,\frac{\partial t}{\partial\theta}\right) \quad \text{(B)}$$

(c) For direction ϕ :

(i) Rate of heat entering the volume,

$$(q_\phi)_E = -k\,(r d\theta\,dr)\,\frac{1}{r\sin\theta}\,\frac{\partial t}{\partial\phi}$$

(ii) Rate of heat leaving the volume,

$$(q_\phi)_L = (q_\phi)_E + \frac{1}{r\sin\theta}\,\frac{\partial}{\partial\phi}\,(q_\phi)_E\,(r\sin\theta\,d\phi)$$

(iii) Heat storage rate,

$$= (q_\phi)_E - (q_\phi)_L$$

$$= kr^2\,d\theta\,dr\sin\theta\,d\phi\,\frac{1}{r^2\sin^2\theta}\,\frac{\partial^2 t}{\partial\phi^2} \quad \text{(C)}$$

(d) Total heat storage rate is the summation of Eqs. (A), (B) and (C) and will be given by :

$$kr^2\sin\theta\,d\theta\,d\phi\,dr\left[\frac{1}{r^2}\,\frac{\partial}{\partial r}\left(r^2\,\frac{\partial t}{\partial r}\right) + \frac{1}{r^2}\,\frac{1}{\sin\theta}\,\frac{\partial}{\partial\theta}\left(\sin\theta\,\frac{\partial t}{\partial\theta}\right) + \frac{1}{r^2\sin^2\theta}\,\frac{\partial^2 t}{\partial\phi^2}\right]$$

(e) Rate of heat production

$$= Q'\,r^2\sin\theta\,d\theta\,d\phi\,dr$$

(f) Rate of change of energy within the elemental volume,

$$= \rho\,c\,(r^2\sin\theta\,d\theta\,d\phi\,dr)\,\frac{\partial t}{\partial\tau}$$

Since the rate of change of energy within the volume must equal the total heat storage plus the heat production rate, we get ;

$$\rho\,(r^2\sin\theta\,d\theta\,d\phi\,dr)\,\frac{\partial t}{\partial\tau}$$

$$= k\,(r^2\sin\theta\,d\theta\,d\phi\,dr)\left[\frac{1}{r^2}\,\frac{\partial}{\partial r}\left(r^2\,\frac{\partial t}{\partial r}\right) + \frac{1}{r^2\sin\theta}\,\frac{\partial}{\partial\theta}\left(\sin\theta\,\frac{\partial t}{\partial\theta}\right) + \frac{1}{r^2\sin^2\theta}\,\frac{\partial^2 t}{\partial\phi^2}\right] + Q'\,r^2\sin\theta\,d\theta\,d\phi\,dr$$

or $$\frac{1}{\alpha}\frac{\partial t}{\partial \tau} = \left[\frac{1}{r^2}\frac{\partial}{\partial r}\left(r^2\frac{\partial t}{\partial r}\right) + \frac{1}{\sin\theta}\frac{\partial}{\partial \theta}\left(\sin\theta\frac{\partial t}{\partial \theta}\right) + \frac{1}{\sin^2\theta}\frac{\partial^2 t}{\partial \phi^2}\right] + \frac{Q'}{k} \tag{2.17}$$

Eq. (2.17) is the *general heat conduction equation* in spherical coordinates.

2.5 LINEAR ONE DIMENSIONAL STEADY STATE HEAT CONDUCTION

We shall now discuss the steady state heat conduction *without heat generation.* The cases in which heat is generated within the system will be discussed in Section 2.19.

One dimensional steady state heat conduction cases include heat flow through a wall or along an insulated rod.

For one dimensional, steady state heat conduction through a solid with no heat generation, Eq. (2.15) gives :

$$\frac{\partial^2 t}{\partial x^2} = 0 \tag{2.18}$$

The particular solution to the equation will give the value of the temperature t which will represent the temperature field of the system.

Integrating Eq. (2.18) we have ;

$$\frac{\partial t}{\partial x} = \text{C (a const.)}$$

again integrating,

$$t = Cx + D \tag{a}$$

The values of C and D can be obtained by imposing initial and boundary conditions of the system (Fig. 2.3).

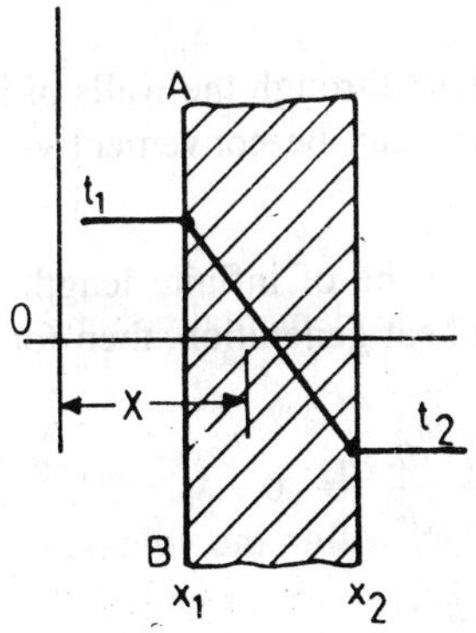

Fig. 2.3 One Dimensional Heat Conduction

Let $$t = t_1 \text{ at } x = x_1$$

$$t = t_2 \text{ at } x = x_2$$

Substituting these values in equation (a) we get ;

$$C = \frac{t_1 - t_2}{x_1 - x_2}$$

$$D = \frac{x_2 t_1 - x_1 t_2}{x_2 - x_1}$$

so the temperature distribution in the solid is given by :

$$t = \frac{t_2 - t_1}{x_2 - x_1} x + \frac{x_2 t_1 - x_1 t_2}{x_2 - x_1}$$

or in dimensionless form we have ;

$$\frac{t - t_1}{t_2 - t_1} = \frac{x - x_1}{x_2 - x_1} = \frac{x}{l} \tag{2.19}$$

(where l is the thickness of solid and origin lies in the face AB).

Now, the heat flow,

$$q = \frac{dQ}{d\tau} = -kA\frac{\partial t}{\partial x} \tag{1.1}$$

$$= -kA\frac{\partial}{\partial x}\left\{\frac{t_2 - t_1}{x_2 - x_1} x + \frac{x_2 t_1 - x_1 t_2}{x_2 - x_1}\right\}$$

$$= -kA\frac{t_2 - t_1}{x_2 - x_1}$$

$$q = \frac{k}{l} A (t_1 - t_2) \tag{1.2}$$

(where $l = x_2 - x_1$; the thickness of solid)

2.6 RADIAL HEAT CONDUCTION THROUGH TUBES

The case of conduction of heat through the walls of hollow cylinders or tubes usually encountered in heat exchangers, can be conveniently treated in cylindrical coordinates instead of cartesian coordinates.

Let us assume the cylinder to be of infinite length and the heat flow to be only in radial direction. If there is no heat generation, then for steady state we obtain from Eq. (2.16),

$$\frac{d^2 t}{dr^2} + \frac{1}{r}\frac{dt}{dr} = 0 \tag{2.20}$$

or

$$\frac{1}{r}\left\{\frac{d}{dr}\left(r\frac{dt}{dr}\right)\right\} = 0$$

since

$$\frac{1}{r} \neq 0 ; \frac{d}{dr}\left(r\frac{dt}{dr}\right) = 0$$

$$r\frac{dt}{dr} = C \text{ (a const.)}$$

or

$$\frac{dt}{dr} = \frac{C}{r}$$

Integrating further we have,

$$t = C \log_e r + D \tag{2.21}$$

Boundary conditions are (say),

(i) at $r = r_1$; $t = t_1$

(ii) at $r = r_2$; $t = t_2$

Substituting these values in Eq. (2.21), the values of C and D are obtained as :

$$C = -\frac{t_1 - t_2}{\log_e (r_2 / r_1)}$$

$$D = \frac{(t_1 \log_e r_2) - t_2 (\log_e r_1)}{\log_e (r_2 / r_1)}$$

The temperature distribution in hollow cylinder will thus be given by ;

$$\text{t} = \frac{(t_1 \log_e r_2) - t_2 (\log_e r_1)}{\log_e (r_2 / r_1)} - \frac{t_1 - t_2}{\log_e (r_2 / r_1)} (\log_e r)$$

or $$t \log_e \frac{r_2}{r_1} = (t_1 \log_e r_2) - (t_2 \log_e r_1) - (t_1 - t_2) \log_e r$$

Subtracting $t_1 \log_e (r_2/r_1)$ from both sides we have ;

$$(t - t_1) \log_e \frac{r_2}{r_1} = (t_2 - t_1) \log_e \frac{r}{r_1}$$

or $$\frac{t - t_1}{t_2 - t_1} = \frac{\log_e (r / r_1)}{\log_e (r_2 / r_1)} \qquad (r_1 \le r \le r_2) \tag{2.22}$$

Further, the rate of heat flow is given by ;

$$q = \frac{dQ}{d\tau} = -k A \frac{\partial t}{\partial r} \tag{1.1}$$

$$= -k A \frac{C}{r}$$

$$= -\frac{k A (t_1 - t_2)}{r \log_e (r_2 / r_1)}$$

or, $$q = \frac{2\pi kl (t_1 - t_2)}{\log_e (r_2 / r_1)} \tag{2.23}$$

(where $l = A/2\pi r$ is length of tube)

Aliter

Eq. (2.23) can also be derived directly with the help of the linear Fourier's law Eq. (1.1).

Consider a cross-section of the hollow cylinder or of the tube as shown in Fig. 2.4. Since there is only radial flow of heat, the rate q will be same through the inner and outer surfaces as well as through any cylindrical surface in between.

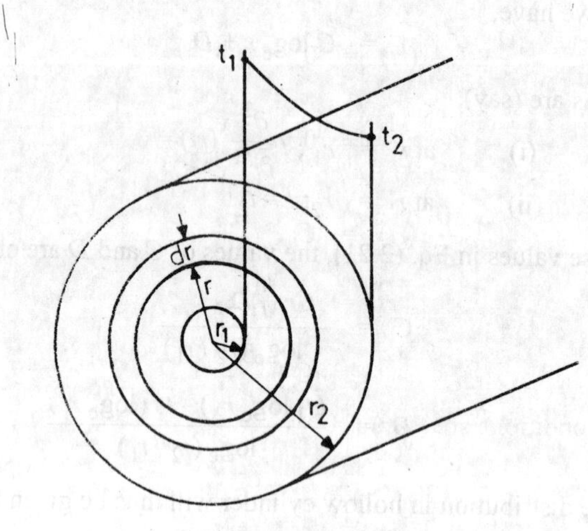

Fig. 2.4 Radial Heat Flow in a Tube

Consider an elementary thin cylinder of radius r and thickness dr, then

$$q = -kA\frac{dt}{dr} = -k(2\pi rl)\frac{dt}{dr} \qquad (1.1)$$

Integrating this for the same boundary condition as for Eq. (2.21) we have ;

$$q\int_{r_1}^{r_2}\frac{dr}{r} = -(k\,2\pi l)\times\int_{t_1}^{t_2}dt$$

or
$$q\log_e\frac{r_2}{r_1} = (k\,2\pi l)(t_1-t_2)$$

or
$$q = \frac{(2\pi kl)(t_1-t_2)}{\log_e(r_2/r_1)} \qquad (2.23)$$

Sometimes it is advantageous to write this equation in the same form as for heat flow through a plane wall. To do this we set ;

$$q = \frac{kA_m(t_1-t_2)}{(r_2-r_1)}$$

where
$$A_m = \frac{2\pi l(r_2-r_1)}{\log_e(r_2/r_1)}$$

$$= \frac{(A_2-A_1)}{\log_e(A_2/A_1)} \qquad (2.24)$$

where A_1, A_2 = inner and outer surface areas

A_m = logarithmic mean area of the tube

2.7 RADIAL HEAT CONDUCTION THROUGH SPHERICAL SHELL

For steady state radial heat flow in a spherical shell the use of Eq. (2.17) is simpler. The same is reduced for these conditions to the form ;

$$\frac{1}{r}\frac{\partial^2}{\partial r^2}(rt) = 0 \tag{2.25}$$

since $$\frac{1}{r} \neq 0 ; \frac{\partial^2}{\partial r^2}(rt) = 0$$

so $$\frac{\partial}{\partial r}(rt) = C$$

or $$rt = Cr + D$$

or $$t = C + \frac{D}{r} \tag{2.26}$$

If the boundary conditions are (say) ;

at $$r = r_1 ; t = t_1$$

$$r = r_2 ; t = t_2$$

then after substitutions we get ;

$$C = t_1 - \frac{1}{r_1}\left\{\frac{t_1 - t_2}{(1/r_1) - (1/r_2)}\right\}$$

$$D = \frac{t_1 - t_2}{(1/r_1) - (1/r_2)}$$

The temperature distribution is obtained by substituting these values in Eq. (2.26),

$$\frac{t - t_1}{t_2 - t_1} = \frac{(1/r) - (1/r_1)}{(1/r_2) - (1/r_1)} \qquad (r_2 \geq r \geq r_1) \tag{2.27}$$

Also, the rate of heat flow,

$$q = \frac{dQ}{d\tau} = -kA\frac{dt}{dr} = kA\frac{D}{r^2}$$

$$= \frac{k\,4\pi r^2}{r^2} D = 4\pi k \frac{t_1 - t_2}{(1/r_1) - (1/r_2)}$$

or $$q = 4\pi k \frac{t_1 - t_2}{r_2 - r_1}(r_1 r_2) \tag{2.28}$$

This expression could also be found by integrating the fundamental Fourier Law by similar method as was used in the alternative solution for the case of hollow cylinders.

Again, if Eq. (2.27) is written in the same form as that for heat flow through a plane wall, the mean area A_m of the spherical shell may be evaluated ;

$$q = 4\pi k \frac{t_1 - t_2}{r_2 - r_1}(r_1 r_2)$$

$$= k A_m \left(\frac{t_1 - t_2}{r_2 - r_1}\right)$$

so $$A_m = 4\pi r_1 r_2 = \sqrt{A_1 A_2} \tag{2.29}$$

where A_1 and A_2 are the inner and outer surface areas of the shell.

2.8 COMPOSITE STRUCTURES

There are a few systems of considerable practical utility which are made up of two or more layers of different materials. For example, cold storage walls have a layer of bricks, a layer of thick insulation and plaster on both sides. Similarly steam pipes have the cylindrical steel wall of the pipe, a layer of insulating material and then a layer of protecting plaster. The treatment of conductive heat flow in such structures may be considered as an extension of the single wall structures discussed in earlier sections.

In such composite structures each layer has a different value of resistance, so the overall cumulative effect of these resistances has to be studied. Usually the resistances of the free surfaces also have to be accounted for as additional resistances in series, in the calculations. This is due to the fact that the actual surface temperatures, which are different than the ambient values, are not known.

It is generally assumed that heat flow is steady and one-dimensional, there are no heat sources in the structure and that the resistance due to interface contact is negligible.

Interface contact depends upon total force exerted on the contacting surfaces, physical nature of the bond, the quality of the fluid trapped in the contacting surface etc. The mechanism of heat transfer across the interface contact is therefore complicated. However the values of resistances at interface contact obtained experimentally are so small as to justify the above assumption. We shall now consider a few cases of composite structures.

(a) Composite plane wall

The general method of tracking problems of heat flow by conduction through a series of resistances has been dealt with in Section 1.11 and illustrated by Example 1.4.

On similar lines three cases of composite plane wall structures will be considered.

(i) Multiple layers in series : Fig. 2.5 shows a system consisting of three layers of different materials placed in series so as to make a composite plane structure. The convective films on both sides of the wall are also shown. In the lower portion of the diagram thermal resistances are given corresponding to different layers.

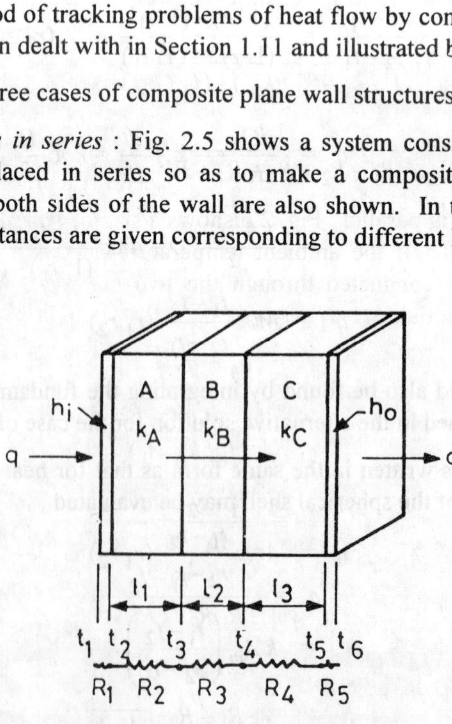

Fig. 2.5 Conduction Through Three Layers in Series

The rate of heat conducted through each layer is the same. However the temperatures at each interface are different hence the temperature gradient in each layer is different. Further the resistance offered by films on the exposed surfaces will be treated along with other resistances in series.

Let q be the rate of heat conduction and the temperatures of ambient air on two sides of the wall be t_1 and t_6.

hence

$$q = h_i A (t_1 - t_2) \quad \text{through left side film}$$

$$= \frac{k_A A}{l_1} (t_2 - t_3) \quad \text{through material A}$$

$$= \frac{k_B A}{l_2} (t_3 - t_4) \quad \text{through material B}$$

$$= \frac{k_C A}{l_3} (t_4 - t_5) \quad \text{through material C}$$

$$= h_o A (t_5 - t_6) \quad \text{through right side film}$$

eliminating temperatures t_2, t_3, t_4 and t_5 from the above we get,

$$q = \frac{(t_1 - t_6) A}{[1/h_1 + (l_1/k_A) + (l_2/k_B) + (l_3/k_C) + 1/h_o]}$$

$$= UA (t_1 - t_6)$$

hence the overall heat transfer coefficient, U is given by ;

$$U = \frac{1}{[1/h_1 + (l_1/k_A) + (l_2/k_B) + (l_3/k_C) + 1/h_o]} \quad (2.30)$$

$$= \frac{1}{R_1 + R_2 + R_3 + R_4 + R_5} = \frac{1}{\Sigma R} \quad (2.31)$$

(ii) Two materials in parallel : Fig. 2.6 shows a slab of two different materials of area A_1 and A_2 in parallel. If the ambient temperature on the two sides of the wall are uniform, heat will be conducted through the two materials along the separate paths between the same potential, $\Delta t = t_1 - t_2$.

$$q = q_1 + q_2$$

$$= \frac{t_1 - t_2}{(l_1/k_A A_1)} + \frac{t_1 - t_2}{(l_2/k_B A_2)}$$

$$= \frac{t_1 - t_2}{R_1} + \frac{t_1 - t_2}{R_2}$$

or

$$q = \frac{t_1 - t_2}{R_1 R_2/(R_1 + R_2)}$$

$$= UA (t_1 - t_2) \quad (2.32)$$

hence

$$UA = \frac{R_1 + R_2}{R_1 R_2} \quad (2.33)$$

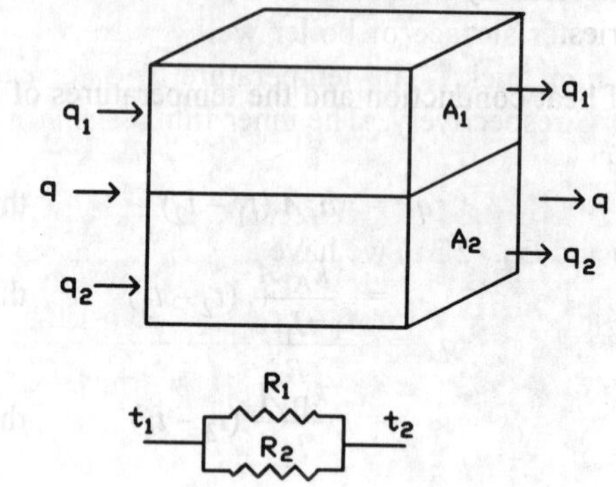

Fig. 2.6 Conduction Through Wall Section, with Two Material Paths in Parallel

(iii) Walls in series and in parallel : A more complex application of thermal network approach is illustrated in Fig. 2.7 where heat is transferred through a composite structure involving thermal resistances in series and in parallel.

For this system the resistance of middle layer R_2 is given by

$$R_2 = \frac{R_B R_C}{R_B + R_C} \tag{2.34}$$

and
$$q = \frac{\Delta t}{\sum R}$$

where the resistances of individual layers are as shown in Fig. 2.7.

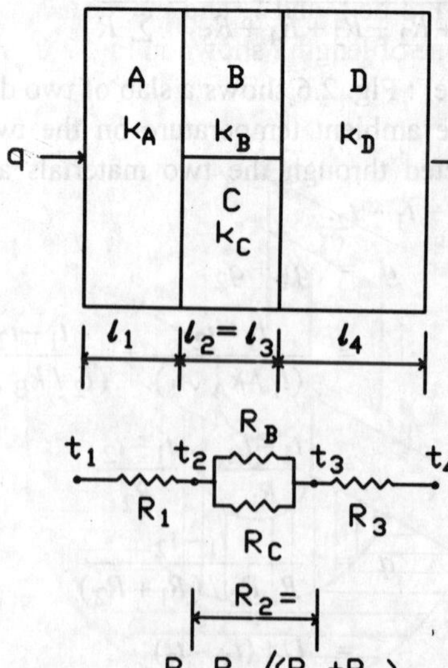

Fig. 2.7 Conduction Through a Wall Consisting of Material-Paths in Series and Parallel

Example 2.1

Determine the rate of heat flow through a boiler wall made of 20 mm thick steel (k = 58 W/m K). The outer surface of boiler wall is covered with asbestos insulation (k = 0.116 W/m K), 5 mm thick. The temperature of outer surface and that of fluid inside are 50°C and 300°C respectively. The inner film resistance is 0.0023 K/W.

Solution :

According to Eq. (1.17) and Eq. (2.31) we have ;

$$q = \frac{t_1 - t_4}{\sum R} = \frac{300 - 50}{\left(R_1 + \frac{x_2}{k_2} + \frac{x_3}{k_3}\right)}$$

$$= \frac{250}{0.0023 + \frac{0.02}{58} + \frac{0.005}{0.116}}$$

$$= 5540 \text{ W/m}^2$$ Ans.

The temperature of the inner surface of steel may be determined as follows :

$$t_2 = t_1 - q\,R_1 = 300 - (5540 \times 0.0023)$$

$$= 287.3°\text{C}$$ Ans.

(b) Composite cylinder

Let us assume, for a general case, a composite cylindrical tube made of two layers having thermal conductivity k_1 and k_2 respectively. A practical application of such a system is usually found in industrial installations where one fluid at a certain temperature flows inside the tube while the outer surface is exposed to some other fluid at different temperature. Let these temperature be t_i and t_o and the surface conductances be h_1 and h_o respectively in an insulated pipe of length *l* shown in Fig. 2.8.

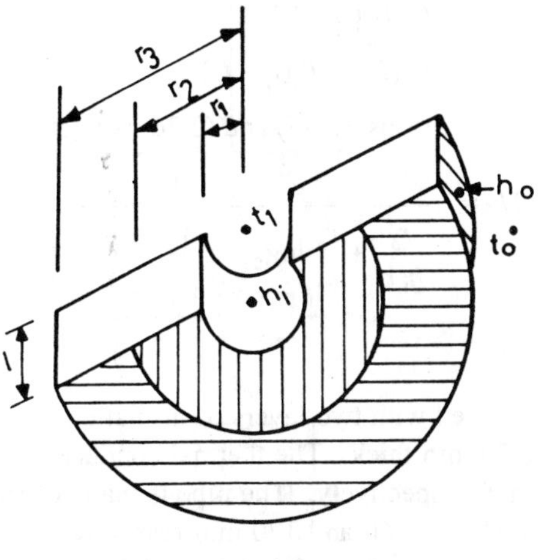

Fig. 2.8 Composite Cylindrical Wall

If the cylinder is long then the heat flow may be assumed to be radial only. In the steady state the rate of heat flow is same through each section and is given by ;

(i) For inner surface

$$q = 2\pi r_1 l\, h_i (t_i - t_1)$$

(ii) For inner cylindrical wall, by Eq. (2.23)

$$q = \frac{2\pi k_1 l}{\log_e (r_2 / r_1)} (t_1 - t_2)$$

(iii) For outer cylindrical wall, by Eq. (2.23)

$$q = \frac{2\pi k_2 l}{\log_e (r_3 / r_2)} (t_2 - t_3)$$

(iv) For outer surface

$$q = 2\pi r_3 l\, h_o (t_3 - t_o)$$

By adding the expressions for temperature differences in the above relations we have ;

$$(t_i - t_1) + (t_1 - t_2) + (t_2 - t_3) + (t_3 - t_0) = (t_i - t_o)$$

$$= \frac{q}{2\pi r_1 l h_i} + \frac{q \log_e (r_2 / r_1)}{2\pi k_1 l} + \frac{q \log_e (r_3 / r_2)}{2\pi k_2 l} + \frac{q}{2\pi r_3 l h_o}$$

or

$$q = \frac{(t_i - t_o)}{\frac{1}{2\pi l}\left\{\frac{1}{h_i r_1} + \frac{\log_e (r_2/r_1)}{k_1} + \frac{\log_e (r_3/r_2)}{k_2} + \frac{1}{h_o r_3}\right\}}$$

$$= \frac{t_i - t_o}{\sum R_n} \qquad (2.35)$$

Since it is easier to measure the outer area, the overall heat transfer coefficient may be obtained on the basis of area A_o ;

$$q = U A_0 (t_i - t_o)$$

$$= U (2\pi r_3 l)(t_i - t_o) \qquad (2.36)$$

Equating the expressions of Eqs. (2.35) and (2.36) we get ;

$$U = \frac{1}{\frac{r_3}{h_i r_1} + \frac{r_3}{k_1}\log_e\left(\frac{r_2}{r_1}\right) + \frac{r_3}{k_2}\log_e\left(\frac{r_3}{r_2}\right) + \frac{1}{h_o}} \qquad (2.37)$$

Example 2.2

A steam pipe is covered with two layers of insulation. The inner layer is 30 mm thick and the outer one is 50 mm thick. The thermal conductivities of insulating materials are 0.17 and 0.023 W/m K respectively. The pipe is made of steel (k = 58 W/m K) and has inner and outer diameter of 160 and 170 mm respectively. The temperature of steam is 300 C and the ambient air is at 50 C. If the inside and outside film coefficients are 30 and 5.8 W/m^{2}K respectively, calculate the heat lost per unit length of pipe.

Solution :

We have ;

$$r_1 = 80 \text{ mm} \; ; r_2 = 85 \text{ mm} \; ; r_3 = 115 \text{ mm and } r_4 = 165 \text{ mm}$$

Now
$$q = \frac{t_i - t_0}{R_1 + R_2 + R_3 + R_4}$$

$$q = \frac{2\pi \times 1 \times (300 - 50)}{\frac{1}{30 \times 0.08} + \frac{\ln(85/80)}{58} + \frac{\ln(115/85)}{0.17} \times \frac{\ln(165/115)}{0.093} + \frac{1}{5.8 \times 0.15}}$$

$$= 200 \text{ W/m}$$ Ans.

(c) Composite Spherical Shells

A system consisting of a number of spherical layers of different materials will also have radial flow of heat due to conduction if the unit surface conductances and the ambient temperature are uniform.

With the same nomenclature and by similar treatment as for composite cylindrical tubes; the heat flow equation for this case will be (assuming two layers);

$$q = \frac{4\pi(t_i - t_0)}{\left\{\frac{1}{h_i r_1^2} + \frac{r_2 - r_1}{k_1 r_1 r_2} + \frac{r_3 - r_2}{k_2 r_2 r_3} + \frac{1}{h_o r_3^2}\right\}} \qquad (2.38)$$

and overall heat transfer coefficient based on outer surface of the spherical shell will be obtained as ;

$$U = \frac{1}{\frac{(r_3/r_1)^2}{h_i} + \frac{(r_2 - r_1)(r_3^2/r_1 r_2)}{k_1} + \frac{(r_3 - r_2)(r_3^2/r_3 r_2)}{k_2} + \frac{1}{h_0}} \qquad (2.39)$$

2.9 CRITICAL INSULATION THICKNESS FOR PIPES

The addition of insulation on plane wall does not affect the face area of heat transfer, while it increases the outer area in case of cylindrical or spherical walls. Therefore, the rate of heat transfer will always decrease due to increase in thickness (by application of insulation) of a plane wall, but it may not be so in case of cylinders or spheres.

According to equation (2.22), the value of q decreases with increasing value of r_2/r_1 or with increasing wall thickness if the inside radius r_1 is kept constant. However, the process of conduction through the cylinder wall is usually accompanied by convective heat transfer at the inside and outside surfaces. As the wall thickness or the value of r_2 increases the convective heat transfer is also increased. On the other hand the rate of heat transfer by conduction through the wall decreases. If the overall heat transfer from inside surface to outside is considered, the rate of heat flow may first increase with increasing r_2 and after passing through a maximum may start decreasing. This argument, when applied to insulated pipes or cables, shows that :

(a) The heat transfer rate from the cylindrical surface increases by addition of insulation (increase in outer radius) upto a certain value (r_c).

(b) At this value of the outer radius (r_c) called *critical radius* the rate of heat transfer by conduction (equal to rate of heat transfer by convection at the surface) is maximum.

(c) The heat transfer rate decreases if further insulation is applied to the surface beyond critical r_c, (Fig. 2.9).

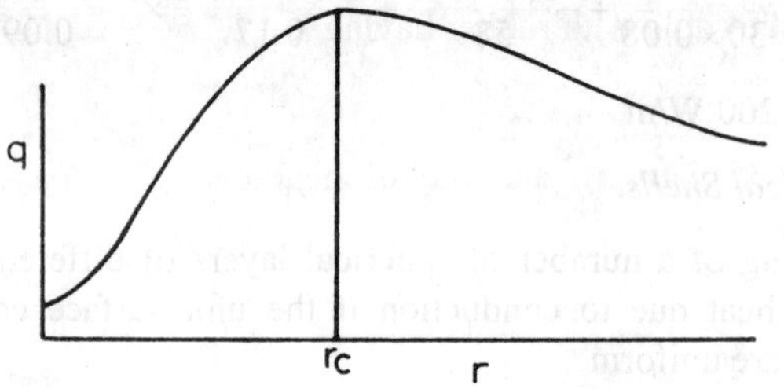

Fig. 2.9 Critical Radius of Insulated Pipe

Therefore in case of pipes the outside radius of pipe after applying insulation is kept greater than r_c while in case of current carrying wires and cables it is kept lower than or equal to r_c (as this will dissipate more heat than a bare pipe). The insulation of electrical wires is therefore advantageous in as much as it increases the dissipation of heat generated within the conductor.

In order to calculate the value of critical radius consider a cylinder having inner and outer radii r_1 and r_2. The ambient fluids are at t_i and t_0, film coefficients are h_i and h_0 respectively. Then by equation (2.35) ;

$$q = \frac{(t_i - t_o)}{\frac{1}{2\pi}\left\{\frac{1}{h_i r_1} + \frac{ln(r_2/r_1)}{k} + \frac{1}{h_o r_2}\right\}}$$

Keeping r_1 fixed ; maximum heat flow will occur at a particular value of r_2 so that $dq/dr_2 = 0$; thus ;

$$\frac{dq}{dr_2} = \frac{-(2\pi l)(t_i - t_o)\left(\frac{1}{kr_2} - \frac{1}{h_o r_2^2}\right)}{\left\{\frac{1}{h_i r_1} + \frac{1}{k} ln\left(\frac{r_2}{r_1}\right) + \frac{1}{h_o r_2}\right\}^2} = 0$$

The value of r_2 obtained from this equation is the *critical radius*, r_c given by ;

$$r_c = k/h_0 \qquad (2.40)$$

(The critical radius in case of spherical shell can be evaluated on the same lines as $2k/h_0$)

In practice, however, the selection of insulation will be based on the cost analysis and on the compromise between desirability of heat dissipation and the cost involved.

Cases will arise when $r_c < r_1$. The addition of insulation in such cases will always reduce the heat transfer. For electrical conductors therefore r_1 (inner radius of insulation or outer radius of wire) should always be chosen so that $r_1 < r_c$. This will eliminate the anomaly of the necessity of application of insulation against electrical current on one hand and decrease in the heat dissipation rate on the other, if $r_1 > r_c$.

Example 2.3

A 10 mm cable is to be laid in atmosphere of 20 C, (h_0 = 8.5 W/m^{2}K). The surface temperature of the cable is likely to be 65 C due to heat generated within. Discuss the effect of insulating the cable with rubber having k = 0.155 W/mK.

Solution :

The equation (2.35), giving the rate of heat dissipation per unit length, can be simplified for this case as ;

$$q = \frac{(t_i - t_o)}{\frac{1}{2\pi}\left\{\frac{\log_e(r_2/r_1)}{k} + \frac{1}{h_0 r_2}\right\}}$$

$$q = \frac{2\pi(65-20)}{\frac{\log_e r_2/5}{0.155} + \frac{1000}{8.5 r_2}} = \frac{43.8}{\log_e \frac{r_2}{5} + \frac{18.4}{r_2}}$$

(where r_2 is in mm)

This will have a maximum value at critical radius ;

$$r_c = \frac{k}{h_0} = \frac{0.155}{8.5} \times 1000 = 18.4 \text{ mm}$$

or $$(q)_{max} = \frac{43.8}{\log_e(18.4/5)+1} = 19 \text{ W/m}$$ Ans.

If the pipe were bare the heat dissipation would have been :

$$(q)_{bare} = \frac{(t_i - t_0)2\pi}{1/h_0 r_2}$$

$$= \frac{(65-20)\times 2\pi \times 8.5 \times 5}{1000}$$

$$= 12 \text{ W/m}$$ Ans.

It may be noted that the bare pipe could dissipate heat @ 12 W/m. On application of insulation upto a thickness of (18.4 – 10/2) = 13.4 mm the heat dissipation is increased to 19 W/m i.e. an increase of 36 %.

Example 2.4

A 60 watt lamp is buried in soil (k = 0.0084 J/s cm°C) at 0°C and burned until steady state is reached. Find the temperature 30 cm away if 1 watt lamp produces 1 J/s heat.

Solution :

From Eq. (2.28)

$$q = \frac{4\pi k(t_1 - t_2) r_1 r_2}{r_2 - r_1}$$

$$= \frac{4\pi k(t_1 - t_2)}{\frac{1}{r_1} - \frac{1}{r_2}}$$

In the present case the boundary conditions are :

$$t = t_1 \text{ at } r = r_1$$

$$t = 0 \text{ at } r = \infty$$

so $$q = 4\pi k(t_1) r_1$$

or $$t_1 = \frac{q}{4\pi k r_1}$$

$$q = 60 \times 1 \text{ J/s}$$

or $$t = \frac{60 \times 1}{4\pi \times 0.0084 \times 30}$$

$$= 19°\text{C}$$ Ans.

Example 2.5

Calculate the rate of radial heat flow from blading to the center for a turbine rotor as shown in Fig. 2.10. The dimensions are : R = 50 cm, y_0 = 4 cm, y_c = 14 cm. The conductivity of the material is 0.37 J/s cm°C. The temperatures are 600°C at r_2 = R = 50 cm, and 320°C at r_1 = 10 cm.

Solution :

Although the cooling of a turbine rotor is mainly by gas convection, a certain amount of conduction cooling is also present. To compute this amount it may be assumed that the heat flows radially inwards from the bladed periphery of the rotor disk and is carried away at the center by conduction along the axle, or by liquid cooling in the axle.

Considering the rotor shown in Fig. 2.10, let

y_c = thickness of the rotor disc at center (r = 0)

y_o = thickness of the rotor disc where the blading begins (r = R)

y = thickness of the rotor disk at radius r

Then, from the geometry of the rotor disc,

$$y = y_c - \frac{y_c - y_o}{R} r$$

or $$y = y_c - mr$$

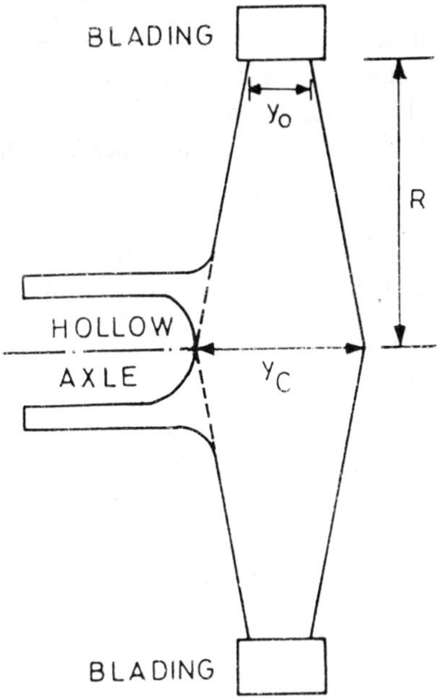

Fig. 2.10 Example 2.5

where $$m = \frac{y_c - y_o}{R}$$

since $$q = \frac{t_1 - t_2}{\text{Resistance}} \quad (1.7)$$

so $$q = \frac{t_2 - t_1}{\int \frac{dx}{kA}} = k\,(t_2 - t_1) \div \int_{r_1}^{r_2} \frac{dr}{2\pi r y}$$

$$= k\,(t_2 - t_1) \div \int_{r_1}^{r_2} \frac{dr}{2\pi r\,(y_c - mr)}$$

but $$\int_{r_1}^{r_2} \frac{dr}{r\,(y_c - mr)} = \frac{1}{y_c} \log_e \frac{r_2\,(y_c - mr_1)}{r_1\,(y_c - mr_2)}$$

so $$q = \frac{2\pi\, y_c\, k\,(t_2 - t_1)}{\log_e \frac{r_2\,(y_c - mr_1)}{r_1\,(y_c - mr_2)}}$$

now $$m = \frac{y_c - y_o}{R} = \frac{14 - 4}{50} = 0.2$$

so $$q = \frac{2\pi\,(14)(0.37)(600 - 320)}{\log_e \frac{50(14 - 0.2 \times 10)}{10(14 - 0.2 \times 50)}}$$

$$q = 3420 \text{ J/s} = 3420 \text{ W}$$ Ans.

Example 2.6

A hollow cone of metal (k = 0.15 W/cm°C) 1 mm thick connects a pipe of 6 cm diameter with the exterior metal sheath of an insulated vessel. The base of the cone is 20 cm in diameter and its length, measured along the cone is 24 cm. If the pipe is at 200°C and the base of the cone at 0°C, calculate the rate of heat loss through the cone.

Solution :

A hollow cone may be considered to be made from the sector of circle. Hence, we can use for this case Eq. (2.23) with appropriate modification.

$$q = \frac{2\pi l k(t_1 - t_2)}{\log_e (r_2 / r_1)} \times p \qquad (2.23)$$

Here l will be the thickness of the cone. i.e., $l = 1$ mm = 0.1 cm. To compute r_1 and r_2 we proceed as follows. Let p be the fraction of the circle which the sector represents.

Then $\quad 2\pi r_1 p = 6 \times \pi$

$$2\pi r_2 p = 20 \times \pi$$

or $\quad r_2/r_1 = \frac{10}{3}$

also $\quad r_2 - r_1 = 24$

so $\quad r_1 = 10.3$ cm

$$r_2 = 34.3 \text{ cm and } p = 0.292$$

Substituting in Eq. (2.23) we get :

$$q = \frac{(2\pi)(0.292)(0.1)(0.15)(200)}{\log_e (34.3/10.3)}$$

$$= 4.6 \text{ W}$$

Ans.

2.10 EFFECT OF VARIABLE CONDUCTIVITY

In all the cases considered so far we assumed that k is constant within the temperature range in question. There are, however, cases where its variation may cause an appreciable change in heat flow. For all practical purposes the conductivity k may be expressed as a linear function of the temperature,

$$k = k_0 (1 + \alpha t)$$

where k_0 is the thermal conductivity at 0 C and α a constant which is negative for metallic conductors and positive for insulating materials.

Case 1 : Plane Wall

The temperature profile and the heat flow expression corresponding to Eqs. (2.19) and (1.2) for the case of plane wall may be obtained as follows :

$$q = -k A \frac{dt}{dx}$$

$$= - k_0 (1 + \alpha t) A \frac{dt}{dx} \tag{2.41}$$

The temperature profiles inside the wall for different values of α may be plotted with this equation and are shown in Fig. 2.11. It will be observed that :

If α = positive ; curve is convex.

α = 0 ; curve is linear.

α = negative ; curve is concave.

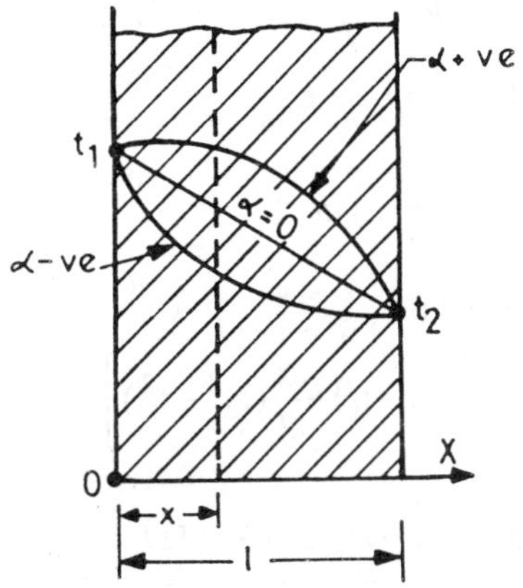

Fig. 2.11 Effect of Variable Conductivity

Separating variables in Eq. (2.41) and integrating ;

$$\int_0^x q\,dx = \int_{t_1}^{t} - \{k_o A (1 + \alpha t)\}\, dt$$

or

$$q = -\frac{k_o A}{x}\left\{\left(t + \frac{\alpha t^2}{2}\right) - \left(t_1 + \frac{\alpha t_1^2}{2}\right)\right\} \tag{2.42a}$$

Now at $x = l$; $t = t_2$

so

$$q = -\frac{k_o A}{l}\left\{\left(t_2 + \frac{\alpha t_2^2}{2}\right) - \left(t_1 + \frac{\alpha t_1^2}{2}\right)\right\}$$

so

$$q = \frac{k_o A}{l}\left\{1 + \frac{\alpha}{2}(t_1 + t_2)\right\}(t_1 - t_2) \tag{2.42b}$$

Equating Eqs. (2.42a) and (2.42b) we get ;

$$\frac{x}{l}\left\{\left(t_2 + \frac{\alpha t_2^2}{2}\right) - \left(t_1 + \frac{\alpha t_1^2}{2}\right)\right\} = \left\{\left(t + \frac{\alpha t^2}{2}\right) - \left(t_1 + \frac{\alpha t_1^2}{2}\right)\right\}$$

or

$$t = \frac{1}{\alpha}\left[(1+\alpha t_1)^2 \{(1+\alpha t_1)^2 - (1+\alpha t_2)^2\}\frac{x}{l}\right]^{\frac{1}{2}} - \frac{1}{\alpha} \tag{2.43}$$

Thus equations (2.42b) and (2.43) give the expression for heat flow and temperature distribution through a plane wall.

Case 2 : Tube

For the case of heat conduction through a long tube made of material having linearly varying conductivity the corresponding expressions may be obtained as follows :

$$q = -kA\frac{dt}{dr}$$

$$= -kL\,2\pi r\frac{dt}{dr}$$

$$= -[k_o(1+\alpha t)]\,L\,2\pi r\frac{dt}{dr}$$

or $$q\frac{dr}{r} = -2\pi L k_o(1+\alpha t)\,dt \qquad \text{(a)}$$

Integrating Eq. (a) between r_2 and r_1,

$$q\log_e\frac{r_2}{r_1} = -2\pi L k_o\left[(t_2-t_1)+\frac{\alpha}{2}(t_2^2-t_1^2)\right]$$

or $$q = \frac{2\pi Lk_o\left[1+\frac{\alpha}{2}(t_1+t_2)\right](t_1-t_2)}{\log_e(r_2/r_1)} \qquad \text{(b)}$$

Integrating Eq. (a) between r and r_1,

$$q = \frac{2\pi Lk_0\left[1+\frac{\alpha}{2}(t_1+t)\right](t_1-t)}{\log_e(r/r_1)} \qquad (2.44)$$

From Eq. (b) and Eq. (2.44), we get ;

$$\frac{\left[1+\frac{\alpha}{2}(t_1+t)\right](t_1-t)}{\log_e(r/r_1)} = \frac{\left[1+\frac{\alpha}{2}(t_1+t_2)\right](t_1-t_2)}{\log_e(r_2/r_1)}$$

or $$\frac{\alpha}{2}t_1^2-\frac{\alpha}{2}t^2+t_1-t-\frac{\log_e(r/r_1)}{\log_e(r_2/r_1)}\left[1+\frac{\alpha}{2}(t_1+t_2)\right](t_1-t_2)=0$$

or $$t^2+\frac{2}{\alpha}t-t_1^2-\frac{2}{\alpha}t_1+\frac{2}{\alpha}\frac{\log_e(r/r_1)}{\log_e(r_2/r_1)}\left[\frac{(1+\alpha t_1)^2-(1+\alpha t_2)^2}{2\alpha}\right]=0$$

so $$t = -\frac{1}{\alpha}\pm\left[\frac{1}{\alpha^2}+t_1^2+\frac{2}{\alpha}t_1-\frac{2}{\alpha}\frac{\log_e(r/r_1)}{\log_e(r_2/r_1)}\left\{\frac{(1+\alpha t_1)^2-(1+\alpha t_2)^2}{2\alpha}\right\}\right]^{\frac{1}{2}}$$

$$= -\frac{1}{\alpha}\pm\frac{1}{\alpha}\left[1+\alpha^2t_1^2+2\alpha t_1-\frac{\log_e(r/r_1)}{\log_e(r_2/r_1)}\{(1+\alpha t_1)^2-(1+\alpha t_2)^2\}\right]^{\frac{1}{2}}$$

Since $t = t_2$ where $r = r_2$, only the positive sign of the square root can be used.

Hence, $$t = \frac{1}{\alpha}\left[(1+\alpha t_1)^2 - \frac{\log_e (r/r_1)}{\log_e (r_2/r_1)}\{(1+\alpha t_1)^2 - (1+\alpha t_2)^2\}\right]^{\frac{1}{2}} - \frac{1}{\alpha} \quad (2.44a)$$

Case 3 : Sphere

The procedure for a sphere is entirely analogous to that for the cylinder. We start from the equation,

$$q = -kA\frac{dt}{dr}$$

$$= -k\,4\pi r^2\frac{dt}{dr}$$

or $$q\frac{dr}{r^2} = -4\pi\,[k_o(1+\alpha t)]\,dt \quad (c)$$

Integrating Eq. (c) between r_2 and r_1,

$$q\left[\frac{1}{r_1} - \frac{1}{r_2}\right] = -4\pi k_0\left[(t_2 - t_1) + \frac{\alpha}{2}(t_2^2 - t_1^2)\right]$$

or $$q = \frac{4\pi k_0 r_1 r_2}{(r_2 - r_1)}\left[1 + \frac{\alpha}{2}(t_1 + t_2)\right](t_1 - t_2) \quad (d)$$

Integrating Eq. (c) between r and r_1.

$$q = \frac{4\pi k_0 r_1 r}{(r - r_1)}\left[1 + \frac{\alpha}{2}(t_1 + t)\right](t_1 - t) \quad (2.45)$$

From Eqs. (d) and (2.45) we get,

$$\frac{r_2}{(r_2 - r_1)}\left[1 + \frac{\alpha}{2}(t_1 + t_2)\right](t_1 - t_2) = \frac{r}{(r - r_1)} \times \left[1 + \frac{\alpha}{2}(t_1 + t)\right](t_1 - t)$$

or $$\frac{\alpha}{2}(t_1^2 - t^2) + (t_1 - t) - \left(\frac{r - r_1}{r_2 - r_1}\right)\left(\frac{r_2}{r}\right) \times \left[\frac{\alpha}{2}(t_1^2 - t_2^2) + (t_1 - t_2)\right] = 0$$

or $$t^2 + \frac{2}{\alpha}t - t_1^2 - \frac{2}{\alpha}t_1 + \frac{2}{\alpha}\left(\frac{r - r_1}{r_2 - r_1}\right)\frac{r_2}{r} \times \left[\frac{(1+\alpha t_1)^2 - (1+\alpha t_2)^2}{2\alpha}\right] = 0$$

so, $$t = -\frac{1}{\alpha} \pm \frac{1}{\alpha}\left[(1+\alpha t_1)^2 - \left(\frac{r - r_1}{r_2 - r_1}\right)\left(\frac{r_2}{r}\right) \times \{(1+\alpha t_1)^2 - (1+\alpha t_2)^2\}\right]^{\frac{1}{2}}$$

since $t = t_2$ at $r = r_2$, only the positive sign of the square root can be used.

Hence,

$$t = \frac{1}{\alpha}\left[(1+\alpha t_1)^2 - \left(\frac{r - r_1}{r_2 - r_1}\right)\left(\frac{r_2}{r}\right) \times \{(1+\alpha t_1)^2 - (1+\alpha t_2)^2\}\right]^{\frac{1}{2}} - \frac{1}{\alpha} \quad (2.45a)$$

Example 2.7

Heat flows through an annular wall of inside radius R_0 and outside radius R_1. The thermal conductivity varies linearly with temperature from K_0 at T_0 to K_1 at T_1.

(a) Develop an expression for heat flow through the wall at $r = R_0$. The temperatures of the wall at $r = R_0$ and $r = R_1$ are T_0 and T_1, respectively.

(b) Show how the expression for heat flow may be simplified if $(R_1 - R_0)$ is very small.

Solution :

(a) $$q = -k\,(2\pi r\,L)\,\frac{dt}{dr} \qquad (1.1)$$

or $$q\,\frac{dr}{r} = -(2\pi\,L)\,k\,dt$$

Since k varies linearly with temperature from K_0 at T_0 to K_1 at T_1 we can write ;

$$k = K_0 + \left(\frac{K_1 - K_0}{T_1 - T_0}\right)(t - T_0)$$

Hence, $$q\,\frac{dr}{r} = -(2\pi\,L)\left[K_0 + \left(\frac{K_1 - K_0}{T_1 - T_0}\right)(t - T_0)\right]dt$$

Integrating between R_1 and R_0 ;

$$q\log_e\frac{R_1}{R_0} = -(2\pi\,L)\left[K_0(T_1 - T_0) + \left(\frac{K_1 - K_0}{T_1 - T_0}\right)\left\{\frac{T_1^2 - T_0^2}{2} - T_0(T_1 - T_0)\right\}\right]$$

$$= (2\pi\,L)(T_0 - T_1)\left[K_0 + \frac{K_1 - K_0}{2}\right]$$

$$= (2\pi\,L)\left(\frac{K_0 + K_1}{2}\right)(T_0 - T_1)$$

or $$q = (2\pi\,L)\left(\frac{K_0 + K_1}{2}\right)(T_0 - T_1)\,[\log_e(R_1/R_0)]^{-1} \qquad \text{Ans.}$$

(b) If $(R_1 - R_0)$ is very small, we can write

$$\log_e\frac{R_1}{R_0} = \log_e\left(1 + \frac{R_1 - R_0}{R_0}\right)$$

$$\approx \frac{R_1 - R_0}{R_0}$$

hence, $$q = (2\pi\,L)(T_0 - T_1)\left(\frac{K_0 + K_1}{2}\right)\times\left(\frac{R_0}{R_1 - R_0}\right) \qquad \text{Ans.}$$

Example 2.8

(a) Develop an expression for the steady-state heat transfer rate through the walls of spherical container of inner radius R_0 and outer radius R_1. The temperatures are T_0 and T_1 at R_0 and R_1 respectively. Assume that the thermal conductivity of the wall varies as

$$k = k_0 + (k_1 - k_0)\left(\frac{T - T_0}{T_1 - T_0}\right)$$

(b) Estimate the rate of evaporation of liquid oxygen from a spherical container 1.8 m inside diameter covered with 30 cm of asbestos insulation. The temperatures at the inner and outer surfaces of the insulation are −183 C and 0 C respectively. The boiling point of oxygen is −183 C and the latent heat of vaporization is 212.5 kJ/kg. The thermal conductivity of the insulation is 0.157 and 0.125 W/mK at 0 C and −183 C respectively.

Solution :

(a)
$$q = -k\,(4\pi r^2)\,\frac{dt}{dr} \tag{1.1}$$

or
$$q\,\frac{dr}{r^2} = -4\pi k\, dt$$

$$= -4\pi\left[k_0 + (k_1 - k_0)\left(\frac{T - T_0}{T_1 - T_0}\right)\right] dt$$

Integrating between R_1 and R_0

$$q\left[\frac{1}{R_0} - \frac{1}{R_1}\right] = -4\pi\left[k_0 + (T_1 - T_0) + \left(\frac{k_1 - k_0}{T_1 - T_0}\right) \times \left\{\frac{T_1^2 - T_0^2}{2} - T_0(T_1 - T_0)\right\}\right]$$

$$= 4\pi\,(T_0 - T_1)\left(\frac{k_0 + k_1}{2}\right)$$

or
$$q = 4\pi R_0 R_1\left(\frac{k_0 + k_1}{2}\right)\left(\frac{T_0 - T_1}{R_1 - R_0}\right) \qquad \text{Ans.}$$

(b) Substituting the numerical values we get,

$$q = 4\pi\left(\frac{1.8}{2}\right)(1.2)\left(\frac{0.157 + 0.125}{2}\right) \times \left(\frac{-183 - 0}{1.2 - 0.9}\right)$$

$$= -1175 \text{ J/s}$$

so rate of heat flow into the container, $= 1175$ W

or rate of evaporation of oxygen, $= \dfrac{1175 \times 3600}{212.5 \times 1000} = 19.8$ kg/hr Ans.

2.11 EXTENDED SURFACES

In practice, problems are encountered in which solid walls of small areas are required to transfer relatively large quantities of heat. Normally, *heat exchangers* have plane or cylindrical walls which separate two *heat exchanging* fluids at different temperatures. The thermal conductivity of the surface film on one side may be less than that on the other. In order to increase the rate of heating or cooling, the heat flow area is increased by the attachment of *fins* or solid material to the wall of the heat exchanger on that side for which the coefficient h is small. These fins serve the purpose of increasing the surface transmittance.

Fins are widely used in engineering heat transfer equipments. Electrical apparatus like transformers and motors in which the generated heat should be efficiently transferred, are provided with fins on the outside surface. Similarly fins are provided on the cylinders of air cooled I. C. engines (like that of a motor cycle), and on a large variety of heat exchangers.

The extended surface located on the heat exchanging wall may be either in the form of individual rods known as *pin fins or spines*, or in the form of continuous plates known as fins. On tubes the *fins* may be located either longitudinally or circumferentially. The cross sectional shape of the extended surface in a plane normal to the wall is known as the *profile* of the spine or fin.

While increasing cooling performance is most important in selection and design of fins it is also desirable that they should require minimum cost of production, offer minimum resistance to fluid flow surrounding the surface and be easy to manufacture.

It may be observed that a fin is to be regarded as a system in which a metallic rod or plate is transferring heat to a fluid and is connected at its base to a heated wall. The surface coefficient *h* enters the basic equation ; however the attention has to be focussed mainly on heat conduction through the fin, in order to study the temperature distribution which varies from the base to the tip. The analysis of fins therefore is actually a heat conduction problem. Within the scope of the present work only elementary cases of fins will be considered.

2.12 HEAT TRANSFER FROM A SPINE OF UNIFORM CROSS-SECTION

As a simple illustration consider a thin rod protruding from a surface at a temperature higher than the surrounding air. Let the surface temperature be t_s and that of the ambient air be t_∞. Also let the heat transfer coefficient of air film at the surface of the spine be *h*.

Further, because of high thermal conductivity of the metal, we can assume that the temperature at any cross section is uniform. This assumption reduces the problem to unidirectional heat flow.

Consider an infinitesimal element of the spine, of length *dx* and cross sectional area *A* at a distance *x* from the base wall as shown in Fig. 2.12.

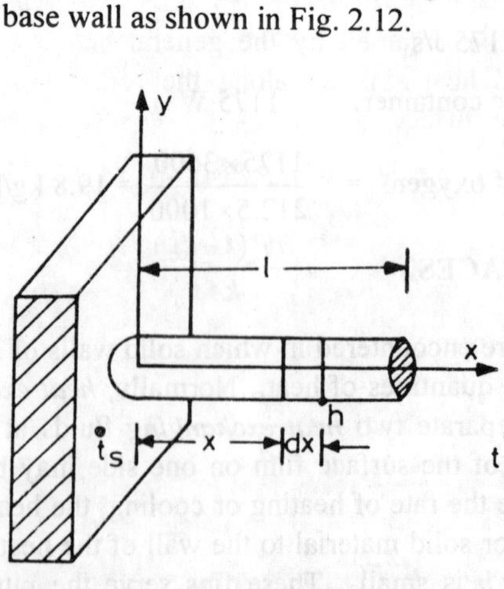

Fig. 2.12 Heat Transfer From A Spine

Under steady state the rate of heat flow into the element is ;

$$q_x = -kA\frac{dt}{dx} \tag{1.1}$$

Rate of heat flowing out of the element is the sum of heat flow by conduction at x + dx and that lost to surrounding air by convection ;

so $$q_x = \left\{-kA\frac{dt}{dx}+\frac{d}{dx}\left(-kA\frac{dt}{dx}\right)dx\right\}+\{h\,P\,dx\,(t-t_\infty)\}$$

where P = perimeter of the element,

t = temperature ; $t\,(x)$

Equating the two expressions we get ;

$$\frac{d}{dx}\left(-kA\frac{dt}{dx}\right)dx + h\,P\,dx\,(t-t_\infty) = 0$$

or $$-\frac{d^2t}{dx^2}+\frac{hP}{kA}(t-t_\infty) = 0$$

or $$\frac{d^2t}{dx^2} = m^2\,(t-t_\infty) \tag{2.46}$$

where $m = \sqrt{\dfrac{hP}{kA}}$

The constant m has the physical significance that ;

$$ml = \sqrt{\frac{hPl}{kA/l}}$$

i.e. it is the ratio of the surface conductance to the internal conductance.

Aliter

Eq. (2.46) may also be obtained by the general heat conduction equation (2.14) considering a distributed heat sink all along the rod, consuming heat equal to the convective heat loss at the surface.

$$-Q'\,(A\,dx) = h\,P\,dx\,(t-t_\infty)$$

or $$Q'/k = \frac{h\,P(t-t_\infty)}{kA}$$

Thus Eq. (2.14) takes the form ;

$$\nabla^2 t = \frac{h\,P}{kA}(t-t_\infty)$$

or for a unidirectional case ;

$$\frac{d^2t}{dx^2} = m^2\,(t-t_\infty) \tag{2.46}$$

The general solution of this second order linear differential equation may be obtained by assuming ;

$$T = (t - t_\infty)$$

so that Eq. (2.46) takes the form ;

$$\frac{d^2 T}{dx^2} = m^2 T$$

The solution of this equation will be ;

$$T = t - t_\infty = C_1 e^{mx} + C_2 e^{-mx} \qquad (2.47)$$

The constants C_1 and C_2 may be obtained by suitable boundary conditions. The boundary conditions may be assumed differently in three cases as below :

(a) Case 1 : (i) at $x = 0$; $t = t_s$

(ii) the rod is infinitely long ($l = \infty$) i.e. $x = \infty$; $t = t_\infty$

Applying these conditions to Eq. (2.47) we have respectively,

$$t_s - t_\infty = C_1 + C_2$$

and

$$t_\infty - t_\infty = 0 = C_1 e^{\infty} + C_2 e^{-\infty}$$

Solving these equations we get ;

$$C_1 = 0 \text{ and } C_2 = t_s - t_\infty$$

Thus the temperature distribution in the spine is given by ;

$$\frac{t - t_\infty}{t_s - t_\infty} = e^{-mx} \qquad (2.48)$$

The rate of heat flow through the fin is given by ;

$$(q)_{fin} = -kA\left(\frac{dt}{dx}\right)_{x=0}$$

$$= kA\{-m(t_s - t_\infty)e^{-mx}\}_{x=0}$$

$$= kA\,m(t_s - t_\infty)$$

$$= \sqrt{PhkA}\,(t_s - t_\infty) \qquad (2.49)$$

Aliter

The same result as Eq. (2.49) can also be obtained by integrating the expression for heat dissipated by the infinitesimal element due to convection ;

$$(q)_{fin} = \int_0^\infty h\,P\,dx\,(t - t_\infty)$$

$$= \int_0^\infty \{hP(t_s - t_\infty)e^{-mx}\}\,dx$$

$$= \sqrt{hPkA}\,(t_s - t_\infty) \qquad (2.49)$$

(b) Case 2 : (i) at $x = 0$; $t = t_s$

(ii) the end of rod is loosing heat i.e. heat conducted to the rod at $x = l$ is the same as convected by the end to the ambient air, or

$$= -k\left(\frac{dt}{dx}\right)_{x=l} = h_l\,(t_l - t_\infty)$$

Therefore Eq. (2.47) will result in the following :

$$t_s - t_\infty = C_1 + C_2$$

and
$$\left\{-k\frac{d}{dx}(C_1 e^{mx} + C_2 e^{-mx})\right\}_{x=l} = h_1\,(C_1 e^{ml} + C_2 e^{-ml})$$

or
$$mk\,(C_2 e^{-ml} - C_1 e^{ml}) = h_1\,(C_1 e^{ml} + C_2 e^{-ml})$$

Solving the two equations we get ;

$$C_1 = \frac{t_s - t_\infty}{1-\sigma e^{2ml}}$$

$$C_2 = \frac{(t_s - t_\infty)\sigma}{(\sigma - e^{-2ml})}$$

$$\text{where} \quad \sigma = \frac{h_1 + mk}{h_1 - mk}$$

Substituting these constants into Eq. (2.47) we get the dimensionless temperature distribution in the spine as :

$$\frac{t-t_\infty}{t_s - t_\infty} = \frac{exp\left\{ml\left(\frac{x}{l}\right)\right\} - exp\left\{-ml\left(\frac{x}{l}-2\right)\right\}}{1-\sigma e^{2ml}} \tag{2.50}$$

This may be expressed in terms of hyperbolic functions as ;

$$\frac{t-t_\infty}{t_s - t_\infty} = \frac{cosh\; m(l-x) + (h_1/mk)\,sinh\; m(l-x)}{cosh\;\; ml + (h_1/mk)\,sinh\; ml} \tag{2.51}$$

The rate of heat transfer from the fin is given by ;

$$(q)_{fin} = -kA\left(\frac{dt}{dx}\right)_{x=0}$$

$$= \sqrt{PhAk}\left\{\frac{\sigma e^{ml}+1}{\sigma e^{ml}-1}\right\} \tag{2.52}$$

$$= \sqrt{PhAk}\left\{\frac{(h_1/mk) + tanh\; ml}{1+(h_1/mk)\,tanh\; ml}\right\}(t_s - t_\infty) \tag{2.53}$$

(c) Case 3 : (i) at $x = 0$; $t = t_s$

(ii) the end of the rod is insulated i.e. there is no heat transfer at the tip, i.e.

$$\text{at} \qquad x = l; \frac{dt}{dx} = 0$$

Applying these conditions to equation (2.47) we get ;

$$t_s - t_\infty = C_1 + C_2$$

$$\text{and} \qquad C_1 e^{ml} - C_2 e^{-ml} = 0$$

Solving these equations we get ;

$$C_1 = \frac{t_s - t_\infty}{1 + e^{2ml}}$$

$$C_2 = \frac{t_s - t_\infty}{1 + e^{-2ml}}$$

Substituting these constants in Eq. (2.47) we get the temperature distribution in the spine for this case as ;

$$\frac{t - t_\infty}{t_s - t_\infty} = \frac{e^{mx}}{1 + e^{2ml}} + \frac{e^{-mx}}{1 + e^{-2ml}}$$

$$= \frac{\cosh m(l - x)}{\cosh ml} \qquad (2.54)$$

The rate of heat flow from the fin is given by ;

$$(q)_{fin} = -kA\left(\frac{dt}{dx}\right)_{x=0}$$

$$= \sqrt{PhkA}\,(t_s - t_\infty)\tanh ml \qquad (2.55)$$

The results obtained in Eqs. (2.54) and (2.55) can be verified by putting $h_1 = 0$ in Eqs. (2.51) and (2.53) respectively.

2.13 HEAT TRANSFER FROM A STRAIGHT FIN (PLATE) OF UNIFORM CROSS SECTION

In many applications straight fins are used all along the length of the wall, instead of a number of spines. Such fins are usually of rectangular profile and are called *straight rectangular fins* (Fig. 2.13)

The heat transfer from such a fin can be analysed by the same method as discussed in Section 2.12 for the spines. The results obtained will be similar for all the three cases.

If width of the plate fin is b and the thickness is δ then :

$$A = b\,\delta$$

$$P = 2(b + \delta)$$

$$= 2b \quad (b >> \delta)$$

$$\text{so} \qquad (m)_{rect} = \sqrt{\frac{hP}{kA}} = \sqrt{\frac{2h}{k\delta}} \qquad (2.56)$$

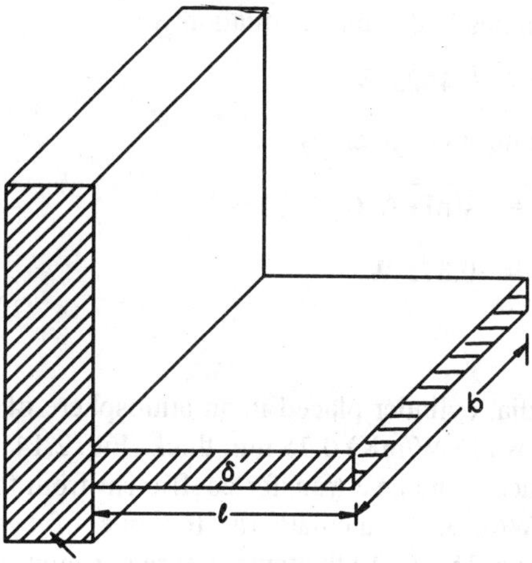

Fig. 2.13 Plate Fin

For a circular spine of radius r ;

$$(m)_{cir} = \sqrt{\frac{h2\pi r}{k\pi r^2}} = \sqrt{\frac{2h}{kr}} \tag{2.57}$$

It may also be pointed out that the results for straight fins attached to plane walls can also be made applicable to such fins attached to tubes if $\delta << D$ where D is the tube diameter.

Example 2.9

A pin fin 2.5 mm diameter is made of copper (k = 396 W/mK). It protrudes from a wall maintained at 95°C and placed in 25°C air. The convective heat transfer coefficient over the fin is 10 W/m^2K. Calculate the heat loss for the two cases as below :

(a) fin length 25 mm

(b) fin length infinite

Solution :

(a) Heat loss from a finite fin is given by

$$q = \sqrt{PhAk}\cdot\left\{\frac{(h_1/mk)+\tanh ml}{1+(h_1/mk)\tanh ml}\right\}(t_s - t_\infty) \tag{2.53}$$

$$P = \pi d = \pi \times (0.0025)\text{ m}$$

$$h = 10\text{ W/m}^2\text{k} = h_1 \text{ (say)}$$

$$A = \frac{\pi}{4}(0.0025)^2\text{ m}^2$$

$$k = 396\text{ W/mK}$$

$$m = \sqrt{hP/kA}$$

$$l = 0.025\text{ m}$$

Substituting these values in the above equation we get

$$q = 0.4425 \text{ W} \qquad \text{Ans.}$$

(b) Heat loss from infinite fin is given by

$$q = \sqrt{hPkA}\ (t_s - t_\infty) \qquad (2.49)$$

$$= 0.875 \text{ W} \qquad \text{Ans.}$$

Example 2.10

A 1 m long, 5 cm dia, cylinder placed in an atmosphere of 40 C is provided with 12 longitudinal straight fins (75 W/mK) 0.75 mm thick (Fig. 2.14). The fins protrude 2.5 cm from the cylinder surface. The heat transfer coefficient from the cylinder and fins to the ambient air is 23.3 W/m^{2}K. Calculate (a) the rate of heat transfer if the surface temperature of cylinder is 150°C (b) the temperature at center of fin.

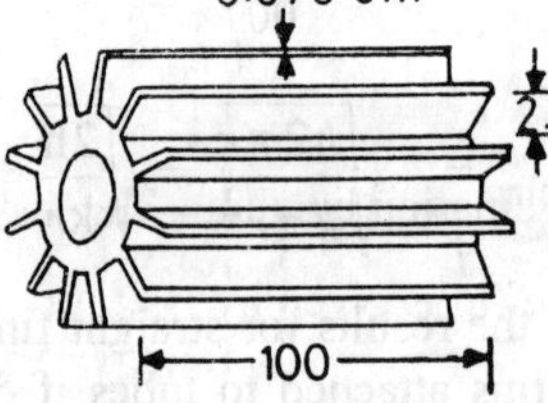

Fig. 2.14 Example 2.10

Solution :

(a) It will not be appropriate to assume that the fins are infinitely long. Let us therefore apply boundary conditions of case 3 mentioned in Section 2.12 i.e. heat loss from the tip is negligible

$$(q)_{fin} = \sqrt{PhAk}\ (t_s - t_\infty) \tanh (ml)$$

$$P = 2b = 2 \times 1 = 2 \text{ m}$$

$$A = \frac{1 \times 0.75}{1000} = 0.00075 \text{ m}^2$$

$$k = 75 \text{ W/mK}$$

Hence $$\sqrt{PhAk} = \sqrt{2 \times 23.3 \times 0.00075 \times 75} = 1.6$$

Also $$ml = \sqrt{\frac{Ph}{Ak}} \times \left(\frac{2.5}{100}\right)$$

$$= \sqrt{\frac{2 \times 23.3}{0.00075 \times 75}} \times \frac{2.5}{100} = 0.71$$

So $$(q)_{fin} = 1.6 \times (150 - 40) \times \tanh (0.71)$$

$$= 1.6 \times 110 \times 0.61 = 110 \text{ W}$$

For 12 fins the heat dissipation will be ;

$$(q)_F = 12 \times 110 = 1320 \text{ W}$$

Further, the area of the cylinder not occupied by the fins will also dissipate heat :

$$(q)_c = h\,A\,\Delta t$$

$$= 23.3 \times \left\{\left(\pi \times \frac{5}{100} \times 1\right) - (12 \times 0.00075 \times 1)\right\} \times (150 - 40)$$

$$= 23.3 \times 0.147 \times 110 = 375 \text{ W}$$

Hence the total heat transfer ;

$$q = (q)_F + (q)_c = 1320 + 375 = 1695 \text{ W} \qquad \text{Ans.}$$

If the cylinder was bare the heat transfer would have been

$$q = h\,A\,\Delta t = 23.3 \times \left(\pi \times \frac{5}{100} \times 1\right) \times 110$$

$$= 398 \text{ W}$$

Thus there is an increase in heat dissipation by about 300 %.

(b) The temperature at the center of the fin is given by :

$$\frac{t - t_\infty}{t_s - t_\infty} = \left(\frac{\cosh m(l - x)}{\cosh ml}\right)_{x=1.25}$$

$$\frac{t - 40}{150 - 40} = \frac{\cosh 28.4(1.25/100)}{\cosh 0.71} = \frac{\cosh 0.355}{\cosh 0.71}$$

$$= \frac{1.063}{1.2628} = 0.79$$

or $$t = (40 + 0.79 \times 110) = 126.9°\text{C} \qquad \text{Ans.}$$

2.14 ERROR OF MEASUREMENT OF TEMPERATURE IN A THERMOMETER WELL

Another important application of the above theory is found in estimating the error in the value of temperature measured by a thermometer dipped in a *thermometer well.* A thermometer well is a small tube welded radially into another tube through which the fluid, the temperature of which is required to be measured, is flowing. The well is filled with some other liquid and a thermometer is dipped into this liquid.

The temperature recorded by the thermometer will be that of the fluid in the bottom of the well. Since the wall temperature of the tube will be less than the fluid temperature flowing within (if hotter than ambient), the heat will flow from the bottom of the well towards the wall of the tube all along the well (Fig. 2.15). Thus the bottom of the well will have a tendency to become colder than the fluid flowing around. Hence the indicated temperature will not be true fluid temperature. The error in the recorded temperature may be calculated as illustrated in the following example.

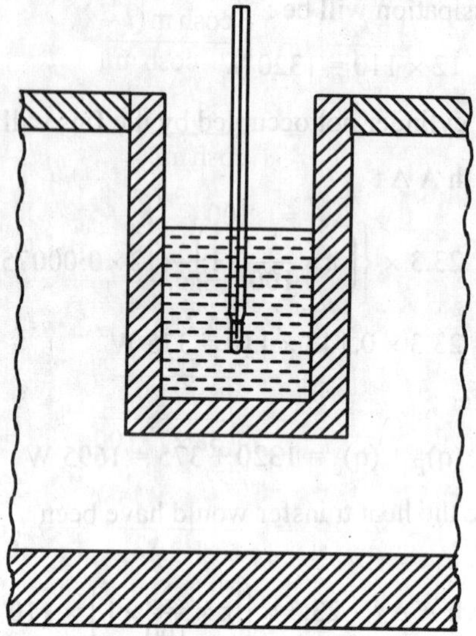

Fig. 2.15 A Thermometer Well

Example 2.11

A mercury thermometer placed in a well filled with oil is required to measure the temperature of compressed air flowing in a pipe. The well is 140 mm long and is made of steel (k = 58 W/mK) 1.0 mm thick. The temperature recorded by the well is 100 C while the pipe wall temperature is 50 C. The air film conductance outside the wall may be assumed to be 29 W/m²K. Estimate the true temperature of the air.

Solution :

Consider the well to be a spine protruding 140 mm from the wall of the pipe (at 50 C). Fig. 2.16 shows the temperature distribution in the well.

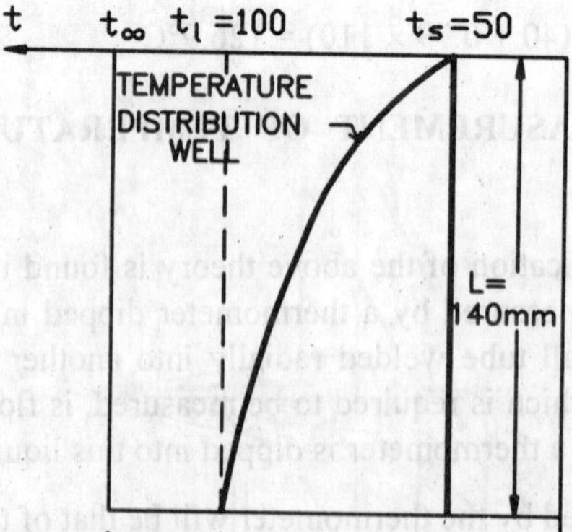

Fig. 2.16 Temperature Distribution in Thermometer Well

As an approximation we may assume that there is no heat flow at the tip of the well, hence Eq. (2.54) may be used ;

$$\frac{t-t_\infty}{t_s-t_\infty} = \left(\frac{\cosh m(l-x)}{\cosh ml}\right)_{x=l}$$

$$= \frac{1}{\cosh ml}$$

Now $t_l = 100\text{ C} ;\ t_s = 50\text{ C} ;\ l = 140\text{ mm}$

so $$ml = \sqrt{\frac{Ph}{kA}} \times l = \sqrt{\frac{2\pi rh}{k(2\pi r\delta)}} \times l$$

$$= \sqrt{\frac{29}{58 \times 1/1000}}\left(\frac{140}{1000}\right) = 3.14$$

so $$\frac{100-t_\infty}{50-t_\infty} = \frac{1}{\cosh 3.14} = \frac{1}{11.6}$$

or $t_\infty = 104.5°\text{C}$ Ans.

Hence the true temperature of air is 104.5°C and the measurement error = 4.5°C.

In order to reduce the temperature error the following steps are necessary.

(a) The k value of the metal of well should be small.

(b) The well should be long and as thin as possible.

(c) The value of h should be large or preferably the well should be finned from the outside.

2.15 GENERALISED EQUATION FOR FINS OR SPINES

Equation (2.46) was developed in Section 2.12 for the case of a spine of uniform section. In case the profile of fin or spine is non- uniform, and the area and perimeter at any section are functions of x, then a generalized equation is helpful in solving the problem.

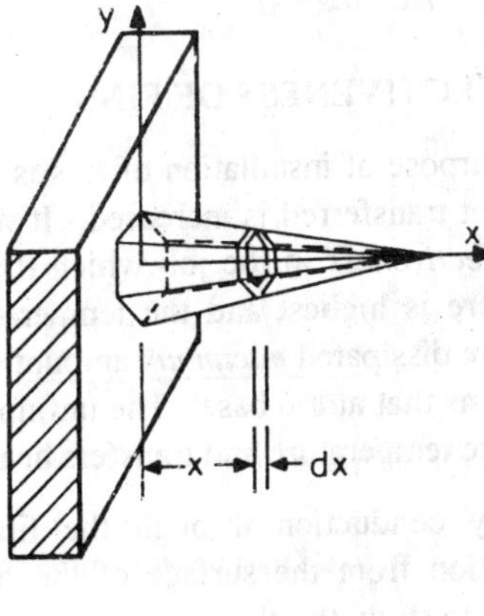

Fig. 2.17 A Fin of a General Profile

In this case also the same assumptions, as mentioned in Section 2.12 will be made. Further, let the contour of the spine be such that its cross-sectional area be given by $A_x = \phi(x)$, and the perimeter be $P_x = \psi(x)$. Then the heat balance equation for the infinitesimal element of thickness dx at a distance x will be given by :

$$-kA_x \frac{dt}{dx} = \left\{ -kA_x \frac{dt}{dx} + \frac{d}{dx}\left(-kA_x \frac{dt}{dx} \right) dx \right\} + [h\, P_x\, dx\, (t - t_\infty)]$$

$$\frac{d}{dx}\left(A_x \frac{dt}{dx} \right) = \frac{h}{k} P_x (t - t_\infty)$$

or
$$A_x \frac{d^2 t}{dx^2} + \frac{d\,A_x}{dx}\frac{dt}{dx} - \frac{h}{k} P_x (t - t_\infty) = 0$$

or
$$x^2 \frac{d^2 t}{dx^2} + \left(\frac{d\,A_x / dx}{A_x} x^2 \right) \frac{dt}{dx} - \frac{h x^2 P_x}{k\,A_x} (t - t_\infty) = 0 \qquad (2.58)$$

This equation is the generalized equation applicable to all fins of any *profile* or cross-section. It will be noted that it is a modified *Bessel equation,* hence its solution may be obtained with the help of Bessel functions.

For the case of fins of uniform section, A_x and P_x are not functions of x but are constants, hence Eq. (2.53) takes the form;

$$\frac{d^2 t}{dx^2} - \frac{hP}{kA} (t - t_\infty) = 0 \qquad (2.46)$$

A generalized equation similar to Eq. (2.58) may also be set up for fins of symmetrical sections along the central lines whose profile contours are of the form $y_x = f(x)$. By the substitution of contour equation we have ;

$A_x = 2\, y_x$ per unit length ; $P_x = 2 + 4\, y_x \approx 2$ as y_x is very smali

or
$$2y_x \frac{d^2 t}{dx^2} + 2\, \frac{dy_x}{dx} \frac{dt}{dx} - \frac{h}{k} 2\, (t - t_\infty) = 0$$

or
$$y_x \frac{d^2 t}{dx^2} + \frac{dy_x}{dx} \frac{dt}{dx} - \frac{h}{k} (t - t_\infty) = 0 \qquad (2.59)$$

2.16 EFFICIENCY OR EFFECTIVENESS OF FINS

As already discussed the purpose of installation of fins is to increase the heat transfer area, so that the amount of heat transferred is increased. It will, however, be interesting to study their *efficiency* or *effectiveness* in the job which they are required to perform. Evidently, the base temperature is highest and the temperature along the fin goes on decreasing. The fin would have dissipated *maximum* amount of heat if its temperature all along the length was the same as that at the base. The unfinned portion at the surface of the heat exchanger is at the base temperature and transfers heat at 100 percent efficiency.

As the heat passes on by conduction through the fin its temperature goes on decreasing. The heat dissipation from the surface of the fin will therefore not be at maximum efficiency. In order to study the *thermal performance* of the fin, we define *fin efficiency* or *effectiveness* as :

$$(\eta)_{fin} = \frac{\{\text{heat transferred across the fin surface}\}}{\left\{\begin{array}{c}\text{theoretical heat transfer across an}\\ \text{equal area held at a constant}\\ \text{temperature equal to that of base}\end{array}\right\}}$$

$$= \frac{(q)_{fin}}{(q)_s}$$

$$= \frac{1}{(t_s - t_\infty)\, Ah} \int_0^A (t - t_\infty)\, h\, dA$$

$$= \frac{1}{(t_s - t_\infty)\, A} \int_0^A (t - t_\infty)\, dA$$

where t_s = base temperature of the fin, °C.

A = area of fin, m^2

t = temperature distribution along the fin, °C

It is assumed that the value of h remains constant throughout.

(a) For the case of a *straight rectangular fin* shown in Fig. 2.13 we have :

$$q_{fin} = \sqrt{PhAk}\, (t_s - t_\infty) \tanh ml \quad (2.55)$$

$$= \sqrt{2b\, h b\delta\, k}\, (t_s - t_\infty) \tanh ml$$

$$= b\sqrt{2h\delta\, k}\, (t_s - t_\infty) \tanh ml$$

and $$(q)_s = 2b\, l\, h\, (t_s - t_\infty)$$

or $$(\eta)_{fin} = \frac{b\sqrt{2h\delta k}\, \tanh ml}{2bhl}$$

$$= \sqrt{\frac{k\delta}{2h}}\, \frac{\tanh ml}{l}$$

or $$(\eta)_{fin} = \frac{\tanh ml}{ml} \quad (2.60)$$

where $$m = \sqrt{\frac{2h}{k\delta}} \quad (2.56)$$

It may be observed from the expression of *fin efficiency*, Eq. (2.60), that higher values of δ, and lower values of h and l, all increase the efficiency of the fin.

(b) An expression for the efficiency of a *spine* or *pin-fin* of uniform cross section can also be derived on similar lines and the result will be :

$$(\eta)_{fin} = \frac{\tanh ml}{ml}$$

where $$m = \sqrt{\frac{2h}{kr}}, \tag{2.57}$$

Standard curves for fin-efficiency are available for :

(i) rectangular and triangular fins, and

(ii) circumferential rectangular fins.

These curves are shown in Fig. 2.18 and Fig. 2.19 respectively. For the former case it has been found convenient to define (i) profile area $A_m = \delta l$ for rectangular fins and $\delta l/2$ for triangular fins (ii) effective length $l_o = l + \delta/2$ for rectangular fins and l for triangular fins.

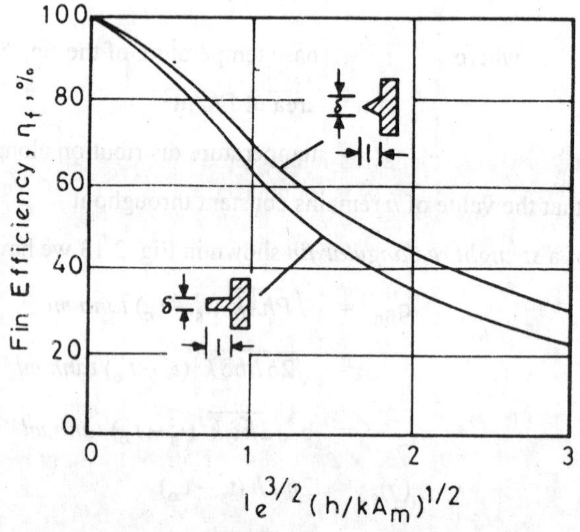

Fig. 2.18 Efficiency of Rectangular and Triangular Fins

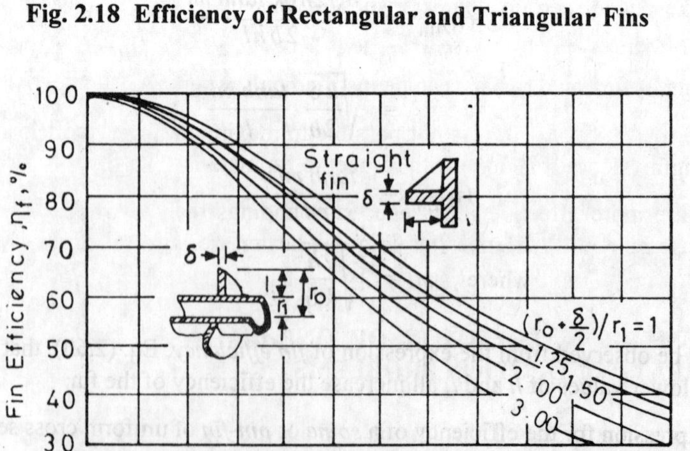

Fig. 2.19 Efficiency of Circumferential Rectangular Fins

Example 2.12

Aluminum circumferential fins (1 mm thick) are soldered to 2.5 cm O. D. pipe. Other data is : fin length = 1.5 cm, pipe wall temperature = 100°C, ambient temperature = 25°C and h = 65 W/m^{2}K. Calculate the rate of heat loss from a fin.

Solution :

The parameters needed to use Fig. 2.19 are :

$$\left(r_o + \frac{\delta}{2} - r_1\right)^{3/2} = \{(2.75 + 0.05 - 1.25)/100\}^{3/2}$$

$$= 0.00193$$

$$[2\, h_o/k\, \delta\, (r_o - r_1)]^{1/2} = [2 \times 65/200 \times 0.001 \times 0.015]^{1/2}$$

$$= 208$$

and
$$\left(r_o + \frac{\delta}{2}\right) \div r_1 = \frac{2.8}{1.25} = 2.24$$

Hence the value of fin efficiency as interpolated on curve of value 2.24 and abscissa of 0.00193 × 208 = 0.402 in Fig. 2.19 is found as ;

$$\eta_{fin} = 0.91$$

hence
$$q = 0.91 \times h \times A\,(t_s - t_\infty)$$

$$= 0.91 \times 65 \times 2\,\pi \left\{\left(r_o + \frac{\delta}{2}\right)^2 - r_i^2\right\}(100 - 75)$$

$$= 0.91 \times 65 \times 2\,\pi\,(7.84 - 1.56) \times 10^{-4} \times 25$$

$$= 17.5 \text{ W}$$ Ans.

From the above, certain characteristics of fins in general can be distinguished as below :

(a) Placing of fins on a surface can be justified only when hA/Pk < < 1.

(b) Since $\eta_{fin} = 1$ at $l = 0$, it is not possible to maximize fin performance with respect to length. Hence it is done with respect to fin material.

(c) Fins are more effective in gaseous surroundings than in the liquid surroundings, because $h_g << h_l$. Hence fins in a gas heater are more effective than in a water heater.

2.17 NUMERICAL SOLUTION OF ONE-DIMENSIONAL STEADY STATE HEAT CONDUCTION PROBLEMS

In earlier sections analytical solutions of the heat conduction equation were given. As an alternative method of solving the problems, a numerical technique can be employed by writing the heat balance equations in finite difference form. For this purpose the continuous domain of the solid is replaced by a *net-work* of preselected points within the domain. The method adopted is quite similar to approximate numerical integration.

The solid in question is subdivided into a number of small volumes of a convenient size. The center of each small volume called *nodal point* is assigned a reference number, 1, 2, 3, ...*n*. The volume size is chosen as small as possible so that its temperature may be assumed to be equal to that at its center. Heat balance equations are then set up for each small volume on the basis of finite difference approximation replacing the derivatives of heat conduction differential equation. There will thus be n simultaneous equations. These equations are solved numerically and the values of unknowns i.e. the temperatures at the n points obtained. The accuracy of the results calculated for these nodal points by this technique will depend upon how closely or rationally these are selected, so that finite difference approximation is accurately applicable.

Let us consider, as an illustration, a long rectangular thin solid bar, in which heat is conducted in one direction only (i.e. surface is considered as insulated). Let the width of the bar be unity and thickness δ as shown in Fig. 2.20. Consider that the bar is divided into small volumes of an arbitrary length δ, and each subvolume is designated with the number 1, 2, 3 and 4 at its nodal points.

We can now replace the whole solid by a network of imaginary wires joining these nodal points as shown in Fig. 2.21.

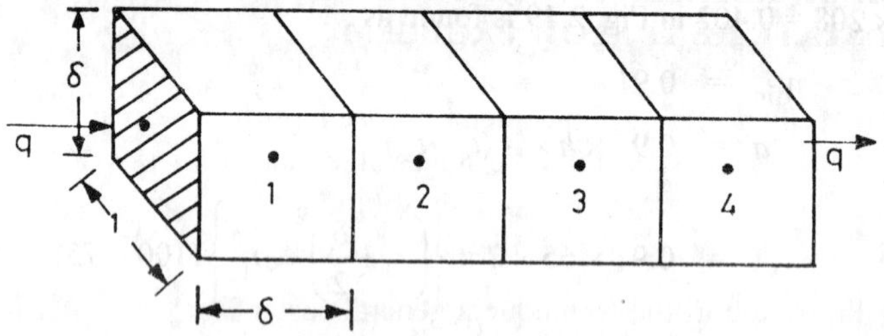

Fig. 2.20 Selection of Nodal Points

Fig. 2.21 Solid Replaced by Imaginary Wires

This network will be a *thermal equivalent* of the original solid in as much as the steady heat flow q in the solid will approximate the steady heat flow by conduction between nodal points of two adjacent subvolumes for which the imaginary wire was substituted. Heat balance equations for these nodal points may now be written in the manner given below :

(i) Consider nodal point 2. For steady state the amount of heat received from nodal point 1 is the same as delivered to nodal point 3 i.e.

$$q_{1-2} - q_{2-3} = 0$$

or

$$q_{1-2} + q_{3-2} = 0$$

or

$$K_{1-2}(t_1 - t_2) + K_{3-2}(t_3 - t_2) = 0$$

where K = thermal conductance.

Since the solid is uniform and homogeneous throughout its length, $K_{1-2} = K_{3-2}$.

or
$$(t_1 - t_2) + (t_3 - t_2) = 0$$
$$t_1 + t_3 - 2\,t_2 = 0 \qquad (2.61)$$

(ii) Similarly heat balance equation for nodal point 3 will be ;

$$t_2 + t_4 - 2\,t_3 = 0$$

(iii) It will be observed that the heat balance equations for points 1 and 4 will require different analysis depending upon their boundary conditions which are unknown in this case. The equations for sub-volumes, however, may be set up as explained later.

(iv) Finally, the four simultaneous equations known as *residual equations*, may be solved for the values of temperature at *nodal points*. We may thus obtain the *temperature distribution* in the solid.

The method discussed above for obtaining heat balance equation may be adopted for two-dimensional heat flow also in a similar manner. In this case each nodal point will be subjected to heat flow from two directions instead of one, as will be discussed later.

2.18 USE OF RELAXATION METHOD FOR SOLVING ONE-DIMENSIONAL STEADY STATE HEAT FLOW PROBLEM

In the preceding section, fundamental procedure of numerical method was discussed for heat conduction. We obtained n linear algebraic equations which can be solved by any of the conventional methods. The solution of a large number of simultaneous equations is most conveniently handled by the Southwell's *relaxation method.*

As an illustration of the technique a general case of one-dimensional heat conduction through a solid is considered in example 2.13.

Example 2.13

A bronze fin with dimensions as shown in Fig. 2.22 is exposed to an ambient fluid on all surfaces except the base. The base is at a uniform temperature of 350 C, while the

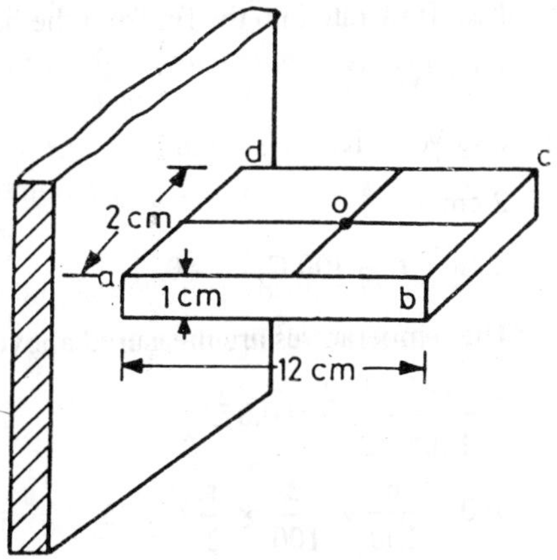

Fig. 2. 22 Example 2.13

ambient air is at 50 C. Calculate numerically the temperature distribution in the fin and also the heat flow from the fin to the ambient fluid. The value of thermal conductivity for bronze is 87.5 W/mK and fin surface conductance is 580 W/m²K.

Solution :

The dimensions of the fin are such that unidirectional flow of heat may be assumed.

Let us divide the fin into subvolumes as shown in Fig. 2.23, with seven nodal points each 2 cm apart. There are five complete subvolumes and two half subvolumes. Let temperatures at nodal points be represented by t_1, t_2, ..., ..., t_7 measured above ambient as datum. Thus, $t_1 = (350 - 50) = 300$ C.

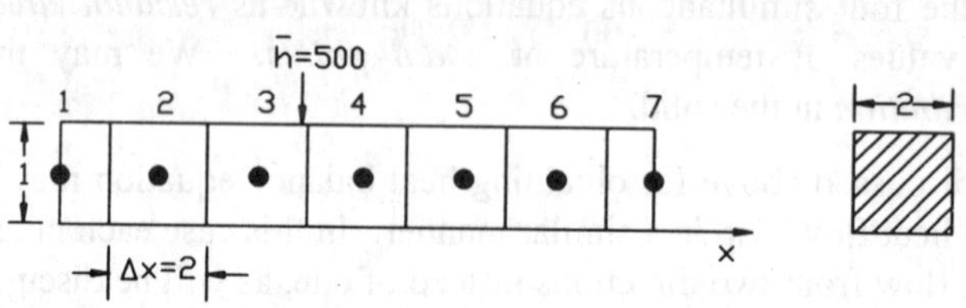

Fig. 2.23 Selection of Nodal Points

If the fin was insulated from the surface then Eq. (2.61) could have been used straight away for writing the heat balance equation. In this case however the heat is being dissipated from the surface ; hence this fact also should be taken into consideration while setting up residual equations.

(i) Nodal point 1

Heat balance equation is :

$$Q_1 = q_b + q_{2-1} + q_{\infty-1}$$

$$= q_b + \frac{kA}{\Delta x}(t_2 - t_1) + hP\frac{\Delta x}{2}(t_\infty - t_1)$$

where Q_1 = residual heat flow rate at point 1 which will be equal to zero for steady state.

q_b = heat flow rate into the fin from the base.

k = 87.5 W/mK

h = 580 W/m²K

Δx = 2 cm

A = 2 cm²; $t_1 = 300$ C; $t_\infty = 0$

(The temperatures are measured above t_∞)

Now $$\frac{kA}{\Delta x} = \frac{87.5 \times 2 \times 100}{100^2 \times 2} = 0.875$$

$$hP\frac{\Delta x}{2} = 580 \times \frac{6}{100} \times \frac{2}{100} \times \frac{1}{2} = 0.35$$

so $$Q_1 = q_b + 0.875\, t_2 - (0.875 + 0.35)\, t_1$$

Let $$q_1 = \frac{Q_1}{0.875}$$

$$= 1.26\, q_b + t_2 - 420 \qquad (2.62)$$

(ii) Nodal Point 2

$$Q_2 = q_{1-2} + q_{3-2} + q_{\infty-2}$$

$$= \frac{k\,A}{\Delta x}(t_1 - t_2) + \frac{k\,A}{\Delta x}(t_3 - t_2) - (h\,P\,\Delta x\, t_2)$$

$$= 0.875\,(t_1 + t_3 - 2\,t_2) - 0.7\,t_2$$

so $$q_2 = \frac{Q_2}{0.875} = 300 + t_3 - 2.8\,t_2 \qquad (2.63)$$

(iii) Nodal Point 3

$$Q_3 = q_{2-3} + q_{4-3} + q_{\infty-3}$$

$$q_3 = t_2 + t_4 - 2.8\,t_3 \qquad (2.64)$$

The equations for points 4, 5 and 6 will be similar.

(iv) Nodal Point 7

$$Q_7 = q_{6-7} + q_{\infty-7}$$

$$= \frac{k\,A}{\Delta x}(t_6 - t_7) - \left(h\,P\frac{\Delta x}{2}\right) t_7 - (h\,A\,t_7)$$

or $$q_7 = t_6 - \left(1 + \frac{0.35}{0.875} + \frac{580 \times 2/100^2}{0.875}\right) t_7$$

$$= t_6 - 1.413\,t_7$$

Therefore, the residual equations for various nodal points may be summarized as below :

Nodal Point	*Residual Equation*
1	$q_1 = 1.26\,q_b + t_2 - 420$
2	$q_2 = 300 + t_3 - 2.8\,t_2$
3	$q_3 = t_2 + t_4 - 2.8\,t_3$
4	$q_4 = t_3 + t_5 - 2.8\,t_4$
5	$q_5 = t_4 + t_6 - 2.8\,t_5$
6	$q_6 = t_5 + t_7 - 2.8\,t_6$
7	$q_7 = t_6 - 1.413\,t_7$

It will be observed from Eq. (2.61) that the values of *residuals* q_1, q_2, ..., q_7 are zero in *steady state* heat flow. The above residual equations equated to zero could then be solved for the values of temperatures by relaxation method as given in Tables 2.1 and Table 2.2.

Table 2.1 Operation Table - Example 2.13

Operation	Δq_1	Δq_2	Δq_3	Δq_4	Δq_5	Δq_6	Δq_7
$\Delta t_2 = 1$	+ 1	– 2.8	+ 1	0	0	0	0
$\Delta t_3 = 1$	0	+ 1	– 2.8	+ 1	0	0	0
$\Delta t_4 = 1$	0	0	+ 1	– 2.8	+ 1	0	0
$\Delta t_5 = 1$	0	0	0	+ 1	– 2.8	+ 1	0
$\Delta t_6 = 1$	0	0	0	0	+ 1	– 2.8	+ 1
$\Delta t_7 = 1$	0	0	0	0	0	+ 1	– 1.413
$\Delta t_2 = 1$, $\Delta t_3 = 1$	+ 1	– 1.8	– 0.8	– 0.8	– 0.8	– 0.8	– 0.413

The numerical solution is carried on in Table 2.2 using initial block relaxation and over relaxation. The final results have been checked with those obtained by analytical solution.

2.19 ONE DIMENSIONAL STEADY STATE HEAT CONDUCTION WITH UNIFORM HEAT GENERATION

There are some practical systems where heat is generated internally at a uniform rate. The conduction of heat in such cases may be considered analytically, in similar manner, as dealt with in Section 2.5 and 2.6 with the difference that the Eq. (2.14) is used as the fundamental equation.

Case 1 : Heat generation in Plane Wall

For one dimensional, steady state heat conduction through a wall with internal heat generation Q' W/m^3, the Eq. (2.14) takes the form :

$$\frac{d^2 t}{dx^2} + \frac{Q'}{k} = 0 \qquad (2.66)$$

Integrating this equation, we have ;

$$\frac{dt}{dx} = -\frac{Q'}{k} x + C$$

or

$$t = -\frac{Q'}{k}\frac{x^2}{2} + Cx + D \qquad (2.67)$$

The values of C and D are obtained as below by applying the boundary conditions that $t = t_1$ at $x = x_1$ and $t = t_2$ at $x = x_2$

$$C = -\frac{t_1 - t_2}{x_2 - x_1} + \frac{Q'}{2k}(x_1 + x_2)$$

$$D = \frac{t_1 x_2 - t_2 x_1}{x_2 - x_1} - \frac{Q'}{2k} x_1 x_2$$

Hence the temperature distribution is given by ;

$$t = -\frac{Q'}{2k} x^2 + \frac{Q'}{2k}\{(x_1 + x_2)x - x_1 x_2\} - \left\{\frac{t_1 - t_2}{x_2 - x_1} x - \frac{t_1 x_2 - t_2 x_1}{x_2 - x_1}\right\} \qquad (2.68)$$

Table 2.2 Relaxation Table 2.13

No.	Operation	1		2		3		4		5		6		7	
		t_1	q_1	t_2	q_2	t_3	q_3	t_4	q_4	t_5	q_5	t_6	q_6	t_7	q_7
1	Assume	300 (fixed)		200		100		50		25		15		5	
2	Residuals				–160		–30		–15		–5		–12		7.935
3	$\Delta t_2 = -60$			140											
4	Residuals				8		–90		–15		–5		–12		7.935
5	$\Delta t_3 = -40$					60									
6	Residuals				–32		22		–55		–5		–12		7.935
7	$\Delta t_2 = -15$			125											
8	Residuals				10		7		–55		–5		–12		7.935
9	$\Delta t_4 = -20$							30							
10	Residuals				10		13		1		–25		–12		7.935
11	$\Delta t_5 = -15$									10					
12	Residuals				10		–13		–14		17		–27		7.935
	Temperature Check	300		125	10	60	–13	30	–14	10	17	15	–27	5	7.935

Table 2.2(Contd.) **Relaxation Table- Example 2.13**

No.	Operation	1		2		3		4		5		6		7	
		t_1	q_1	t_2	q_2	t_3	q_3	t_4	q_4	t_5	q_5	t_6	q_6	t_7	q_7
13	$\Delta t_3 = -7$					53									
14	Residuals				3		6.6		–21		17		–27		7.935
15	$\Delta t_4 = -8$							22							
16	Residuals				3		–1.4		1.4		9		–27		7.935
17	$\Delta t_6 = -10$											5			
18	Residuals				3		–1.4		+1.4		–1		1		2.07
19	$\Delta t_2 = +1.2$			126.2											
20	Residuals				–0.36		–0.2		+1.4		–1		1		52.07
21	$\Delta t_7 = -1.75$													3.25	
	Residuals				–0.36		–0.2		+1.4		–1		–0.75		0.41
	Temperature	300		126.2		53		22		10		5		3.25	
	Check				–0.36		–0.2		+1.4		–1		–0.75		0.41
22	$\Delta t_4 = 0.5$							22.5							
	Residuals			–0.36			0.3		0		–0.05		–0.75		0.41
23	$\Delta t_3 = 0.1$					53.1									
	Residuals				0.26		0		0.1		–0.5		0.75		0.41
	Final temps.	300		126.2		53.1		22.5		10		5		3.25	
	Calculated analytically	300		126.1		53.1		22.4		9.65		4.62		3.25	

If it is assumed that the origin lies on face AB of Fig. 2.3 and that two faces of the wall, distance l apart, are at the same temperature t_1 then this equation reduces to,

$$t - t_1 = \frac{Q' l^2}{2k}\left\{\left(\frac{x}{l}\right) - \left(\frac{x}{l}\right)^2\right\} \qquad (2.68a)$$

Thus in this case the temperature distribution is a parabola with apex at the center of the wall ($x = l/2$). Further since the two faces are at the same temperature we should expect that there is a symmetry of heat flow at the center of wall. Hence at the center line ($x = l/2$) the temperature gradient should be zero. This could be verified as,

$$\left(\frac{dt}{dx}\right)_{x=l/2} = \left\{\frac{Q' l^2}{2k}\left(\frac{1}{l} - \frac{2x}{l^2}\right)\right\}_{x=l/2} = 0$$

Further the rate of heat flow is :

$$q = \frac{dQ}{d\tau} = -kA\frac{dt}{dx} \qquad (1.1)$$

$$= -kA\left(-\frac{Q'}{k}x + C\right)$$

$$= AQ'x - \frac{Q'A}{2}(x_1 + x_2) + \frac{kA(t_1 - t_2)}{x_2 - x_1} \qquad (2.69)$$

Example 2.14

Obtain the dimensionless temperature distribution in a plain wall of uniform thickness 2a meter, generating heat @ Q' (W/m^3). The wall is exposed at both surfaces to a gas at uniform temperature t_g. The surface film conductance is assumed as h (W/m^{2}K).

Solution :

Let the wall (Fig. 2.24) have infinite face area and the heat flow be unidirectional along the thickness.

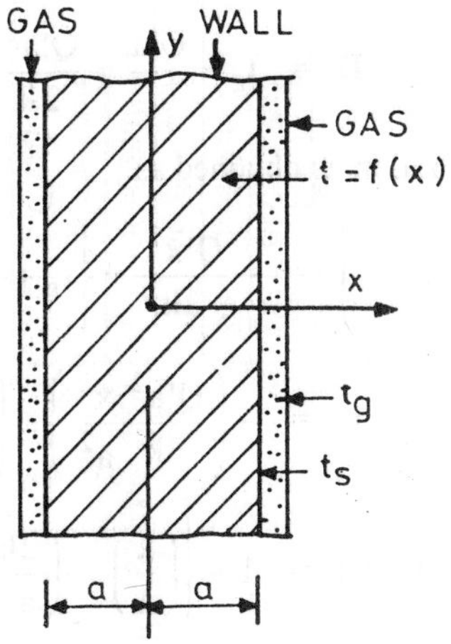

Fig. 2.24 Example 2.14

The heat flow equation is :

$$\frac{d^2t}{dx^2} + \frac{Q'}{k} = 0 \qquad (2.66)$$

Boundary conditions are ;

(i) at $x = 0$, dt/dx = 0 due to symmetry.

(ii) The rate of heat generated (W/m^2) in the wall is equal to that conducted across both the face surfaces, i.e.

$$2\,h\,(t_s - t_g) = Q'\,(2\,a)$$

{where t_s = surface temperature (C)}

or $$t_s = t_g + \frac{Q'a}{h} \quad (\text{at } x = \pm a)$$

Integrating Eq. (2.66), we get

$$\frac{dt}{dx} = -\frac{Q'}{k}x + C$$

Applying boundary condition (i) at $x = 0$; $\frac{dt}{dx} = 0$

we obtain $$C = 0\,;\ \text{so}\ \frac{dt}{dx} = -\frac{Q'}{k}x$$

On further integration we have

$$t = -\frac{Q'x^2}{2k} + D$$

Now, applying boundary condition (ii) ; at $x = \pm a$; $t_s = t_g + \frac{Q'a}{h}$

$$t_g + \frac{Q'a}{h} = \frac{Q'a^2}{2k} + D$$

or $$D = t_g + \frac{Q'a}{h} + \frac{Q'a^2}{2k}$$

Hence the temperature distribution is obtained as ;

$$t = -\frac{Q'x^2}{2k} + \left\{t_g + \frac{Q'a}{h} + \frac{Q'a^2}{2k}\right\}$$

or $$t - t_g = -\left(\frac{Q'a^2}{2k}\frac{x^2}{a^2}\right) + \left(\frac{Q'a}{h} + \frac{Q'a^2}{2k}\right)$$

$$\frac{2k}{a^2Q'}(t - t_g) = 1 - \left\{\left(\frac{x}{a}\right)^2 - \left(\frac{2k}{ah}\right)\right\}$$

where $(-a \le x \le a)$ Ans.

Example 2.15

The two ends of a cylindrical metal wire of length 2L, which is perfectly insulated against radial heat flow are held at the same temperature t_0. The left end and the middle of the wire are connected with a source of electricity so that a current flows through the left half of the wire and produces Joulean heat at the uniform rate of Q'. Assuming steady state conditions and k of wire to be independent of temperature :

(a) Express the temperature as a function of the distance from left hand end (Two different equations will be found for the left and right half of the wire)

(b) Will the temperature have a maximum somewhere between the two ends ? If so, at what distance from the left end ?

(c) Express the maximum temperature, if any, in mathematical form.

Solution :

(a) Referring to Fig. 2.25,

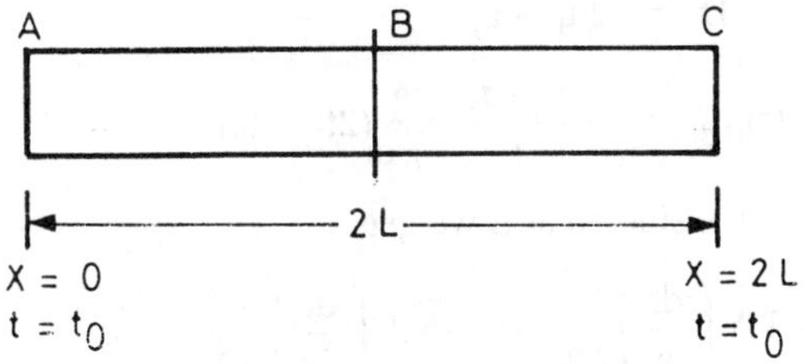

Fig. 2.25 Example 2.15

(i) For portion AB,

$$\frac{d^2t}{dx^2} = -\frac{Q'}{k}$$

so
$$t = -\frac{Q'}{k}\frac{x^2}{2} + Ax + B$$

where A, B are constants of integration.

Now, At
$$x = 0, \quad t = t_0$$
$$x = L, \quad t = t_L$$

so
$$B = t_0$$

and,
$$t_L = -\frac{Q'}{k}\frac{L^2}{2} + AL + t_0$$

or
$$AL = t_L - t_0 + \frac{Q'}{k}\frac{L^2}{2}$$

$$A = \frac{t_L - t_0}{L} + \frac{Q'L}{2k}$$

$$(t)_{AB} = -\frac{Q'}{k}\frac{x^2}{2} + \frac{(t_L - t_0)x}{L} + \frac{Q'Lx}{2k} + t_0$$

$$= \frac{Q'}{2k}(L - x)\,x + \frac{t_L - t_0}{L}\,x + t_0 \qquad \text{(a)}$$

Before eliminating t_L, let us write the differential equation for BC. It is :

$$\left(\frac{d^2t}{dx^2}\right)_{BC} = 0$$

Let $\quad t = Cx + D$

at $\quad x = L,\ t = t_L;$ at $x = 2L,\ t = t_0$

so

$$t_L = CL + D$$

$$t_0 = 2CL + D$$

or

$$C = -\frac{t_L - t_0}{L}$$

and

$$D = 2\,t_L - t_0$$

thus

$$(t)_{BC} = -\frac{t_L - t_0}{L}\,x + (2t_L - t_0) \qquad \text{(b)}$$

Now, if the heat flow is considered at B we get,

$$-kA\left(\frac{dt}{dx}\right)_{\substack{AB\\x=L}} = -kA\left(\frac{dt}{dx}\right)_{\substack{BC\\x=L}}$$

From Eqs. (a) and (b) we get,

$$\frac{Q'L}{2k} - \left(\frac{Q'x}{k}\right)_{x=L} + \left(\frac{t_L - t_0}{L}\right) = \frac{t_0 - t_L}{L}$$

or

$$-\frac{Q'L}{2k} = \frac{t_0 - t_L}{L} - \frac{t_L - t_0}{L}$$

$$= \frac{2(t_0 - t_L)}{L}$$

$$(t_0 - t_L) = -\frac{Q'L^2}{4k}$$

or

$$t_L = t_0 + \frac{Q'L^2}{4k}$$

So from Eq. (a), the temp distribution in AB is given by :

$$(t)_{AB} = \frac{Q'}{2k}(L - x)\,x + \frac{Q'L}{4k}\,x + t_0$$

$$= \frac{Q'}{4k}\left[Lx + 2\,Lx - 2x^2\right] + t_0$$

$$= \frac{Q'}{4k}\left[x\,(3\,L - 2x)\right] + t_0$$

(ii) For portion BC,

$$(t)_{BC} = -\frac{Q'L^2}{4kL}x + 2\left(t_0 + \frac{Q'L^2}{4k}\right) - t_0$$

$$= \frac{Q'L}{4k}(2L - x) + t_0$$

Hence,

$$(t)_{AB} = \frac{Q'}{4k}x(3L - 2x) + t_0 \qquad (x = 0 \text{ to } x = L)$$

$$(t)_{BC} = \frac{Q'L}{4k}(2L - x) + t_0 \qquad (x = L \text{ to } x = 2L)$$

(b) For maximum or minimum ;

$$\frac{dt}{dx} = 0$$

$$\left(\frac{dt}{dx}\right)_{AB} = \frac{Q'}{4k}[3L - 4x] = 0$$

or

$$3L = 4x$$

$$x = \frac{3}{4}L$$

and

$$\left(\frac{dt}{dx}\right)_{BC} = \frac{Q'L}{4k}(-1)$$

Evidently temperature distribution curve BC has no maximum or minimum. However, curve AB has a maximum at x = 3L/4 which is seen easily;

$$\left(\frac{d^2t}{dx^2}\right)_{AB} = \frac{Q'}{4k}(-4) = -\text{ve}$$

(c)

$$(t)_{max} = \frac{Q'}{4k}\left[3L - 2\times\frac{3}{4}L\right]\frac{3}{4}L + t_0$$

$$= \frac{Q'}{4k}\left(\frac{3L}{2}\right)\left(\frac{3L}{4}\right) + t_0$$

or

$$(t)_{max} = \frac{9}{32}\frac{Q'}{k}L^2 + t_0 \qquad \left(\text{at } x = \frac{3L}{4}\right) \qquad \textbf{Ans.}$$

Case 2 : Heat Generation in a Hollow Cylinder

For the case of a hollow cylinder the conduction equation in cylindrical coordinates is more appropriate. Hence we shall use Eq. (2.16) which for the case of steady-state one-dimensional heat flow reduces to :

$$\frac{d^2t}{dr^2} + \frac{1}{r}\frac{dt}{dr} + \frac{Q'}{k} = 0$$

or

$$\frac{d}{dr}\left(r\frac{dt}{dr}\right) = -\frac{Q'}{k}(r)$$

Integrating this equation twice we have :

$$r\frac{dt}{dr} = -\frac{Q'}{k}\frac{r^2}{2} + C_1$$

and

$$t = -\frac{Q'}{k}\frac{r^2}{4} + C_1 \ln r + C_2 \qquad (2.70)$$

where C_1 and C_2 are the constants of integration which may be evaluated by the boundary conditions. Let us assume that the boundary conditions are :

(i) The outer surface is insulated i.e.,

$$\left(\frac{dt}{dr}\right)_{r_2} = 0 \qquad (2.71)$$

(ii) The inner surface is cooled by a fluid having temperature t_g. This means that the heat generated inside the cylindrical wall by the internal heat sources is completely transferred through this surface to the cooling fluid. Thus the rate of heat generated per unit length :

$$Q = Q'\pi(r_2^2 - r_1^2) = -k\left(\frac{dt}{dr}\right)_{r_1} \times (2\pi r_1)$$

$$= h(2\pi r_1)(t_{s1} - t_g)$$

or

$$t_{s1} = \frac{Q'(r_2^2 - r_1^2)}{2hr_1} + t_g \qquad (2.72)$$

where r_1 and r_2 are inner and outer radii of the cylinder.

From the first boundary condition we get ;

$$C_1 = \frac{Q'}{k}\frac{r_2^2}{2}$$

so From Eq. (2.70),

$$t = -\frac{Q'}{k}\frac{r^2}{4} + \frac{Q'}{k}\frac{r_2^2}{2}(\ln r) + C_2 \qquad (2.73)$$

at

$$r = r_1 \,;\, t = t_{s1}$$

so

$$t_{s1} = -\frac{Q'}{k}\frac{r_1^2}{4} + \frac{Q'}{k}\frac{r_2^2}{2}(\ln r_1) + C_2 \qquad (2.74)$$

Thus from Eqs. (2.72) and (2.74) we have ;

$$\frac{Q'}{2hr_1}(r_2^2 - r_1^2) + t_g = \frac{Q'}{k}\left\{\frac{r_2^2(\ln r_1)}{2} - \frac{r_1^2}{4}\right\} + C_2$$

or

$$C_2 = t_g + \frac{Q'}{2hr_1}(r_2^2 - r_1^2) - \frac{Q'}{2k}\left\{r_2^2(\ln r_1) - \frac{r_1^2}{2}\right\}$$

Hence, the temperature distribution in the cylinder is given by ;

$$t = -\left(\frac{Q'}{k}\frac{r^2}{4}\right) + \left(\frac{Q'}{k}\frac{r_2^2}{2}\ln r\right) + \left[t_g + \frac{Q'}{2hr_1}(r_2^2 - r_1^2) - \frac{Q'}{2k}\{r_2^2 \ln r_1 - \tfrac{1}{2}r_1^2\}\right]$$

$$= t_g + \frac{Q' r_1^2}{4k}\left[\frac{2k}{hr_1}\left\{\left(\frac{r_2}{r_1}\right)^2 - 1\right\} + 1 - \left(\frac{r}{r_1}\right)^2 + 2\left(\frac{r_2}{r_1}\right)^2 \ln\frac{r}{r_1}\right] \quad (2.75)$$

The temperature at the outer surface can be obtained by putting $r = r_2$ in Eq. (2.75) ;

$$t_0 = t_g + \frac{Q' r_1^2}{4k}\left[\left(\frac{2k}{hr_1} - 1\right)\left\{\left(\frac{r_2}{r_1}\right)^2 - 1\right\} + 2\left(\frac{r_2}{r_1}\right)^2 \ln\frac{r_2}{r_1}\right] \quad (2.76)$$

and that at inner surface is obtained by putting $r = r_1$ in Eq. (2.75)

$$t_i = t_g + \frac{Q' r_1}{2h}\left\{\left(\frac{r_2}{r_1}\right)^2 - 1\right\} \quad (2.77)$$

A special case of the above is a heat generating solid cylinder cooled from the outside. Here $r_1 = 0$, and the fluid at temperature t_g flows at the outer surface with a film coefficient, h. The temperature distribution can be shown to be given by ;

$$t = t_g + \frac{Q' r_2^2}{4k}\left\{\frac{2k}{h r_2} + 1 - \left(\frac{r}{r_2}\right)^2\right\} \quad (2.78)$$

and the temperature at the center line is given by ;

$$t_C = t_g + \frac{Q' r_2^2}{4k}\left(\frac{2k}{h r_2} + 1\right) \quad (2.79)$$

Example 2.16

A thin hollow tube with 3 mm outer diameter and 2 mm inner diameter is subjected to a current of 1000 amp while water at 30 C circulates inside the same. The resistivity of the material is 0.1 ohm-mm^2/m. The heat transfer coefficient of the water film may be estimated as 35,000 W/m^{2}K and k = 18 W/mK. Calculate the surface temperature of the tube.

Solution :

Due to the flow of current in the tube there will be heat generation at a uniform rate given by :

$$Q = \frac{I^2 R_e}{A l}$$

where I = current, amp

R_e = electrical resistance, ohms

$$A = \text{cross sectional area of conductor, m}^2$$

$$l = \text{length, m}$$

so $$Q' = \frac{1000^2}{\pi(3^2 - 2^2)\times 10^{-6}\times 1} \times \{0.1 \div \pi (3^2 - 2^2) \times 1\}$$

$$= 4.05 \times 10^8 \text{ W/m}^3$$

Now the outside and inside surface temperatures may be obtained by using Eqs. (2.76) and (2.77), and the data :

$$r_2 = 3 \text{ mm}; \quad r_1 = 2 \text{ mm}; \quad t_g = 30 \text{ C, and}$$

$$h = 35{,}000 \text{ W/m}^2\text{K}; \quad Q' = 4.05 \times 10^8 \text{ W/m}^3$$

so $$t_0 = 30 + \frac{4.05\times 10^8 \times 2^2 \times 10^{-6}}{4\times 18}\left[\left\{\left(\frac{3}{2}\right)^2 - 1\right\} \times \left(\frac{2\times 18}{35{,}000\times 2\times 10^{-3}} - 1\right) + 2\left(\frac{3}{2}\right)^2 \ln\frac{3}{2}\right]$$

$$= 57.8 \text{ C}$$ **Ans.**

and $$t_i = 30 + \frac{4.05\times 10^8 \times 2\times 10^{-3}}{2\times 35{,}000}\left\{\left(\frac{3}{2}\right)^2 - 1\right\}$$

$$= 44.6 \text{ C}$$ **Ans.**

Case 3 : Heat conduction in a wire carrying electrical current

Consider an electric wire of radius R and electrical conductivity k_e ohm^{-1} cm^{-1}. The wire carries an electric current of density I amp/cm^2. The surface of the wire is maintained at temperature t_o. It is desired to know the radial temperature distribution within the wire.

This case is very similar to Case 2, only the boundary conditions are different. The Eq. (2.16) simplifies to the form ;

$$\frac{d^2t}{dr^2} + \frac{1}{r}\frac{dt}{dr} + \frac{Q'}{k} = 0$$

because it is a steady-state system and heat flow is only in the radial direction.

The Eq. (2.16) may be written as ;

or $$\frac{d}{dr}\left(r\frac{dt}{dr}\right) = -\frac{Q'}{k}(r)$$

Integrating the above equation :

$$r\frac{dt}{dr} = -\frac{Q' r^2}{2k} + C_1$$

Now, the heat flux q_r' (i.e. the rate of heat flow per unit area at r = r) is given by ;

$$q_r' = -k\frac{dt}{dr}$$

$$= \frac{Q'r}{2} - \frac{C_1 k}{r}$$

To determine the constant of integration C_1 we use the boundary condition :

at $r = 0$, q_r' is not infinite

this gives $C_1 = 0$,

hence, $$q_r' = \frac{Q'r}{2}$$

and $$\frac{dt}{dr} = -\frac{Q'r}{2k}$$

Integrating again we get,

$$t = -\frac{Q'r^2}{4k} + C_2$$

The second boundary condition is ;

at $r = R, t = t_o$

Using this boundary condition,

$$t_0 = -\frac{Q'R^2}{4k} + C_2$$

or $$C_2 = t_0 + \frac{Q'R^2}{4k}$$

hence, $$t = -\frac{Q'r^2}{4k} + t_0 + \frac{Q'R^2}{4k}$$

or $$(t - t_0) = \frac{Q'R^2}{4k}\left(1 - \frac{r^2}{R^2}\right) \tag{2.80}$$

Since the heat generated per unit volume Q', is

$$Q' = \frac{I^2}{k_e}$$

the temperature distribution inside the wire, Eq. (2.80) is given by :

$$t - t_0 = \frac{I^2}{k_e}\frac{R^2}{4k}\left(1 - \frac{r^2}{R^2}\right) \tag{2.81}$$

The maximum temperature occurs at the center of the wire (r = 0) and is given by ;

$$t_{max} - t_0 = \frac{I^2}{k_e}\frac{R^2}{4k} = Q'\frac{R^2}{4k} \tag{2.82}$$

The average temperature $\bar{t}$ is given by ;

$$\bar{t} - t_0 = \frac{\int_0^{2\pi}\int_0^R (t - t_0)\, r\, dr\, d\theta}{\int_0^{2\pi}\int_0^R r\, dr\, d\theta} = \frac{\frac{I^2}{k_e}\frac{R^2}{4k}\int_0^{2\pi}\int_0^R \left(1 - \frac{r^2}{R^2}\right) r\, dr\, d\theta}{(2\pi R^2/2)}$$

$$= \frac{I^2}{k_e}\,\frac{R^2}{4k}\,\frac{2\pi R^2}{4} \times \frac{2}{2\pi R^2}$$

or
$$\bar{t} - t_0 = \frac{I^2}{k_e}\,\frac{R^2}{8k} = \frac{Q' R^2}{8k} \qquad (2.83)$$

The rate of heat flow at the surface of wire ($r = R$) per unit length is ;

$$q_R = (2\pi R)\,(1)\,(q'')_{r=R}$$

$$= (2\pi R)\,\frac{Q' R}{2}$$

or
$$q_R = \pi R^2 Q'$$

$$= \text{heat generated per unit length} \qquad (2.84)$$

This means that all the Joulean heat produced must leave through the wire surface under steady state conditions.

Example 2.17

A copper wire has a radius of 1 mm and a length of 10 m. For what voltage drop would the temperature rise at the wire axis be 10°C, if the surface temperature of the wire is 20°C. The electrical and thermal conductivities of copper are 5.1×10^5 ohm^{-1} cm^{-1}, and 384 W/mK respectively.

Solution :

From Eq. (2.82)

$$t_{max} - t_0 = \frac{I^2}{k_e}\,\frac{R^2}{4k}$$

Now,
$$R = 0.1 \text{ cm}$$

$$k_e = 5.1 \times 10^5 \text{ ohm}^{-1}\text{ cm}^{-1}$$

$$k = 384 \text{ W/mK} = 3.84 \text{ watt/cm°C}$$

so
$$I^2 = \frac{(t_{max} - t_0)(k_e)(4k)}{R^2}$$

$$= \frac{10 \times 5.1 \times 10^5 \times 4 \times 3.84}{0.01}$$

$$= 77.5 \times 10^8$$

so
$$I = 8.8 \times 10^4 \text{ amp/cm}^2$$

If V is the voltage drop over length L, and R_e is the electrical resistance in ohms, then the current IA is given by

$$IA = \frac{V}{R_e}$$

$$R_e = \frac{L}{k_e A}$$

or
$$V = IA\,R_e = \frac{IL}{k_e}$$

$$= \frac{8.8\times10^4\times(10\times100)}{5.1\times10^5}$$

$$= 172 \text{ Volt}$$ Ans.

Example 2.18

A copper wire 0.1 cm in diameter is insulated with plastic to an outer diameter of 0.3 and is exposed to an environment at t_e = 40°C. The heat transfer coefficient from the outer surface of the plastic to the surroundings is h_e = 8.75 W/m²K. What is the maximum steady current in amperes that this wire can carry without heating any part of the plastic above 95°C. The thermal conductivities of plastic and copper are 0.35 and 384 W/mK respectively. The electrical conductivity of copper is 5.1×10^5 ohm^{-1} cm^{-1}.

Solution :

Let r_1 and r_2 denote the inside and outside radii of the plastic insulation. Then, the heat flow rate per unit length of the wire is given by,

$$q = \frac{2\pi(t_i - t_e)}{\frac{\log_e r_2/r_1}{k} + \frac{1}{r_2 h_e}}$$

$$= \frac{2\pi(95-40)}{\frac{\log_e(0.15/0.05)}{0.35} + \frac{1}{(0.15\times10^{-2}\times8.75)}}$$

$$= \frac{2\pi\times55}{3.14+76.2}$$

$$= 4.47 \text{ watt per m length}$$

In the steady state all the heat produced on account of current flow must be transferred to the surroundings. Here the rate of heat production is 4.47 watt per m length of the wire. The heat production per unit volume is therefore ;

$$Q' = \frac{4.47}{\pi(0.05\times10^{-2})^2\times1}$$

$$= 5.7\times10^6 \text{ watt/m}^3$$

Since $\quad Q' = \dfrac{I^2}{k_e}$

$$I^2 = Q'\,k_e$$

$$= 5.7 \times 10^6 \times 5.1 \times 10^7$$

$$I = 1.705 \times 10^7 \text{ amp/m}^2$$

or $\quad \text{Current} = (1.705 \times 10^7)(\pi)(0.05 \times 10^{-2})^2$

$$= 13.4 \text{ amp}$$

Ans.

Case 4 : Heat conduction with a nuclear heat source

Let us consider a nuclear fuel element of spherical form consisting of a sphere of fissionable material surrounded by a spherical shell of aluminum cladding (Fig. 2.26). Inside the fissionable material thermal energy is generated owing to the slowing down of the neutrons as a result of collision with the atoms of the fissionable material. Let the rate of heat energy generation per unit volume be given by ;

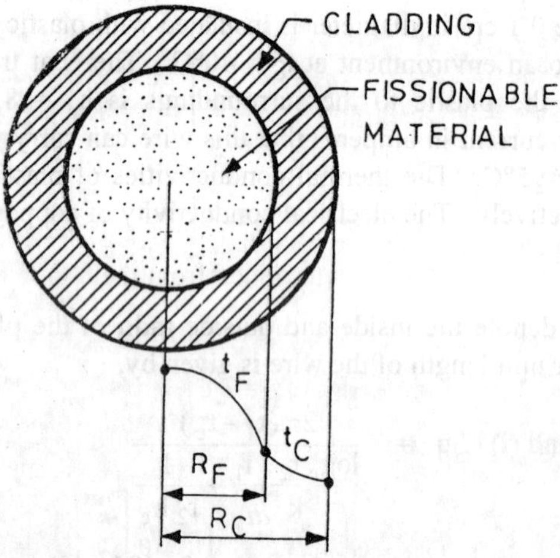

Fig. 2.26 Nuclear Heat Source

$$Q' = Q_0\left[1 - b\left(\frac{r}{R_F}\right)^2\right]$$

where Q_0 is the rate of heat production per unit volume at r = 0 and b is a constant. Note that Q' is a function of the position r.

In order to determine the temperature distribution within the fissionable material and the cladding we start from Eq. (2.17). For steady-state, one-dimensional heat conduction in the radial direction, Eq. (2.17) simplifies to ;

$$\frac{1}{r^2}\frac{d}{dr}\left(r^2\frac{dt}{dr}\right) + \frac{Q'}{k} = 0 \tag{2.85}$$

or

$$\frac{d}{dr}\left(r^2\frac{dt}{dr}\right) = -\frac{Q'}{k}r^2 \tag{2.86}$$

Since the heat flux q' is given by ;

$$q' = -k\frac{dt}{dr}$$

Eq. (2.86) may also be written as ;

$$\frac{d}{dr}(r^2 q') = Q' r^2 \tag{2.87}$$

where k has been assumed to be constant.

If suffix F and C denote the fissionable material and the cladding respectively, we get:

$$\frac{d}{dr}(r^2 q_F') = Q_0 r^2\left[1-b\left(\frac{r}{R_F}\right)^2\right] \tag{a}$$

$$\frac{d}{dr}(r^2 q_c') = 0 \tag{b}$$

Integrating Eqs. (a) and (b),

$$q'_F = Q_0\left[\frac{r}{3}-\frac{b}{5}\frac{r^3}{R_F^2}\right]+\frac{C_1}{r^2} \tag{c}$$

$$q'_c = \frac{C_2}{r^2} \tag{d}$$

The constants C_1 and C_2 are determined from the boundary conditions :

at $r = 0$, q'_F is not infinite (i)

at $r = R_F$, $q'_F = q'c$ (ii)

From Eqs. (c) and (i) $C_1 = 0$

so $$q'_F = \left(-k_F\frac{dt_F}{dr}\right) = Q_0\left[\frac{r}{3}-\frac{b}{5}\frac{r^3}{R_F^2}\right] \tag{e}$$

From Eqs. (d) and (ii).

$$C_2 = Q_0 R_F^3\left(\frac{1}{3}-\frac{b}{5}\right)$$

so $$q'c = \left(-k_c\frac{dt_c}{dr}\right) = \frac{Q_0 R_F^3}{r^2}\left(\frac{1}{3}-\frac{b}{5}\right) \tag{f}$$

Integration of Eqs. (e) and (f) gives.

$$t_F = -\frac{Q_0}{k_F}\left[\frac{r^2}{6}-\frac{b}{20}\frac{r^4}{R^2{}_F}\right]+C_3 \tag{g}$$

$$t_c = +\frac{Q_0 R_F^3}{k_C}\left(\frac{1}{3}-\frac{b}{5}\right)\frac{1}{r}+C_4 \tag{h}$$

The constants C_3 and C_4 are determined from the boundary conditions :

at $$r = R_F, t_F = t_c \quad \text{(iii)}$$

$$r = R_c, t_c = t_0 \quad \text{(iv)}$$

From Eqs. (h) and (iv).

$$C_4 = t_0 - \frac{Q_0 R_F^3}{k_c R_c}\left(\frac{1}{3} - \frac{b}{5}\right)$$

so $$t_c = t_0 + \frac{Q_0 R_F^3}{k_c}\left(\frac{1}{3} - \frac{b}{5}\right)\left(\frac{1}{r} - \frac{1}{R_c}\right)$$

or $$t_c = t_0 + \frac{Q_0 R_F^2}{k_c}\left(\frac{1}{3} - \frac{b}{5}\right)\left(\frac{R_F}{r} - \frac{R_F}{R_c}\right) \quad \text{(j)}$$

From Eqs. (g), (iii) and (j),

$$-\frac{Q_0}{k_F}\left[\frac{R_F^2}{6} - \frac{b}{20} R_F^2\right] + C_3 = t_0 + \frac{Q_0^2 R_F^2}{k_c}\left(\frac{1}{3} - \frac{b}{5}\right)\left(1 - \frac{R_F}{R_c}\right).$$

$$\text{or } C_3 = t_0 + \frac{Q_0 R_F^2}{k_c}\left(\frac{1}{3} - \frac{b}{5}\right)\left(1 - \frac{R_F}{R_c}\right) + \frac{Q_0 R_F^2}{k_F}\left(\frac{1}{6} - \frac{b}{20}\right)$$

Thus, from Eq. (g) we get ;

or $$t_F = t_0 + \frac{Q_0 R_F^2}{k_c}\left(\frac{1}{3} - \frac{b}{5}\right)\left(1 - \frac{R_F}{R_c}\right) + \frac{Q_0 R_F^2}{k_F} \times \left(\frac{1}{6} - \frac{b}{20}\right) - \frac{Q_o}{k_F} r^2\left(\frac{1}{6} - \frac{b}{20}\frac{r^2}{R_F^2}\right)$$

$$= t_0 + \frac{Q_0 R_F^2}{3 k_c}\left(1 - \frac{3}{5} b\right)\left(1 - \frac{R_F}{R_c}\right) + \frac{Q_0 R_F^2}{6 k_F}\left(1 - \frac{r^2}{R_F^2}\right) - \frac{Q_0 R_F^2 b}{20 k_F}\left(1 - \frac{r^4}{R_F^4}\right) \quad (2.88)$$

The maximum temperature will occur in the fissionable material at r = 0.

$$(t_F)_{max} = t_0 + \frac{Q_0 R_F^2}{3 k_c}\left(1 - \frac{3}{5} b\right)\left(1 - \frac{R_F}{R_c}\right) + \frac{Q_0 R_F^2}{6 k_F}\left(1 - \frac{3}{10} b\right) \quad (2.89)$$

Example 2.19

A long nuclear fuel rod is surrounded by annular layer of aluminum cladding. The rate of heat generation per unit volume by fission within the fuel rod is given by ;

$$Q' = Q_0\left[1 - b\left(\frac{r}{R_F}\right)^2\right]$$

where Q_0 is the rate of heat generation per unit volume at r = 0 and r is the distance from the axis of the fuel rod. Calculate the temperature distribution and maximum temperature in the fuel rod if the outer surface of the cladding is in contact with a liquid coolant at temperature T_L. The heat transfer coefficient from cladding surface to coolant may be taken as h_L. The thermal conductivities of the rod and the cladding are k_F and k_c.

Solution :

For steady state heat flow in radial direction only, we have,

$$\frac{d}{dr}\left(kr\frac{dt}{dr}\right) = -Q'\,r \text{ (since k = constant)}$$

Let t_F and t_c denote the temperature in the fuel rod and cladding respectively. We then obtain, for the fuel rod

$$\frac{d}{dr}\left(k_F\,r\frac{dt_F}{dr}\right) = -Q_0\left[1-b\left(\frac{r}{R_F}\right)^2\right]r$$

Integrating the above,

$$k_F\,r\frac{dt_F}{dr} = -Q_0\left[\frac{r^2}{2}-\frac{b}{4}\frac{r^4}{R_F^2}\right]+C_1$$

$$k_F\frac{dt_F}{dr} = -\frac{Q_0\,r}{2}\left(1-\frac{b}{2}\frac{r^2}{R_F^2}\right)+\frac{C_1}{r}$$

so

$$q_F' = -\left(k_F\frac{dt_F}{dr}\right)$$

$$= \frac{Q_0\,r}{2}\left(1-\frac{b}{2}\frac{r^2}{R_F^2}\right)-\frac{C_1}{r}$$

To evaluate C_1 we use the boundary condition :

at $r = 0$, heat flux q_F' is not infinite

or $C_1 = 0$

hence,

$$q_F' = -\left(k_F\frac{dt_F}{dr}\right) = \frac{Q_0\,r}{2}\left(1-\frac{b}{2}\frac{r^2}{R_F^2}\right) \qquad \text{(a)}$$

Similarly, for the cladding, since Q' = 0,

$$\frac{d}{dr}\left(k_c\,r\frac{dt_c}{dr}\right) = 0$$

Integrating this equation,

$$k_c\frac{dt_c}{dr} = -\frac{C_2}{r}$$

or $$q'_c = -k_c \frac{dt_c}{dr} = -\frac{C_2}{r}$$

To evaluate C_2 we use the boundary condition :

at $$r = R_F, \; q'_F = q'_c$$

Thus, $$-\frac{C_2}{R_F} = \frac{Q_0 R_F}{2}\left(1 - \frac{b}{2}\right)$$

or $$C_2 = -\frac{Q_0 R_F^2}{2}\left(1 - \frac{b}{2}\right)$$

hence, $$q_c' = -k_c \frac{dt_c}{dr} = \frac{Q_0 R_F^2}{2r}\left(1 - \frac{b}{2}\right) \qquad \text{(b)}$$

Integrating Eq. (a),

$$t_F = -\frac{Q_0}{k_F}\frac{1}{2}\left(\frac{r^2}{2} - \frac{b}{8}\frac{r^4}{R_F^2}\right) + C_3 \qquad \text{(c)}$$

or $$t_F = -\frac{Q_0 r^2}{4k_F}\left(1 - \frac{b}{4}\frac{r^2}{R_F^2}\right) + C_3$$

Similarly, integration of Eq. (b) gives,

$$t_c = \frac{Q_0 R_F^2}{2k_c}\left(1 - \frac{b}{2}\right)\log_e r + C_4 \qquad \text{(d)}$$

we can determine the constant C_4 by the condition that ;

at $$r = R_c, \; q_c' = h_L(t_c - t_L)$$

or at $$r = R_c, \; t_c = t_L + \frac{q_c'}{h_L}$$

This gives,

$$-\frac{Q_0 R_F^2}{2k_c}\left(1 - \frac{b}{2}\right)\log_e R_c + C_4 = t_L + \frac{Q_0 R_F^2}{2R_c h_L}\left(1 - \frac{b}{2}\right)$$

or $$C_4 = t_L + \frac{Q_0 R_F^2}{2k_c}\left(1 - \frac{b}{2}\right)\left(\log_e R_c + \frac{k_c}{R_c h_L}\right)$$

The temperature distribution in the cladding is then given by

$$t_c = t_L + \frac{Q_0 R_F^2}{2k_c}\left(1 - \frac{b}{2}\right)\left(\log_e \frac{R_c}{r} + \frac{k_c}{R_c h_L}\right) \qquad \text{(e)}$$

The integration constant C_3 is now obtained from the boundary condition :

$$r = R_F, \; t_F = t_c$$

Substitution of this condition in Eqs. (c) and (e) gives ;

$$-\frac{Q_0 R_F^2}{4k_F}\left(1-\frac{b}{4}\right)+C_3 = t_L+\frac{Q_0 R_F^2}{2k_c}\left(1-\frac{b}{2}\right)\times\left(\log_e\frac{R_c}{R_F}+\frac{k_c}{R_c h_L}\right)$$

$$\text{or}\quad C_3 = \frac{Q_0 R_F^2}{4k_F}\left(1-\frac{b}{4}\right)+t_L+\frac{Q_0 R_F^2}{2k_c}\left(1-\frac{b}{2}\right)\times\left(\log_e\frac{R_c}{R_F}+\frac{k_c}{R_c h_L}\right)$$

Hence, the temperature distribution in the fuel rod is ;

$$t_F = t_L+\frac{Q_0 R_F^2}{4k_F}\left(1-\frac{b}{4}\right)+\frac{Q_0 R_F^2}{2k_c}\left(1-\frac{b}{2}\right)\times\left(\log_e\frac{R_c}{R_F}+\frac{k_c}{R_c h_L}\right)-\frac{Q_0 r^2}{4k_F}\times\left(1-\frac{b}{4}\frac{r^2}{R_F^2}\right)$$

The maximum temperature occurs in the fuel rod at r = 0.

Hence,

$$t_{F,max} = t_L+\frac{Q_0 R_F^2}{4k_F}\left(1-\frac{b}{4}\right)+\frac{Q_0 R_F^2}{2k_c}\left(1-\frac{b}{2}\right)\times\left(\log_e\frac{R_c}{R_F}+\frac{k_c}{R_c h_L}\right)$$

Ans.

Example 2.20

An oil acts as a lubricant between two cylindrical surfaces as shown in Fig. 2.27. The outer cylinder has a radius of 5 cm and it rotates at 9000 r.p.m. The clearance between the cylinders is 0.025 cm. What is the maximum temperature in the oil if both wall temperatures are kept at 70°C. The physical properties of oil are :

$$\mu = 0.95 \text{ poise}$$

$$k = 0.023 \text{ J/sec-cm-C}$$

$$\rho = 1.22 \text{ g/cm}^3$$

Solution :

The rate of heat generation per unit volume, Q' due to viscous dissipation may be assumed to be :

$$Q' = \mu\left(\frac{dv_y}{dy}\right)^2$$

Since the clearance between the cylinders is small compared to the radius of the outer cylinder, we may simplify the system of Fig. 2.27 (a) by that shown in Fig. 2.27 (b). For this system, Eq (2.66) may be written in the form :

$$\frac{d^2t}{dy^2}+\frac{Q'}{k} = 0 \qquad \text{(a)}$$

or

$$\frac{d^2t}{dy^2} = -\frac{Q'}{}$$

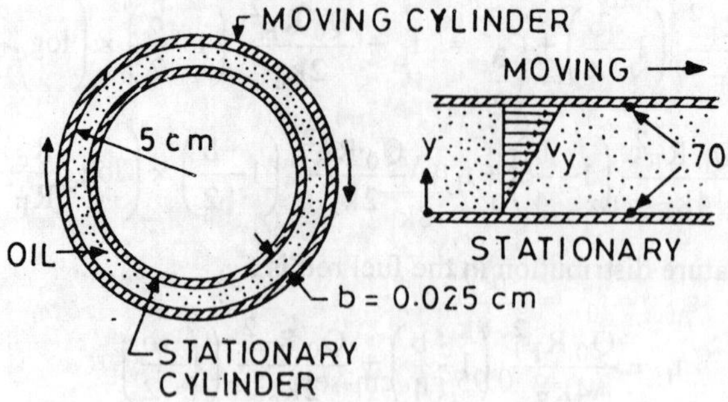

Fig. 2.27 Example 2.20

Now, for the steady laminar flow of a fluid with constant viscosity in a slit, as shown in Fig. 2.27 (b), the velocity varies linearly with distance, i.e,

$$V_y = \left(\frac{y}{b}\right) V$$

or
$$\frac{dV_y}{dy} = \frac{V}{b}$$

thus
$$Q' = \mu \frac{V^2}{b^2}$$

Hence Eq. (a) becomes
$$\frac{d^2t}{dy^2} = -\frac{\mu}{k}\frac{V^2}{b^2} \tag{b}$$

Integrating Eq. (b) twice,

$$t = -\frac{\mu}{k}\left(\frac{V^2}{b^2}\right)\frac{y^2}{2} + C_1 y + C_2 \tag{c}$$

The boundary condition are

(i) at $y = 0, \; t = t_0$

(ii) at $y = b, \; t = t_0$

Eqs. (c) and (i) give ;

$$C_2 = t_0$$

Eqs. (c) and (ii) then give ;

$$C_1 = \frac{\mu V^2}{2kb}$$

so
$$t = t_0 + \frac{\mu V^2}{2kb}\left[y - \frac{y^2}{b}\right]$$

For t to be maximum,

$$\frac{dt}{dy} = 0$$

or $$y = \frac{b}{2}$$

thus $$t_{max} = t_0 + \frac{\mu V^2}{8k}$$

or $$t_{max} - t_0 = \frac{\mu V^2}{8k}$$

now $$\mu = 0.95 \text{ gm/cm-sec}$$

$$V = \frac{2\pi \times 9000}{60} \times 5 = 4710 \text{ cm/sec}$$

so $$\mu V^2 = (0.95)(4710)^2 \text{ (g/cm s) (cm}^2\text{/s}^2)$$

$$\mu = 21.1 \times 10^6 \text{ erg/cm s}$$

$$= (21.1 \times 10^6)(10^{-7}) = 2.11 \text{ J/cm s}$$

or $$\frac{\mu V^2}{4k} = \frac{2.11}{8 \times 0.023}\left(\frac{J}{\text{cm} - \text{sec}}\right)\left(\frac{\text{cm} - \text{sec} - C}{J}\right)$$

$$= 11.55°C$$

so $$t_{max} = t_0 + 11.55$$

$$= 70 + 11.55 = 88.55°C$$ Ans.

Example 2.21

A chemical reaction is being carried out at constant pressure in a packed bed between two coaxial cylinders with radii 1.14 cm and 1.27 cm. The entire inner wall is at a uniform temperature of 500°C and there is almost no heat transfer through this surface.

The reaction releases heat at a uniform rate of 560,000 W/m^3 throughout the reactor. The effective thermal conductivity of the packed bed may be taken as 0.525 W/mK. What is the temperature of the outer wall.

Solution :

For this case,

$$\frac{1}{r}\left\{\frac{d}{dr}\left(r\frac{dt}{dr}\right)\right\} = -\frac{Q'}{k}$$

or $$\frac{dt}{dr} = -\frac{Q'r}{2k} + \frac{C_1}{r}$$

Since $$\frac{dt}{dr} = 0 \text{ at } r = R_0 \text{ (inside radius)}$$

$$C_1 = \frac{Q'R_0^2}{2k}$$

so $$\frac{dt}{dr} = \frac{Q'R_0^2}{2kr} - \frac{Q'r}{2k}$$

Thus $$t = \frac{Q'}{2k}\left(R_0^2 \log_e r - \frac{r^2}{2}\right) + C_2$$

Since $$t = T_o \text{ at } R = R_o$$

$$C_2 = T_0 - \frac{Q'R_0^2}{2k}\left(\log_e R_0 - \frac{1}{2}\right)$$

$$t = T_0 + \frac{Q'R_0^2}{2k}\left(\log_e \frac{r}{R_0}\right) - \frac{Q'R_0^2}{4k}\left(\frac{r^2}{R_0^2} - 1\right)$$

Substituting the numerical values,

$$t = 500 + \frac{560{,}000 \times (1.14 \times 10^{-2})^2}{2 \times 0.524} \times \left(\log_e \frac{1.27}{1.14}\right)$$

$$- \frac{560{,}000 \times (1.14 \times 10^{-2})^2}{4 \times 0.524}\left(\frac{1.27^2}{1.14^2} - 1\right)$$

$$= 500 + 7.28 - 8.38$$

so $$t = 498.9°\text{C}$$ Ans.

2.20 TWO DIMENSIONAL STEADY STATE HEAT CONDUCTION

In the preceding sections we dealt with heat conduction problems where the heat flow could be treated as function of a *single variable* i.e. one space coordinate. There may be many engineering applications where it is actually so ; in other cases it may be possible that the heat flow in other directions is ignored. However, when the boundaries of the system are *irregular* or when the temperature along a boundary is *nonuniform*, a one-dimensional treatment of the problem may no longer be satisfactory. In such cases the permanent state of the temperature is a function of *two* or even *three* space coordinates. Here we shell examine only a few simple cases of *steady state two dimensional heat flow*.

The various methods for the solution of two dimensional heat conduction problems may be divided into four categories.

(i) Analytical

(ii) Graphical

(iii) Analogical

(iv) Numerical

A complete *analytical solution* may not be possible for all practical problems. Even in simple and regular solids, such a method requires the knowledge of Fourier series, Bessel and Legendre functions, Laplace transform and Complex variables theory. We will deal with only two simple cases in Sections 2.21 and 2.22. Then in section 2.23 the *graphical method* will be explained by discussing the potential field plotting technique.

The *analogical method* is based on the principle explained in Section 1.12, and may be applied to even complicated problems. Finally, in Section 2.24 the *numerical method* will be taken up.

Among the simple cases which will be discussed in the succeeding Sections are the cases of *semi-infinite* and *infinite* solids. The two types of solids are illustrated in Fig. 2.28. It will be observed that at least one dimension is finite in both cases. In the case of *semi-infinite* solids the other two dimensions may both be infinite but one of the surfaces (normal to the axis of finite dimension) lies in a finite location. The *infinite solid,* however, will have no surface, (normal to the finite dimension), located in finite position.

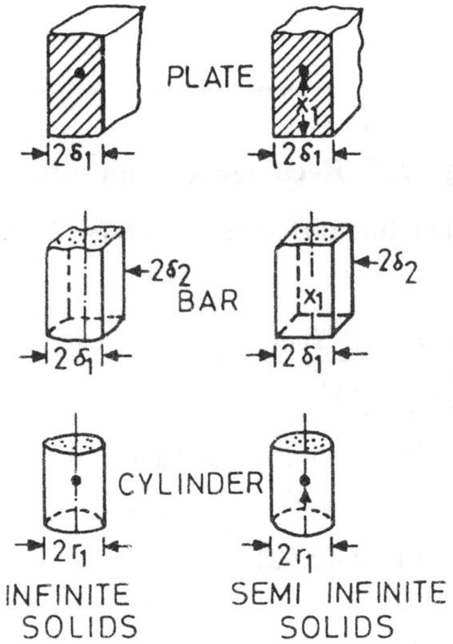

Fig. 2.28 Infinite and Semi-infinite Solids

2.21 STEADY STATE HEAT CONDUCTION IN A RECTANGULAR ADIABATIC PLATE

The general heat conduction Equation (2.12) may be solved analytically for the boundary conditions of the simple problem of a *thin rectangular plate*, free of internal heat sources and insulated at two faces. The plate is shown in Fig. 2.29. The significance of thin plate is that $\partial t/\partial z = 0$ and the temperature is a function of x and y only .

The Eq. (2.12) is reduced to the form ;

$$\frac{\partial^2 t}{\partial x^2} + \frac{\partial^2 t}{\partial y^2} = 0 \tag{2.90}$$

This equation will now be solved for the initial and boundary conditions :

$$\begin{aligned} t &= t_1 \text{ at } x = 0, l \\ t &= t_1 \text{ at } y = 0 \\ t &= t_m \sin(\pi x / l) \text{ at } y = b \end{aligned} \tag{2.91}$$

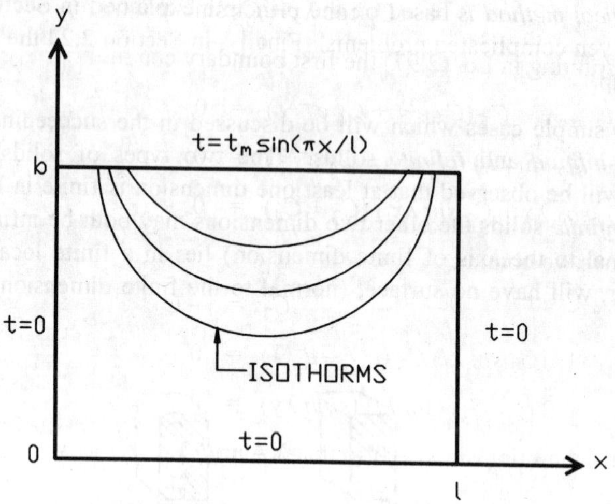

Fig. 2. 29 Rectangular Thin Adiabatic Plate

Let us define the temperature difference, $(t - t_1) = T$, so that Eqs. (2.90) and (2.91) are reduced to the form ;

$$\frac{\partial^2 T}{\partial x^2} + \frac{\partial^2 T}{\partial y^2} = 0 \tag{2.92}$$

$$\left.\begin{array}{ll} \text{(i)} & T = 0 \text{ at } x = 0, l \text{ and } y = 0 \\ \text{(ii)} & T = T_m \sin(\pi x / l) \text{ at } y = b \end{array}\right\} \tag{2.93}$$

Let $T = X\,Y$ be a solution of Eq. (2.92)

where $X = X(x)$, and $Y = Y(y)$

Substituting this in Eq. (2.92) we get ;

$$\frac{1}{X}\frac{d^2X}{dx^2} = -\frac{1}{Y}\frac{d^2Y}{dy^2} = -\lambda^2 \text{ (say)}$$

This equality gives two ordinary total differential equations each having one variable and λ as unknown ;

$$\frac{d^2X}{dx^2} + \lambda^2 X = 0 \tag{2.93}$$

and

$$\frac{d^2Y}{dy^2} - \lambda^2 Y = 0 \tag{2.94}$$

The general solutions of Eqs. (2.93) and (2.94) are ;

$$X = C_1 \cos \lambda x + C_2 \sin \lambda x \tag{2.95}$$

$$Y = C_3 e^{\lambda y} + C_4 e^{-\lambda y} \tag{2.96}$$

(where C_1, C_2, C_3, C_4 are constants)

The general solution of Eq. (2.92) is ;

$$T = (C_1 \cos \lambda x + C_2 \sin \lambda x)(C_3 e^{\lambda y} + C_4 e^{-\lambda y}) \tag{2.97}$$

Now, substituting in Eq. (2.97) the first boundary condition of Eq. (2.92) we get ;

$$C_1 (C_3 e^{\lambda y} + C_4 e^{-\lambda y}) = 0$$

$$(C_1 \cos \lambda l + C_2 \sin \lambda l)(C_3 e^{\lambda y} + C_4 e^{-\lambda y}) = 0$$

and $$(C_1 \cos \lambda x + C_2 \sin \lambda x)(C_3 + C_4) = 0$$

From the above we obtain ;

$$C_3 = -C_4 ; C_1 = 0$$

and $$(C_2 \sin \lambda l) C_3 (e^{\lambda y} - e^{-\lambda y}) = 0$$

or $$2 (C_2 C_3)(\sin \lambda l)(\sinh \lambda y) = 0 \tag{2.98}$$

To satisfy this condition $\sin \lambda l = 0$ so that $\lambda = n\pi/l$

Hence $$T = \sum_{n=1}^{\infty} C_n \left(\sin \frac{n\pi x}{l}\right)\left(\sinh \frac{n\pi y}{l}\right) \tag{2.99}$$

Substituting the second boundary condition of Eq. (2.92) we get ;

$$\sum_{n=1}^{\infty} C_n \sin\left(\frac{n\pi x}{l}\right)\left(\sinh \frac{n\pi b}{l}\right) = T_m \sin \frac{\pi x}{l}$$

This indicates that only the first term with $n = 1$ is needed in the total series solution ; accordingly ;

$$C_n = C_1 = T_m \div \left(\sinh \frac{\pi b}{l}\right)$$

The final solution therefore becomes

$$T(x, y) = t - t_1$$

$$= T_m \frac{\sinh(\pi y / l)}{\sinh(\pi b / l)} \left(\sin \frac{\pi x}{l}\right) \tag{2.100}$$

The corresponding isotherms are shown by lines in Fig. 2.29.

2.22 STEADY STATE HEAT CONDUCTION IN A SEMI-INFINITE PLATE

We next consider the case of *semi-infinite* plate as shown in Fig. 2.30. Here again, no heat sources are assumed to be existing.

Let the plate be so oriented that its two edges lie in the first x – y quadrant on the x and y axes. Let the three edges $x = 0$; $x = l$ and $y = \infty$ be held at steady and uniform temperature t_1. The fourth edge $y = 0$ is maintained at a steady but arbitrary temperature distribution $t = f(x)$.

Now, Eq. (2.12) is reduced to the form (for steady state) ;

$$\frac{\partial^2 t}{\partial x^2} + \frac{\partial^2 t}{\partial y^2} = 0 \tag{2.90}$$

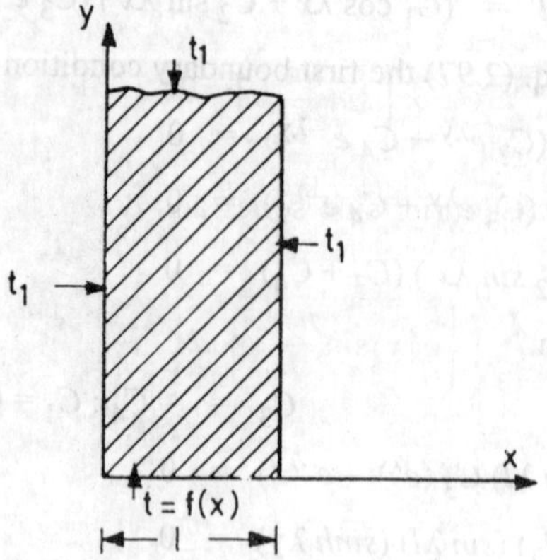

Fig. 2.30 Semi-infinite Adiabatic Plate

This equation has to be solved for the conditions ;

$$\left.\begin{array}{ll}\text{(i)} & t=t_1 \text{ at } x=0, l \\ \text{(ii)} & t=t_1 \text{ at } y=\infty, \\ \text{(iii)} & t=f(x) \text{ at } y=0,\end{array}\right\} \tag{2.101}$$

Let us define the temperature difference $(t - t_1) = T$; so that the conditions are reduced to ;

$$\left.\begin{array}{ll}\text{(i)} & T=0 \text{ at } x=0, l \\ \text{(ii)} & T=0 \text{ at } y=\infty, \\ \text{(iii)} & T=\phi(x) \text{ at } y=0,\end{array}\right\} \tag{2.102}$$

Further, Eq. (2.90) also reduces to ;

$$\frac{\partial^2 T}{\partial x^2} + \frac{\partial^2 T}{\partial y^2} = 0 \tag{2.92}$$

Thus the general solution of Eq. (2.92) is given by ;

$$T = (C_1 \cos \lambda x + C_2 \sin \lambda x)(C_3 e^{\lambda y} + C_4 e^{-\lambda y}) \tag{2.97}$$

In order to evaluate the constants let the boundary conditions of Eq. (2.102) be applied one by one ;

(i) when $T = 0$ at $x = 0$; $C_1 = 0$

(ii) when $T = 0$ at $y = \infty$; $C_3 = 0$

Hence Eq. (2.97) reduces to ;

$$T = C_2 C_4 e^{-\lambda y} (\sin \lambda x) \tag{2.103}$$

or $T = C e^{-\lambda y} (\sin \lambda x)$ where $(C_2 C_4 = C$ say)

(iii) Now if $T = 0$ at $x = l$ then $\sin \lambda l = 0$

or $$\lambda = \frac{n\pi}{l} ; n = 1, 2, 3, \ldots\ldots$$

$$T = \sum_{n=1}^{\infty} C_n e^{-(n\pi/l)y} \times \left(\sin \frac{n\pi}{l} x \right)$$

(iv) Finally if T = ϕ (x) at y = 0 we have ;

$$\phi(x) = \sum_{n=1}^{\infty} C_n \times \left(sin \frac{n\pi}{l} x \right)$$

By expanding ϕ (x) in a Fourier sine series and comparing this expression with the Fourier expansion we get ;

$$C_n = \frac{2}{l} \int_0^l \left\{ \phi(x) \sin \frac{n\pi}{l} x \right\} dx$$

Hence ;

$$T = t - t_1$$

$$= \frac{2}{l} \sum_{n=1}^{\infty} e^{-(n\pi/l)y} \times \left(sin \frac{n\pi}{l} x \right) \times \int_0^l \left\{ \phi(x) \sin \frac{n\pi}{l} x \right\} dx \qquad (2.104)$$

Eq. (2.104) gives the temperature distribution in the semi-infinite plate and also illustrates how similar cases may be dealt with.

2.23 GRAPHICAL SOLUTION OF STEADY STATE HEAT CONDUCTION; POTENTIAL FIELD PLOTTING

As already mentioned, the analytical solution in many cases may be too complex or even impossible ; in others, the order of accuracy required in the evaluation of the temperature field may not justify a detailed analytical investigation . In such cases, an *approximate* solution of the two-dimensional heat conduction equation can be obtained graphically by plotting the temperature field freehand. This graphical method is known as *potential field plotting.*

This method is particularly simple in systems with isothermal boundaries, and we shall discuss here these systems only.

It may be shown with the help of the theory of *complex variables* that the locus of isothermal and adiabatic points in a Laplacean field, represented by Laplace Eq. (2.15), must be orthogonal to each other. This means that the approximate graphical method will consist of the filling up of the given region with a net of *isothermal* and *adiabatic* lines. The construction of lines will be carried out by trial and error until the criterion that these lines meet at right angles at all points of intersection is satisfied. It will be observed that such a construction will be similar to the *flow nets* in fluid mechanics where the *stream function* and *velocity potential* lines are orthogonal to each other.

By this method the temperature distribution is directly obtained, and the heat flow rate can be calculated. The principle as applied to the case of plane wall is illustrated in Fig. 2.31. It may be assumed that the temperatures t_1 and t_2 are constant over the faces of the wall. Here the *isotherms* run parallel to the faces while the *heat-flow* lines run perpendicular to them.

Actually, the heat-flow lines analogous to stream lines of fluid flow, are tangential to the direction of heat flow at any point. Thus no heat can flow across any heat flow lines. If the spacing of *heat-flow* lines is so adjusted that they are at equal distance apart, then the same amount of heat will flow through each *flow tube* formed between two adjacent heat flow lines.

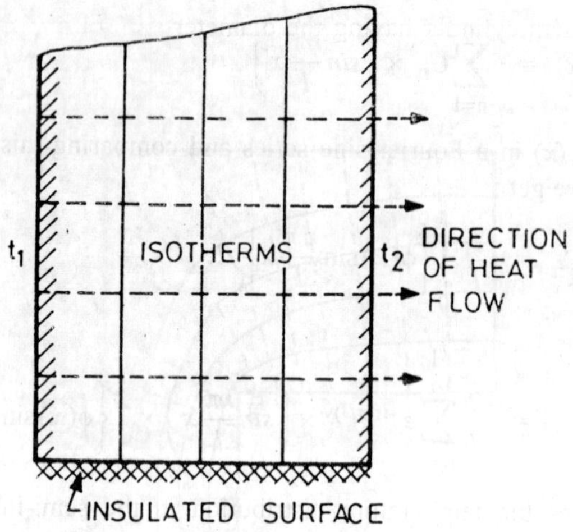

Fig. 2.31 Isotherms and Heat Flow Lines in a Plane Wall

Thus the total rate of heat flow is given by ;

$$q = n \times (\Delta q) \tag{2.105}$$

where n = number of flow tubes

Δq = heat flow rate per tube, W

The following procedure may be adopted to construct the *flux plot* or *potential field plot.*

(i) Choose a certain scale and draw the cross section of the solid.

(ii) Arbitrarily select the temperature intervals, and draw the corresponding *isotherms*; just by intuition. It may be advisable to choose a small number of intervals at the start and then to fill in between them later. Note that the isotherms are not necessarily equidistant (Fig. 2.31)

(iii) Draw *heat flow* lines, so that they intersect the isotherms at right angles everywhere. The two types of lines will form small *curvilinear quadrilaterals*, which should approximately be curvilinear squares. This may be done by suitably adjusting the distance between heat flow lines.

In making the flux plot, with the above method, the curvilinear squares will be obtained by a careful trial and error technique. The problem will be solved if the *network* satisfies the following boundary conditions :

(a) Flow-lines are perpendicular to isothermal boundaries.

(b) Isotherms are perpendicular to insulated boundaries.

(c) Flow-lines leading to a corner of an isothermal boundary bisect the angle between the surfaces of the boundary at the corner.

As an illustration, Figs. 2.32 (a) and (b) show the *heat-flow network* for the case of heat conduction :

(i) across a corner of a building having isothermal faces *A B C* and *D E F*, and,

(ii) across a solid semicylinder having the diameter-surface and the outer surface as isothermals.

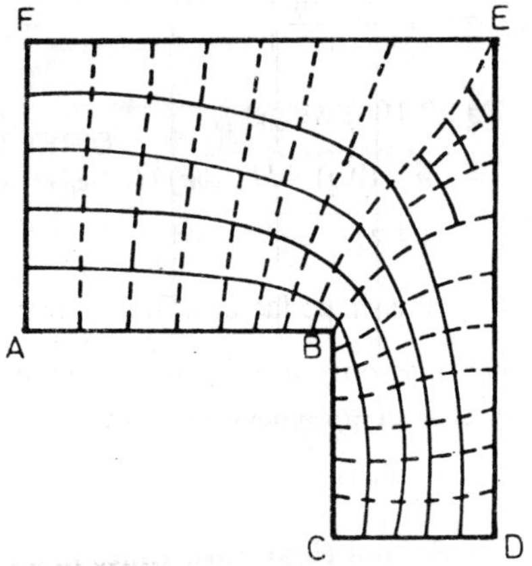

Fig. 2.32 (a) Potential Field Plot for Corner

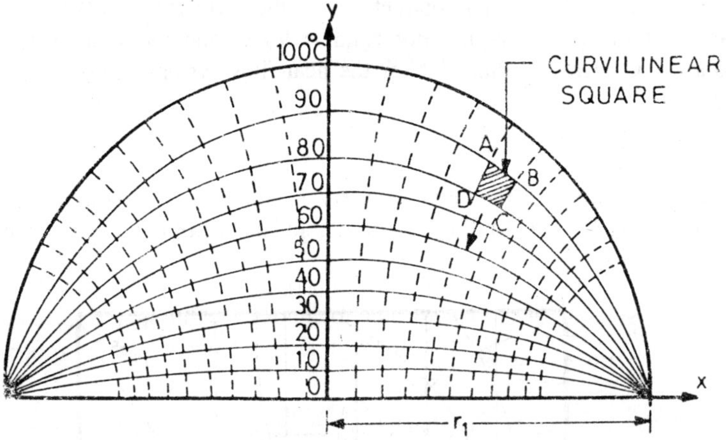

Fig. 2.32 (b) Potential Field Plot for a Solid Semi-Cylinder

The rate of heat flow through any curvilinear square *A B C D* of Fig. 2.32 (b) may be evaluated as :

$$\Delta q = -kl\,\Delta x\left(\frac{\Delta t}{\Delta r}\right) \tag{2.106}$$

where l = length normal to the diagram, m

$\Delta x, \Delta r$ = dimensions of curvilinear square, m

Δt = temperature interval between isothermals, C

$= \left(\frac{t_1 - t_0}{m}\right)$

m = number of temperature increments, or the number of curvilinear squares in one channel of heat flow.

and k = thermal conductivity, W/mK

Thus,
$$\Delta q = kl\left(\frac{t_1 - t_0}{m}\right) \quad (2.107)$$

Substituting Eq. (2.107) in Eq. (2.105) we get ;

$$q = n \times (\Delta q) = kl\,(n/m)\,(t_1 - t_0)$$
$$= k\,l\,S\,(t_1 - t_0) \quad (2.108)$$

where S = the shape factor of the system.

It may be noted that since S is only a ratio of the number of flow tubes to that of temperature increments, the size of the curvilinear squares makes no difference in the result.

Example 2.22

A 15 cm outer diameter pipe, 10 m long is buried in ground with its center line 75 cm below the surface. The thermal conductivity of the earth is 0.346 W/mK. The surface temperature of the pipe carrying hot fluid is 90 C and the ground surface may be assumed at 10 C. Draw a flux plot for the heat-flow system and determine the rate of heat loss.

Solution :

Arbitrarily we choose 8 temperature intervals and 18 heat flow tubes around the pipe as shown in Fig. 2.33. Only half of the flow field has been shown due to symmetry.

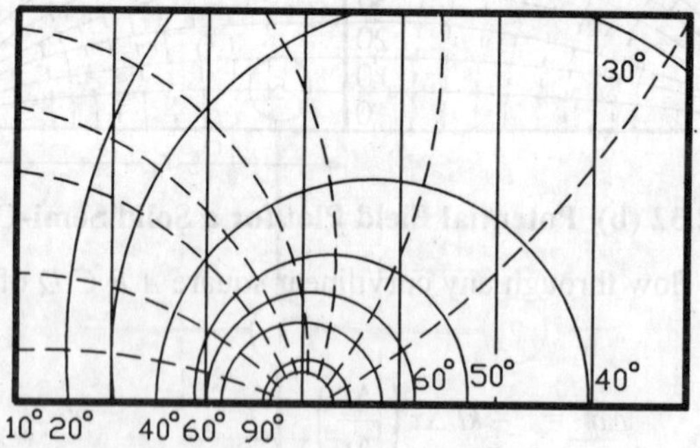

Fig. 2. 33 Example 2.22

now, $m = 8$; $n = 18$; $k = 0.346$ W/mK; $l = 10$ m; $t_1 = 90$ C; $t_0 = 10$ C;

Hence
$$q = k\,l\,S\,(t_1 - t_0) \quad (2.108)$$
$$= 0.346 \times 10 \times \frac{18}{8} \times (90 - 10)$$
$$= 62 \text{ W}$$

Ans.

2.24 THE NUMERICAL SOLUTION OF TWO-DIMENSIONAL STEADY STATE HEAT CONDUCTION PROBLEMS

The numerical method of solution of heat conduction problems for one dimensional flow discussed in Section 2.17 may be extended to two dimensional flow as well. The *network* for selection of *nodal points* in this case, however, will not be as simple as shown in Fig. 2.20.

For instance, let us consider a solid of thickness b where no variation in temperature exists along the axis representing thickness. Hence the system may be considered as a *two dimensional* one. Let us divide the whole solid into small volumes equal to $(\Delta l\ \Delta l\ b)$, as shown in Fig. 2.34. The center of each volume may be considered as a nodal point.

Accordingly, the material between nodal points may be replaced by imaginary rods having same conductance as the material. This is shown in Fig. 2.35.

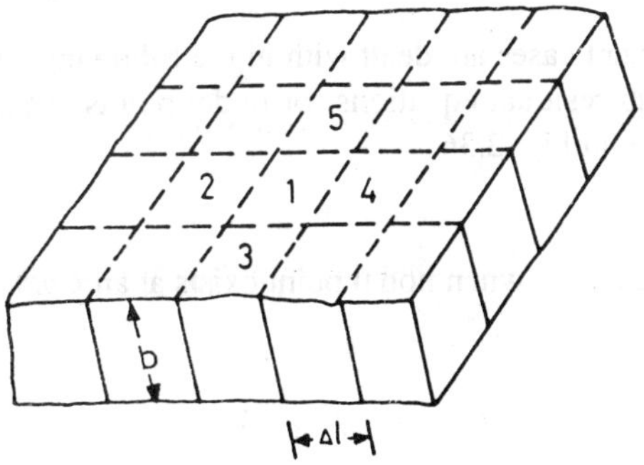

Fig. 2.34 A Solid Divided into Small Volumes

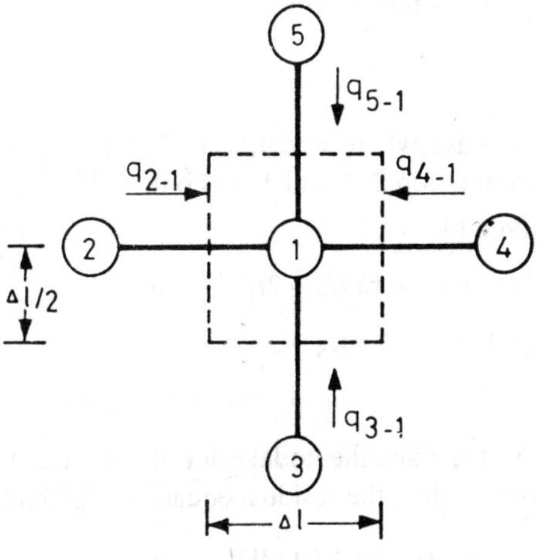

Fig. 2.35 Heat Flow Between Nodal Points

The steady state heat balance on the nodal point 1 is given by :

$$q_{2-1} + q_{3-1} + q_{4-1} + q_{5-1} = 0$$

$$\text{or} \quad K_{2-1}(t_2 - t_1) + K_{3-1}(t_3 - t_1) + K_{4-1}(t_4 - t_1) + K_{5-1}(t_5 - t_1) = 0$$

Now

$$K_{2-1} = \frac{k(\Delta l)b}{\Delta l} = kb = K_{3-1} = K_{4-1} = K_{5-1}$$

Hence the residual equation for the interior nodal point 1 is given by :

$$kb\,(t_2 + t_3 + t_4 + t_5 - 4t_1) = 0$$

or

$$t_2 + t_3 + t_4 + t_5 - 4t_1 = 0 \qquad (2.109)$$

Eq. (2.109) holds for all the *interior nodal points*. However nodal points at the *outer boundary* will have residual equations which depend upon the geometry and other condition of the boundary. Therefore, no simple and general equation may be written for these *boundary nodal points* each of which requires individual consideration.

Some of the important cases are dealt with in the following paragraphs to illustrate the technique of setting up residual equations for nodal points at the boundary. Eight cases so discussed are shown in Fig. 2.36.

(a) Case 1

Fig. 2.36 (a) shows a case when nodal point exists at an *isothermal boundary*.

$$t_3 = t_1 = t_5$$

so

$$q_{3-1} = q_{5-1} = 0$$

Hence residual equation is :

$$q\,(\Delta l b) + kb\,(t_4 - t_1) = 0$$

or

$$q\left(\frac{\Delta l}{k}\right) + (t_4 - t_1) = 0 \qquad (2.110)$$

(b) Case 2

Fig. 2.36 (b) shows the case where nodal point lies on an *insulated surface*, hence $q_{3-1} = 0$. Further, the area between 2–1 and 4–1 is only half of that between 5–1. Hence the residual equation is given by ;

$$(kb/2)\,(t_2 + t_4) + kb\,(t_5 - 2t_1) = 0$$

or

$$(½)\,(t_2 + t_4) + (t_5 - 2t_1) = 0 \qquad (2.111)$$

(c) Case 3

Fig. 2.36 (c) shows the case when the nodal point lies on a surface which is in contact with a fluid at temperature t_∞. Here the residual equation is given by :

$$(kb/2)\,(t_3 + t_5) + kb\,(t_2 - 2t_1) + (t_\infty - t_1)\,h\,(\Delta l b) = 0$$

$$\text{or} \quad \left(\frac{1}{2}\right)(t_3 + t_5) + t_2 + t_\infty\left(\frac{h\Delta l}{k}\right) - t_1\left(2 + \frac{h\Delta l}{k}\right) = 0 \qquad (2.112)$$

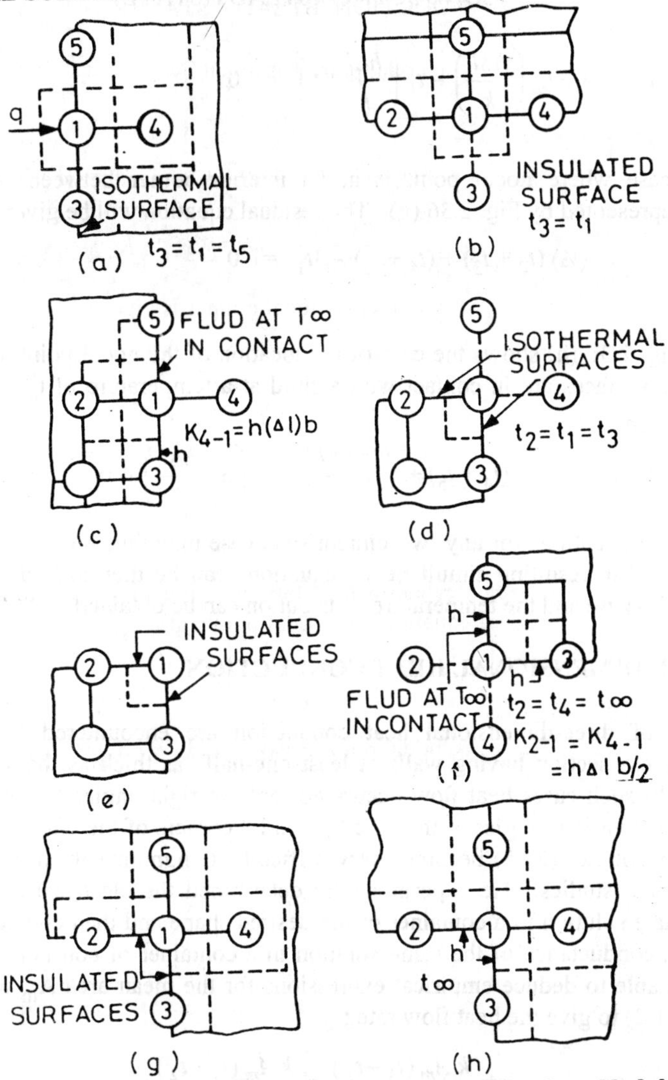

Fig. 2. 36 Eight different Cases of Heat Flow for Boundary Nodal Points

(d) Case 4

Fig. 2.36 (d) shows a case when nodal point forms a corner between *isothermal surfaces*. This case has no residual equation as no heat flows through the nodal point.

(e) Case 5

Fig. 2.36 (e) shows a case when the nodal point forms a corner between two insulated surfaces. In this case the heat flows only between 2–1 and 3–1, and that too only through area ($\Delta l b/2$). The residual equation will be given by :

$$(½)\,(t_2 + t_3) - t_1 \;=\; 0 \tag{2.113}$$

(f) Case 6

This is a case of the nodal point located at a corner which is in contact with a fluid at t_∞ and is shown by Fig. 2.36 (f). The residual equation may be obtained as given by :

$$(\tfrac{1}{2})(t_4 + t_5) + t_\infty \left(\frac{h\Delta l}{k}\right) - t_1\left(\frac{h\Delta l}{k} + 1\right) = 0 \qquad (2.114)$$

(g) Case 7

Another case where nodal point is at an interior corner between two insulated surfaces, is represented by Fig. 2.36 (g). The residual equation will be given by :

$$(\tfrac{1}{2})(t_2 + t_3) + (t_4 + t_5) - 3t_1 = 0 \qquad (2.115)$$

(h) Case 8

Finally, Fig. 2.36 (h) shows the case of the location of the nodal point at the *interior corner* whose surfaces are in contact with a fluid at a temperature of t_∞. The residual equation is given by :

$$(\tfrac{1}{2})(t_2 + t_3) + (t_4 + t_5) + t_\infty \left(\frac{h\Delta l}{k}\right) - t_1\left(3 + \frac{h\Delta l}{k}\right) = 0 \qquad (2.116)$$

The residual equations for any two-dimensional case may thus be set up for different nodal points. The resulting simultaneous equations can be then solved by *relaxation method* or *otherwise* and the temperature distribution can be obtained.

2.25 THREE-DIMENSIONAL HEAT CONDUCTION

Problems of three-dimensional heat conduction are encountered in furnaces of rectangular cross section having walls at least one-half as thick as the shortest inside dimension. In such cases heat flow occurs not only at right angles to the entire inside area, but also at various angles to the 12 edges and 8 corners of the outside walls. Here, the integration of Eq. (2.15) becomes very difficult. Langmuir solved the problem by means of model studies. He experimentally determined the electrical conductance of copper sulphate solution in a container of the desired shape and then compared this with the electrical conductance of the same solution in a container of constant cross section. He was thus able to deduce empirical expressions for the mean area A_m which may be used in Eq. (1.2) to give the heat flow rate ;

$$q = \frac{k\,A_m\,(t_1 - t_2)}{(x_2 - x_1)} = \frac{k\,A_m\,(t_1 - t_2)}{\Delta x} \qquad (2.117)$$

We shall consider empirical results for four cases dealing with only those bodies which are bounded by right rectangular parallelepipeds, whose wall thickness is constant throughout. All faces intersect at right angles

Let y = length of any inside dimension, (i.e, length, breadth or height), m

Δx = wall thickness, m

A_1 = total inside surface area, m^2

A_2 = total outside surface area, m^2

Case 1:

Suppose that all inside dimensions exceed one-fifth of the wall thickness i.e. $y > \Delta x/5$. In this case A_m is obtained by adding to the total inside surface area A_1, the quantity 0.54 $(\Delta x)(\Sigma y)$ to account for the 12 edges and the quantity 0.15 $(\Delta x)^2$ for each of the corners.

$$A_m = A_1 + 0.54\,\Delta x\,(\Sigma y) + 8 \times 0.15\,(\Delta x)^2$$

or
$$A_m = A_1 + 0.54\,\Delta x\,(\Sigma y) + 1.2\,(\Delta x)^2 \quad (2.118)$$

Case 2:

Suppose one inside dimension is less than $\Delta x/5$. In this case the length of the four inside edges, less than $\Delta x/5$, are neglected in computing Σy. Then A_m is computed from:

$$A_m = A_1 + 0.465\,\Delta x\,\Sigma y + 0.35(\Delta x)^2 \quad (2.119)$$

where Σy is the sum of the lengths of the remaining 8 edges, each of which is grater than $\Delta x/5$.

Case 3:

Suppose two inside dimensions are each less than $\Delta x/5$.

Here,

$$A_m = \frac{2.78\,y_{max}\,\Delta x}{\log_{10}(A_2 / A_1)} \quad (2.120)$$

where y_{max} is the largest interior dimension.

Case 4:

Suppose all three inside dimensions are less than $\Delta x/5$.

Then,

$$A_m = 0.79\,(A_1 A_2)^{1/2} \quad (2.121)$$

Sometimes Eq. (2.117) is expressed in the form

$$q = kS\,(t_1 - t_2) \quad (2.122)$$

where S is called the shape factor. It is easily seen that the shape factor has dimensions of length and is equal to $A_m/\Delta x$

The shape factors for the above four cases are :

Case 1 : $S = A_1/\Delta x + 0.54\,\Sigma y + 1.2\,\Delta x$ (2.123)

Case 2 : $S = A_1/\Delta x + 0.465\,\Sigma y + 0.35\,\Delta x$ (2.124)

Case 3 : $S = 2.78\,y_{max}/\log_{10}(A_2/A_1)$ (2.125)

Case 4 : $S = 0.79\,(A_1 A_2)^{1/2}/\Delta x$ (2.126)

If follows from the above discussion that shape factor S_E for one edge of length y is

$$S_E = 0.54\,y \quad (2.127)$$

Similarly, the shape factor S_C, for a corner, (for wall thickness Δx) is

$$S_C = 0.15\,\Delta x \quad (2.128)$$

Shape factors for a cylindrical pipe and a spherical shell are found to be

$$S\text{ (cylindrical pipe)} = 2\pi l/\log_e(r_2/r_1) \quad (2.129)$$

$$S\text{ (spherical shell)} = 4\pi\,r_1\,r_2/(r_2 - r_1) \quad (2.130)$$

Example 2.23

The inside dimensions of a furnace are 3 m × 2.5 m × 2 m. The walls are 20 cm thick and have k = 1.4 W/mK. If the inner and outer surface temperatures are 220°C and 90°C compute the heat loss per hour.

Solution :

Since all inside dimensions are greater then one-fifth wall thickness, this comes under Case 1. Hence

$$A_m = A_1 + 0.54\,\Delta x\,(\Sigma y) + 1.2\,(\Delta x)^2 \qquad (2.118)$$

$$A_1 = 2\,(3 \times 2.5 + 2.5 \times 2 + 3 \times 2) = 37 \text{ m}^2$$

$$\Sigma y = 4\,(3 + 2.5 + 2) = 30 \text{ m}$$

$$\Delta x = 0.2 \text{ m}$$

so

$$A_m = 37 + (0.54 \times 0.2 \times 30) + 1.2\,(0.2)^2$$

$$= 40.288 \text{ m}^2$$

or

$$q = \frac{k A_m (t_1 - t_2)}{\Delta x} = \frac{(1.4)(40.288)(200-50)}{0.2}$$

$$= 42{,}300 \text{ W} \qquad \text{Ans.}$$

2.26 THERMAL CONDUCTIVITY

Thermal conductivity is a physical property of a substance and characterises its ability to transfer heat by conduction. In general, the numerical value of the thermal conductivity is highest for solids, the lowest for gases, and it has intermediate value for liquids.

The conductivity of gases ranges from 0.005 to 0.5 W/mK. It increases with the temperature and does not vary with pressure with the exception of very high (2000 *atm*) or very low (20 mm *Hg*) pressures.

Conductivity of liquids ranges from 0.08 to 0.6 W/mK. It decreases for most of the liquids with increase in temperature, with the exception of water and glycerine.

Conductivity of solids ranges from 0.04 to 300 W/mK. It decreases with temperature increase in the case of some solids while increases in the case of others.

It will be observed that the value of thermal conductivity in case of liquids and gases is so small that heat transfer due to conduction may be assumed to be negligible as compared to other modes viz convection. Hence further discussion of conductivity of gases and liquids is not considered necessary. For the case of solids, however, conduction is the only mode of heat transfer ; hence the study of *thermal conductivity* of solids is more important.

2.27 THERMAL CONDUCTIVITY OF SOLIDS

There are a large number of solid materials, each differing in its behaviour with others. They may, however, be divided broadly into two groups :

(a) Non-metallic substances

(b) Metallic substances

Non-metallic solids do not conduct heat as much as metallic solids do. Table 2.3 gives a few examples of each of the two groups of solids along with the value of the *thermal conductivity* for each of them.

Table 2.3 Thermal conductivity of solid substances

Group	*Type of Substance*	*Name*	*k (W/mK)*
Non Metallic solids	Insulating materials	Aluminum Foil	0.046
		Asbestos Fibre	0.11
		Cellotex	0.046
		Cork sheet	0.04
		Mineral wool	0.046
	Refractory materials	Fire brick	0.14
		Coal	0.18
		Earth dry	0.14
		Clay	1.00
	Amorphous crystaline materials	Glass	0.75
		Mica	0.60
		Porcelain	1.00
		Sand dry	0.32
		Slate	1.48
	Building materials	Brick	0.23
		Concrete	1.13
		Gravel	0.37
		Linoleum	0.19
		Marble	1.32
		Plaster	0.77
		Cement	0.30
	Other materials	Chalk	0.92
		Gypsum	0.29
		Ice	2.24
		Leather	0.15
		Rubber	0.16
		Saw dust	0.07
Metallic solids	Pure metals	Aluminum	200.00
		Copper	380.00
		Iron	63.00
		Lead	35.00
		Silver	420.00
		Mercury	8.00
	Alloys	Brass	85.50
		Bronze	64.00
		Steel (0.3% C)	45.00

2.28 THERMAL CONDUCTIVITY OF NON-METALLIC SUBSTANCES

The value of thermal conductivity for non-metallic substances ranges from 0.04 to 2.5 W/mK. It increases, for all cases, with temperature. It also depends upon the structure of the material, and also on its porosity and the humidity. If the material is damp the value of k may increases considerably. For example the conductivity of brick increases to 0.93 W/mK after it is dipped in water for some time. Fig. 2.37 shows the variation of k with temperature for non-metallic solids.

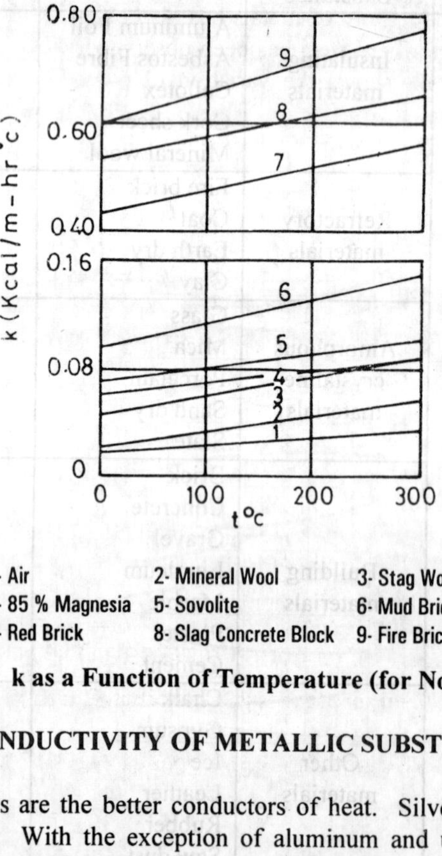

Fig. 2.37 k as a Function of Temperature (for Non-metals)

2.29 THERMAL CONDUCTIVITY OF METALLIC SUBSTANCES

Of all solids, metals are the better conductors of heat. Silver is the best conductor among known metals. With the exception of aluminum and uranium the value of *k* decreases with increasing temperature.

Thermal conductivity of metals depends upon the purity of the substance. It drops abruptly in presence of other elements or compounds. For instance ;

(i) *k* for pure copper = 380 W/mK

k for copper with traces of arsenic = 142 W/mK

(ii) *k* for cast iron = 63 W/mK

k for steel (0.1% C) = 52 W/mK

k for steel (1.5% C) = 36 W/mK

Also it is observed that the conductivity of an alloy is much less as compared to the value for any of its constituents. In some cases it may be as low as 20% of the value for any of the constituents. For instance :

k for nickel = 93 W/mK

k for copper = 380 W/mK

k for Monel metal = 24.4 W/mK

(30% nickel ; 70% copper)

Further, the thermal conductivity of metals has also been found to depend upon the grain structure. The value for hard steel has been observed to be 10 to 25 % lower than that for soft steel. Fig. 2.38 shows the variation of conductivity with purity of the substance and its temperature.

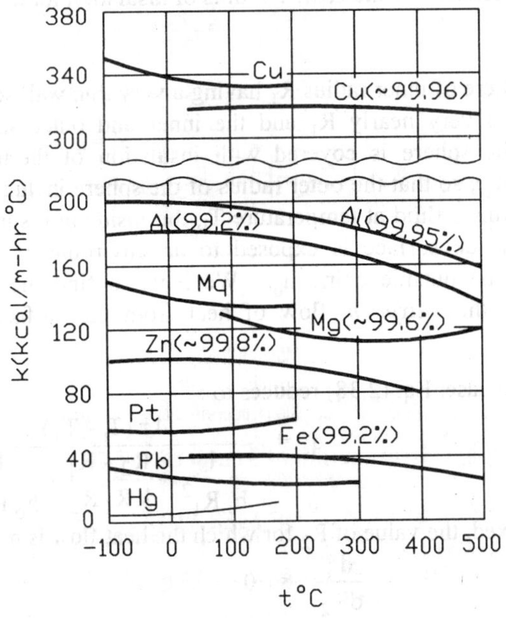

Fig. 2.38 k as a Function of Composition (for Metals)

It may be mentioned that some grained materials have nonuniform thermal conductivity. Its value differs in x and y direction. Such materials are known as *anisotropic*.

2.30 RELATIONSHIP BETWEEN THERMAL AND ELECTRICAL CONDUCTIVITY

In 1853 Wiedeman and Franz observed that the ratio of the thermal conductivity k to the electrical conductivity k_e is approximately constant for all metals. In 1872, however, Lorenz correlated this ratio with temperature of the metal and enunciated the *Weideman-Franz law* :

$$\frac{k}{k_e T} = L \tag{2.131}$$

where T = Absolute temperature, °K

L = Lorenz number = 3 (B^2/ε^2)

= 2.45×10^8 (volt/C)^2

B = Boltzmann constant

ε = charge on the electron

Since it is easier to measure the electrical conductivity than the thermal conductivity, *Lorenz number* may be used to assess the value of thermal conductivity of a metal.

2.31 ILLUSTRATIVE EXAMPLES

Example 2.24

Derive an expression for the critical radius of insulation for a sphere.

Solution :

Consider a sphere of inner radius R_1 having a very thin wall so that the outer radius of the sphere is also very nearly R_1 and the inner and outer surfaces are at the same temperature. The sphere is covered with insulation of thermal conductivity k and thickness ($R_2 - R_1$), so that the outer radius of the sphere including the insulation is R2. The sphere contains a fluid at temperature T_i, the inside unit surface conductance being h_i, while the outside surface is exposed to an environment at T_o, the unit surface conductance on the outside being h_o. We have to find the thickness of insulation corresponding to the *maximum* flow of heat from the surface of the sphere to the environment.

For the present case, Eq. (2.38) reduces to :

$$q = \frac{4\pi(T_i - T_o)}{\frac{1}{h_i R_1^2} + \frac{R_2 - R_1}{k R_1 R_2} + \frac{1}{h_o R_2^2}}$$

Keeping R_1 fixed, the value of R_2 for which the heat flow is maximum is given by

$$\frac{dq}{dR_2} = 0$$

so

$$\frac{4\pi(T_i - T_o)(-1)\left[\frac{1}{k R_2^2} - \frac{2}{h_o R_2^3}\right]}{\left[\frac{1}{h_i R_1^2} + \frac{R_2 - R_1}{k R_1 R_2} + \frac{1}{h_o R_2^2}\right]^2} = 0$$

or

$$\frac{1}{k} = \frac{2}{h_o R_2} \quad \text{or} \quad R_2 = \frac{2k}{h_o}$$

Hence, critical radius of insulation is given by

$$r_c = 2k/h_o$$

Ans.

Example 2.25

An orange, 10 cm in outer diameter, is placed in a refrigerator in which the air temperature at a particular instant is –2°C. The surface temperature of the orange is 18° C and the convection coefficient, h, between surface and air is 15 W/m^{2}K. If the orange peel is 3 mm thick and its thermal conductivity is 0.40 W/mK, determine the temperature at the interface between the peel and the remainder part of the orange. The emissivity, ε of the orange surface may be assumed to be 0.7.

Solution :

The rate of heat transfer by conduction across the orange peel equals the rate of heat transfer by convection and radiation from the surface of the orange to the air in the refrigerator and its walls.

i.e. $$q_{cond} = q_{conv} + q_{rad}$$

(i) Rate of convective heat transfer from orange surface to air,

$$q_{conv} = h\,A_s\,(t_s - t_a)$$

where suffix s denotes orange surface

$$A_s = \pi\,D_s^2 = \pi\,(0.1)^2 = 0.0314\ \text{m}^2$$

so $$q_{conv} = (15)\,(0.0314)\,(18+2) = 9.42\ \text{W}$$

(ii) Now, assuming the walls of the refrigerator to be at the air temperature

$$q_{rad} = A_s\,(F_g)_{12}\,\sigma\,(T_s^4 - T_a^4)$$

$$(F_g)_{12} = \varepsilon_1 = 0.7 \quad \text{(from Table 6.4)}$$

so $$q_{rad} = (0.0314)\,(0.7)\,(5.67\times10^{-8})\,(291^4 - 271^4)$$

$$= 0.391\ \text{W}$$

thus $$q_{conv} + q_{rad} = 9.42 + 0.391 = 9.811\ \text{W}$$

If t_1 denotes the temperature at the interface between the peel and the remaining part of the orange, the rate of heat transfer by conduction is given by,

$$q_{cond} = \frac{4\pi k R_1 R_s (t_1 - t_s)}{R_s - R_1} \qquad (2.28)$$

where $$R_1 = \frac{10}{2} - 0.3 = 4.7\ \text{cm}$$

$$q_{cond} = q_{conv} + q_{rad} = 9.811$$

$$= \frac{(4\pi)(0.4)(0.047)(0.05)(t_1 - t_s)}{3\times10^{-3}}$$

so $$t_1 - t_s = 2.49$$

or $$t_1 = t_s + 2.49$$

$$= 18 + 2.49$$

$$= 20.49°\text{C}$$ Ans.

Example 2.26

A cylindrical tank 1.2 m in diameter and 6 m long has hemispherical ends and is used to store liquid oxygen which has a boiling point of –183°C. The heat of vaporization of liquid oxygen is 213.6 kJ/kg. Determine the thermal conductivity of the insulation required if the thickness of the insulation has to be 7.5 cm and the boil-off rate not to exceed 10 kg/hr. Assume that the temperature of the outer surface of the insulation is 17°C.

Solution :

The tank may be considered to be composed of a central cylindrical section of length 4.8 m and two equal hemispherical portions of diameter 1.2 m at both ends. It follows from Eq. (2.28) and Eq. (2.23) that the rate of heat flow is :

$$q = \frac{4\pi k R_1 R_2 (T_1 - T_2)}{R_2 - R_1} + \frac{2\pi k L(T_1 - T_2)}{ln(R_2 / R_1)}$$

where $R_1 = 0.6$ m; $R_2 = 0.6 + 0.075$

$= 0.675$ m and $T_1 - T_2 = -200°C$

Hence, $$q = \frac{(4\pi k)(0.6)(0.675)(-200)}{0.075} + \frac{(2\pi k)(4.8)(-200)}{\ln(0.675/0.6)}$$

$$= -(13{,}564.8\ k + 51{,}185.6\ k) = -64{,}750.4\ k$$

The heat transfer rate *to liquid oxygen* must not exceed :

$$-q_{max} = \frac{10 \times 213.6 \times 1000}{3600} = 593.3 \text{ W}$$

so, $q = -593.3 = -64{,}750.4\ k$

or $k = 0.00916$ W/m K   Ans.

Example 2.27

A rocket motor chamber consists of a sphere 40 cm ID. The wall consists of an inner layer of refractory 1 cm thick (k = 1.8 W/m K) followed by a 3 cm thick layer of steel having a thermal conductivity (W/mK) given by ;

$$k = 25.023\,(1 + 0.002\,t)$$

where t is in °C. Under steady state conditions, the gases inside the chamber are at 3200°C, and the unit surface conductance at the inner wall, h_i, is 60 W/m K. Determine the amount of heat transferred through the wall (kW) if the outer surface temperature is 100°C.

Solution :

Let R_1, R_2, R_3 denote respectively the inner radius of the sphere, the radius including the refractory layer and the outer radius of the sphere. Then $R_1 = 0.2$ m, $R_2 = 0.21$ m, $R_3 = 0.24$ m. Also let T_1, T_2, T_3 denote the temperature of the hot gases inside the chamber, interface between refractory and steel layers, and outer surface of the chamber, respectively. $T_1 = 3200$ C, $T_3 = 100$ C.

The rate of heat transfer from the hot gases to the outer surface of the refractory is,

$$q_1 = \frac{T_1 - T_2}{\left[\dfrac{1}{4\pi h_1 R_1^2} + \dfrac{R_2 - R_1}{4\pi k_1 R_1 R_2}\right]} \tag{2.38}$$

$$= \frac{3200 - T_2}{\left[\dfrac{1}{4\pi \times 60 \times 0.2^2} + \dfrac{0.01}{4\pi \times 1.8 \times 0.2 \times 0.021}\right]}$$

$$= \frac{3200 - T_2}{(0.033174 + 0.0105314)}$$

i.e. $$q_1 = \frac{3200 - T_2}{0.0437054}\ \text{W} \tag{a}$$

The rate of heat transfer across the steel layer is

$$q_2 = \frac{(T_2 - T_3)}{(R_3 - R_2)/\{4\pi k_o R_2 R_3[1+(\alpha/2)(T_2+T_3)]\}} \tag{2.45}$$

$$= \frac{(T_2 - 100)}{(0.03)/\{(4\pi)(25.023)(0.21)(0.24)[1+0.001(T_2+100)]\}}$$

i.e. $$q_2 = \frac{(T_2 - 100)[1+0.001(T_2+100)]}{0.00189392}\ \text{W} \tag{b}$$

Under steady state $q_1 = q_2$, hence by equating expressions (a) and (b) we obtain by trial,

$$T_2 = 200\ \text{C}$$

Hence $$q_1 = q_2 = 68641\ \text{W}$$

$$= 68.641\ \text{kW}$$ Ans.

Example 2.28

Two long rods of the same diameter, one made of brass (k = 85 W/m K) and the other made of copper (k = 375 W/m K) have one of their ends inserted into a furnace. Both rods are exposed to the same environment. At a section 10.5 cm away from the furnace end, the temperature of the brass rod is 120°C. At what distance from the furnace end, the same temperature would be reached in the copper rod ? (UPSC '93)

Solution :

The temperature distribution in a spine is given by,

$$\frac{t - t_\infty}{t_s - t_\infty} = e^{-mx} \tag{2.48}$$

where $$m = \sqrt{\left(\frac{hP}{kA}\right)}$$

For the same value of temperature in the two rods,

$$m_b x_b = m_c x_c$$

(suffix b denotes brass and c denotes copper)

So,
$$x_c = x_b \frac{m_b}{m_c} = x_b \left[\frac{hP}{k_b A} \times \frac{k_c A}{hP} \right]^{1/2}$$

$$= x_b \left(\frac{k_c}{k_b} \right)^{1/2} = (10.5)\,(375/85)^{1/2}$$

$$= 22.05 \text{ cm}$$ **Ans.**

Example 2.29

A composite pin fin is made up of two materials. The inner material of diameter 10 mm has thermal conductivity of 18 W/m K while the outer material of outside diameter 20 mm has thermal conductivity 54 W/m K. The length of the fin is 200 mm. Calculate the effectiveness of the fin if $h = 10$ W/m^2 K.

Solution :

It can be verified that the governing differential equation, which corresponds to Eq. (2.46), for the temperature distribution in the composite fin may be written as :

$$(k_1 A_1 + k_2 A_2) \frac{d^2 t}{dx^2} = h P (t - t_\infty)$$

The suffix 1 and 2 denote the inner and the outer material. It follows that,

$$m = \sqrt{\frac{hP}{k_1 A_1 + k_2 A_2}}$$

here,
$$h = 10 \text{ W/m}^2\text{ K}; \quad P = 20\,\pi \times 10^{-3} \text{ m}; \; k_1 = 18 \text{ W/m K};$$
$$k_2 = 54 \text{ W/m K}; \quad A_1 = (\pi/4)\,(10)^2 \times 10^{-6} \text{ m}^2$$
$$= 0.785 \times 10^{-4} \text{ m}^2;$$
$$A_2 = (\pi/4)\,[(20)^2 - (10)^2] \times 10^{-6} = 2.355 \times 10^{-4} \text{ m}^2.$$

so,
$$m = \sqrt{\frac{10 \times 20\pi \times 10^{-3}}{18 \times 0.785 \times 10^{-4} + 54 \times 2.355 \times 10^{-4}}} = 6.666$$

or
$$mL = 6.666 \times 0.2 = 1.333$$

From Eq. (2.56), the effectiveness of the fin is,

$$\eta_f = \frac{\tanh(mL)}{mL} = \frac{\tanh(1.333)}{1.333}$$

$$= \frac{0.869}{1.333} = 0.652$$ **Ans.**

Example 2.30

Ten fins of rectangular cross section, 3 mm thick and 25 mm long (k = 50 W/m K), are attached circumferentially to a cylinder of 20 cm outer diameter. The fin spacing is 1.25 cm. Determine the increase in the amount of heat transferred from the cylinder to air due to fins. The cylinder wall and ambient air temperatures are 300°C and 30°C, respectively, and the convective heat transfer coefficient is 15 W/m^2 K.

Solution :

Let us first calculate the efficiency of the fins,

$$\eta_f = \frac{\tanh(mL)}{mL} \tag{2.56}$$

$$\text{where} \quad m = \sqrt{\frac{hP}{kA}}$$

In the present case, at any radius R,

$$\frac{P}{A} = \frac{2\pi R + 2\pi R + 2\delta}{2\pi R \delta} \; ; \; (\delta = \text{fin thickness})$$

$$\approx \frac{2}{\delta} \text{ since } \delta << R$$

Hence,

$$m = \sqrt{\frac{15 \times 2}{50 \times 3 \times 10^{-3}}} = 14.14 \text{ m}^{-1}$$

$$mL = 14.14 \times 25 \times 10^{-3} = 0.3535$$

so

$$\eta_f = \frac{\tanh(0.3535)}{0.3535} = \frac{0.3394}{0.3535} = 0.96$$

Surface area of 1 fin,

$$= 2\pi(R_2^2 - R_1^2) = 2\pi(12.5^2 - 10^2)\, 10^{-4}$$

$$= 353.25 \times 10^{-4} \text{ m}^2$$

Total surface area of 10 fins,

$$A_f = 10 \times 353.25 \times 10^{-4} = 0.3533 \text{ m}^2$$

The exposed surface area of cylinder wall without fins, i.e., other than $A_{\text{fin base}}$ covered by fins,

$$A_{wf} = 10 \times 1.25 \times 10^{-2} \times \pi \times 20 \times 10^{-2}$$

$$= 0.0785 \text{ m}^2$$

Heat transfer rate is given by,

$$q = (A_f \eta_f + A_{wf})\, h\, (t_s - t_\infty)$$

$$= \{(0.3533 \times 0.96 + 0.0785) \times 15 \times 270\}$$

$$= 1691.6 \text{ W}$$

If no fins are provided, the rate of heat transfer from the cylinder wall will be

$$q_1 = A\,h\,(t_s - t_\infty)$$

where $A = A_{wf} + A_{fin\ base}$

$$= [(0.0785) + (10 \times \pi \times 0.2 \times 0.003)]\,(15)\,(270)$$

$$= 394.23 \text{ W}$$

Hence, increase in heat transfer rate due to fins

$$= 1691.60 - 394.23 = 1297.4 \text{ W}$$

$$= \frac{1297.4}{394.23} \times 100 = 329\ \%$$ **Ans.**

Example 2.31

A copper tube 2.858 cm OD has fins consisting of a metal strip spirally wound around it. Each fin has a thickness of 0.0635 cm and a tip-to-tip diameter of 5.56 cm. The thermal conductivity of the fin is 43 W/m K. The value of the convection coefficient h is 4 W/m²K. If the ambient temperature is 30°C and the tube wall temperature at a given section is 95°C, determine the heat transferred by one fin. What is the efficiency of the fin ?

Solution :

We may use Fig. 2.19 to determine the efficiency of the circumferential fin :

$$r_1 = \frac{2.858}{2} \times 10^{-2} = 1.429 \times 10^{-2} \text{ m}$$

$$r_o = \frac{5.56}{2} \times 10^{-2} = 2.78 \times 10^{-2} \text{ m}$$

$$\delta = 0.0635 \times 10^{-2} \text{ m}$$

$$\frac{r_o + \delta/2}{r_1} = \frac{2.78 + 0.0635/2}{1.429} = 1.97$$

$$(r_o + \delta/2 - r_1)^{3/2}\left[\frac{2h}{k\,\delta\,(r_o - r_1)}\right]^{1/2} = \left[\left(2.78 + \frac{0.0635}{2} - 1.429\right) \times 10^{-2}\right]^{3/2} \times \left[\frac{2 \times 4}{43 \times 0.0635 \times 10^{-2}\,(2.78 - 1.429)10^{-2}}\right]^{1/2}$$

From Fig. 2.19, we get ;

$$= 1.626 \times 10^{-3} \times 147.264 = 0.2395$$

$$\eta_f = 0.965$$ **Ans.**

Heat transfer from one fin,

$$q_f = \eta_f A_f h\,(T_s - T_\infty)$$

$$A_f = 2\left[\pi\left\{\left(r_o + \frac{\delta}{2}\right)^2 - r_1^2\right\}\right]$$

$$= 2\pi\,[(2.81175)^2 - (1.429)^2] \times 10^{-4}\ \text{m}^2$$

$$= 36.825 \times 10^{-4}\ \text{m}^2$$

$$T_s = 95°\text{C}\ ;\ T_\infty = 30°\text{C}$$

So, $$q_f = 0.965 \times 36.825 \times 10^{-4} \times 4\,(95 - 30)$$

$$= 0.924\ \text{W}$$ Ans.

PROBLEMS

2.1 Reduce Eq. (2.16) for the case when the temperature variation is symmetrical about the axis of a tube and there is no heat generation.

2.2 Reduce Eq. (2.17) for the case when the temperature distribution is such that concentric spherical surfaces are isothermal.

2.3 Show that the heat conduction equation for a sphere having steady temperature and radial heat flow only is given by ;

$$\frac{1}{r^2}\frac{d}{dr}\left(r^2\frac{dt}{dr}\right) + \frac{Q'}{k} = 0$$

2.4 Show that for the case of no heat source, Eq. (2.16) can be reduced to

$$\frac{1}{r}\frac{\partial}{\partial r}\left(r\frac{\partial t}{\partial r}\right) + \frac{1}{r^2}\frac{\partial^2 t}{\partial \theta^2} + \frac{\partial^2 t}{\partial z^2} = \frac{1}{\alpha}\frac{\partial t}{\partial \tau}$$

2.5 If $t = A + B \ln r + Cr^2$. where A, B and C are constants, show that the equation of problem 2.4 is satisfied only if t is independent of θ, r and z.

2.6 A pipe carries condensing steam at 9.17 bar. The unit surface conductance of the steam side film is 5650 and that of outside film at pipe is 8.15 W/m^{2}K. The thermal resistance of the pipe per unit area is 1.72×10^{-4} m^2 K/W. If the pipe is laid in a room at temperature of 25 C, calculate the temperature at outer surface of the pipe. [Ans. 174.5 C]

2.7 A steam pipe with I D and O D as 100 and 170 mm is covered with two layers of insulation, 30 mm and 50 mm thick. Thermal conductivities of the insulating materials are 0.175 and 0.093 W/mK respectively while that of steel is 50 W/mK. The inner surface of the pipe is at 300 C while the outer layer surface is at 50 C. Determine the heat loss from the pipe and the layer contact temperatures.

[Ans. 280 W/m ; 300 C; 223 C]

2.8 The integrated form of one-dimensional Fourier Law leads to a mean conducting area for a plane wall as ;

$A_m = A$. Show that :

(a) The mean conducting area for cylindrical shell is

$$A_m = \frac{2\pi l\,(r_2 - r_1)}{\ln(r_2 / r_1)}$$

(b) The mean conducting area for spherical shell is ;

$$A_m = 4\pi\, r_1\, r_2$$

2.9 Determine the percentage error in the rate of heat flow if the value of A_m in problem 2.8 is taken as the arithmetic mean : $\pi/\,(r_2 + r_1)$, for ratios of r_2/r_1 = 1.5, 2.0, 3 and 3.5. Plot the results.

2.10 A 15 cm O. D. pipe is covered with 5 cm of insulation of k = 0.093 W/mK and 4 cm of insulation of k = 0.07 W/mK. Calculate the heat loss from 100 m of pipe if the inner and outer surface temperatures of the insulation are 400°C and 30°C respectively. What is the heat loss per 100 m^2 of outer pipe surface ?

2.11 The outside and inside surface temperatures of a 20 cm O. D. and 18 cm I. D. pipe (k = 40 W/mK) are 400 °C and 406°C respectively. Calculate the reduction in heat loss if 5 cm layer of insulation (k = 0.06 W/m K) is put on the pipe. Assume that the inner and outer surface temperatures of the insulation are 405°C and 140°C. What is the inside surface temperature of this pipe in this case ?

2.12 A standard 5 cm steel pipe carrying steam is insulated with 7.5 cm layer of glass wool. The outer surface temperature of the pipe is 200 C and that of the covering is 65 C. Estimate the heat loss from pipe (Assume the value of k_{steel}).

2.13 A 30 cm outside diameter pipe is covered with two layers of insulation (k_1 = 0.09, k_2 = 0.06 W/mK). The better insulating material is on the outside and it is 4 cm thick. The other insulating material is of 5 cm thickness. The inner and outer surface temperatures of the insulation are 350 C and 50 C. Compute :

(a) Heat loss per m length per hr.

(b) Heat loss per m^2 of outer insulation surface.

(c) Heat loss per m^2 of pipe surface.

(d) Temperature of the surface between the two layers of insulation.

2.14 A 5 cm outer diameter pipe is to be covered with two layers of insulation each 5 cm thick. If the thermal conductivity of one material is 4 times that of the other, and the inner and outer surface temperatures of the composite insulation are fixed, what will be the ratio of the heat transfer when the better insulating material is next to the pipe to that when the better insulating material forms the outer layer.

2.15 A steel locker is used for storing currency notes which get damaged at the temperature of 180 C. Suggest a suitable thickness of insulation (by making necessary assumptions) that will keep the notes safe even in case of fire having broken out. The flame temperature may be assumed to be 2000 C. k_{steel} = 42 W/mK; k_i = 0.8 W/mK.

2.16 Air at 90 C flows in a copper tube 3.75 cm inner diameter and with 6 mm thick wall which is heated from outside by water at 125 C. The thermal conductivity of scale deposit 1.575 mm thick is 1.5 W/mK. Air and water side unit surface conductances are 175 and 3000 W/m^2K. Calculate (a) water to air heat exchange and (b) temperature drop across the scale.

2.17 A 15 cm steam main has inside diameter of 12 cm and an outside diameter of 16 cm. It is insulated at the outside with asbestos. The steam temperature is 150 C and the air temperature is 20 C, h (steam side) = 100 W/m^2K, h (air side) = 30 W/m^2K, k (asbestos) = 0.8, k (steel) = 42 W/mK. How thick should the asbestos be provided in order to limit the heat loss to 5 W/m^2.

2.18 Determine the heat transfer between two fluids separated by a copper pipe 3 mm thick, 20 mm outside diameter and 1.5 meter long, if the inner fluid (water) temperature is 15 C and the outer fluid (steam) temperature is 100 C. The water side coefficient is 1,200 W/m^2 K, and the steam side coefficient is 10,000 W/m^2K and the conductivity for copper is 300 W/mK. Determine the log mean area and the arithmetic mean area of conducting surface and show if arithmetic mean area can be used in place of log mean area.

2.19 The rear window of an automobile is defogged by means of a thin transparent film type heating element to its inner surface. By electrically heating this element a uniform heat flux may be established at the inner surface. What is the electric power that must be provided per unit of window area in order to maintain the outer surface at 5°C when the interior air temperature and convection coefficient are 25° C and 10 W/mK, respectively, and the exterior (ambient) air temperature and convection coefficient are –10°C and 65 W/m^2K. The window glass is 4 mm thick (k = 0.78 W/mK). [Ans. 1118.75 W/m^2; t_i = 8.75°C]

2.20 The wind chill which is experienced on a cold windy day is due to increased heat transfer from exposed human skin to the surrounding atmosphere. Consider a layer of fatty tissue (k = 0.2 W/mK) 3 mm thick whose interior surface is kept at 36°C. On a calm day the convection heat transfer coefficient at the outer surface is 25 W/m^2K but with 50 kmph wind it reaches 65 W/m^2K. If the ambient air temperature is –15°C in both cases :

(a) What is the ratio of heat loss per unit area from the skin for the calm day to that for the windy day.

(b) What will be the skin outer surface temperature for the calm day and for the windy day.

(c) What temperature would the calm air have to attain to produce the same heat loss as on the windy day.

[Ans. (a) 0.542; (b) 21.36°C, 10.82°C; (c) –56.32°C]

2.21 The door of a domestic refrigerator has an area of 0.7 m^2. It consists of a metal sheet 3 mm thick (k = 40W/m C) with a 25 mm thick layer of insulation (k = 0.25 W/m C) on the inside. h_i = h_o = 10 W/m^2 C. Determine the heat flow rate through the door and the temperatures of the surfaces of the metal sheet if the refrigerator and room temperatures are 0°C and 20°C, respectively.

2.22 Derive the heat conduction equation for a hollow spherical shell, taking the heat flux at the inner surface and the temperature at the outer surface to have constant values. Solve the resulting equation for the temperature distribution in the shell.

2.23 Calculate the heat loss per m^2 of outside area of a heated sphere, 20 cm in diameter covered with 5 cm of insulation (k = 0.06 W/mK). The inside and outside surface temperatures of the insulation are 300°C and 50°C respectively.

2.24 A hollow sphere whose inner and outer diameters are 28 cm and 32 cm, is heated by means of a 20 ohm coil placed inside the sphere. Calculate the current required to keep the two surface at a constant temperature difference of 50°C and calculate the heat supplied. (k = 30 W/mK).

2.25 A hollow sphere 10 cm radius is covered with a 5 cm thick layer of insulation (k = 0.06 W/m K). An electric heater is located inside the sphere. What wattage is required to maintain a temperature difference of 20°C between the inner and outer surfaces of the insulation.

2.26 A wire whose resistance per cm length is 0.1 ohm is embedded along the axis of a cylindrical cement tube of radii 0.05 cm and 1.0 cm. A current of 5 amp is found to keep a steady difference of 125°C between the inner and outer surfaces. What is the conductivity of the cement and how much heat must be supplied per cm length ? [Ans. 0.0096 W/cm K; 2.5 W]

2.27 A hollow sphere (k = 0.347 W/cm K) has radii of 0.5 and 5 cm. It is heated by means of 10 ohm coil placed in the cavity. Find the current required to keep the two surfaces at a temperature difference of 5°C. Find the rate at which heat is supplied. [Ans. 1.1 amp; 12 W]

2.28 A turbine blade 6 cm long and having a cross sectional area A of 4.65 cm^2, perimeter P of 12 cm, is made of stainless steel (k = 23.4 W/mK). The temperature at the root is 500 °C. The blade is exposed to a hot gas at 870 C. The unit surface conductance is 42 W/m^2K. Determine the temperature distribution and rate of heat flow at the root of the blade. Assume the tip to be insulated.

2.29 Calculate the rate of heat flow in the following cases if the thermal conductivity of the material is 0.595 W/cm K.

(i) a circular disk 0.1 cm thick and 10 cm diameter with a central hole 1 cm diameter and with 100°C temperature difference between hole and edge.

(ii) a hollow cone, 0.1 cm thick, 20 cm long, 1 cm and 10 cm diameters at the two ends with 100°C temperature difference between the ends,

(iii) a solid cone of the same dimensions and same temperature difference as in (ii)

[Ans. 16.1 W; 3.62 W; 23.6 W]

2.30 A 2 mm diameter wire with resistance of 0.07 ohm per m carries a 15 ampere current. It is to be insulated with a material for which k = 0.0875 W/mK. If h = 23.3 W/m^2K, and the wire temperature is uniform, calculate the thickness of insulation which will result in the lowest operating temperature of the wire.

2.31 An electrical cable of 1 cm O.D. is to be insulated with rubber (k = 0.058 W/mK). The cable is to be located in air (h = 7 W/m^2 K) at 18 C. Investigate the effect of insulation thickness on heat dissipation assuming the cable surface temperature of 85 C. Determine the percentage increase in heat flow by providing the critical thickness of insulation. [Ans. 16.3 W/m^2, 14.7 W/m^2, 10.3%]

2.32 Compute the heat loss from the surface of a furnace wall 25 cm thick. The inner and outer surface temperatures are 300 C and 50 C respectively. The conductivity of the wall material is given by $k = 0.002\,T - 10^{-6}\,T^2$ where T is in °C and k in W/mK. Also determine the average value of conductivity in this range.

2.33 Compute the heat loss per m^2 of surface area of a wall 25 cm thick. The inner and outer surfaces of the wall are at 300°C and 40°C respectively. The variation of k with temperature is given by

$$k = 0.03 + 5 \times 10^{-6}\, t^2$$

where t is in °C and k is in W/mK.

2.34 Compute the heat loss per m^2 of surface area of a wall 40 cm thick having surface temperatures of 300°C and 50°C if the thermal conductivity of the wall material is given by

$$k = 0.005\, t - 5 \times 10^{-6}\, t^2$$

2.35 The thermal conductivity of a material is given by :

$$k = A + Bt + Ct^2$$

where *t* is temperature and A, B, C are constants.

Derive an expression giving the heat loss per m length of a hollow cylinder made of this material and having inner radius R_1 and outer radius R_2. Consider radial heat flow only.

2.36 A steel tube is placed in quiet air. The tube is heated at one end at a given constant rate of heat flow and perfectly insulated at the other end. Develop expressions for the temperatures at either end of the tube, assuming that no appreciable heat flow occurs in the air inside the tube and that radial temperature difference in the steel is negligible. Assume, further, that the thermal conductivity of steel is independent of the temperature and that the coefficient of heat transfer is the same all over the outer tube surface. The result should be given in the form of mathematical formulae and for the following special case :

Tube length	=	0.0254 m
Outer diameter of tube	=	0.3048 m
Inner diameter of tube	=	0.019 m
Thermal conductivity of steel	=	51 W/mK
Temperature of air outside the tube	=	23.9 C
Coefficient of heat transfer on the outside	=	8 W/m^2 K
Rate of heat flow	=	2.94 W

[Ans. 65 C, 31.9 C]

2.37 A circular ring 12.7 cm mean diameter and rectangular cross section (2.54 cm high, 0.0508 cm thick) is located in air at 16.2 C and is held at 100 C in one cross sectional area. What will be the temperature in the coldest cross sectional area (diametrically opposite to the plane of heating), when a steady state of heat exchange has been reached, (a) if the ring consists of glass (k = 0.364 W/mK), (b) if the ring consists of copper (k = 298 W/mK).

Assume that any influence of the curvature of the ring and temperature differences in any cross sectional area can be neglected. The coefficient of heat transfer between the ring and air may be taken as 9.78 W/m^{2}K at each point of the surface.

2.38 Two vessels are kept at 132 C and 49 C, respectively, in a room at 19 C air temperature. The vessels are connected by 4 non-insulated bolts, each 1.27 cm in diameter, 0.3048 m long. The thermal conductivity of the nickel steel is 41.5 W/mK and the coefficient of heat transfer at surface of rods is 5.7 W/m^{2}K. How much heat energy per hour is conducted away from the warmer vessel and how much is supplied to the colder vessel by the connecting rods ?

[Ans. 15 W, –0.01 W]

2.39 A thermometer well, 2 cm in diameter and 0.06 cm thick, is made of steel (k = 30 W/mK) and it is to be used to measure the temperature of steam flowing through a pipe. Find the minimum length of the well so that the error is less than 0.5% of the difference between the pipe wall and fluid temperature. Assume that the steam temperature is 260°C and h = 98 W/m^{2}K.

2.40 Brass fins are welded longitudinally on a 5 cm diameter 30 cm long brass cylinder which is surrounded by air at 26.7°C. If 12 uniformly spaced fins are used, each 0.075 cm thick and extending 1.25 cm radially from the cylinder, calculate the rate of heat transfer from the cylinder to the air. The cylinder surface temperature is 137.7°C. Assume h = 17 W/m^2 K and for brass k = 120 W/mK. Calculate also the heat transfer per fin only. [Ans. 264 W, 14.8 W]

2.41 The tip of a soldering iron consists of a 6 mm OD copper rod (k = 401 W/mK), 75 mm long. If the tip must attain a temperature of 475 K to effect the soldering what must be the base temperature and heat flow into the base in Watts. h = 23 W/m^{2}K and ambient temperature is 295 K.

[Ans. 465 K; 6.14 W]

2.42 Two long copper rods of 10 mm diameter are to be soldered together end to end with a solder having a melting point of 650°C. The rods are in air at 25°C. h = 10 W/m^2 K. What is the minimum power input needed for soldering ?

[Ans. 124.3 W]

2.43 An experimental arrangement for measuring thermal conductivity, involves the use of two long rods that are equivalent in every respect expect that one is fabricated from a material of known k (k_A) while the other from a material whose k (k_B) is unknown. Both rods are attached at one end to a heat source at T_b, are exposed to a fluid at T_∞ and are provided with thermocouples to measure the temperature at distance x_1 from the heat source. If k_A = 200 W/mK; T_A = 75°C and T_B = 60 °C at x_1, T_b = 100°C and T_∞ 25°C, what is the value of k_B. Assume h = 20 W/m^{2}K.

[Ans. 56.6 W/mK]

2.44 A 1.25 cm diameter iron rod (k = 40 W/mK), 3 cm long is giving up heat to the environment at 20°C. The ends of the rod are connected to two heat sources at 130°C. Determine :

(a) the temperature distribution in the rod,

(b) the temperature at the middle of the rod,

(c) the heat flowing out of each heat source,

(d) the heat flowing at the middle of the rod.

Assume $h = 20 \text{ W/m}^2\text{K}$

2.45 A 1.25 cm diameter iron rod (k = 40 W/mK) 15 cm long protrudes out of a heat source at 130°C into environment at 20°C. The free end of the rod is insulated. If $h = 20 \text{ W/m}^2 \text{ K}$, find :

(a) the temperature distribution in the rod,

(b) the temperature at the insulated end,

(c) the heat flowing out of the heat source.

2.46 A 1.25 cm diameter 15 cm long iron rod (k = 40 W/mK) protrudes out from a heat source at 130°C into environment at 20°C. If the value of h is $20 \text{ W/m}^2\text{K}$ determine :

(a) the temperature distribution in the rod,

(b) the temperature at the free end,

(c) the heat flow out of the source,

(d) the heat flow at the free end.

2.47 A fin 0.6 cm thick and 5 cm long has its base on a plane plate at 100°C. The environment is at 20°C. The conductivity of fin material is 60 W/mK. Value of h is $50 \text{ W/m}^2 \text{ K}$. Calculate :

(a) temperature at the end of fin.

(b) temperature at middle of the fin.

(c) total heat dissipated by the fin.

(d) effectiveness of the fin.

2.48 Consider a circular pin fin 5 cm dia and 25 cm long which is attached to a wall at 300°C. Its surface is exposed to a gas at 100 C. The average film conductance is $20 \text{ W/m}^2\text{K}$. The fin is made of material such that k = 10 W/mK. Determine the temperature distribution along the fin by using the relaxation method, and then determine the amount of heat dissipation by the fin. Compare the result with that obtained analytically.

2.49 One end of a long rod is inserted into a furnace while the other projects into ambient air. Under steady state, the temperature of the rod is measured at two points 7.5 cm apart and found to be 125°C and 88.5°C when the ambient temperature is 20°C. If the rod is 2.5 cm in diameter and $h = 20 \text{ W/m}^2\text{K}$, find the thermal conductivity of the rod material. [Ans. 100 W/mK]

2.50 A thin rod of copper (k = 348 W/mK), 1.22 cm diameter connects two parallel plates 15 cm apart. Air flows in the space between the plates ($h = 57 \text{ W/m}^2\text{K}$). The plates are at 70°C while the air temperature is 15°C. Find the temperature at the midpoint of the rod. [Ans. 62.4°C]

2.51 Calculate the length of 0.159 cm diameter spines which will give the maximum heat transfer rate per unit weight of tube from a 1.59 cm OD, 1.41 cm ID brass tube under following conditions :

Tube surface temperature	=	177°C
Air temperature	=	24°C
Number of spines per cm^2 of outer tube area	=	12
h	=	10.4 W/m^{2}K

[Ans. 1.9 cm]

2.52 What is the temperature distribution in a plane wall 7.5 cm thick in which heat is generated uniformly at the rate of 1,050 W/m^3, one surface of the wall is maintained at 88 K and the other at 100 C. Assume k = 8.5 W/mK. Also determine the magnitude and direction of heat flow through each face per unit area.

[Ans. $t = -61.8x^2 - 3795.4x + 100$, 32,261 W/m^2 32,340 W/m^2]

2.53 Heat is generated at the interface between two slabs both 50 mm thick, one is of steel of which k = 35 W/mK and the other is brass of k = 76 W/mK. The temperature of the outside of the steel slab is 95 C, while that of outside of the brass is 47 C. When the rate of the heat generation is 17,500 W/m^2 of contact area, calculate the flow rate through the outer surfaces and the interface temperature. Also show how such a condition can be obtained.

[Ans. 1700 W, 70 C]

2.54 A heating element is constructed from carbon in the shape of bar 7.5 cm wide and 1.25 cm thick and 90 cm long. When a potential of 12 V is applied to ends of the bar, its surface reaches a uniform temperature of 760°C. What is the temperature at the center of the bar ? The electrical resistivity of the bar is 39 × 10^{-4} ohm cm, and its thermal conductivity is 5 W/mK. [Ans. 830°C]

2.55 A flat plate electric resistance heater 30 cm × 60 cm × 1.25 cm is used to keep the temperature of a bath at 100°C. If the temperature at the mid-plane of the heater is maintained at 115 C, calculate the rate of heat flow through both faces. Assume that k = 4.5 W/mK and that heat is generated uniformly per unit volume. Neglect the temperature difference between heater surface and bath liquid, and also heat flow through the edges.

2.56 The outer surfaces of a flat plate electric resistance heater, 1.25 cm thick, of effective k = 15 W/mK, are at 300°C. Calculate the temperature at the centre of the heater if the electrical power input is 1200 kW/m^3. Neglect heat flow through the edges.

2.57 Determine the general expression for the temperature distribution in wire of radius r_o in meter, thermal conductivity k in W/mK, and resistance R in ohms per meter of length, when a current of I ampers flows through it.

[Ans. $t = t_\infty + I^2 R A \left\{1 + \frac{2k}{hr_0} - \left(\frac{r}{r_0}\right)^2\right\}$]

2.58 Show that the temperature distribution in long solid tubes, insulated on the outside, cooled on the inside with uniform heat within the solid, is given by ;

$$t(r) - t_0 = -\frac{Q'}{4k}(r^2 - r_0^2) + \frac{Q' r_0^2}{2k} \ln\left(\frac{r}{r_0}\right)$$

2.59 Show that the temperature distribution in a solid cylinder of radius r_0 in which heat is generated homogeneously is given by ;

$$t = t_0 + \frac{Q'}{4k}(r_0^2 - r^2)$$

where t_o is temperature at the outer surface.

The outer surface temperature of a conductor 1.25 cm in diameter is 150°C. The I^2R loss is 4.187 kJ/cm^3 and k = 15 W/mK. Determine :

(i) Temperature difference between the center and outer surface of the conductor.

(ii) Heat flow from the outer surface per m length.

2.60 A coil 3 cm thick and 30 cm inside diameter is placed in an oil bath. At current of 0.01 amp and voltage of 0.12 volt,the surface temperature remained unchanged. Calculate the maximum temperature inside the coil when the current was 2.8 amp. Assume $\varepsilon = 0.0043\ C^{-1}$.

2.61 Determine the center line temperature of bare steel wire (k = 22.6 W/mK) of 1.9 cm diameter carrying a current of 1000 amp. The surface heat transfer coefficient is 28.4 W/m^{2}K and the air temperature is 21°C. The electrical resistance is 3.28×10^{-4} ohm per m length. [Ans. 215.5°C]

2.62 Show that the steady state temperature distribution in a sphere in which heat is produced at a uniform rate per unit volume, Q' is given by ;

$$(t - t_o) = \frac{Q' R_0^2}{6k}\left[1 - \left(\frac{R}{R_0}\right)^2\right]$$

where R_0 = outer radius and t_0 = outer surface temperature.

2.63 The average heat production during ripening of oranges is estimated to be 320 W/m^3. If the orange is assumed to be homogeneous with k = 0.15 W/mK, compute the temperature at the center of the orange and the heat flow from the outer surface. Assume that its diameter is 8.5 cm and the external surface temperature is 10°C.

2.64 A long cylindrical rod of radius 100 mm consists of a nuclear material (k = 0.5 W/m K) generating 24,000 W/m^3 uniformly. It is surrounded by another cylinder whose outer radius is 200 mm (k = 4 W/mK). The outer surface is cooled by a fluid at 100°C (h = 20 W/m^{2}K). Find the temperature at the interface between the two cylinders and at the outer surface. [Ans. 150.8°C, 130°C]

2.65 An electric current of 700 amp flows through a stainless steel cable of 5 mm diameter and electrical resistance of 6×10^{-4} Ω/m. The cable is in an environment of temperature 30°C and the total coefficient of heat transfer (convection and radiation combined) between cable and environment is 25 W^2/mK. What is the surface temperature if the cable is bare? If a very thin coating of electrical insulation is applied to the cable with a contact resistance of 0.02 m^2K/W what are the insulation and cable surface temperatures?

[Ans. 788°C; 788°C;1153°C]

2.66 In a cylindrical fuel rod of a nuclear reactor, the heat generation is,

$$Q' = Q_0 \{1 - (r/r_o)^2\}$$

where Q' = local heat generation per unit volume at radius r

r_o = outside radius

Q_o = rate of heat generation per unit volume at the center line.

Calculate the temperature drop from the center line to the surface of a 2.5 cm OD rod having a thermal conductivity of 30 W/mK, if the heat removal from its surface is 2,500,000 W/m^2K.

2.67 Deduce expressions for the shape factors for :

(a) A composite plane wall made up of two materials,

(b) A hollow sphere having inner and outer radii of R_1 and R_2.

2.68 The inside dimensions of an oven made of brick (k = 1.2 W/mK) are 2 m by 1.8 m by 0.6 m. The wall is 10 cm thick. The temperatures of the inner and outer surfaces are 600°C and 100°C. Calculate the heat loss per hour.

(General Note : 1 Nm = 1 Ws = 1 J)

CHAPTER
3

Heat Transfer by Conduction (unsteady state)

3.1 UNSTEADY STATE HEAT CONDUCTION

In the previous chapter attention was restricted to the systems in which the temperature did not vary with time and it was a function of space coordinates only. However, before such a steady state is reached, some time must elapse after the heat transfer process is initiated. During this period the temperature is a function of time also, i.e.,

$$t = f(x, y, z, \tau) \tag{3.1}$$

Thus heat transfer occurs under *transient conditions* and the temperature at any location varies from time to time. Such systems are called *unsteady state* systems. Once the unsteady or transient state disappears the system temperature is no more a function of time and the system is supposed to have attained the *steady state*.

In most engineering applications, transient systems are encountered; for instance, heat flow to boiler tubes, cooling the I. C. engine cylinders, and heat gain or loss in buildings are transient heat flow processes. We intend to discuss only a few simple cases of transient state heat flow in this chapter.

All transient state systems may be broadly classified into two types.

(a) *Non-periodic heat flow systems* where the temperature at any point within the system changes as some non-linear function of time.

(b) *Periodic heat flow systems* where the temperature within the system undergoes periodic changes which may be *regular* or *irregular* but definitely cyclic.

In general, however, for purpose of analytically solving the problems of *unsteady state heat flow* we may subdivide *non-periodic* systems into the following classes :

(i) Heating or cooling of bodies with known temperature distribution.

(ii) Heating or cooling of bodies whose internal thermal resistance is negligible as compared to the surface resistance.

(iii) Heating or cooling of bodies whose surface temperature is either known or whose surface resistance is negligible as compared to internal thermal resistance.

(iv) Heating or cooling of bodies where both surface and internal thermal resistance are appreciable.

3.2 UNSTEADY STATE HEATING OR COOLING OF BODIES WITH KNOWN TEMPERATURE DISTRIBUTION

In this section we will treat simple cases of one-dimensional unsteady heat conduction in which the temperature distribution across the body is known at some instant of time τ and it is desired to know the rate of heat flow, q (x, τ) or the rate of change of temperature with time at various points within the body. The solution of such problems requires the application of equation (2.12) which, for one-dimensional unsteady state conduction without heat generation, simplifies to

$$\frac{1}{\alpha}\frac{\partial t}{\partial \tau} = \frac{\partial^2 t}{\partial x^2} \tag{3.2}$$

The method of solution can be illustrated by means of an example.

Example 3.1

The temperature distribution through a larger wall 50 cm thick, heated from one side, at a certain time τ is found to be given by ;

$$t = 90 - 80x + 16x^2 + 32x^3 - 25x^4$$

where *x* is in meter and *t* in °C. If the area of the wall is 10 m^2 compute :

(a) the heat entering and leaving the wall in unit time ;

(b) the heat energy stored in the wall in unit time ;

(c) the rate of temperature change with time at x = 0, x = 0.5 m ;

(d) the location where the rate of heating (or cooling) is maximum.

(Assume k = 58 W/mK; α = 0.02 m^2/hr)

Solution :

Now, $$\frac{\partial t}{\partial x} = -80 + 32x + 96x^2 - 100x^3$$

$$\frac{\partial^2 t}{\partial x^2} = 32 + 192x - 300x^2$$

$$q = -kA\left(\frac{\partial t}{\partial x}\right)$$

so $$q_o = -kA\left(\frac{\partial t}{\partial x}\right)_{x=0}$$

or $$q_o = (5.8 \times 10)(-80) = 4640 \text{ W}$$

$$= 16{,}800 \text{ kJ/hr}$$ Ans.

Similarly, $$q_L = -kA\left(\frac{\partial t}{\partial x}\right)_{x=L=0.5}$$

$$q_L = -(5.8 \times 10)\,[-80 + 32\,(0.5) + 96\,(0.5)^2 - 100\,(0.5)^3]$$
$$= -(58)\,(-52.5) = 3050 \text{ W}$$
$$= 11{,}000 \text{ kJ/hr}$$ Ans.

Heat storage rate,

$$q_{stored} = q_o - q_L = 4640 - 3050$$
$$= 1590 \text{ W} = 5800 \text{ kJ/hr}$$ Ans.

Aliter

Heat storage rate may also be calculated by ;

$$q_{stored} = \int_{x=0}^{L} c\rho A\,dx \frac{\partial t}{\partial \tau} = c\,\rho\,A \int_{x=0}^{L} \alpha \frac{\partial^2 t}{\partial x^2}\,dx$$

or

$$q_{stored} = kA \int_{x=0}^{L} \alpha \frac{\partial^2 t}{\partial x^2}\,dx$$
$$= (5.8 \times 10) \int_{x=0}^{0.5} (32 + 192\,x - 300\,x^2)\,dx$$
$$= 58\,(32 \times 0.5 + 96 \times 0.25 - 100 \times 0.125)$$
$$= 58\,(27.5) \text{ W} = 1590 \text{ W} = 5724 \text{ kJ/hr}$$ Ans.

(c) Since

$$\frac{\partial t}{\partial \tau} = \alpha \frac{\partial^2 t}{\partial x^2} = \alpha\,(32 + 192\,x - 300\,x^2)$$

$$\left(\frac{\partial t}{\partial \tau}\right)_{x=0} = 0.02 \times 32 = 0.64\ °\text{C/hr}$$ Ans.

$$\left(\frac{\partial t}{\partial \tau}\right)_{x=0.25} = 0.02\,(32 + 192 \times 0.25 - 300 \times 0.0625)$$
$$= 0.02\,(61.25) = 1.225\ °\text{C/hr}$$ Ans.

$$\left(\frac{\partial t}{\partial \tau}\right)_{x=0.5} = 0.02\,(32 + 192 \times 0.5 - 300 \times 0.25)$$
$$= 0.02\,(53) = 1.06\ °\text{C/hr}$$ Ans.

(d) To find the depth at which the rate of heating or cooling is maximum we should set ;

$$\frac{\partial}{\partial x}\left(\frac{\partial t}{\partial \tau}\right) = 0$$

$$\frac{\partial}{\partial x}\left(\alpha \frac{\partial^2 t}{\partial x^2}\right) = 0\ ; \quad \frac{\partial^3 t}{\partial x^3} = 0$$

Hence, $$\frac{\partial^3 t}{\partial x^3} = 192 - 600\,x = 0$$

so $$x = 0.32 \text{ m}$$ Ans.

Example 3.2

At a certain instant, the temperature distribution in a long cylindrical fire-tube (inner radius 30 cm, outer radius 50 cm) is given by

$$t = 800 + 1000\,r - 5000\,r^2$$

where *t* is in °C and *r* is in meter. The thermal conductivity and thermal diffusivity of the tube material are 58 W/mK and 0.004 m^2/hr, respectively. Find :

(a) Rate of heat flow at the inside and outside surfaces per m length.

(b) Rate of heat storage per m length

(c) Rate of change of temperature with time at the inner and outer surfaces.

Solution :

The differential equation to be used is the simplified form of Eq. (2.16) for radial heat flow without heat generation,

ie, $$\frac{\partial t}{\partial \tau} = \alpha\left[\frac{\partial^2 t}{\partial r^2} + \frac{1}{r}\frac{\partial t}{\partial r}\right] \tag{3.3}$$

Now, from the given temperature distribution,

$$\frac{\partial t}{\partial r} = 1000 - 10{,}000\,r$$

$$\frac{\partial^2 t}{\partial r^2} = -10{,}000$$

(a) $$q_{inner} = -k\,A_i\left(\frac{\partial t}{\partial r}\right)_{r=0.3}$$

$$= -(58)\,(2\pi \times 0.3 \times 1)\,(1000 - 10{,}000 \times 3)$$

$$= 21.9 \times 10^4 \text{ W}$$ Ans.

$$q_{outer} = -k\,A_o\left(\frac{\partial t}{\partial r}\right)_{r=0.5}$$

$$= -(58)\,(2\pi \times 0.5 \times 1)\,(1000 - 10{,}000 \times 0.5)$$

$$= 73 \times 10^4 \text{ W}$$ Ans.

(b) Rate of heat storage

$$= q_i - q_o$$

$$= (21.9 - 73) \times 10^4$$

$$= -51.1 \times 10^4 \text{ W}$$ Ans.

This means that the tube is losing heat at the rate of 51.1×10^4 W at the given instant.

(c)
$$\frac{\partial t}{\partial \tau} = \alpha \left[\frac{\partial^2 t}{\partial r^2} + \frac{1}{r}\frac{\partial t}{\partial r}\right]$$

$$\left(\frac{\partial t}{\partial \tau}\right)_{r=0.3} = 0.004\left(-10{,}000 + \frac{1000}{0.3} - 10{,}000\right)$$

$$= -0.004 \times 16{,}667$$

$$= -66.67\ °\text{C/hr}$$ Ans.

$$\left(\frac{\partial t}{\partial \tau}\right)_{r=0.5} = 0.004\left(-10{,}000 + \frac{1000}{0.5} - 10{,}000\right)$$

$$= -0.004 \times 18{,}000$$

$$= -72\ °\text{C/hr}$$ Ans.

3.3 UNSTEADY STATE SYSTEMS WITH NEGLIGIBLE INTERNAL THERMAL RESISTANCE (k = ∞)

Although there are no materials in nature that possess an infinite thermal conductivity, still many transient heat conduction problems can be solved easily by making such an assumption.

Consider a simple case of *cooling* (or *heating*). Let the initial temperature of an object be t_i and let it be cooled in an ambient atmosphere at t_a. Since the vale of $k = \infty$, there are no temperature gradients within the object at any instant τ. Such systems are also known as *lumped heat capacity* systems.

Rate of heat gain by the object ;

$$q = c \rho V \frac{dt}{d\tau} = -h A_1 (t - t_a) \tag{3.4}$$

where
c = specific heat of solid, kJ/kg K
ρ = density of solid, kg/m^3
V = volume, m^3
A_1 = surface area, m^2
h = surface film conductance, W/m^{2}K
t_a = ambient temperature; °C.

The Eq. (3.4) may be rewritten as

$$\frac{dt}{d\tau} + \frac{hA_1}{c\rho V}(t - t_a) = 0$$

or $$\int \frac{dt}{t-t_a} = \int -\frac{h A_1}{c\rho V} d\tau$$

or $$ln\,(t - t_a) = -\frac{h A_1}{c\rho V} \tau + c_1$$

at $$\tau = 0 \,;\, t = t_i$$

so $$c_1 = ln\,(t_i - t_a)$$

hence $$ln\,(t - t_a) = -\frac{h A_1}{c\rho V} \tau + ln\,(t_i - t_a)$$

or $$\frac{t-t_a}{t_i - t_a} = e^{-(hA_1/c\rho V)\tau} \tag{3.5}$$

$$= e^{-(hA_1/c\omega)\tau}$$

(where $\omega = \rho$ V is the total mass of solid)

Eq. (3.5) gives the dimensionless *temperature history* of the object. It indicates that during *newtonian cooling* (or *heating*) the temperature falls or rises exponentially with time. The cooling curve is shown in Fig. 3.1. It further indicates that it will take an infinite amount of time for the object to attain steady state conditions. This is however just a mathematical necessity. The difference between t and t_a becomes extremely small after a short time so that we can count that $t \approx t_a$ beyond that period.

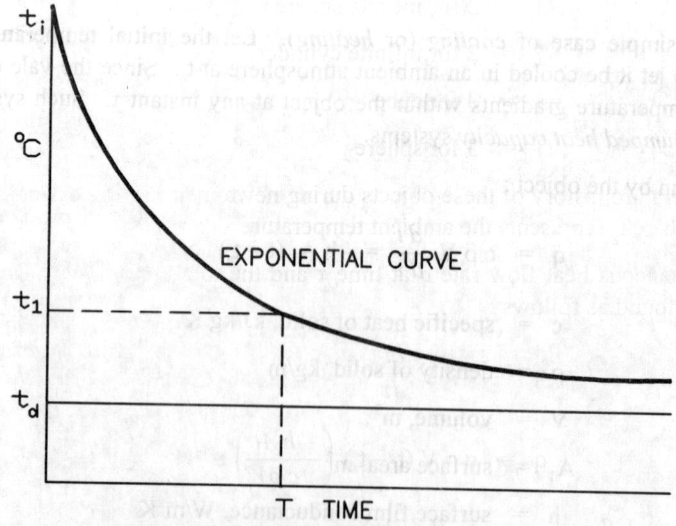

Fig. 3.1 Unsteady State Cooling when k = ∞

Let us now introduce the following dimensionless numbers :

$$\text{Biot number, } Bi = \frac{hS}{k} \tag{3.6}$$

$$\text{Fourier number, } Fo = \frac{\alpha\tau}{S^2} = \frac{k\,\tau}{\rho c S^2} \tag{3.7}$$

where S is a *semi-thickness* or a *surface radius*,

then, $$\frac{h A_1}{c\rho V}\tau = \left(\frac{hS}{k}\right)\left(\frac{k}{c\rho}\right)\left(\frac{A_1 S}{V}\right)\left(\frac{\tau}{S^2}\right)$$

$$= (Bi)\left(\frac{\alpha\tau}{S^2}\right)\left(\frac{A_1 S}{V}\right) = \left(\frac{A_1 S}{V}\right)(Bi)\times(Fo)$$

Hence Eq. (3.5) yields

$$\frac{t-t_a}{t_i-t_a} = e^{-\{(A_1 S/V)\,Bi\times Fo\}}$$

$$\frac{t-t_a}{t_i-t_a} = e^{-(G.Bi.Fo)} \tag{3.9}$$

where $$G = \left(\frac{A_1 S}{V}\right) \tag{3.10}$$

Note that the dimensionless quantity G = (A_1 S/V) depends only on the geometry of the object. For example,

G = 1 for infinite plate

G = 2 for infinite square rod

G = 2 for infinite cylinder

G = 3 for cube

G = 3 for sphere

The temperature history of these objects during newtonian heating or cooling is given by Fig. 3.2 where t_g represents the ambient temperature.

The instantaneous heat flow rate q at time τ and the total heat flow Q from τ = 0 to τ = τ may be found as follows :

$$q = c\,\rho\,V\,\frac{dt}{d\tau}$$

or $$q = c\,\rho\,V\,(t_i - t_a)\left(-\frac{h A_1}{c\rho V}\right) e^{-(hA_1/c\rho V)\tau}$$

hence $$\frac{q}{h A_1 (t_a - t_i)} = e^{-(hA_1/c\rho V)\tau} \tag{3.11}$$

or $$\frac{q}{h A_1 (t_a - t_i)} = e^{-(G.Bi.Fo)} \tag{3.12}$$

Since, $$Q = \int_{\tau=0}^{\tau} q\,d\tau = \int_0^{\tau} h\,A_1\,(t_a - t_i)\,e^{-(hA_1/c\rho V)\tau}\,d\tau$$

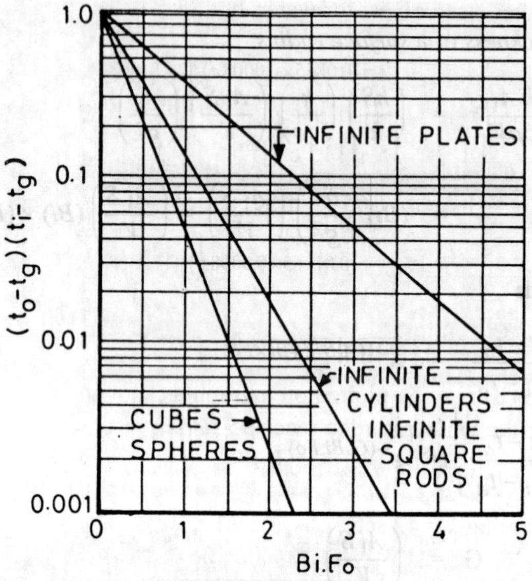

Fig. 3.2 Temperature History during Newtonian Cooling

or $$Q = -\frac{h A_1 (t_a - t_i)}{h A_1 / c\rho V}\left(e^{-(hA_1/c\rho V)\tau} - 1\right)$$

$$= c\,\rho V\,(t_a - t_i)\left(1 - e^{-G.Bi.Fo}\right)$$

or $$\frac{Q}{c\rho V (t_a - t_i)} = 1 - e^{-G.Bi.Fo} \qquad (3.13)$$

Note that the values of q and Q can be obtained directly from Fig. 3.2 by interpreting the ordinate as $[(q/h\,A_1\,(t_a - t_i)]$ and $[1 - Q/c\,\rho\,V\,(t_a - t_i)]$ respectively.

Example 3.3

(a) A copper wire 0.8 mm diameter, at 150 C, is suddenly dipped into water at 35 C. Assume unit surface film coefficient of heat transfer, $(h)_{water}$ = 85.5 W/m^{2}K and estimate the time required to cool the wire upto 95 C.

(b) If the wire is placed in air, $(h)_{air}$ = 11.65 W/m^{2}K what is the time required to cool it to 95 C. (For copper : k = 373 W/mK; c = 0.38 kJ/kg K ; ρ = 9000 kg/m^3)

Solution :

Let us consider one m length of wire ;

$$A_1 = (\pi \times D \times 1) = 0.0008\,\pi \text{ m}^2$$

$$w = \rho V = 9000 \times (\pi/4)\, D^2 \times 1$$

$$= 9000 \times (\pi/4)\,(0.8/1000)^2$$

or $$cw = 0.38 \times 9000 \times (\pi/4) \times (0.8/1000)^2$$

$$= 17.2 \times 10^{-4} \text{ kJ/K}$$

(a) Thus Eq. (3.5) when used gives; for water, h = 85.5 W/m^{2}K

$$\frac{t-35}{150-35} = e^{-\tau(0.0855\times 0.0008\pi)/(17.2\times 10^{-4})}$$

or $$t = 35 + 115\, e^{-0.125\tau} \quad \text{(a)}$$

(b) for air ; $$h = 11.65 \text{ J/s m}^2\text{K}$$

$$\frac{t-35}{150-35} = e^{-\tau(0.01165\times 0.0008\pi)/(17.2\times 10^{-4})}$$

or $$t = 35 + 115\, e^{-0.017\tau} \quad \text{(b)}$$

Equations (a) and (b) may be plotted on a graph as shown in Fig. 3.3. It is noted that the time required for the temperature to reach 95 °C is ;

(a) for water surrounding = 5.7 sec Ans.

(b) for air surroundings = 42 sec Ans.

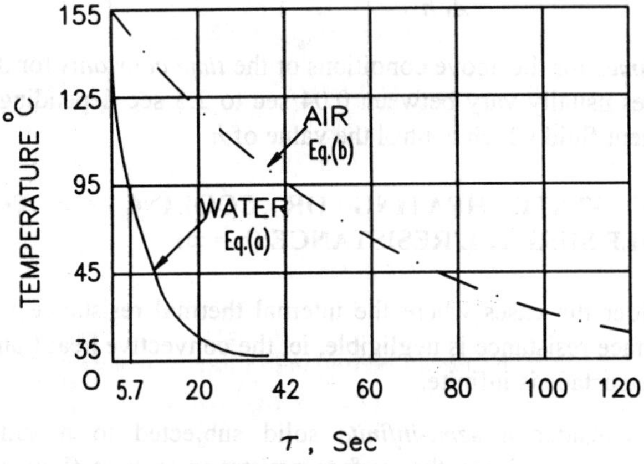

Fig. 3.3 Example 3.5

3.4 TEMPERATURE - TIME RESPONSE OF THERMOCOUPLES

The most important example of heat conduction with infinite thermal conductivity is that of a *thermocouple* which is discussed below.

From the results of Example 3.3 it is observed that if a wire (of 0.8 mm diameter) were used in thermocouple to measure rapid temperature changes of air, there would be a considerably longer time lag, as compared to the water surroundings, for the temperature difference between the thermocouple and the surroundings to become zero. The thermocouple wire will be heated or cooled due to changes in ambient temperature. Thus the temperature-time response of the thermocouple will be more rapid if the cooling or heating of the thermocouple wire, as the case may be, is fast. The exponential term of Eq. (3.5) may fast approach zero by increasing the value of h (as for the case of water in Example 3.3) or by *decreasing the wire diameter*. Hence a very thin wire is recommended for use in thermocouples to ensure a rapid response.

However, due to the asymptotic nature of the exponential curve, it is difficult to determine precisely just when the temperature of the wire has reached the temperature of the surroundings. Therefor for practical purpose, it is desirable to specify the response of thermocouple, by the *time* required for it to indicate a certain percentage of the initial temperature difference.

As a usual practice, the time required by a thermocouple to reach 63.2 % of the value of initial temperature change indicates the *sensitivity* of the thermocouple and is known as the *time constant* of the thermocouple. This point is reached when ;

$$\frac{t-t_a}{t_i-t_a} = (1-0.632) = 0.368$$

$$= e^{-1} \text{ (approx.)}$$

or

$$\left(\frac{A_1 h}{c\rho V}\right)\tau = 1 \text{ (say)}$$

or the *time constant* is given by ;

$$\tau_1 = \frac{V\rho c}{A_1 h} \tag{3.14}$$

The *response times* for the above conditions or the *time constants* for different sizes of thermocouple wires usually vary between 0.04 sec to 2.5 sec depending upon the mass flow rates of ambient fluid which control the value of *h*.

3.5 UNSTEADY STATE HEATING OR COOLING OF BODIES WITH NEGLIGIBLE SURFACE RESISTANCE (h = ∞)

We now consider the cases where the internal thermal resistance is appreciable ie, $k \neq \infty$, but the surface resistance is negligible, ie, the convective heat transfer coefficient, *h* of the film at the surface is infinite.

For instance, consider a *semi-infinite* solid subjected to a sudden change of *environment* temperature. Since the surface resistance to heat flow is negligible, the surface (of the solid) exposed to the environment will attain the environment temperature instantaneously and the surface temperature will stay constant thereafter. However, on account of the finite internal resistance of the solid, the temperature at any point at a distance x from the surface will not attain the surface temperature instantaneously but will vary with time.

In this situation the heat conduction equation (2.12) for unsteady state reduces to the form given below for one dimensional case;

$$\frac{\partial^2 t}{\partial x^2} = \frac{1}{\alpha}\frac{\partial t}{\partial \tau} \tag{3.2}$$

The situation would be exactly the same if the surface resistance is not negligible but the surface temperature is known (by measurement, for instance). In succeeding sections we shall consider the following cases of solids having sudden change of surface (or environment) temperature :

1. An infinitely wide plate of finite thickness (Section 3.6).
2. An infinitely long cylinder (Section 3.7).
3. A sphere (Section 3.8).
4. An infinitely thick plate i.e. semi-infinite solid (Section 3.9).

3.6 SUDDEN TEMPERATURE CHANGES OF THE SURFACE (OR ENVIRONMENT) OF AN INFINITELY WIDE PLATE

(a) Consider the heating or cooling of a large plate (in the $y - z$ plane) of finite thickness L = 2 S. The temperature distribution in the plate t (x, τ) initially (at τ = 0) is some arbitrary function of x ie,

$$t(x, 0) = t_i(x)$$

Now, both surfaces of the plate (x = 0, x = L) are suddenly changed to and then permanently held at temperature t_s (or say t_g) for all τ > 0.

The governing differential equation is

$$\frac{1}{\alpha}\frac{\partial t}{\partial \tau} = \frac{\partial^2 t}{\partial x^2} \qquad (3.2)$$

and the initial and boundary conditions are :

$$t = t_i(x) \text{ at } \tau = 0 ; 0 \leq x \leq L \qquad (3.15)$$

$$t = t_s \text{ at } x = 0; \tau > 0 \qquad (3.16)$$

$$t = t_s \text{ at } x = L ; \tau > 0 \qquad (3.17)$$

Let us define a temperature T by ;

$$T = t - t_s \qquad (3.18)$$

Then, the governing differential equation becomes

$$\frac{1}{\alpha}\frac{\partial T}{\partial \tau} = \frac{\partial^2 T}{\partial x^2} \qquad (3.19)$$

The initial and boundary conditions become

$$T = T_i(x) \text{ at } \tau = 0 ; 0 \leq x \leq L \qquad (3.20)$$

$$T = 0 \text{ at } x = 0 ; \tau > 0 \qquad (3.21)$$

$$T = 0 \text{ at } x = L; \tau > 0 \qquad (3.22)$$

Eq. (3.19) may be solved by the separation of variables method. Let us assume

$$T = X(x)\, Y(\tau)$$

where X is a function of x alone and Y is a function of τ alone.

Then, $$\frac{\partial T}{\partial \tau} = X\frac{dY}{d\tau}$$

and
$$\frac{\partial^2 T}{\partial x^2} = Y \frac{d^2 X}{d x^2}$$

Hence, Eq. (3.19) becomes

$$\frac{1}{\alpha} X \frac{dY}{d\tau} = Y \frac{d^2 X}{d x^2}$$

Dividing by XY,

$$\frac{1}{\alpha} \frac{1}{Y} \frac{dY}{d\tau} = \frac{1}{X} \frac{d^2 X}{dx^2} \qquad \text{(a)}$$

The two sides of the Equation (a) are functions of different independent variables, so that they can be equal only if each side is equal to some constant.

Let this separation constant be $-\lambda^2$. Then, variables can be separated as,

$$\frac{1}{\alpha} \frac{1}{Y} \frac{dY}{d\tau} = -\lambda^2 \qquad \text{(b)}$$

and
$$\frac{1}{X} \frac{d^2 X}{d x^2} = -\lambda^2 \qquad \text{(c)}$$

The solution of Eq. (b) is

$$Y = C' \, e^{-\lambda^2 \alpha \tau}$$

The solution of Eq. (c) is

$$X = C'' \cos \lambda x + C''' \sin \lambda x$$

Hence
$$T_\lambda = XY = C' \, e^{-\lambda^2 \alpha \tau} \, (C'' \cos \lambda \dot{x} + C''' \sin \lambda x)$$

where C′, C″ and C‴ are some constants.

or
$$T = e^{-\lambda^2 \alpha \tau} \, (C_1 \cos \lambda x + C_2 \sin \lambda x) \qquad (3.23)$$

To evaluate C_1, C_2 and λ we have to use the boundary conditions. Since $T = 0$ at $x = 0$ for all values of $\tau > 0$,

$$C_1 \, e^{-\lambda^2 \alpha \tau} = 0$$

This means that $C_1 = 0$. Since $T = 0$ at $x = L$ for all $\tau > 0$, we must have either $C_2 = 0$ or $\sin \lambda L = 0$. Clearly the value of $C_2 = 0$ is not admissible.

Hence,
$$\sin \lambda L = 0$$

or
$$\lambda = \frac{n \pi}{L}; \qquad \text{when } n = 1, 2, 3, \ldots\ldots$$

The solution in Eq. (3.23) therefore takes the form ;

$$T = \sum_{n=1}^{\infty} C_n \left\{ e^{-(n\pi/L)^2 \alpha \tau} \sin\left(\frac{n \pi}{L}\right) x \right\} \qquad (3.24)$$

If Eq. (3.24) is to satisfy the initial condition of Eq. (3.20) we must have

$$T_i(x) = \sum_{n=1}^{\infty}\left\{C_n \sin\left(\frac{n\pi}{L}\right)x\right\} \tag{3.25}$$

The RHS of Eq. (3.25) is a Fourier sine-series expansion of the function T_i (x) for which the values of C_n are given by

$$C_n = \frac{2}{L}\int_o^L T_i(x)\sin\left(\frac{n\pi}{L}\right)x\,dx \tag{3.26}$$

Hence the complete solution of Eq. (3.19) is ;

$$T = \frac{2}{L}\sum_{n=1}^{\infty}\left[e^{-(n\pi/L)^2\alpha\tau}\sin\left(\frac{n\pi}{L}\right)x\right]\times \int_o^L T_i(x)\sin\left(\frac{n\pi}{L}\right)x\,dx \tag{3.27}$$

(b) If the plate is initially at a uniform temperature t_i throughout then t_i (x) = t_i or T_i (x) = T_i. In this case Eq. (3.26) gives,

$$C_n = \frac{2}{L}\int_o^L T_i\sin\left(\frac{n\pi}{L}\right)x\,dx$$

$$= \frac{2T_i}{L}\,\frac{L}{n\pi}\left[-\cos\left(\frac{n\pi}{L}\right)x\right]_0^L$$

$$= \frac{2T_i}{n\pi}(1-\cos n\pi)$$

Hence,

$$C_1 = \frac{4T_i}{\pi};\ C_2 = 0$$

$$C_3 = \frac{4T_i}{3\pi};\ C_4 = 0$$

$$C_5 = \frac{4T_i}{5\pi};\ C_6 = 0, \text{ etc.}$$

Then, from Eq. (3.24),

$$T = \frac{4}{\pi}T_i\left\{e^{-\pi^2\alpha\tau/L^2}\sin\frac{\pi x}{L} + \frac{1}{3}e^{-9\pi^2\alpha\tau/L^2}\sin\frac{3\pi x}{L} + \frac{1}{5}e^{-25\pi^2\alpha\tau/L^2}\sin\frac{5\pi x}{L} + \ldots\right\} \tag{3.28}$$

or

$$\frac{T}{T_i} = \frac{t-t_s}{t_i-t_s} = \frac{4}{\pi}\sum_{n=1}^{\infty}\frac{1}{n}e^{(-\pi n/L)^2\alpha\tau}\sin\left(\frac{n\pi}{L}\right)x \tag{3.29}$$

$$(n = 1, 3, 5, \ldots, \ldots)$$

Since, $$q = -kA\frac{\partial t}{\partial x} = -kA\frac{\partial T}{\partial x}$$

$$q(x, \tau) = 4\left(\frac{kA}{L}\right)(t_s - t_i)\sum_{n=1}^{\infty} e^{-(n\pi/L)^2 \alpha\tau} \cos\left(\frac{n\pi}{L}\right) x$$

$$(n = 1, 3, 5, \ldots, \ldots) \qquad (3.30)$$

Here q (x, τ) is the instantaneous rate of heat flow at time τ = τ across area A of a plane located at x = x. If we integrate Eq. (3.30) with respect to τ from τ = 0 to τ = τ we get the total heat flow across area A of the plane located at x = x from τ = 0 to τ. Thus,

$$Q(x, \tau) = \frac{4}{\pi^2}\left(\frac{KAL}{\alpha}\right)(t_s - t_i)\sum_{n=1}^{\infty}\left[\frac{1}{n^2}\left(1 - e^{-(n\pi/L)^2\alpha\tau}\right)\left(\cos\frac{n\pi}{L}x\right)\right]$$

where n = 1, 3, 5, ... (3.31)

(c) For the mid-plane temperature of the plate (x = L/2) Eq. (3.29) gives

$$\frac{t(L/2,\tau) - t_s}{t_i - t_s} = \frac{4}{\pi}\left[e^{-(\pi^2/L^2)\alpha\tau} - \frac{1}{3}e^{-(9\pi^2/L^2)\alpha\tau} + \frac{1}{5}e^{-(25\pi^2/L^2)\alpha\tau} - \ldots\right] \qquad (3.32)$$

$$= P \text{ (say)}$$

Values of P as a function of $\alpha\tau/S^2$ or $4\alpha\tau/L^2$ where (S = L/2) are listed in Appendix C and plotted as curve 1 in Fig. 3.4.

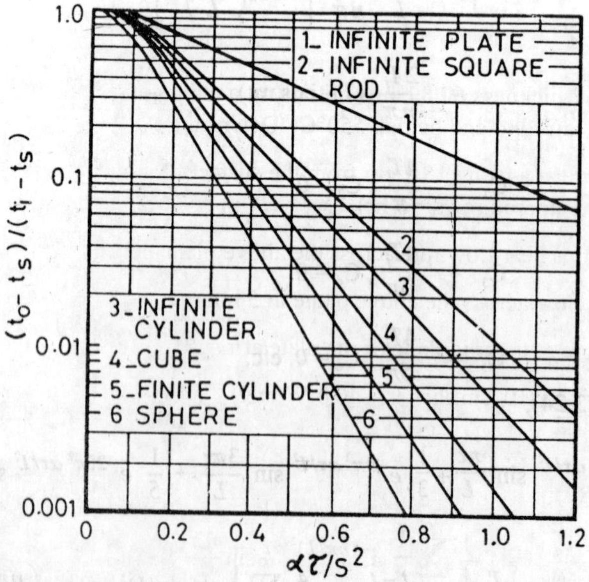

Fig. 3.4 Central Temperature History in Solids

(d) By putting x = 0 in Eqs. (3.30) and (3.31) the values of q and Q at the surface are given by ;

$$\frac{qS}{kA_s(t_s - t_i)} = 2\sum_{n=1}^{\infty}\left\{e^{-(n\pi/2S)^2\alpha\tau}\right\} \tag{3.33}$$

$$\frac{Q}{kA_s(t_s - t_i)S} = \frac{8}{\pi^2}\sum_{n=1}^{\infty}\frac{1}{n^2}\left[1 - e^{-(n\pi/2S)^2\alpha\tau}\right] \tag{3.34}$$

(where n = 1, 3, 5, ...)

(e) When dealing with symmetrical bodies, it is generally more appropriate to count distances from the symmetry plane, axis or centre as zero point. For the plane plate if we take the midplane, as x = 0, then the two surfaces will be at x = – L/2 = – S and x = + L/2 = S and instead of Eq. (3.28) we will obtain ;

$$T = \frac{4}{\pi} T_i \left\{e^{-\pi^2\alpha\tau/L^2}\cos\left(\frac{\pi x}{L}\right) - \frac{1}{3}e^{-9\pi^2\alpha\tau/L^2}\cos\left(\frac{3\pi x}{L}\right) + \frac{1}{5}e^{-25\pi^2\alpha\tau/L^2}\cos\left(\frac{5\pi x}{L}\right) - \ldots\right\}$$

or

$$T = \frac{4}{\pi} T_i \left\{e^{-\pi^2\alpha\tau/4S^2}\cos\left(\frac{\pi x}{2S}\right) - \frac{1}{3}e^{-9\pi^2\alpha\tau/4S^2}\cos\left(\frac{3\pi x}{2S}\right) + \ldots\right\}$$

or

$$\frac{T}{T_i} = \frac{t - t_s}{t_i - t_s} = \frac{4}{\pi}\sum_{n=1}^{\infty}\left[\sin\left(\frac{n\pi}{2}\right)\frac{1}{n}\,e^{-(n\pi/2S)^2\alpha\tau}\cos\left(\frac{n\pi}{2S}\right)x\right]$$

(where n = 1, 3, 5, ...)

Example 3.4

A large plate of thickness 62.8 cm is initially at a temperature of 50°C. Suddenly both of its surfaces are raised to and held at 550°C. Determine :

(a) Temperature at a plane 15.7 cm from the surfaces 5 hours after the sudden change in surface temperature (τ = 5 hr)

(b) Instantaneous heat flow rate across the above plane per m2 at τ = 5 hr.

(c) Total heat flow across the above plane in 5 hours.

(d) Temperature at the midplane of the plate at τ = 5 hr.

(Given $\alpha = 2 \times 10^{-3}$ m²/hr ; k = 46.4 W/mK)

Solution :

(a) From Eq. (3.29)

$$\frac{t - t_s}{t_i - t_s} = \frac{4}{\pi}\sum_{n=1}^{\infty}\frac{1}{n}\,e^{-(n\pi/L)^2\alpha\tau}\sin\frac{n\pi}{L}x \; ; \; (n = 1, 3, 5, \ldots)$$

Now $\left(\frac{n\pi}{L}\right)^2(\alpha\tau) = \left(\frac{n \times 3.14}{0.628}\right)^2 \times (2 \times 10^{-3} \times 5) = 0.25\, n^2$

and
$$\frac{x}{L} = \frac{0.157}{0.628} = \frac{1}{4}$$

so
$$\frac{t-t_s}{t_i-t_s} = \frac{4}{\pi}\left[\frac{1}{1}e^{-0.25}\sin\frac{\pi}{4} + \frac{1}{3}e^{-2.25}\sin\frac{3\pi}{4} + \frac{1}{5}e^{-6.25}\sin\frac{5\pi}{4} + \frac{1}{7}e^{-12.25}\sin\frac{7\pi}{4} + \ldots\right]$$

$$= \left(\frac{4}{\pi}\right)(0.707)\left[\frac{1}{e^{0.25}} + \frac{1}{3e^{2.25}} - \frac{1}{5e^{6.25}} - \frac{1}{7e^{12.25}} \ldots\right]$$

$$= \left(\frac{4}{\pi}\right)(0.707)\left[\frac{1}{1.284} + \frac{1}{3\times 9.5} - \frac{1}{5\times 520} \ldots\right]$$

$$= \left(\frac{4}{\pi}\right)(0.707)\,(0.78 + 0.035 - 0.000384) = 0.734$$

or
$$t = t_s + (t_i - t_s)\,0.734$$
$$= 550 + (50 - 550)\,0.734$$

or
$$t = 183\ °C$$ **Ans.**

(b) From Eq. (3.30),

$$q = \frac{4kA}{L}(t_s - t_i)\sum_{n=1}^{\infty} e^{-(n\pi/L)^2\alpha\tau}\cos\frac{n\pi}{L}x$$

(where n = 1, 3, 5, …)

so
$$q = \frac{4kA}{L}(t_s - t_i)\left[e^{-0.25}\cos\frac{\pi}{4} + e^{-2.25}\cos\frac{3\pi}{4} + e^{-6.25}\cos\frac{5\pi}{4} + \ldots\right]$$

$$= \frac{4kA}{L}(t_s - t_i)\,(0.707)\left[\frac{1}{e^{0.25}} - \frac{1}{e^{2.25}} - \frac{1}{e^{6.25}}\right]$$

$$= \frac{4\times 46.4\times 1}{0.628}(500)\,(0.707)\,(0.673)$$

or
$$q = 7.08\times 10^4\ W$$ **Ans.**

(c) From Eq. (3.31)

$$Q = \frac{4}{\pi^2}\left(\frac{KAL}{\alpha}\right)(t_s - t_i)\sum_{n=1}^{\infty}\frac{1}{n^2}\left[1 - e^{-(n\pi/L)^2\alpha\tau}\right]\cos\frac{n\pi}{L}x$$

(where n = 1, 3, 5, …)

$$= \frac{4}{\pi^2}\left(\frac{KAL}{\alpha}\right)(t_s - t_i)\left[(1-e^{-0.25})\cos\frac{\pi}{4} + \frac{1}{9}(1-e^{-2.25})\cos\frac{3\pi}{4} + \frac{1}{25}(1-e^{-6.25})\cos\frac{5\pi}{4} + \frac{1}{49}(1-e^{-12.25})\cos\frac{7\pi}{4} + \ldots\right].$$

$$= \frac{4}{\pi^2}\left(\frac{KAL}{\alpha}\right)(t_s - t_i)(0.707)\left[(1-0.78) - \frac{1}{9}(1-0.105) - \frac{1}{25}(1-0.00192) + \ldots\right]$$

$$= \frac{4}{\pi^2}\left(\frac{46.4\times 0.628}{2\times 10^{-3}}\right)(500)(0.707)(0.113)$$

or $Q = 23.2 \times 10^4$ W Ans.

(d) $S = L/2 = 0.314$ m

so $\dfrac{\alpha\tau}{S^2} = \dfrac{2\times 10^{-3}\times 5}{(0.314)^2} = 0.101$

From Appendix C of Fig. 3.4, we find that for $\alpha\tau/S^2 = 0.101$, P = 0.9478

so $\dfrac{t - t_s}{t_i - t_s} = 0.9478$

or $t = t_s + (t_i - t_s)(0.9478)$

$= 550 - 500\,(0.9478) = 76$

or $(t)_{L/2} = 76$ °C Ans.

Example 3.5

A large mass of combustible material is piled up against the inside surface of the wall of a large room. The wall is 15 cm thick and is initially at 25 °C. A fire suddenly raises and maintains the outer surface temperature of the wall at 200 °C. How long will it take for the inside surface to attain the ignition temperature (60 °C) of the combustible material. The thermal diffusivity of the wall material is 1.5×10^{-3} m²/hr. The thermal conductivity of the combustible material is low.

Solution :

In this case the temperature of only *one* surface (not both) is raised to and held at t_s. However, on account of the very low thermal conductivity of the combustible material, there will be hardly any heat flow across the *inside* surface. Hence, we may consider the inside surface to be insulated.

Now, suppose we have a wall of thickness 2L and *both* of its surfaces are raised to and held at t_s. Due to symmetry of temperature distribution, there will be *no heat flow* across the mid-plane (x = L). Thus the mid-plane will act as an insulated surface.

It follows that the temperature distribution in a wall of thickness L, originally at uniform temperature t_i, whose one surface is suddenly changed to and held at t_s and whose other surface is insulated ; will be the same as that in a wall of thickness 2L originally at t_i whose both surfaces are suddenly changed to and held at t_s.

Therefore, we can use Eq. (3.32) provided we consider the thickness L of the wall to be 2 × 0.15 = 0.3 m.

Now, $$\frac{t(L/2,\tau)-t_s}{t_i-t_s} = P=\frac{60-200}{25-200}=0.8$$

From Appendix C, for P = 0.8, $\alpha\tau/S^2 = 0.184$

or $$\frac{\alpha\tau}{(L/2)^2} = 0.184$$

$$\tau = \frac{(0.184)(0.15)^2}{1.5\times 10^{-3}}$$

$$= 2.76 \text{ hr}$$ Ans.

3.7 SUDDEN TEMPERATURE CHANGE OF THE SURFACE (OR ENVIRONMENT) OF AN INFINITELY LONG CYLINDER

Consider an infinitely long cylinder initially at a *uniform* temperature t_i throughout. Let the temperature of the surface of the cylinder (r = S where S = radius of cylinder) be suddenly changed to t_s and maintained at this value for all $\tau > 0$. For this case only the results for the expressions for *t* (r, τ), q (r, τ) and Q (r, τ) will be given without derivation.

(i) $$\frac{T}{T_i} = \frac{t(r,\tau)-t_s}{t_i-t_s} = 2\sum_{n=1}^{\infty}\frac{1}{M_n}\, e^{-M_n^2\alpha\tau/S^2}\left(\frac{J_o\, M_n\, r/S}{J_1(M_n)}\right) \quad (3.36)$$

where J_o = the Bessel function of first kind and zero order,

J_1 = the Bessel function of first kind and first order,

M_n = the nth root of the equation $J_o(M_n) = 0$

Value of J_o and J_1 are given in Appendix D and of M_n are given in Appendix E

(ii) $$q(r,\tau) = \left\{4\pi k(t_s-t_i)\frac{r}{s}\right\}\sum_{n=1}^{\infty} e^{-M_n^2\alpha\tau/S^2}\,\frac{J_1(M_n\, r/S)}{J_1(M_n)} \quad (3.37)$$

here q (r,τ) is rate of heat flow per unit length of the cylinder.

(iii) $$Q(r,\tau) = \left\{\frac{4\pi k}{\alpha}(t_s-t_i)\,Sr\right\}\sum_{n=1}^{\infty}\left\{\frac{1}{M_n^2}\left[1-e^{-M_n^2\alpha\tau/S^2}\right]\times\left(\frac{J_1(M_n\, r/S)}{J_1(M_n)}\right)\right\} \quad (3.38)$$

here Q (r, τ) is total heat flow per unit length of the cylinder in time τ.

(iv) From Eq. (3.36) the axial temperature t at r = 0, τ = τ is given by ;

$$\frac{t(0,\tau)-t_s}{t_i-t_s} = 2\left[\frac{e^{-M_1^2\alpha\tau/S^2}}{M_1 J_1(M_1)}+\frac{e^{-M_2^2\alpha\tau/S^2}}{M_2 J_1(M_2)}+\ldots\right] = \text{C} \tag{3.39}$$

Values of C as a function of $\alpha\tau/S^2$ are given in Appendix F and are plotted as curve 3 in Fig. 3.4.

(v) The values of q (r, τ) and Q (r, τ) for the surface (r = S) can be obtained by putting r = S in Eqs. (3.37) and (3.38). We then obtain after some rearrangement ;

$$\frac{q_s S}{k A_s (t_s - t_i)} = 2\sum_{n=1}^{\infty} e^{-M_n^2\alpha\tau/S^2} \tag{3.40}$$

$$\frac{Q_s \alpha}{k A_s S(t_s - t_i)} = 2\sum_{n=1}^{\infty} \frac{1}{M_n^2}\left[1-e^{-M_n^2\alpha\tau/S^2}\right] \tag{3.41}$$

(where A_s = Surface area at x = S)

Example 3.6

A long cylinder of 20 cm radius is initially at a temperature of 500 °C. Its surface temperature is suddenly reduced to 0 °C. Find at 2 hours after the surface temperature change :

(a) Temperature at a distance of 10 cm from the axis.

(b) Rate of heat flow across the cylindrical surface located at 10 cm radius per m length of the cylinder.

(c) Total heat flow across the above surface in 2 hr.

(d) Axial temperature.

(e) Rate of heat flow at the surface (r = 20 cm) per m length.

(f) Total heat flow at the surface (r = 20 cm).

(α = 2 × 10^{-3} m^2/hr; k = 58 W/mK)

Solution :

(a)
$$\frac{t-t_s}{t_i-t_s} = 2\sum_{n=1}^{\infty}\frac{1}{M_n} - e^{-M_n^2\alpha\tau/S^2}\left\{\frac{J_o(M_n r/S)}{J_1(M_n)}\right\} \tag{3.36}$$

Now,
$$\frac{\alpha\tau}{S^2} = \frac{2\times10^{-3}\times2}{(0.2)^2} = 0.1$$

From Appendices D and E we get :

n	M_n	M_n^2	$\frac{M_n^2 \alpha\tau}{S^2}$	$M_n \frac{r}{S}$	$J_o\left(M_n \frac{r}{S}\right)$	$J_1\left(M_n \frac{r}{S}\right)$	$J_1(M_n)$
1	2.405	5.8	0.58	1.20	0.671	0.498	0.52
2	5.52	30.5	3.05	2.76	– 0.168	0.422	– 0.34
3	8.654	75.0	7.50	4.33	– 0.356	– 0.180	0.27
4	11.79	139.0	13.90	5.90	0.122	– 0.295	– 0.23

so
$$\frac{t-t_s}{t_i-t_s} = 2\left[\frac{1}{2.405}\,e^{-0.58}\,\frac{0.6711}{0.52} + \frac{1}{5.52}\,e^{-3.05}\left(\frac{-0.168}{-0.34}\right) + \frac{1}{8.654}\,e^{-7.5}\,\frac{(-0.356)}{0.27} + \ldots\right]$$

$$= 2\left[\frac{1.29}{2.405\times 1.786} + \frac{0.168}{5.52\times 21.1\times 0.34} - \frac{0.356}{8.654\times 1800\times 0.27}\right]$$

$$= 2\,[0.3 + 0.00423 - 0.000085] = 0.6083$$

or
$$t = t_s + (t_i - t_s)(0.6083)$$
$$= 0 + (500)(0.6083)$$
$$= 304°C$$

Ans.

(b) From Eq. (3.37),

$$q = 4\pi k\,(t_s - t_i)\left(\frac{r}{S}\right)\sum_{n=1}^{\infty} e^{-\{M_n^2\alpha\tau/S^2\}}\,\frac{J_1(M_n r/S)}{J_1(M_n)}$$

or
$$q = 4\pi \times 58\,(-500)\left(\frac{10}{20}\right)\Sigma\, e^{-\{M_n^2\alpha\tau/S^2\}} \times \frac{J_1(M_n r/S)}{J_1(M_n)}$$

$$= -18.3\times 10^4\left[e^{-0.58}\,\frac{0.498}{0.52} + e^{-3.05}\,\frac{0.422}{-0.34} + e^{-7.5}\,\frac{(-0.18)}{0.27} + \ldots\right]$$

$$= -18.3\times 10^4\,(0.55 - 0.0588 - 0.00037)$$

$$= -8.93\times 10^4 \text{ W}$$

Ans.

(c) From Eq. (3.38)

$$Q = \frac{4\pi k}{\alpha}\,(t_s - t_i)\,S\,r\sum_{n=1}^{\infty}\frac{1}{M_n^2}\left[1-e^{-M_n^2\alpha\tau/S^2}\right]\times\frac{J_1(M_n r/S)}{J_1(M_n)}$$

$$= \frac{4\pi\times 58}{2\times 10^{-3}}\,(-500)\,(0.2)\,(0.1)\sum_{n=1}^{\infty}\frac{1}{M_n^2}\left[1-e^{-M_n^2\alpha\tau/S^2}\right]\frac{J_1(M_n r/S)}{J_1(M_n)}$$

$$= -1.16\,\pi \times 10^6 \left[\frac{1}{5.8} + (1 - e^{-0.58})\,\frac{0.498}{-0.52} + \frac{1}{30.5} \times (1 - e^{-3.05})\,\frac{0.422}{-0.34} + \frac{1}{75}\,(1 - e^{-7.5})\,\frac{(-0.18)}{0.27} + \frac{1}{139}\,(1 - e^{-13.9})\,\frac{(-0.295)}{(-0.23)} + \ldots\right]$$

$$= -1.16\,\pi\, 10^6\,[0.0729 - 0.0388 - 0.0089 + 0.00924]$$

$$= -1.26 \times 10^5 \text{ W}$$

Ans.

(d) From Eq. (3.39), the axial temperature is given by :

$$\frac{t - t_s}{t_i - t_s} = 2\left[\frac{e^{-M_1^2 \alpha\tau/S^2}}{M_1 J_1(M_1)} + \frac{e^{-M_2^2 \alpha\tau/S^2}}{M_2 J_1(M_2)} + \frac{e^{-M_3^2 \alpha\tau/S^2}}{M_3 J_1(M_3)} + \ldots\right]$$

$$= 2\left[\frac{e^{-0.58}}{2.405 \times 0.52} + \frac{e^{-3.05}}{5.52 \times (-0.34)} + \frac{e^{-7.5}}{8.654 \times 0.27} + \ldots\right]$$

$$= 2\left(\frac{1}{2.405 \times 0.52 \times 1.786} - \frac{1}{5.52 \times 21.1 \times 0.34} + \frac{1}{8.654 \times 0.27 \times 1800}\right)$$

$$= 2\,(0.448 - 0.0254 + 0.000238 + \ldots) = 0.845$$

or

$$t = t_s + (t_i - t_s)\,(0.845)$$

$$= 0 + (500)\,(0.845)$$

$$= 422.85°\text{C}$$

Ans.

Aliter

If we refer to Appendix F,

we find that for $\frac{\alpha\tau}{S^2} = 0.1$, C $(\alpha\tau/S^2) = 0.84836$.

Thus, $t = 500\,(0.84836) = 424.2\ °\text{C}$

Ans.

(e) For r/S = 1, Eq. (3.37) gives

$$q = 4\pi\, k\, (t_s - t_i)\,(1) \sum_{n=1}^{\infty} e^{-M_n^2 \alpha\tau/S^2}$$

$$= -(4\pi)\,(58)\,(500)\,[e^{-0.58} + e^{-3.05} + e^{-7.5} + \ldots]$$

$$= -3.64 \times 10^5\,[0.56 + 0.0474 + 0.000556 + \ldots]$$

$$= -2.27 \times 10^5 \text{ W}$$

Ans.

(f) For r/S = 1, Eq. (3.38) gives

$$Q = \frac{4\pi k}{\alpha}\,(t_s - t_i)\,S^2 \sum_{n=1}^{\infty} \frac{1}{M_n^2}\left(1 - e^{-M_n^2 \alpha\tau/S^2}\right)$$

$$= \frac{4\pi \times 58}{2\times 10^{-3}} (0 - 500)(0.2)^2 \left[\frac{1}{5.8}(1 - e^{-0.58}) + \frac{1}{30.5}(1 - e^{-3.05}) + \frac{1}{75}(1 - e^{-7.5}) + \frac{1}{139}(1 - e^{-13.9}) + \ldots\right]$$

$$= -2.32\pi \times 10^6 [0.0758 + 0.0314 + 0.0133 + 0.0072 + \ldots]$$

$$= -9.35 \times 10^5 \text{ W}$$ **Ans.**

3.8 SUDDEN TEMPERATURE CHANGE OF THE SURFACE (OR ENVIRONMENT) OF A SPHERE

For the case of a solid sphere of radius S, initially at a uniform temperature t_i throughout, whose surface temperature is suddenly changed to and maintained at t_s for all $\tau > 0$, the general and central temperature history and the instantaneous and total heat flows are given by ;

(i) $$\frac{t - t_s}{t_i - t_s} = \frac{2}{n}\left(\frac{S}{r}\right)\sum_{n=1}^{\infty}\frac{(-1)^{n+1}}{n} e^{-(n\pi)^2 \alpha\tau/S^2} \sin\left(n\pi\frac{r}{S}\right) \quad (3.42)$$

(ii) $$\frac{t(0,\tau) - t_s}{t_i - t_s} = 2\sum_{n=1}^{\infty}(-1)^{n+1} e^{-(n\pi)^2 \alpha\tau/S^2} \quad (3.43)$$

The Eq. (3.43) is plotted as curve 6 in Fig. 3.4.

(iii) $$q(r,\tau) = 8k(t_s - t_i) S r^2 \sum_{n=1}^{\infty}\left(\frac{(-1)^{n+1}}{n} e^{-(n\pi)^2 \alpha\tau/S^2}\right) \times \left[\frac{1}{r^2}\sin n\pi\frac{r}{S} - \frac{n\pi}{Sr}\cos n\pi\frac{r}{S}\right] \quad (3.44)$$

(iv) $$Q(r,\tau) = \frac{8k}{\pi^2 \alpha}(t_s - t_i) S^3 r^2 \sum_{n=1}^{\infty}\frac{(-1)^{n+1}}{n}\left[1 - e^{-(n\pi)^2 \alpha\tau/S^2}\right] \times \left(\frac{1}{r^2}\sin n\pi\frac{r}{S} - \frac{n\pi}{Sr}\cos n\pi\frac{r}{S}\right) \quad (3.45)$$

(v) The values of q and Q for the surface (r = S) are given by :

$$\frac{q_s S}{k A_s (t_s - t_i)} = 2\sum_{n=1}^{\infty} e^{-(n\pi)^2 \alpha\tau/S^2} \quad (3.46)$$

$$\frac{Q_s \alpha}{k A_s S (t_s - t_i)} = \frac{2}{\pi^2}\sum_{n=1}^{\infty}\frac{1}{n^2}\left[1 - e^{-(n\pi)^2 \alpha\tau/S^2}\right] \quad (3.47)$$

(where A_s = surface area of the sphere at r = S)

Example 3.7

Calculate the temperature at the centre of a steel sphere 40 cm in diameter 20 minutes after the surface temperature changes from 30 °C to 660 °C. An average value of $\alpha = 24 \times 10^{-3}$ m^2/hr may be assumed. Compute also the instantaneous and cumulative heat flow at the surface of the sphere. (k = 46.4 W/mK).

Solution :

(a) Radius $S = \frac{0.4}{2} = 0.2$ m

so $$\frac{\alpha\tau}{S^2} = (24 \times 10^{-3}) \left(\frac{20}{60}\right) \frac{1}{0.04} = 0.2$$

From Eq. (3.43),

$$\frac{t_o - t_s}{t_i - t_s} = 2 \sum_{n=1}^{\infty} (-1)^{n+1}\, e^{-(n\pi)^2 \alpha\tau/S^2}$$

$$= 2\left[e^{-\pi^2(0.2)} - e^{-4\pi^2(0.2)} + e^{-9\pi^2(0.2)} - \ldots\right]$$

$$= 2\left[\frac{1}{e^{1.97}} - \frac{1}{e^{7.88}} + \frac{1}{e^{17.78}} - \ldots\right] = 0.275$$

Note that same result is obtained from curve 6 of Fig. 3.4

so $$(t_o - 660) = (30 - 660)(0.275)$$

$$t_o = 660 - 630 \times 0.275$$

$$= 486 \text{ °C}$$ Ans.

(b) From Eq. (3.44),

$$q(r, \tau) = 8\,k\,(t_s - t_i)\,Sr^2 \sum_{n=1}^{\infty} \frac{(-1)^{n+1}}{n} e^{-(n\pi)^2 \alpha\tau/S^2} \times \left[\frac{1}{r^2} \sin n\pi \frac{r}{S} - \frac{n\pi}{Sr} \cos n\pi \frac{r}{S}\right]$$

For r = S, we get

$$q(S, \tau) = 8\,k(t_s - t_i)\,S^3 \sum_{n=1}^{\infty} \left[\frac{(-1)^{n+1}}{n} e^{-(n\pi)^2 \alpha\tau/S^2}\right] \times \left[\frac{1}{S^2} \sin n\pi - \frac{n\pi}{S^2} \cos n\pi\right]$$

$$= 8\,k\,S\,(t_s - t_i)\left[\pi e^{-\pi^2 \alpha\tau/S^2} + \frac{(-1)}{2} \times (0 - 2\pi)\, e^{-4\pi^2 \alpha\tau/S^2} + \frac{1}{3}(0 + 3\pi) e^{-9\pi^2 \alpha\tau/S^2} + \ldots\right]$$

$$= 8\pi k S (t_s - t_i) \left[e^{-\pi^2 \alpha\tau/S^2} + e^{-4\pi^2 \alpha\tau/S^2} + e^{-9\pi^2 \alpha\tau/S^2} - \ldots \right]$$

$$= 8 \pi (46.4) (0.2) (630) \left[\frac{1}{e^{1.97}} + \frac{1}{e^{7.88}} + \frac{1}{e^{17.72}} - \ldots \right]$$

$$= 20{,}400 \text{ W}$$ Ans.

(c) From Eq. (3.45)

$$Q(r, \tau) = \frac{8k}{\pi^2 \alpha} (t_s - t_i) S^3 r^2 \sum_{n=1}^{\infty} \frac{(-1)^{n+1}}{n^2} \left[1 - e^{-(n\pi)^2 \alpha\tau/S^2} \right] \times \left[\frac{1}{r^2} \sin n\pi \frac{r}{S} - \frac{n\pi}{Sr} \cos n\pi \frac{r}{S} \right]$$

For r = S, we get,

$$Q(S, \tau) = \frac{8k}{\pi^2 \alpha} (t_s - t_i) S^3 \sum_{n=1}^{\infty} \frac{1}{n^2} \left[1 - e^{-(n\pi)^2 \alpha\tau/S^2} \right] \times (\sin n\pi - n\pi \cos n\pi)$$

$$= \frac{(8 \times 46.4)(630)(0.2)^3}{\pi^2 \times 24 \times 10^{-3}} \left[\left(1 - \frac{1}{e^{1.97}} \right) + \frac{1}{4} \left(1 - \frac{1}{e^{7.88}} \right) + \frac{1}{9} \left(1 - \frac{1}{e^{17.73}} \right) + \ldots \right]$$

$$= (24{,}800)(1.375) = 34{,}150 \text{ W}$$ Ans.

The *central temperature* history in various solids originally at a uniform temperature t_i and suddenly subjected to a surface temperature change to t_s is plotted as function of ($\alpha\tau/S^2$) in Fig. 3.4. Here S is either a semi-thickness or radius of the solid.

3.9 SUDDEN TEMPERATURE CHANGE OF THE SURFACE (OR ENVIRONMENT) OF AN INFINITELY THICK PLATE (SEMI-INFINITE SOLID)

(a) Consider an infinitely thick plate, that is, a plate bounded by the y–z plane only and extending to infinity in the positive x-direction. Such a solid is called a *semi-infinite solid.* The initial temperature distribution in the solid is given by $t = t_i(x)$. At time $\tau = 0$, the temperature of the surface x = 0 is suddenly changed to and held at t_s for all $\tau > 0$

If we define $T = (t - t_s)$ as before, the governing differential equation becomes same as Eq. (3.2) :

$$\frac{1}{\alpha} \frac{\partial T}{\partial \tau} = \frac{\partial^2 T}{\partial x^2} \tag{3.19}$$

with the conditions

$$T = T_i(x) \quad \text{at} \quad \tau = 0;\ 0 \le x \le \infty \tag{3.48}$$

$$T = 0 \quad \text{at} \quad x = 0;\ \tau > 0 \tag{3.49}$$

As found previously, the solution of Eq. (3.19) is ;

$$T = e^{-\lambda^2 \alpha\tau}(C_1 \cos \lambda x + C_2 \sin \lambda x) \tag{3.23}$$

Since $T = 0$ at $x = 0$ for all $\tau > 0$, $C_1 = 0$.

So $$T = C_2 e^{-\lambda^2 \alpha\tau} \sin \lambda x$$

We have no other boundary condition to determine the λ's. Hence we must consider all solutions corresponding to all possible values of λ, integral as well as non-integral. In this case , therefore, instead of the usual series we get an integral of the form:

$$T = \int_{\lambda=0}^{\infty} C(\lambda) e^{-\lambda^2 \alpha\tau} \sin \lambda x \, d\lambda \tag{3.50}$$

If the solution is to satisfy the initial condition that $T = T_i(x)$ at $\tau = 0$,

$$T_i(x) = \int_{\lambda=0}^{\infty} C(\lambda) \sin \lambda x \, d\lambda \tag{3.51}$$

Now, the Fourier sine integral is given by ;

$$f(x) = \frac{2}{\pi} \int_0^{\infty} f(x')\, dx' \times \int_0^{\infty} \sin \lambda x \sin \lambda x' \, d\lambda$$

which for $T_i(x)$ may be written as

$$T_i(x) = \frac{2}{\pi} \int_0^{\infty} T_i(x')\, dx' \times \int_0^{\infty} \sin \lambda x \sin \lambda x' \, d\lambda$$

or $$T_i(x) = \frac{2}{\pi} \int_{\lambda=0}^{\infty} \left[\int_{x'=0}^{\infty} T_i(x') \sin \lambda x' dx' \right] \sin \lambda x \, d\lambda \tag{3.52}$$

Comparing Eqs. (3.51) and (3.52) we get;

$$C(\lambda) = \frac{2}{\pi} \int_{x'=0}^{\infty} T_i(x') \sin \lambda x' \, dx'$$

Hence,

$$T = \int_{\lambda=0}^{\infty} \left[\frac{2}{\pi} \int_{x'=0}^{\infty} T_i(x') \sin \lambda x' dx' \right] e^{-\lambda^2 \alpha\tau} \sin \lambda x \, d\lambda$$

or $$T = \frac{2}{\pi} \int_{x'=0}^{\infty} T_i(x') \left[\int_{\lambda=0}^{\infty} e^{-\lambda^2 \alpha\tau} \sin \lambda x' \sin \lambda x \, d\lambda \right] dx' \tag{3.53}$$

Evaluating the integral in λ,

$$\int_{\lambda=0}^{\infty} e^{-\lambda^2 \alpha\tau} \sin \lambda x' \sin \lambda x \, d\lambda = \frac{1}{2} \int_{\lambda=0}^{\infty} e^{-\lambda^2 \alpha\tau} \cos \lambda (x' - x) \, d\lambda - \frac{1}{2} \int_{\lambda=0}^{\infty} e^{-\lambda^2 \alpha\tau} \cos \lambda (x' + x) \, d\lambda$$

$$= \frac{1}{4} \cdot \sqrt{\frac{\pi}{\alpha\tau}} \left[e^{-(x'-x)^2/4\alpha\tau} - e^{-(x'+x)^2/4\alpha\tau} \right]$$

hence

$$T = \frac{1}{2\sqrt{\pi\alpha\tau}} \int_{x'=0}^{\infty} T_i (x') \left[e^{-(x'-x)^2/4\alpha\tau} - e^{-(x'+x)^2/4\alpha\tau} \right] dx' \qquad (3.53)$$

It is the general solution of Eq. (3.19) for this seminfinite solid. Consider the special case of uniform initial temperature where $t_i(x) = t_i$ or $T_i(x) = T_i$.

Let $\mu = \dfrac{x' \mp x}{2\sqrt{\alpha\tau}}$

then $x' = 2\sqrt{\alpha\tau}\, \mu \pm x$

$dx' = 2\sqrt{\alpha\tau}\, d\mu$

Hence, for $x' = 0, \ \mu = \mp \dfrac{x}{2\sqrt{\alpha\tau}}$

and for $x' = \infty, \ \mu = \infty$

Then Eq. (3.53) becomes,

$$T = \frac{T_i}{\sqrt{\pi}} \left[\int_{-x/2\sqrt{\alpha\tau}}^{\infty} e^{-\mu^2} d\mu - \int_{-x/2\sqrt{\alpha\tau}}^{\infty} e^{-\mu^2} d\mu \right]$$

$$= \frac{T_i}{\sqrt{\pi}} \int_{-x/2\sqrt{\alpha\tau}}^{x/2\sqrt{\alpha\tau}} e^{-\mu^2} d\mu = \frac{2}{\sqrt{\pi}} T_i \int_{0}^{x/2\sqrt{\alpha\tau}} e^{-\mu^2} d\mu$$

Hence, $$\frac{t - t_s}{t_i - t_s} = \frac{T}{T_i} = \frac{2}{\sqrt{\pi}} \int_0^{x/2\sqrt{\alpha\tau}} e^{-\mu^2} d\mu = \frac{2}{\sqrt{\pi}} \int_0^{u} e^{-u^2} du \qquad (3.54)$$

Note that writing e^{-u^2} in place of $e^{-\mu^2}$ does not affect the value of the integral which depends on the value of the upper limit.

or $$\frac{t - t_s}{t_i - t_s} = \frac{T}{T_i} = F(u) = erf(u) \qquad (3.55)$$

where $$u = \frac{x}{2\sqrt{\alpha\tau}} \tag{3.56}$$

$$F(u) = \frac{2}{\sqrt{\pi}} \int_0^u e^{-u^2}\, du = erf(u) \tag{3.57}$$

$$= \frac{2}{\sqrt{\pi}} \left(\frac{u}{1} - \frac{1}{1!}\frac{u^3}{3} + \frac{1}{2!}\frac{u^5}{5} - ... \right) \tag{3.58}$$

F(u) is called the *Gauss's error integral.* Values of this function are given in Appendix G and plotted in Fig. 3.5. Note that $F(u) \approx u$ if $u < 0.75$ and $F(u) \approx 1$ if $u > 1.5$.

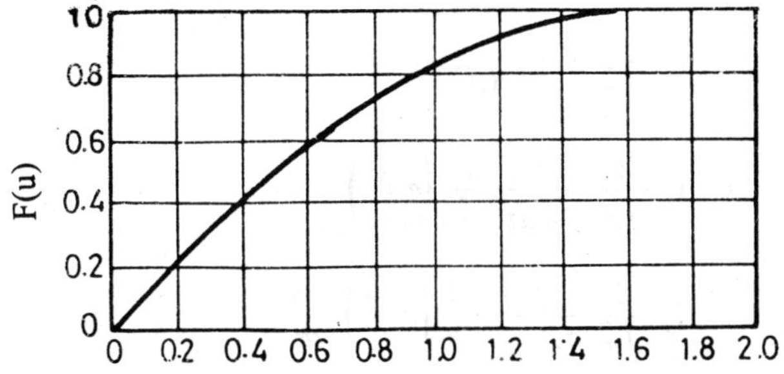

Fig. 3.5 Error function, Eq. (3.57), Appendix G

Aliter

(b) A *somewhat* simpler treatment {having known the result in Eq. (3.59)}, of the case of the semi-infinite solid originally at uniform temperature t_i having its surface (x = 0) suddenly changed to and kept at t_s is given below.

Let us assume that ;

$$T = T_i F(u) = T_i \frac{2}{\sqrt{\pi}} \int_0^u e^{-u^2} \tag{3.54}$$

is the solution of the differential equation ;

$$\frac{\partial t}{\partial \tau} = \alpha \frac{\partial^2 t}{\partial x^2} \tag{3.2}$$

and then prove that Eq. (3.54) satisfies Eq. (3.2), as well as the initial and boundary conditions Eqs (3.48) and (3.49).

(i) Now, $$\frac{\partial t}{\partial \tau} = \frac{\partial T}{\partial \tau} = \frac{\partial}{\partial \tau} [T_i F(u)] = T_i \frac{\partial F(u)}{\partial u} \frac{\partial u}{\partial \tau}$$

$$\frac{\partial F(u)}{\partial u} = \frac{\partial}{\partial u}\left[\frac{2}{\sqrt{\pi}} \int_0^u e^{-u^2} du\right] = \frac{2}{\sqrt{\pi}} e^{-u^2}$$

$$\frac{\partial u}{\partial \tau} = \frac{\partial}{\partial \tau}\left(\frac{x}{2\sqrt{\alpha\tau}}\right) = -\frac{x}{4\tau\sqrt{\alpha\tau}}$$

or
$$\frac{\partial t}{\partial \tau} = -T_i\left\{\frac{2}{\sqrt{\pi}} e^{-u^2} . \frac{x}{4\tau\sqrt{\alpha\tau}}\right\} \quad \text{(a)}$$

Similarly;
$$\frac{\partial t}{\partial x} = \frac{\partial T}{\partial \tau} = \frac{\partial}{\partial x}[T_i F(u)] = T_i \frac{\partial F(u)}{\partial u}\frac{\partial u}{\partial x}$$

$$\frac{\partial u}{\partial x} = \frac{\partial}{\partial x}\left(\frac{x}{2\sqrt{\alpha\tau}}\right) = \frac{1}{2\sqrt{\alpha\tau}}$$

$$\frac{\partial t}{\partial x} = T_i \frac{2}{\sqrt{\pi}} e^{-u^2} \frac{1}{2\sqrt{\alpha\tau}}$$

so
$$\frac{\partial^2 t}{\partial x^2} = T_i \frac{2}{\sqrt{\pi}} \frac{1}{2\sqrt{\alpha\tau}} \frac{\partial}{\partial x}\left(e^{-u^2}\right)$$

$$= T_i \frac{2}{\sqrt{\pi}} \frac{1}{2\sqrt{\alpha\tau}} \frac{\partial}{\partial u}\left(e^{-u^2}\right)\frac{\partial u}{\partial x}$$

$$= T_i \frac{2}{2\sqrt{\pi}} \frac{1}{2\sqrt{\alpha\tau}} (-2u)\, e^{-u^2} \frac{1}{2\sqrt{\alpha\tau}}$$

or
$$\alpha\frac{\partial^2 t}{\partial x^2} = T_i \frac{2}{\sqrt{\pi}}\left\{e^{-u^2} \frac{x}{4\sqrt{\alpha\tau}}\frac{1}{\tau}\right\} \quad \text{(b)}$$

Since the RHS of Eqs. (a)and (b) are equal it follows that Eq. (3.54) satisfies Eq. (3.2). Hence assumption made earlier is true.

(ii) Next we must show that assumed Eq. (3.54) satisfies the boundary conditions Eqs. (3.48) and (3.49) ;

$$T = T_i \text{ at } \tau = 0;\ 0 \le x \le \infty$$

$$T = 0 \text{ at } x = 0\ ;\ \tau > 0$$

From Eqs. (3.56) and (3.58) it follows that F(u) = 0 at x = 0 for all $\tau > 0$. Hence T = 0 at x = 0 for all $\tau > 0$.

(iii) Next we will show that F(u) = 1 at $\tau = 0$ for all x > 0, showing thereby that $T = T_i$, or that initially the plate was at temperature t_i.

Consider *r* as the radial distance of a point from the origin of a two-dimensional rectangular coordinate system in y and z.

Then

$$r^2 = y^2 + z^2$$

Now, from rules of calculus,

$$\int_{y=-\infty}^{y=\infty} \int_{z=-\infty}^{z=\infty} e^{-(y^2+z^2)}\, dy\, dz = \int_{-\infty}^{\infty} e^{-y^2}\, dy \times \int_{-\infty}^{\infty} e^{-z^2}\, dz = \left[\int_{-\infty}^{\infty} e^{-y^2}\, dy\right]^2 \qquad (c)$$

If we use polar coordinates with the surface element, (2π dr) and cover the entire plate by integration ;

$$\int_{0}^{\infty} e^{-r^2}\,(2\pi r)\, dr = \pi \qquad (d)$$

Since the results of Eqs. (c) and (d) must be identical ;

$$\int_{-\infty}^{+\infty} e^{-y^2}\, dy = \sqrt{\pi}$$

or

$$\int_{0}^{\infty} e^{-y^2}\, dy = \frac{\sqrt{\pi}}{2} \qquad (e)$$

Hence Eq. (3.57) gives F(u) = 1 for u = ∞. According to Eq. (3.56), u = ∞ occurs at τ = 0. Thus, both boundary conditions are also satisfied. Thus, Eq. (3.54) is the solution of Eq. (3.2).

Note that there is an unsteadiness at x = 0 and τ = 0 due to sudden change of surface temperature; but for all values of x > 0 and τ > 0 the function F(u) has a steady analytical character.

(c) The heat flow rate at x = x, and τ = τ is given by :

$$q = -kA\,\frac{\partial t}{\partial x}$$

$$= -kA\left(T_i\,\frac{2}{\sqrt{\pi}}\right)\left(e^{-u^2}\right)\frac{1}{2\sqrt{\alpha\tau}}$$

or

$$q = -\frac{k\,A\,T_i\,e^{-x^2/4\alpha\tau}}{\sqrt{\pi\alpha\tau}} \qquad (3.59)$$

The rate of heat entering the plate is obtained by putting x = 0 in Eq. (3.59).

$$q = -\frac{k\,A\,T_i}{\sqrt{\pi\alpha\tau}} \qquad (3.60)$$

The total heat flow into the plate during the time interval τ = 0 to τ = τ then becomes :

$$Q_o = \int_0^{\tau} q_o\, d\tau = -\int_0^{\tau} \frac{k\,A\,T_i}{\sqrt{\pi\alpha\tau}}\, d\tau$$

or

$$Q_o = -2\,k\,A\,T_i\left(\frac{\tau}{\pi\alpha}\right)^{1/2} \qquad (3.61)$$

Example 3.8

Determine the temperature in a plane 18 cm from the surface of a thick wall and also the rate of heat flow into 1 m² of this plane 10 hours after the surface temperature of the wall changes suddenly from 20 °C to 820 °C and remains constant thereafter. Also find the total heat energy taken up by the wall in 10 hr. Given k = 0.87 W/mK and $\alpha = 2.78 \times 10^{-3}$ m²/hr.

Solution :

$$\frac{x}{2\sqrt{\alpha\tau}} = \frac{0.18}{2\,(2.78\times10^{-3}\times10)^{1/2}} = 0.538$$

From Fig. 3.5, for $\frac{x}{2\sqrt{\alpha\tau}} = 0.538,$

$$f\left(\frac{x}{2\sqrt{\alpha\tau}}\right) = F(u) = 0.552$$

Hence, from Eq. (3.55),

$$T = T_i\,F(u) \qquad (3.55)$$

$$(t - 820) = (20 - 820)\,(0.552)$$

or

$$t = 820 - (800 \times 0.552)$$

$$= 378.4\ °C$$

Ans.

Now,

$$q = -\frac{k\,A\,T_i\,e^{-x^2/4\alpha\tau}}{\sqrt{\pi\alpha\tau}} \qquad (3.59)$$

$$= -\frac{(0.87)(1)(-800)\,e^{-[0.18\times0.18/(4\times2.78\times10^{-3}\times10)]}}{\sqrt{3.14\times2.78\times10^{-3}\times10}}$$

$$= \frac{600\times0.745}{0.2955}$$

or

$$q = 1750\ W$$

Ans.

Thus

$$Q_o = -2\,k\,A\,T_i\left(\frac{\tau}{\pi\alpha}\right)^{1/2} \qquad (3.61)$$

$$= -2\times0.87\times1\times(-800)\left(\frac{10}{3.14\times2.78\times10^{-3}}\right)^{1/2}$$

$$= 1390\times100/2.955$$

$$= 47{,}400\ W$$

Ans.

Example 3.9

A large mass of a material for which $\alpha = 0.405$ m²/hr is initially at 100 °C throughout. Suddenly its surface temperature is lowered to and permanently held at 0 °C. Determine :

(a) Temperature 1.5 cm below the surface after 2 sec.

(b) Time required to lower the temperature to 40°C at a point 2.5 cm below the surface.

(c) Time required for the temperature gradient at the surface to reach 4 °C/cm.

(d) The depth at which the rate of cooling is maximum after 1 sec.

Solution :

(a)
$$u = \frac{x}{2\sqrt{\alpha\tau}} = \frac{1.5\times 10^{-2}}{2\sqrt{\left(0.405\times\frac{2}{3600}\right)}}$$

$$= 0.5$$

From Appendix G, $F(u) = 0.52$

or
$$\frac{t-t_s}{t_i-t_s} = 0.52$$

or
$$t = 0 + (100)\,(0.52) = 52\ °C$$ Ans.

(b) In this case

$$\frac{t-t_s}{t_i-t_s} = \frac{40-0}{100-0} = 0.4$$

or $F(u) = 0.4$

For $F(u) = 0.4,\ u = 0.37$

Hence,
$$\frac{x}{2\sqrt{\alpha\tau}} = 0.37$$

or
$$\tau = \frac{x^2}{4\alpha(0.37)^2}$$

or
$$\tau = \frac{(2\,5\times 10^{-2})^2}{4\times 0.405(0.37)^2}$$

$$= 2.82\times 10^{-3}\ \text{hr}$$

$$= 10.15\ \text{sec.}$$ Ans.

(c) From Eq. (3.60)

$$q_0 = -\frac{k\,A\,T_i}{\sqrt{\pi\alpha\tau}}$$

or $$-kA\left(\frac{\partial T}{\partial x}\right)_{x=0} = -\frac{k\,A\,T_i}{\sqrt{\pi\alpha\tau}}$$

so $$\left(\frac{\partial T}{\partial x}\right)_{x=0} = \frac{T_i}{\sqrt{\pi\alpha\tau}} = \frac{t_i - t_s}{\sqrt{\pi\alpha\tau}}$$

Hence, $$\frac{t_i - t_s}{\sqrt{\pi\alpha\tau}} = 4 \times 100$$

$$\frac{100}{\sqrt{\pi \times 0.405 \times \tau}} = 400$$

$$\sqrt{\tau} = \frac{100}{400\sqrt{\pi \times 0.405}}$$

or $$\tau = 0.0493 \text{ hr} = 177 \text{ sec}$$ Ans.

(d) From Eq. (a) of Section 3.9 (b) (i)

$$\frac{\partial t}{\partial \tau} = -\frac{T_i}{2\sqrt{\pi\alpha}} \times \frac{x\,e^{-u^2}}{\tau^{3/2}}$$

or $$\frac{\partial t}{\partial \tau} = -\frac{T_i\,x}{2\sqrt{\pi\alpha}} \times \frac{e^{-x^2/4\alpha\tau}}{\tau^{3/2}}$$

To find the depth x at which the cooling rate $(\partial t/\partial \tau)$ is maxinum at time τ we should differentiate the above equation with respect to x and set the result equal to zero :

$$\frac{\partial}{\partial x}\left(\frac{\partial t}{\partial \tau}\right) = -\frac{T_i}{2\sqrt{\pi\alpha}}\,\frac{1}{\tau^{3/2}}\left[e^{-x^2/4\alpha\tau} + x\left(\frac{-2x}{4\alpha\tau}\right)\times e^{-x^2/4\alpha\tau}\right] = 0$$

or $$\frac{x^2}{2\alpha\tau} = 1$$

or $$x = \sqrt{2\alpha\tau} \text{ (for maximum cooling rate)}$$

Hence, $$x = \sqrt{2 \times 0.405 \times \frac{1}{3600}}$$

or $$x = 1.5 \text{ cm}$$ Ans.

Example 3.10

Water pipes are buried underground. The original soil temperature is 5 °C. Suddenly the temperature of the surface falls to – 5 °C. Calculate the minimum depth at which the pipes be laid so that there is no freezing of water. For soil α = 0.001 m²/hr.

Solution :

The earth may be considered as semi-infinite solid.

Then $$\frac{t - t_s}{t_i - t_s} = \frac{0-(-5)}{5-(-5)} = \frac{5}{10} = 0.5 = F(u)$$

For F(u) = 0.5, u = 0.48 (from Appendix G)

or $$\frac{x}{2\sqrt{\alpha\tau}} = 0.48$$

Let the time interval be assumed as 12 hr during which freezing of water may occur. Then

$$x = 0.48 \times 2\sqrt{12 \times 0.001} = 0.105 \text{ m}$$

or $$x = 10.5 \text{ cm}$$ Ans.

Example 3.11

Soil at –6 °C and containing 3% moisture is to be thawed by heating the surface with fire to 800 °C. How much time will be required for the thawing to reach 50 cm depth. $\alpha = 1.76 \times 10^{-3}$ m²/hr. C_p = 1.68 kJ/kg K.

Solution :

To thaw the soil (i.e. to melt the frozen moisture) we should first raise its temperature to 0°C and then supply enough heat to furnish the latent heat of melting. Considering 1 kg of soil, the heat required to melt the moisture is 1 × 0.03 × 336 = 10.08 kJ/kg.

Now, if we supply 10.08 kJ/kg to dry soil the resulting temperature rise will be 10.06/1.68 = 6°C. Thus, we may say that thawing is equivalent to raising the soil temperature by 6°C above 0°C. The problem then is to find the time when the temperature at a depth of 50 cm reaches 6°C. Now,

$$\frac{t - t_s}{t_i - t_s} = \frac{6 - 800}{-6 - 800} = 0.985 = F(u)$$

For F(u) = 0.985, u = 1.72 (from Appendix G)

so $$\frac{x}{2\sqrt{\alpha\tau}} = 1.72$$

or $$\tau = \frac{x^2}{4\alpha(1.72)^2}$$

$$= \frac{(0.5)^2}{4 \times 1.76 \times 10^{-3} \times (1.72)^2}$$

or $$\tau = 12.1 \text{ hr}$$ Ans.

Example 3.12

A large steel ingot (α = 0.0435 m²/hr) is initially at 800 °C. Its surface (x = 0) is suddenly chilled to 20 °C. Find the time required for the temperature to reach 700 °C at a depth of 1 cm.

Solution :

$$F(u) = \frac{t - t_s}{t_i - t_s} = \frac{700 - 20}{800 - 20} = 0.873$$

$$u = 1.08 = \frac{x}{2\sqrt{\alpha\tau}} \qquad \text{(from Appendix G)}$$

so

$$\tau = \frac{x^2}{4\alpha\,(1.08)^2} = \frac{1\times10^{-4}}{4\times0.0435\times(1.08)^2}\times 3600$$

$$= 1.82$$

Ans.

Example 3.13

A locomotive tire 8 cm thick is to be removed from the wheel by heating its outer surface. If α for tire is 0.0435 m^2/hr, find the time after which the differential expansion of tire and rim will be maximum so that the tire may be easily removed.

Solution :

Since the thickness of the tire is very small compared to the diameter we can treat it as a case of one-dimensional heat transmission and neglect the effect of curvature. It is also reasonable to assume that the differential expansion of tire and rim is maximum when the temperature gradient at the junction of tire and rim is maximum. Therefore we have to find the time at which (∂t/∂x) is maximum 8 cm below the surface.

Now,

$$\frac{\partial t}{\partial x} = \frac{(t_i - t_s)}{\sqrt{\pi\alpha\tau}}\, e^{-x^2/4\alpha\tau} \quad \text{\{as obtained from (3.54)\}}$$

The time when (∂t/∂x) is maximum is given by ;

$$\frac{\partial}{\partial\tau}\left(\frac{\partial t}{\partial x}\right) = 0$$

or

$$\frac{(t_i - t_s)}{\sqrt{\alpha\pi}}\left[-\frac{e^{-x^2/\alpha\tau}}{2\tau^{3/2}} + \frac{1}{\sqrt{\tau}}e^{-x^2/4\alpha\tau}\,\frac{x^2}{4\alpha\tau^2}\right] = 0$$

or

$$\tau = \frac{1}{2}\frac{x^2}{\alpha}$$

whence

$$\tau = \frac{(8\times10^{-2})^2}{2\times0.0435} = 0.0735 \text{ hr} = 4.4 \text{ min}$$

Ans.

Example 3.14

A motor car weighing 1370 kg is travelling at 48 km/hr. It is stopped in 5 sec by 4 brakes with brake-bands of 250 cm^2 area each pressing against steel drums (k = 56 W/mK, α = 0.0445 m^2/hr) of the same area. Assuming that the brake lining and drum surfaces are at the same temperature and that the heat is dissipated by flowing across the surface of the drums (assumed to be very thick), find the maximum temperature rise.

Solution :

Kinetic energy of car dissipated $= \frac{1}{2} mv^2$

$$= \frac{1}{2} (1370) \left(\frac{48 \times 1000}{3600}\right)^2$$

$$= 122 \times 10^3 \text{ J in 5 sec}$$

or Heat flow rate $= \dfrac{122}{5/3600} = 874 \times 10^2$ kJ/hr

Thus, the heat flow rate at the surface (x = 0) is 874×10^2 kJ/hr. From Eq. (3.60),

$$q_o = \frac{kA(t_s - t_i)}{\sqrt{\pi \alpha \tau}}$$

or

$$(t_s - t_i) = \frac{q_o \sqrt{\pi \alpha \tau}}{k\,A}$$

The above equation shows that the maximum temperature rise $(t_s - t_i)$ should be computed on the basis of $\tau = 5$ sec (the largest value of τ). Hence,

$$(t_s - t_i) = \frac{874 \times 10^2}{56 \times 3.6 \times 250 \times 4 \times 10^{-4}} \sqrt{3.14 \times 0.0455 \times 5/3600}$$

$$= 74.6 \text{ °C}$$

So maximum temperature rise = 74.6 °C Ans.

Example 3.15

A very large mass of metal (k = 45.3 W/mK and α = 0.121 cm²/sec) at 100°C with one plane face, is dropped into water at 10°C. What will be the temperature of the surface of contact. How much time will elapse before a point 2m inside the surface will fall in temperature to 95°C. For water, k = 6 W/mK, α = 0.00143 cm²/sec.

Solution :

Suppose that we have two infinite bodies 1 and 2 of thermal conductivities and diffusivities k_1, α_1 and k_2, α_2, each with a single plane surface and these surfaces are placed in contact with each other. Let t_1 and t_2 be the initial temperatures of bodies 1 and 2. Let us determine the conditions under which the contact surface remains at constant temperature t_s where $t_1 > t_s > t_2$

Evidently, the contact surface will remain at a constant temperature if it receives as much heat from one body as it loses to the other.

Now, each unit area of contact surface receives heat from body 1 at the rate given by Eq. (3.60)

$$(q_o)_1 = \frac{k_1 (t_1 - t_s)}{\sqrt{\pi \alpha_1 \tau}}$$

And each unit area of contact surface loses heat to body 2 at the rate given by

$$(q_o)_2 = \frac{k_2(t_s - t_2)}{\sqrt{\pi \alpha_2 \tau}}$$

Hence, the contact surface will neither gain nor loose heat and will remain at a constant temperature if

$$(q_o)_1 = (q_o)_2$$

or
$$\frac{k_1(t_1 - t_s)}{\sqrt{\pi \alpha_1 \tau}} = \frac{k_2(t_s - t_2)}{\sqrt{\pi \alpha_2 \tau}}$$

or
$$\frac{k_1(t_1 - t_s)}{\sqrt{\alpha_1}} = \frac{k_2(t_s - t_2)}{\sqrt{\alpha_2}}$$

or
$$t_s = \frac{(k_1 t_1 / \sqrt{\alpha_1}) + (k_2 t_2 / \sqrt{\alpha_2})}{(k_1 / \sqrt{\alpha_1}) + (k_2 / \sqrt{\alpha_2})}$$

Substituting the numerical values we get ;

$$t_s = \frac{\dfrac{45.3 \times 100}{\sqrt{0.121 \times 10^{-4}}} + \dfrac{6 \times 10}{\sqrt{0.00143 \times 10^{-4}}}}{\dfrac{45.3}{\sqrt{0.121 \times 10^{-4}}} + \dfrac{6}{\sqrt{0.00143 \times 10^{-4}}}}$$

so
$$t_s = 90.2\ °C$$
Ans.

Since the mass of steel is initially at 100°C and its surface temperature is suddenly changed to 90.2 °C, we can find the time required for the temperature at a depth of 2 m to reach 95 °C.

$$F(u) = \frac{t - t_s}{t_i - t_s} = \frac{95 - 90.2}{100 - 90.2} = 0.489$$

For
$$F(u) = 0.489,\ u = 0.465 \quad \text{(from Appendix G)}$$

or
$$\frac{x}{2\sqrt{\alpha \tau}} = 0.465 = \frac{2}{2\sqrt{0.121 \times 10^{-4} \times \tau}}$$

or
$$\tau = 38.5 \times 10^4\ \text{sec} = 4.45\ \text{day}$$
Ans.

3.10 TRANSIENT HEAT CONDUCTION IN BODIES WITH FINITE INTERNAL AND SURFACE RESISTANCE

Finally, we deal with those cases of non-periodic heat flow where both the internal and surface resistance of the solid need consideration. For instance, the case of a realistic cooling or heating of a solid by exposing it to an air stream or by dipping it in a fluid, requires that the thermal conductivity k of the solid, as well as the unit surface conductance h should be taken into account while heat transfer analysis is made.

(a) Consider an infinite plate of uniform thickness 2S undergoing a cooling or heating operation. Initially it is at a uniform temperature of t_i and then it is suddenly exposed to, or immersed in, an infinite fluid at temperature t_g. Let the surface film conductance be h, throughout the heating or cooling ($\tau > 0$)

Since the plate is assumed very long in y and z directions the temperature will be uniform in any y-z plane at a specified value of x.

As in previous cases, the general heat conduction equation reduces to the form ;

$$\frac{1}{\alpha}\frac{\partial t}{\partial \tau} = \frac{\partial^2 t}{\partial x^2} \tag{3.2}$$

This equation yields the temperature history in the solid t = t (x, τ) when solved with the following initial and boundary conditions :

$$\left.\begin{array}{ll} (i) & t = t_i \; at \; \tau = 0 \\ (ii) & \dfrac{\partial t}{\partial x} = 0 \; at \; x = 0 \\ (iii) & \mp\dfrac{\partial t}{\partial x} = \pm\dfrac{h}{k}(t - t_g) \; at \; x = \pm S \end{array}\right\} \tag{3.63}$$

Note that we have taken x = 0 at the mid-plane of the of plate. Boundary condition (iii) expresses the fact that the rate at which heat is given by the fluid is equal to the rate at which heat is taken up by the plate or vice-versa.

Let us define ;

$$T = t - t_g$$

$$T_i = t_i - t_g$$

Then Eq. (3.2) becomes,

$$\frac{1}{\alpha}\frac{\partial T}{\partial \tau} = \frac{\partial^2 T}{\partial x^2} \tag{3.19}$$

and the boundary conditions become

(i) $T = T_i$ at $\tau = 0$

(ii) $\dfrac{\partial T}{\partial x} = 0$ at $x = 0$

(iii) $\mp\dfrac{\partial T}{\partial x} = \pm\dfrac{h}{k}T$ at $x = \pm S$

As shown in Section 3.6, the solution of Eq. (3.19) may be expressed as

$$T_\lambda = e^{-\alpha\lambda^2\tau}\,(A\cos\lambda x + B\sin\lambda x) \tag{3.64}$$

Differentiating Eq. (3.64) with respect to x,

$$\frac{\partial T}{\partial x} = e^{-\alpha\lambda^2\tau} (-A\lambda \sin \lambda x + B\lambda \cos \lambda x)$$

or $$\left(\frac{\partial T}{\partial x}\right)_{x=0} = e^{-\alpha\lambda^2\tau} (B\lambda)$$

In order to satisfy boundary condition (Eq. 3.63) (ii), B must be 0.

Hence $$T = e^{-\alpha\lambda^2\tau} A \cos \lambda x \tag{3.65}$$

also,

$$\left(\frac{\partial T}{\partial x}\right)_{x=S} = -e^{-\alpha\lambda^2\tau} A\lambda \sin \lambda S$$

From boundary condition Eq. (3.63) (iii),

$$-\left(\frac{\partial T}{\partial x}\right)_{x=S} = e^{-\alpha\lambda^2\tau} A\lambda \sin \lambda S = \frac{h}{k} T$$

$$= \frac{h}{k} e^{-\alpha\lambda^2\tau} (A \cos \lambda S)$$

or $$\cot \lambda S = \frac{k}{hS} \lambda S = \frac{1}{Bi} (S\lambda) \tag{3.66}$$

Note that here we have used the *Biot number* which was defined earlier. Also note that the same result is obtained if we use the condition ;

$$+\frac{\partial T}{\partial x} = -\frac{h}{k} T \text{ at } x = -S$$

Eq. (3.66) is a *transcendental equation*, and there are infinite number of values of λ which will satisfy it. These values are called the *eigenvalues* or characteristic values. The solution of this equation is possible by a graphical method as shown in Fig. 3.6, where cot λS is plotted against (k/h) λ or λS/Bi. There are infinite intersections of these curves corresponding to the roots $\lambda_1, \lambda_2, \lambda_3, \ldots$ Therefore the complete solution of Eq. (3.66) is the sum of the solutions corresponding to each eigenvalue.

So from Eq. (3.65)

$$T(x, \tau) = \sum_{n=1}^{\infty} e^{-\alpha\lambda_n^2\tau} (A_n \cos \lambda_n x) \tag{3.67}$$

in which $\lambda_1, \lambda_2, \lambda_3, \ldots, \lambda_n$ have the values determined from Eq. (3.66) and $A_1, A_2, \ldots, A_n$ are selected to satisfy the condition that $T = T_i$ at $\tau = 0$. In order to obtain a particular value of A_n, multiply both sides of Eq. (3.67) by $\cos(\lambda_m x)$ and integrate between O and S ;

$$\int_0^S T_i \cos(\lambda_m x)\, dx = \sum_{n=1}^{\infty} A_n \int_0^S \cos(\lambda_m x) \cos(\lambda_n x)\, dx$$

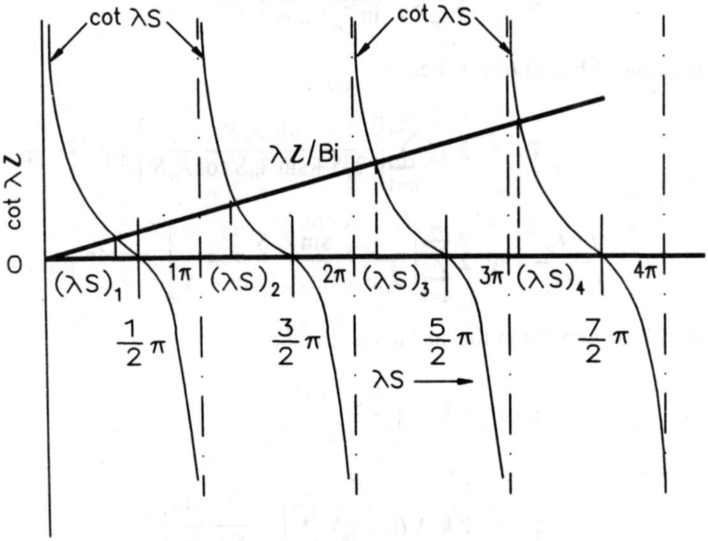

Fig. 3.6 Graphical Solution of Eq. (3.66)

Now $$\int_0^S \cos \lambda_m x \, dx = \frac{1}{\lambda_m} \sin \lambda_m S$$

The RHS integral gives ;

$$\int_0^S \cos \lambda_n x \cos \lambda_m x \, dx = \frac{\lambda_m \sin \lambda_m S \cos \lambda_n S - \lambda_n \cos \lambda_m S \sin \lambda_n S}{\lambda_m{}^2 - \lambda_n{}^2}$$

Since from Eq. (3.66),

$$\lambda_m \sin \lambda_m S = \frac{h}{k} \cos \lambda_m S$$

and $$\lambda_n \sin \lambda_n S = \frac{h}{k} \cos \lambda_n S$$

so $$\lambda_m \sin \lambda_m S \cos \lambda_n S = \lambda_n \cos \lambda_m S \sin \lambda_n S$$

Therefore, the RHS integral is zero if $m \neq n$.

If $$m = n$$

$$\int_0^S \cos^2 \lambda_n x \, dx = \left[\frac{S}{2} + \frac{\sin \lambda_n S \cos \lambda_n S}{2\lambda_n}\right]$$

So $$\frac{T_i}{\lambda_n} \sin \lambda_n S = A_n \left[\frac{S}{2} + \frac{\sin \lambda_n S \cos \lambda_n S}{2\lambda_n}\right]$$

or
$$A_n = \frac{2\,T_i \sin\lambda_n S}{(\lambda_n S + \sin\lambda_n S\cos\lambda_n S)} \tag{3.68}$$

The final solution of Eq. (3.67) is then ;

$$T = 2\,T_i \sum_{n=1}^{\infty}\left[\frac{\sin\lambda_n S}{\lambda_n S + \sin\lambda_n S\cos\lambda_n S}\right] e^{-\alpha\lambda_n^2\tau}\cos\lambda_n x$$

or
$$\frac{t-t_g}{t_i-t_g} = 2\sum_{n=1}^{\infty}\left[\frac{\sin\lambda_n S}{\lambda_n S + \sin\lambda_n S\cos\lambda_n S}\right] e^{-\alpha\lambda_n^2\tau}\cos\lambda_n x \tag{3.69}$$

(b) The rate of heat transfer at the surface is given by :

$$q_s = -k\,A\left(\frac{\partial(t-t_g)}{\partial x}\right)_{x=S}$$

or
$$q_s = 2\,k\,A\,(t_i-t_g)\sum_{n=1}^{\infty}\left[\frac{\sin\lambda_n S}{\lambda_n S + \sin\lambda_n S\cos\lambda_n S}\right] e^{-\alpha\lambda_n^2\tau} \times (\lambda_n \sin\lambda_n S) \tag{3.70}$$

The total heat transfer at the surface during interval τ is;

$$Q = \int_0^{\tau} q_s\, d\tau$$

or
$$Q = \left[\frac{2\,k\,A(t_i-t_g)}{\alpha}\sum_{n=1}^{\infty}\left(\frac{\sin^2\lambda_n S}{\lambda_n^2 S + \lambda_n \sin\lambda_n S\cos\lambda_n S}\right)\times\left(1-e^{-\lambda_n^2\alpha\tau}\right)\right] \tag{3.71}$$

Example 3.16

A rocket nozzle is constructed of a special alloy steel 6.35 mm thick. The thermal conductivity and diffusivity of steel are 26.8 W/mK and 0.0232 m^2/hr respectively. The convective heat transfer coefficient at the flame side may be assumed as 7,330 W/m^2 K and the uniform flame temperature as 2000 C. The nozzle temperature is initially 40 C throughout. Calculate the permissible combustion duration so that the safe operating temperature for this type of steel, 1000 C, is not exceeded.

Solution :

The nozzle diameter is large as compared to the thickness of its walls, hence its behaviour may be assumed to be similar to that of a flat plate 6.35 mm thick. Further, to simplify, we may assume the outer surface of nozzle to be insulated against heat loss from its surface so the axis x = 0 may be chosen at outer surface. Then the plate will be 12.7 mm thick where S = 6.35 mm at the flame side.

Now, $$\frac{t-t_g}{t_i-t_g} = \frac{1000-2000}{40-2000} = 0.51$$

$$\text{Bi} = \frac{hS}{k} = \frac{7330 \times 6.35}{1000 \times 26.8} = 1.74$$

So Eq. (3.66) takes the form;

$$1.74 = \lambda S \tan(\lambda S)$$

From mathematical tables we may obtain the first two eigenvalues for this equation ;

$$\lambda_1 S = 1.06 \text{ and } \lambda_2 S = 3.56$$

So $$\lambda_1 = 1.06 \times \frac{1000}{6.35} = 167$$

and $$\lambda_2 = 3.56 \times \frac{1000}{6.35} = 562$$

Limiting the computation to two terms, we obtain from Eq. (3.69) ;

$$\frac{t-t_g}{t_i-t_g} = 2\sum_{n=1}^{\infty}\left\{\frac{\sin\lambda_n S}{S\lambda_n + \sin\lambda_n S\cos\lambda_n S}\right\} e^{-\alpha\lambda_n^2\tau} \cos\lambda_n S$$

$$= 2\sum_{n=1}^{\infty}\left\{\frac{1}{1+(2\lambda_n S/\sin 2\lambda_n S)}\right\} e^{-\alpha\lambda_n^2\tau}$$

$$= 2\left\{\frac{1}{1+(2\times 1.06/\sin 2.12)}\right\} e^{-(0.0232\times 167^2)\tau} +$$

$$2\left\{\frac{1}{1+(2\times 3.56/\sin 7.12)}\right\} e^{-(0.0232\times 562^2)\tau}$$

$$= 2\,[0.288\, e^{-646\tau} + 0.094\, e^{-7295\tau}]$$

or $$0.51 = 0.576\,(e^{-646\tau}) + 0.188\,(e^{-7295\tau})$$

By trial and error the solution of this equation is found to be $\tau = 1$ sec. Thus the nozzle will attain a temperature of 1000 °C after 1 sec of firing. Ans.

3.11 TRANSIENT HEAT CONDUCTION CHARTS

In the foregoing section, we dealt with the analytical method of solving problems where a solid undergoes heating or cooling. The temperature variation with time and location in an infinite plate can be determined by evaluating appropriate number of terms of Eq. (3.69). If the series is found to converge rapidly then only a few terms may be considered. The resulting solution, however, will be applicable to the particular plate under the particular condition. Further, variation in the values of S, h and k would lead to different sets of eigenvalues, which require re-evaluation of the results. Therefore a more general method for the determination of the temperature history is considered desirable.

One method of presenting the solution of Eq. (3.69) and of similar equations for other solids, like cylinder or sphere, is to give the solution in the form of charts. Evidently, the charts must be prepared with dimensionless coordinates so that they are of universal usage. Such charts or graphical plots for the evaluation of temperature distribution have been made in a variety of forms for practical use. We shall, however, discuss in this section only those charts which are due to *Heisler*.

During the discussion of transient state heat conduction the various parameters which have been used are ;

$$t, t_i, t_g, \alpha, \tau, S, h, k \text{ and } x$$

It is found that all these variables can be correlated by an expression of the form ;

$$\frac{t - t_g}{t_i - t_g} = f\left(\frac{\alpha\tau}{S^2}, \frac{k}{hS}, \frac{x}{S}\right)$$

$$= f(\text{Fo}, 1/\text{Bi}, x/S) \tag{3.72}$$

The left hand side of Eq. (3.72) is the *dimensionless temperature distribution* in solid, the expression for which is to be evaluated. The right hand side term is a function of three dimensionless parameters given below :

(a) *Fourier number* Fo ; ($\alpha\tau/S^2$). It signifies the degree of penetration of the heating or *cooling effect* through the solid. For small values of α/S^2, for instance a large τ is required to obtain a significant temperature change.

(b) *Biot number* Bi ; (hS/k). It signifies the ratio of the resistance to heat transfer at the surface of a solid to the internal resistance of the body. It is somewhat similar to *Nusselt number* used in convective heat transfer.

(c) *Dimensionless length parameter* ; (x/S). It indicates the distance of any location from an arbitrary origin as compared to the plate thickness. In case of solids having circular cross section this parameter will have the value (r/R).

(i) The *Heisler Charts* presented herein are drawn on the basis of the dimensionless numbers discussed above. Figures 3.7, 3.8 and 3.9 show three Heisler Charts for determining the temperature history. In these charts the dimensionless temperature, $T_0/T_i = (t_0 - t_g)/(t_i - t_g)$, where the suffix 0 refers to x = 0, is plotted versus Fo, i.e. $\alpha\tau/S^2$ for various values of (1/Bi) < 100. In the first chart for infinite plate, the curve for (1/Bi) = 0 corresponds to zero surface resistance ($h = \infty$). This is equivalent to the case where the temperature of the exposed surface is suddenly raised and held at the fluid temperature, t_g.

The charts of Figs. 3.7 to 3.9 give the temperature history of the solid at its centre. Temperatures at other locations are obtained by multiplying by *correction factors* as obtained from Figs. 3.10, 3.11 and 3.12 which show the companion *Heisler charts* for the plate, the cylinder and the sphere, respectively. The values of $T(x, \tau)/T_0$ are plotted against 1/Bi for different values of x/S or r/S. It will be observed from these figures that the temperature throughout the solid is nearly uniform for values of (1/Bi) > 10, say. It is therefore permissible as a first approximation to consider that the heating or cooling process in these cases is being controlled solely by the surface resistance and the uniform temperature history is then given by Eq. 3.5.

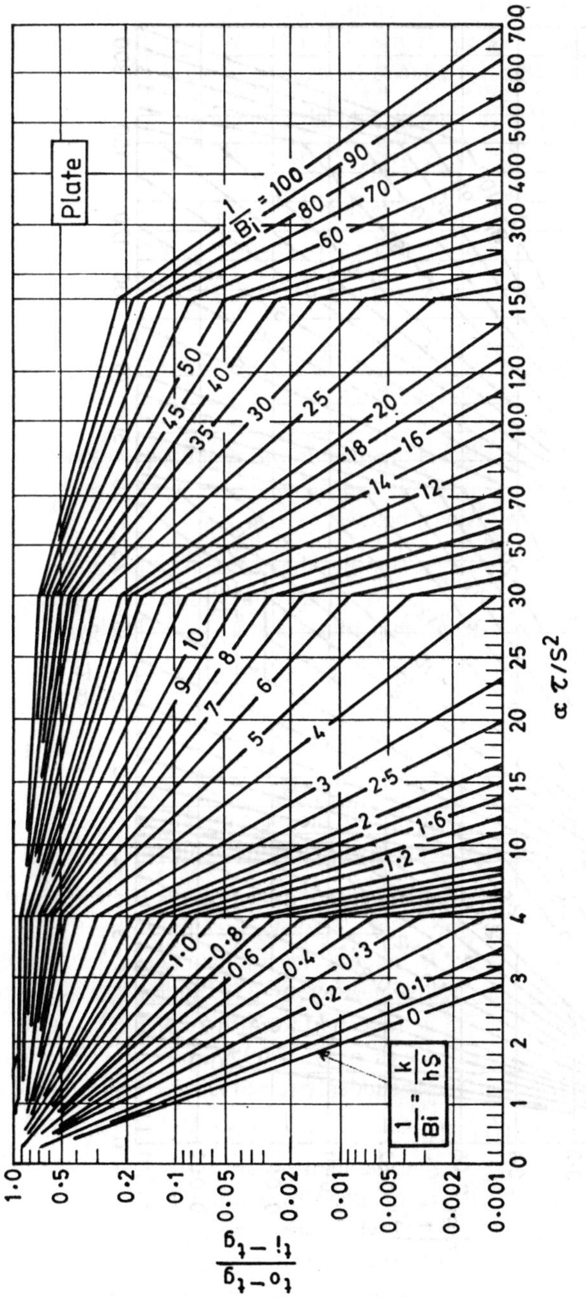

Fig. 3.7 Hesiler Chart for Temperature History (i) at insulated surface of an Infinite Flat Plate ($x/S = 0$) or (ii) at Centre of a plate of thickness 2S with both sides subjected to similar heating or cooling

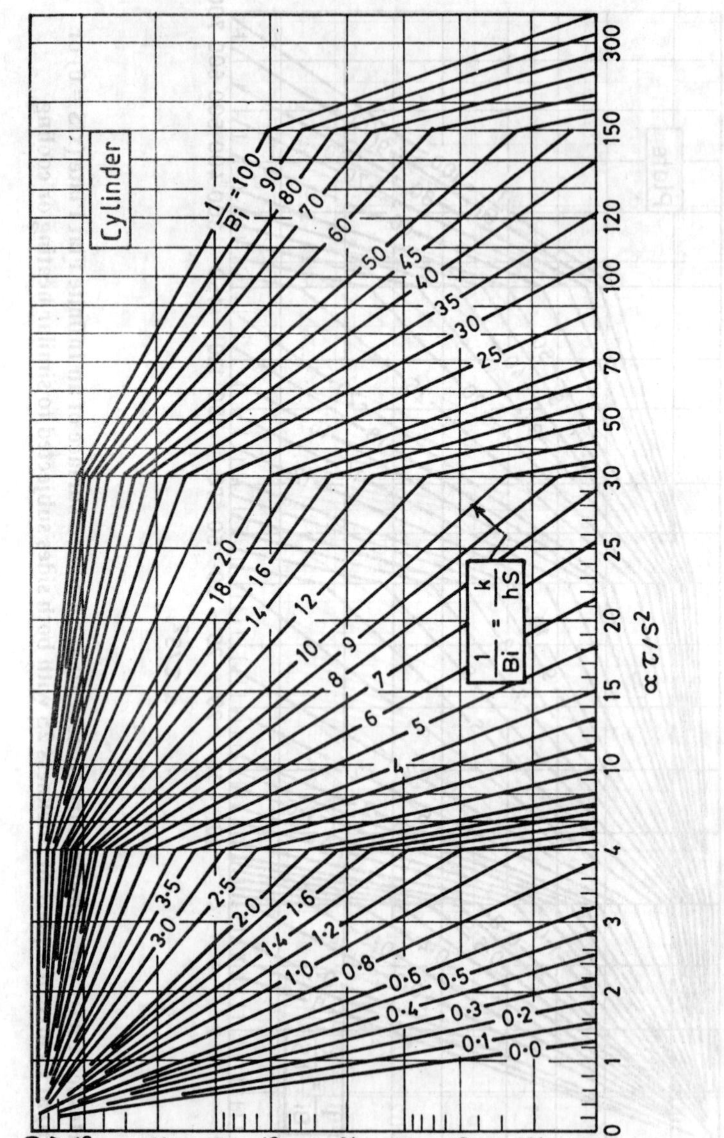

Fig. 3.8 Hesiler Chart for Central Temperature History in Cylinder

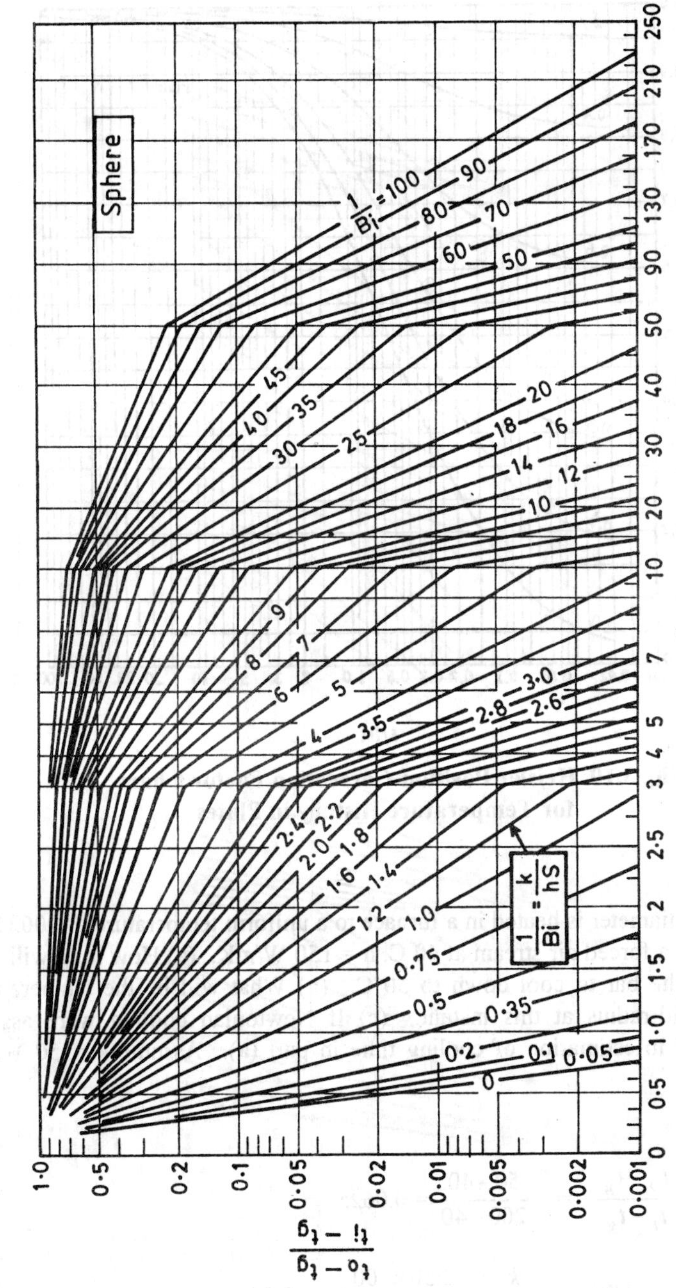

Fig. 3.9 Hesiler Chart for Central Temperature History in Sphere

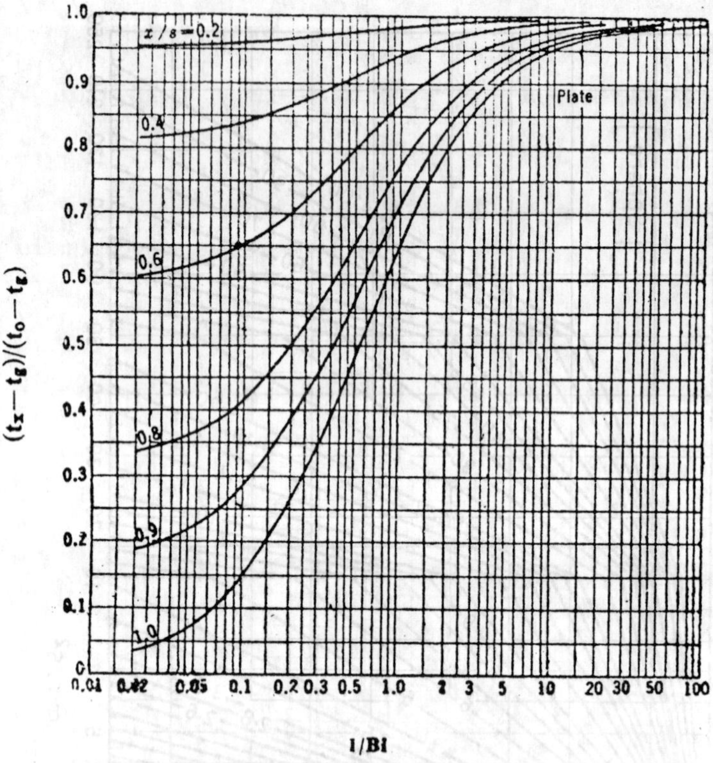

Fig. 3.10 Heisler Position-Correction Factor Chart for Temperature History in Plates

Example 3.17

A bar 12.5 cm diameter is heated in a furnace to a uniform temperature of 200 C. It is then left to cool in a forced air stream at 40 C, h = 150 W/mK. (a) How long will it take for the centre of the bar to cool down to 50 C. (b) What will be the temperature of surface and at mid-radius at this instant. (c) If Newtonian cooling were assumed, calculate the error in estimation of cooling time in part (a). Assume k = 50 W/mK ; $\alpha = 0.0725$ m²/hr.

Solution :

(a) Now, $$\frac{t_o - t_g}{t_i - t_g} = \frac{50-40}{200-40} = 0.0625$$

and $$1/\text{Bi} = \frac{k}{hS} = \frac{50 \times 100}{150 \times 12.5/2} = 5.34$$

From the chart of Fig. 3.8 for the above value we get the value of Fo = 8.2 = $\alpha\tau/S^2$

or $$\tau = \frac{8.2 \times (12.5/200)^2}{0.0725} \times 60$$

$$= 26.57 \text{ min} = 26 \text{ min } 34 \text{ sec.}$$ Ans.

(b) Now, from Fig. 3.11, for 1/Bi = 5.34 and r/S = 0.5, we find the value of correction factor = 0.98.

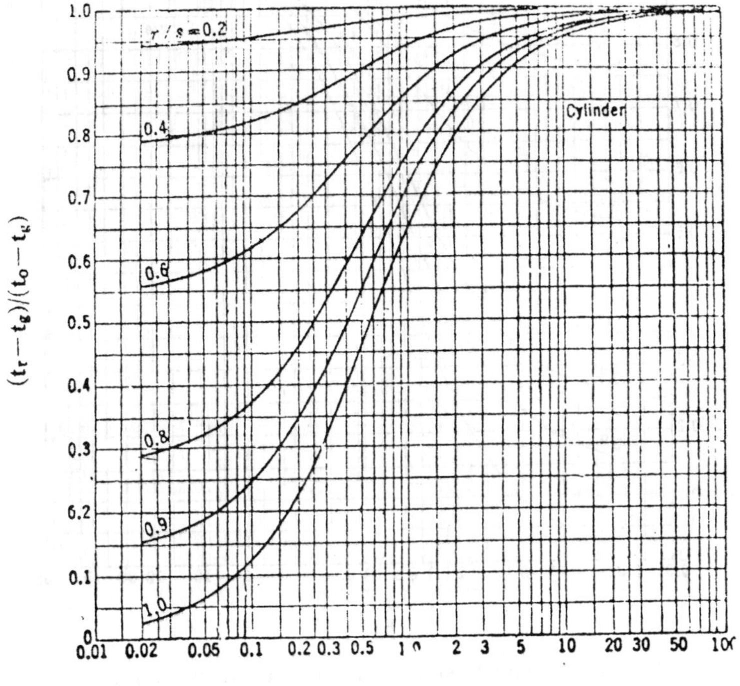

Fig. 3.11 Heisler Position-Correction Factor Chart for Temperature History in Cylinders

So the temperature distribution is given by ;

$$\frac{t-40}{200-40} = 0.98 \times 0.0625$$

$$= 0.0612$$

The temperature at (r/S) = 0.5 i.e. at mid-radius, is given by ;

$$(t)_{R/2} = 40 + 0.0612\,(160)$$

$$= 49.9 \text{ C}$$ Ans.

Similarly the correction factor at r/S = 1, is found to be 0.92 and therefore the surface temperature is given by ;

$$(t)_{surface} = 40 + (0.92 \times 0.0625) \times (160)$$

$$= 49.2\text{C}$$ Ans.

(c) For Newtonian cooling ; Eq. (3.5) is applicable

i.e. $$\frac{t-t_g}{t_i-t_g} = e^{-(hA_1/c\rho v)\tau}$$

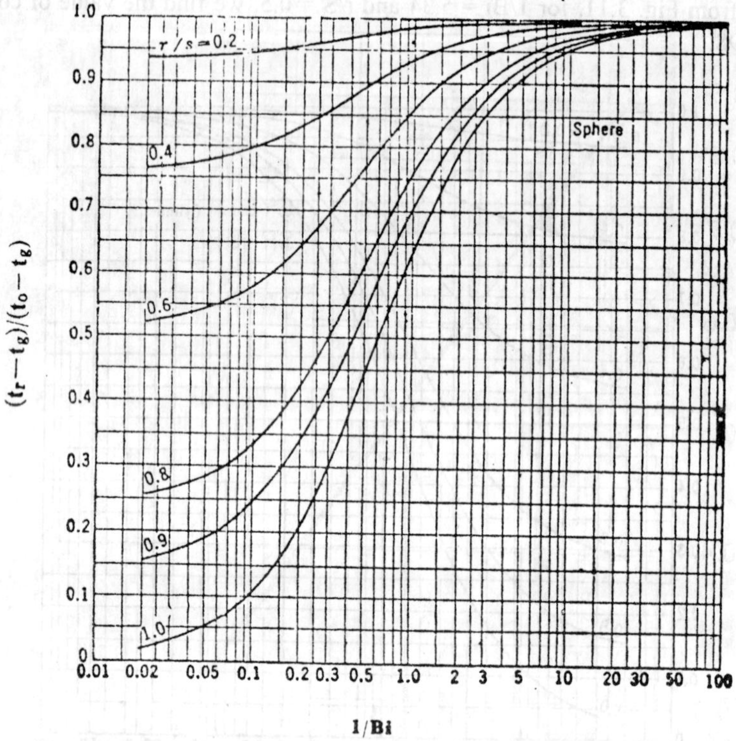

Fig. 3.12 Heisler Position Correction Factor Chart for Temperature History in Spheres

or $$0.0625 = e^{-(h\alpha/k)(A_1/v)\tau} \quad \text{(since } \alpha = k/\rho c)$$

or $$\log_e (1/0.0625) = \frac{150 \times 0.0725}{50}\left(\frac{2 \times 100}{12.5/2}\right)\tau \quad \text{(since } A_1/v = 2/R \text{ for cylinder)}$$

$$= 6.956\ \tau$$

or $$\tau = \frac{\ln 16}{6.956} = 0.4 \text{ hr} = 24 \text{ min}$$

$$\text{Thus error in calculations} = \frac{26.57 - 24}{26.57}$$

$$= 9.5\ \%$$ Ans.

(ii) Heisler further suggested that for problems in which k/hS or 1/Bi > 100 and Fo > 0.2 ; the internal resistance may be considered small as compared to surface resistance. The results in such cases may be satisfactorily represented by Eq. (3.5) or Eq. (3.9). It follows that the temperature throughout the solid at any time differs negligibly from the temperature at x = 0 and is a function of time only. He proposed the charts as given in Figs. 3.13 (a) and 3.13 (b) for evaluation of temperature history for problems having this range of Bi and Fo.

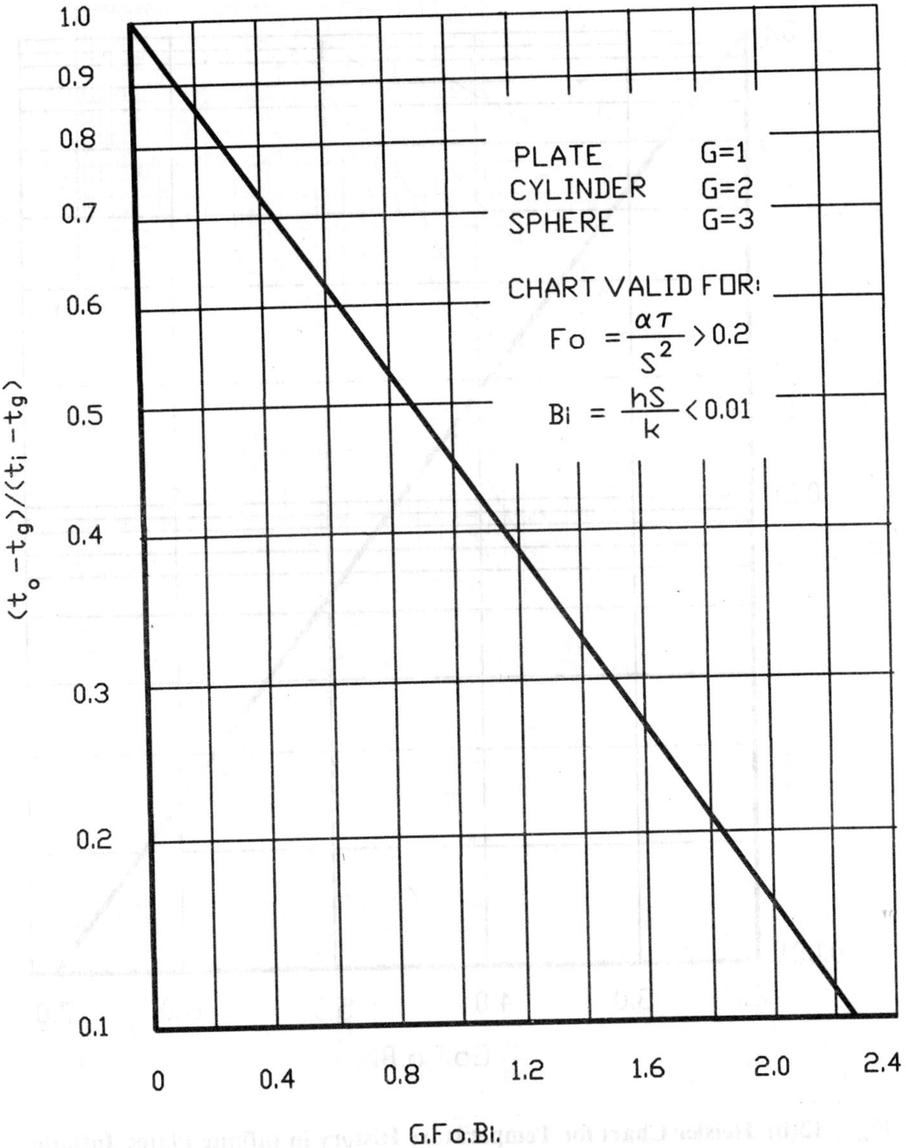

Fig. 3.13(a) Heisler Chart for Temperature History in Infinite Plates, Infinite Cylinders and Spheres for (1/Bi) > 100 and (G.Fo.Bi) < 2.2.

There are many situations when the heating or cooling time is small e.g. quenching in heat treatment of metals. Such operations are called *short-time heating or cooling methods*. Apparantly the charts of Figs. 3.7 to 3.12 are not useful for such cases where the value of Fo is small. The series solutions in this range converge very slowly and hence detailed temperature calculations are very tedious and inaccurate unless a large number of terms are computed. Heisler, therefore developed *short time charts* also, as given in Figs. 3.14 to 3.19. These charts were developed to cover the range of 1/Bi from 0.05 to ∞ and that of Fo from 0 to 0.2.

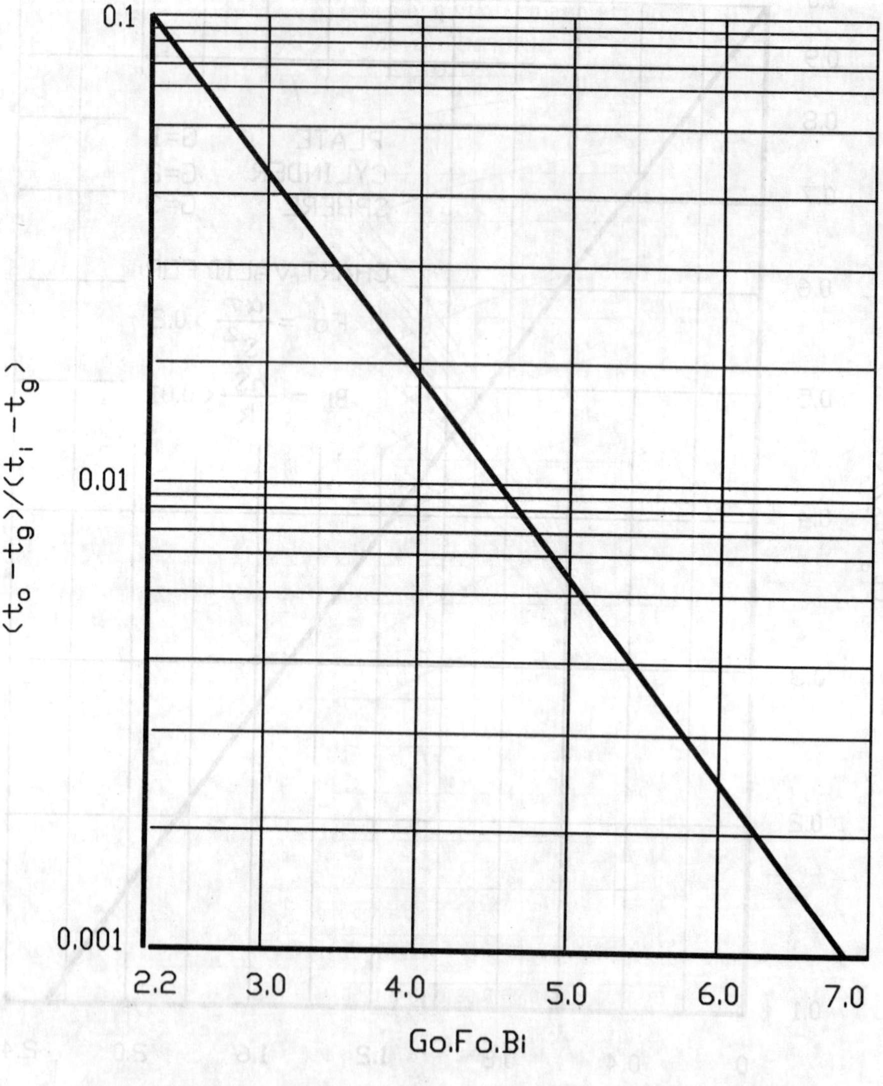

Fig. 3.13(b) Heisler Chart for Temperature History in Infinite Plates, Infinite Cylinders and Spheres for (1/Bi) > 100 and (G.Fo.Bi) > 2.2.

Example 3.18

A long metal bar 10 cm × 5 cm (l_x = 5 cm ; l_y = 2.5 cm) is heated to a uniform temperature 250 C and then suddenly dipped in a fluid at 50 C. The unit surface conductance h = 1,200 W/m^{2}K. Calculate (a) central temperature in the bar after 3 sec (b) surface temperature at the centre of its short face (x = 5 cm ; y = 0) and (c) surface temperature at the centre of its long face (x = 0 ; y = 2.5 cm). Assume that ;

	x - direction	y-direction
k (W/mK)	30	7.5
α (m^2/hr)	0.515	0.0925

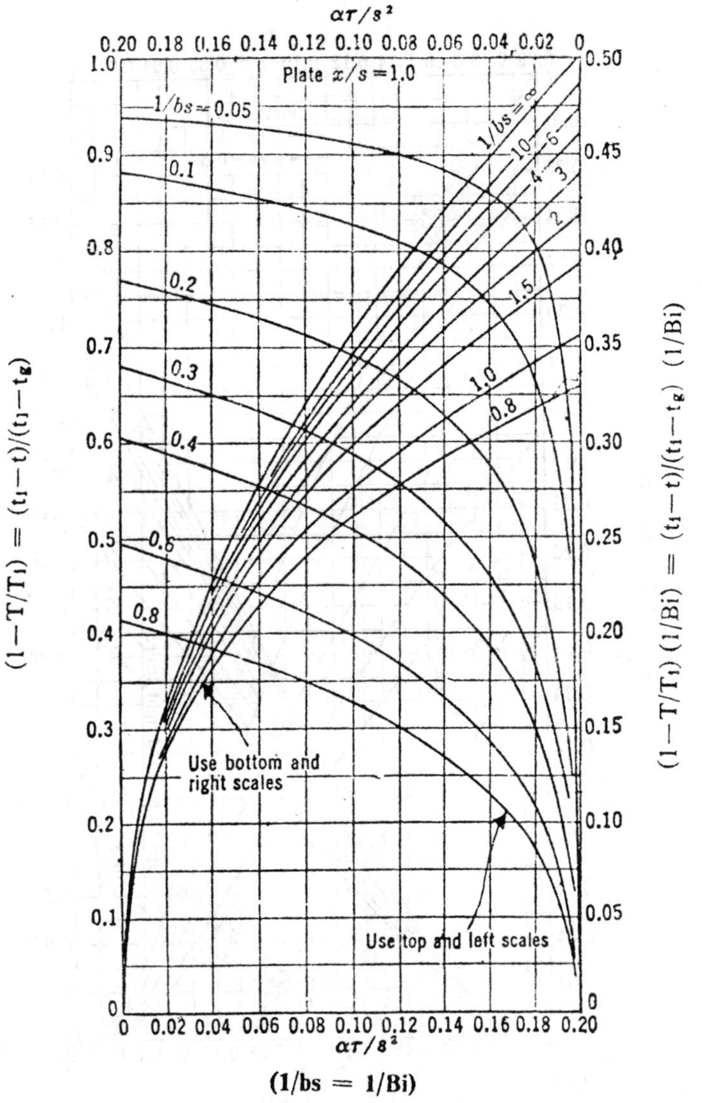

Fig. 3.14 Heisler Short Time Chart for Surface Temperature History in Infinite Plates

Solution :

(a) Here, the two dimensional temperatures may be computed as the product of one-dimensional solutions.

Now,

$$(1/\mathrm{Bi})_x = \frac{k_x}{hs_x} = \frac{30}{1200 \times 5/100} = 0.5$$

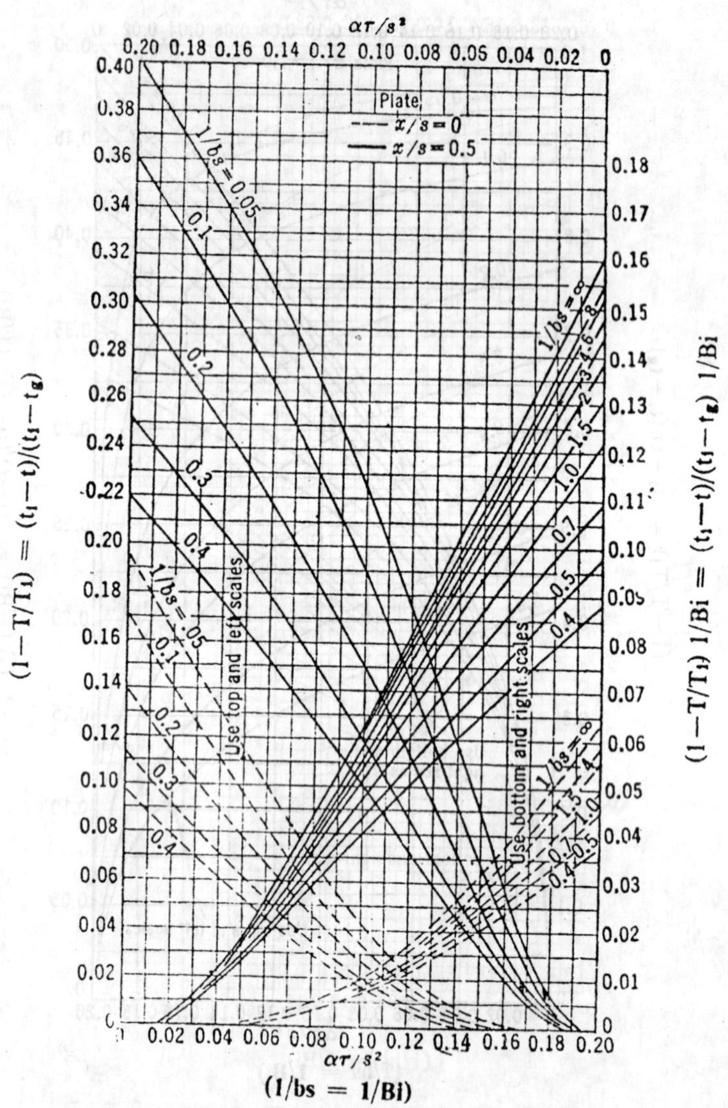

Fig. 3.15 Heisler Short Time Chart for Interior Temperature History in Infinite Plates

$$(1/Bi)_y = \frac{k_y}{hs_y} = \frac{7.5}{1200 \times 2.5/100} = 0.25$$

$$(Fo)_x = \frac{\alpha_x \tau}{s_x^2} = \frac{0.515 \times 3/3600}{(5/100)^2} = 0.168$$

$$(Fo)_y = \frac{\alpha_y \tau}{s_y^2} = \frac{0.0925 \times 3/3600}{(2.5/100)^2} = 0.123$$

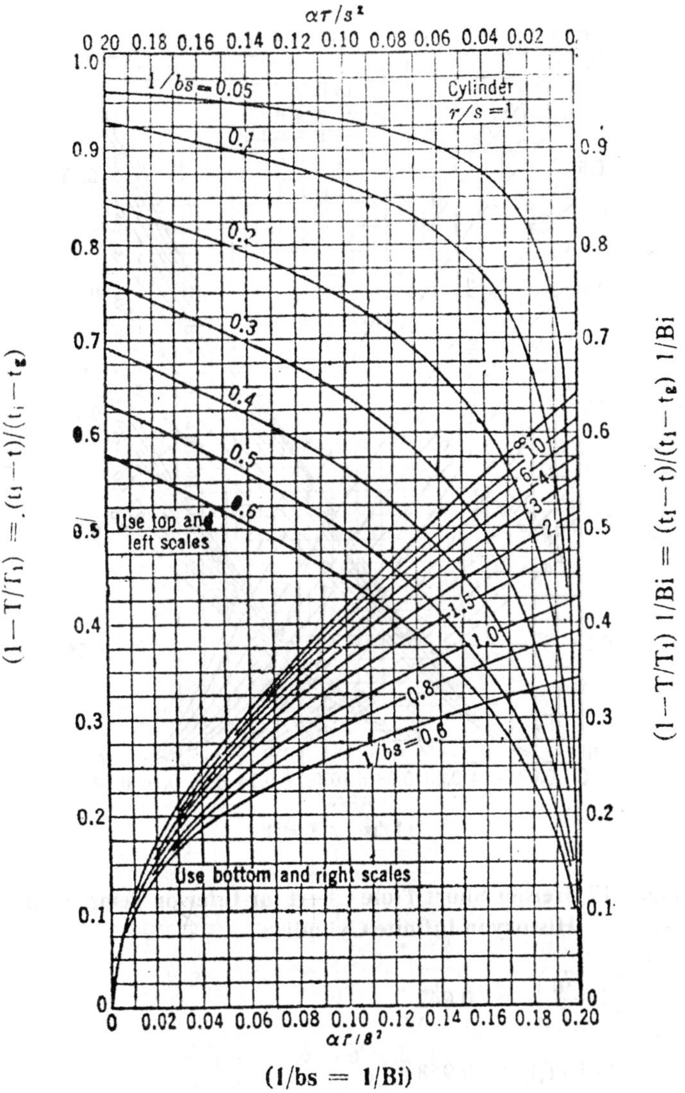

Fig. 3.16 Heisler Short Time Chart for Surface Temperature History in Cylinders

Here both values of Fo are less than 0.2, hence Fig. 3.15 will be used. The central temperature of the plate is given by dotted curves on the chart.

For $(1/Bi)_x = 0.5$ and $(Fo)_x = 0.168$;

$$\left(1-\frac{T_0}{T_i}\right)/Bi = 0.029$$

or $(T_0/T_i) = 1 - 0.029 \times (1/0.5) = 0.942$

and for $(1/Bi)_y = 0.25$ and $(Fo)_y = 0.123$;

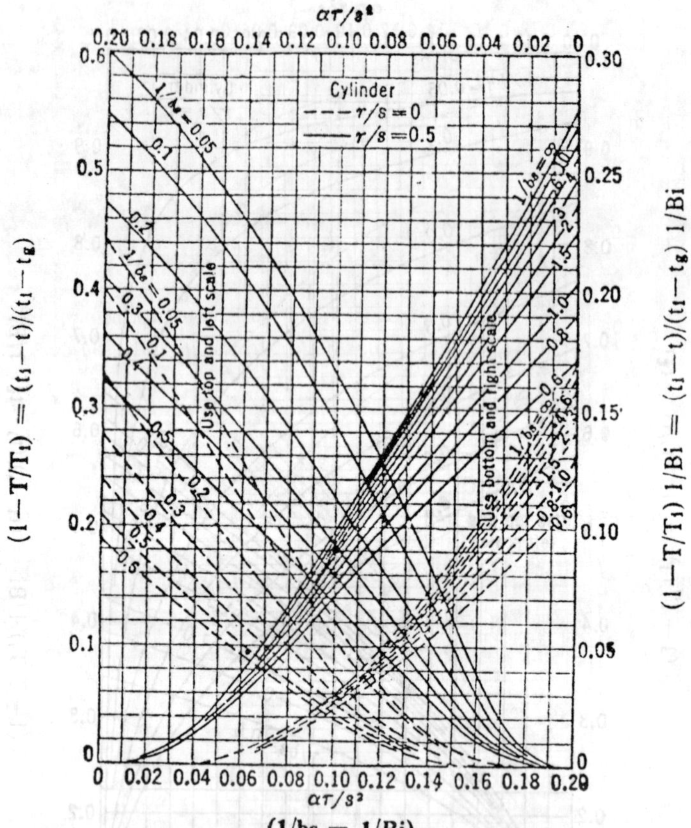

Fig. 3.17 Heisler Short Time Chart for Interior Temperature History in Infinite Cylinders

$$\left(1-\frac{T_0}{T_i}\right) = 0.042$$

or $\quad (T_0/T_i) = 0.958$

hence,

$$\frac{T_0}{T_i} = \frac{t_0-50}{250-50} = 0.942 \times 0.958 = 0.904$$

or $\quad t_0 = 231$ **Ans.**

(b) In order to evaluate $(t)_{x,y}$ we use the product solutions for x and y directions. Now using Fig. 3.14,

$$\left(1-\frac{T_0}{T_i}\right)_x = 0.53 \text{ or } (T_0/T_i)_{x=5} = 1 - 0.53 = 0.47$$

also $\quad (T_0/T_i)_{y=0} = 0.958$

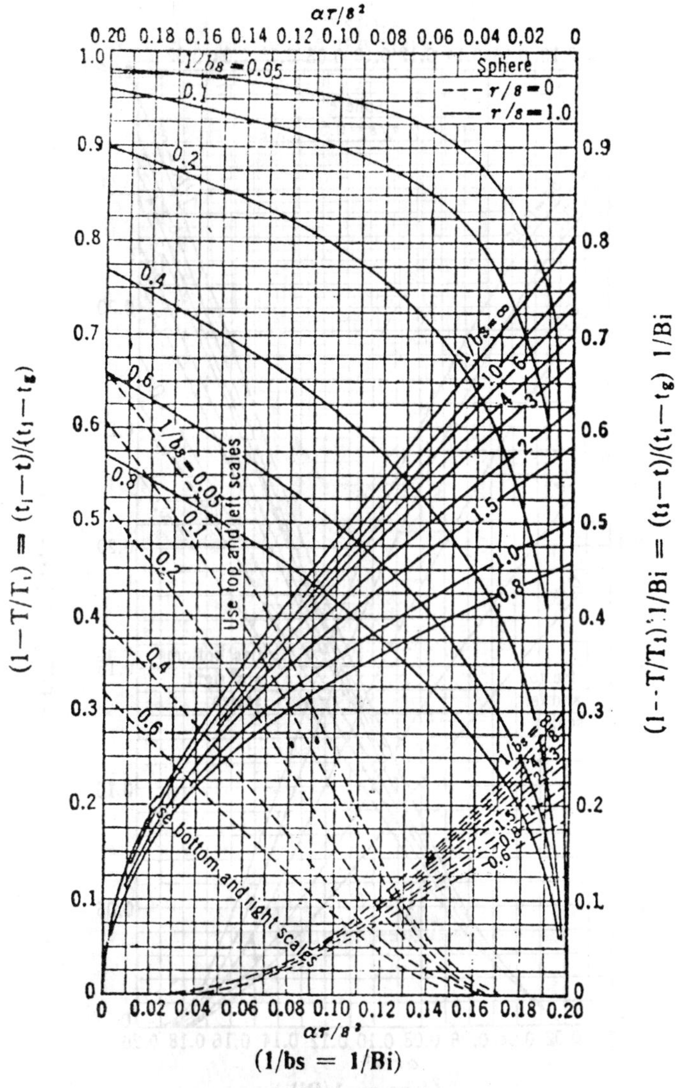

Fig. 3.18 Heisler Short Time Chart for Surface Temperature History in Spheres

hence $$\frac{(t)_{5,0}-50}{250-50} = 0.47 \times 0.958 = 0.45$$

or $$(t)_{5,0} = 140\text{ C}$$ Ans.

(c) By the same procedure and from the same chart we get ;

$$\left(1-\frac{T_0}{T_i}\right)_y = 0.67 \text{ or } \left(\frac{T_0}{T_i}\right)_{y=2.5} = 1 - 0.67 = 0.33$$

also $$\left(\frac{T_0}{T_i}\right)_{x=0} = 0.942$$

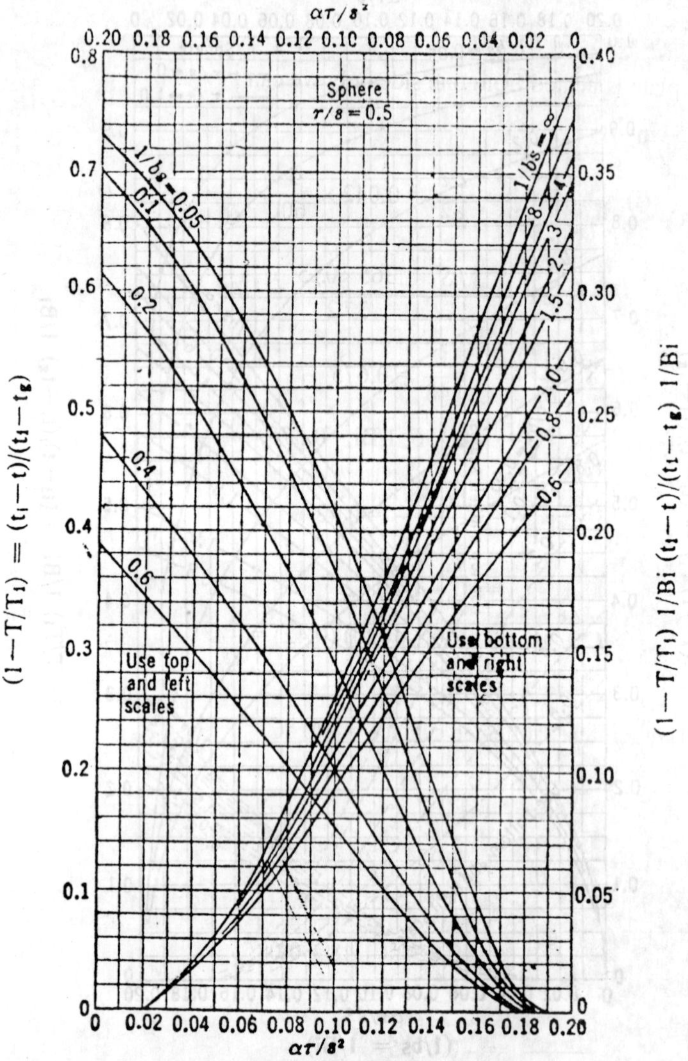

(1/bs = 1/Bi)

Fig. 3.19 Heisler Short Time Chart for Interior Temperature History in Spheres

hence $$\frac{(t)_{0,2.5}-50}{250-50} = 0.33 \times 0.942 = 0.312$$

or $$(t)_{0,2.5} = 112.5 \text{ C}$$ Ans.

Example 3.19

A 15 cm thick plate initially at 20 °C is suddenly put into a 1020 °C furnace and is heated from one side only. The values of k, α and h are 37.2 W/mK, 0.042 m²/hr, and 365 W/m²K. Calculate :

(a) Temperatures at the two surfaces and at center after 5 min of heating.

(b) Temperature at a distance of 9 cm from the heated surface, after 1 hr.

Solution :

Since the plate is heated from one side only we can treat the other surface as insulated.

Hence S = 15 cm

(a) (i) $$\text{Fo} = \frac{\alpha\tau}{S^2} = 0.042 \times \frac{5}{60} \times \frac{1}{(0.15)^2} = 0.156$$

$$\text{Bi} = \frac{hS}{k} = \frac{365 \times 0.15}{37.2} = 1.48$$

$$\frac{1}{Bi} = \frac{1}{1.48} = 0.675$$

From Fig. 3.14, for Fo = 0.156, 1/Bi = 0.675,

$$\left(1 - \frac{T}{T_i}\right) = \frac{t_i - t}{t_i - t_g} = 0.42$$

$$t_i - t = (t_i - t_g)\ 0.42$$

$$(20 - t) = (20 - 1020)\ (0.42) = -420$$

or $$t = 440\ °C$$

Hence, Temperature at heated surface = 440 °C Ans.

(ii) From Fig. (3.15), at x/S = 0,

$$\left(1 - \frac{T}{T_i}\right)\frac{1}{Bi} = 0.024$$

$$\left(\frac{T_i - T}{T_i}\right) = 0.024 \times 1.48 = 0.0355$$

$$\frac{t_i - t}{t_i - t_g} = 0.0355$$

or $$t = 55.5\ °C$$

Hence, temperature at the unheated surface = 55.5 °C. Ans.

(iii) From Fig. 3.15 at x/S = 0.5

$$\left(1 - \frac{T}{T_i}\right)\frac{1}{Bi} = 0.078$$

or $$\frac{t_i - t}{t_i - t_g} = 0.078 \times 1.48 = 0.16$$

$$t = 136\ °C$$

Hence, temperature at x/S = 0.5 is equal to 136 °C Ans.

(b) $$\text{Fo} = \frac{\alpha\tau}{S^2} = 0.042 \times 1 \times \frac{1}{(0.15)^2} = 1.8$$

At a distance of 9 cm from heated surface (where x = S)

$$x = S - 9 = 15 - 9 = 6 \text{ cm}$$

or $$\frac{x}{S} = \frac{6}{15} = 0.4$$

From Fig. 3.7, for Fo = 1.8, 1/Bi = 0.675,

$$\frac{T_0}{T_i} = \frac{t_0 - t_g}{t_i - t_g} = 0.2$$

or $$(t_0 - t_g) = 0.2 \times (-1000) = -200\ °C$$

From Fig. 3.10,

$$\frac{t_x - t_g}{t_0 - t_g} = 0.925$$

$$t_x - t_g = -(0.925 \times 200) = -185$$

$$t_x = 1020 - 185 = 835\ °C$$ Ans.

3.12 TRANSIENT HEAT CONDUCTION IN A SEMI-INFINITE PLATE WITH FINITE INTERNAL AND SURFACE RESISTANCE

In Section 3.5 we obtained the *error function* solution of the equation for transient heat conduction in a *semi-infinite* solid with zero surface resistance. Here we consider a similar case with the exception that the surface resistance $1/hA$, will also be taken into account. Thus the conduction equation;

$$\frac{\partial^2 t}{\partial x^2} = \frac{1}{\alpha}\frac{\partial t}{\partial \tau} \tag{3.2}$$

will be solved for the initial and boundary conditions as below ;

(i) at $\tau \le 0;\ t = t_i\ ;\ x \ge 0$

(ii) at $\tau > 0;\ \frac{\partial t}{\partial x} = \frac{h}{k}(t - t_a);\ x = 0$

Here the semi-infinite solid is suddenly exposed to an ambient atmosphere at temperature t_a and the heat transfer coefficient at the surface of solid is h.

The solution of Eq. (3.2) with these boundary conditions requires more advanced mathematical treatment, then envisaged for this text. The result may however be obtained as;

$$\frac{t - t_a}{t_i - t_a} = erf(u) + e^{[(hx/k)+(h^2/k^2)\alpha\tau]}\ \{erfc\,(u + h\sqrt{\alpha\tau}/k)\} \tag{3.73}$$

$$= erf(u) + \left\{ e^{Bi\sqrt{Fo}\,(2u + Bi\sqrt{Fo})} \right\} \times erfc\,(u + Bi\,\sqrt{Fo})$$

where
$$u = \frac{x}{2\sqrt{\alpha\tau}}$$

$$erfc = 1 - erf = \textit{Error function complementary};$$

$$Bi\sqrt{Fo} = \frac{h}{k}\sqrt{\alpha\tau} = \text{Boundary Fourier Modulus} \qquad (3.74)$$

The numerical evaluation of temperatures by using Eq. (3.73) is cumbersome; hence the solution is represented by charts as given in Fig. 3.20. These charts are known as *Boelter charts*. Here the dimensionless temperature ratio is plotted against local Biot number, hx/k for different values of Fourier modulus squared, $h^2\alpha\tau/k^2$. The use of this chart may be illustrated by the following example.

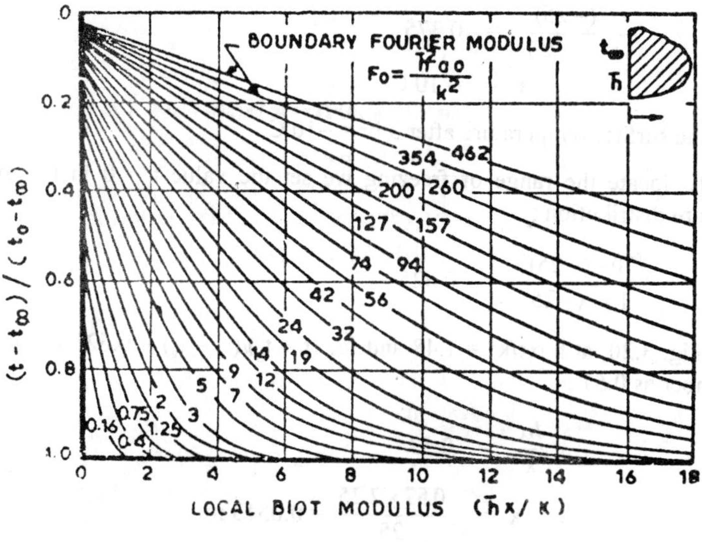

Fig. 3.20 Temperature Distribution in Semi-infinite Slab, h ≠ ∞

Example 3.20

During a winter day in Kashmir, a sudden cold wave reduced the ambient air temperature to –25 °C. The temperature of earth was uniform at 15 C before the onset of cold wave. Calculate (a) the surface temperature of earth after 3 hours (b) the depth in earth upto which freezing effect will be noted in this period. Assume; h = 25 W/m²K; k = 2.25 W/mK, and α = 0.004 m²/hr for wet soil.

Solution :

The basic difference in this problem and Example 3.10 is that here the heat flow is facing resistance through the surface film on the earth also. The heat in that case was flowing due to the conductivity of earth soil only and there was no resistance to flow of heat at the surface of pipes.

Treating the earth as a semi-infinite solid and neglecting the latent heat effects of freezing of the moisture in soil, the temperature distribution will be given by Eq. (3.73), as plotted in Fig. 3.20.

(a) Now,

$$\left\{\frac{h}{k}\sqrt{\alpha\tau}\right\}^2 = \left(\frac{25}{2.25}\right)^2 \times (0.004 \times 3) = 1.48$$

also at the surface ; $x = 0$; $Bi = hx/k = 0$

From Fig. 3.20; at $Bi = 0$ and $\frac{h^2 \alpha\tau}{k^2} = 1.48$

we find $$\frac{(t - t_g)}{(t_o - t_g)} = 0.375$$

or $$\frac{t-(-25)}{15-(-25)} = 0.375$$

or $$t = -10\ C$$

Thus the surface temperature after 3 hr = –10 C Ans.

(b) To locate the range of freezing we set the value t = 0 in the expression for temperature distribution ;

$$\frac{0-(-25)}{15-(-25)} = 0.625$$

so from Fig. 3.20, at $h^2\alpha\tau/k^2 = 1.48$ and for $(t - t_g)/(t_o - t_g) = 0.625$ we get the value of Biot number as 0.57 ;

$$\frac{hx}{k} = 0.53$$

or $$x = \frac{0.57 \times 2.25}{25} = 0.0513\ m$$

Thus the freezing temperature will have penetrated upto a depth of 5.13 cm after 3 hours. Ans.

3.13 GRAPHICAL METHODS OF SOLUTION OF TRANSIENT HEAT CONDUCTION PROBLEMS

Another simple and flexible method of solving unsteady state heat conduction problems is a graphical method due to *Schmidt*. This technique, known as *Schmidt plot*, is a graphical representation of the finite difference method applied to transient heat flow, on the lines as discussed in Chapter 2 for steady state heat conduction. Thus, it requires only the simple arithmetic and *graphical averaging*. Another advantage of the technique is that it can be extended without difficulty to problems where surrounding fluid temperature varies with time, the solution of which is extremely difficult otherwise. Further, it has now been established that the method may be used for cylindrical and spherical solids as well. The principle of the graphical method is discussed in what follows.

Consider a one-dimensional heat flow problem and assume the system network similar to the one as given by Fig. 2.16. Let the $t - x$ curve be represented by Fig. 3.21. The solid wall is divided into a number of parallel layers of thickness, Δx. The dividing planes have been labelled 1, 2, 3, ... n. The centre of the layers were called *nodal points* in Fig. 2.21. We may, for analogy, consider these planes to be passing through nodal points of layers of thickness Δx, as shown by broken vertical lines in the Fig. 3.21.

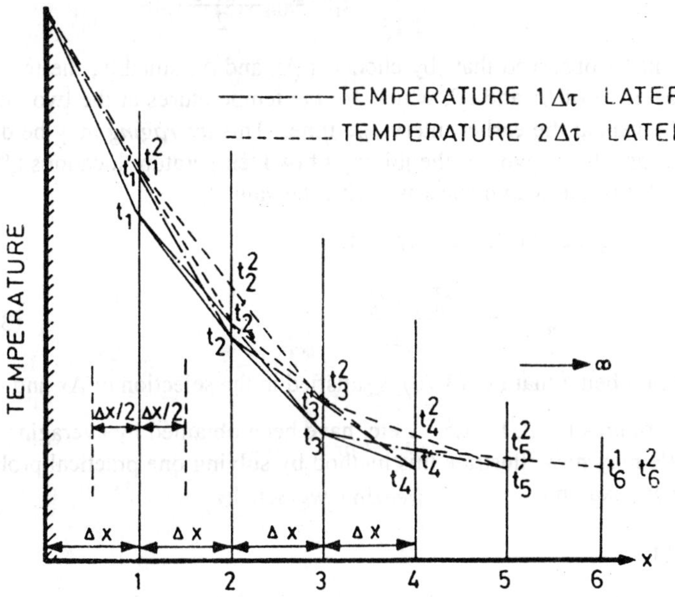

Fig. 3.21 Principle of Schmidt Plot

In the temperatures indicated in Fig. 3.21, the subscript refers to plane number while the superscript refers to time interval Δτ. For instance, t_3^2, refers to temperature at plane 3 after time 2 Δτ.

Let $t_s°, t_1°, t_2°, \ldots t_n°$ be the initial temperatures at various planes at $\tau = 0$. During the first time interval Δτ heat will flow across the plane 1, (nodal plane of layer extending Δx/2 to the left and right);

$$\begin{Bmatrix}\text{Heat flow}\\ \text{towards plane 1}\end{Bmatrix} - \begin{Bmatrix}\text{Heat flow from}\\ \text{plane 1 to plane 2}\end{Bmatrix} = \begin{Bmatrix}\text{Change in internal}\\ \text{energy of the layer}\end{Bmatrix}$$

or $$q_{s-1} - q_{1-2} = \Delta U$$

or $$K_{s-1}\,(t_s° - t_1°)\,\Delta\tau - K_{1-2}(t_1° - t_2°)\,\Delta\tau = c\rho\,\Delta x\,(t_1^1 - t_1°)$$

Dividing by 2 K Δτ; where $K = K_{s-1} = K_{1-2} = (k/\Delta x)$

$$\tfrac{1}{2}\{(t_s° - t_1°) - (t_1° - t_2°)\} = \frac{c\rho\,\Delta x}{2K\Delta\tau}\,(t_1^1 - t_1°)$$

or $$\frac{t_s° + t_2°}{2} - t_1° = (t_1^1 - t_1°)\,\frac{c\rho(\Delta x)^2}{2k\Delta\tau} \qquad (3.75)$$

If we choose for convenience, Δx and Δτ in a way that ;

$$\frac{c\rho(\Delta x)^2}{2\,k\,\Delta\tau} = \frac{(\Delta x)^2}{2\alpha\,\Delta\tau} = 1 \tag{3.76}$$

The Eq. (3.75) reduces to ;

$$t_1^{\,1} = \frac{t_s^{\,o} + t_2^{\,o}}{2} \tag{3.77}$$

Thus it will be observed that by choosing Δx and Δτ suitably, the temperature $t_1^{\,1}$ is simply the average of $t_s^{\,o}$ and $t_2^{\,o}$, which are the temperatures at the two nodal planes of the adjacent layers at the earlier interval to time. This *averaging* may be done easily by graphical means which involves the joining of two temperature locations $t_s^{\,o}$ and $t_2^{\,o}$. The intersection of this line with plane 1 will give the value $t_1^{\,1}$.

Likewise, as a general rule we may write ;

$$t_n^{\tau+1} = \frac{t_{n-1}^{\tau} + t_{n+1}^{\tau}}{2} \tag{3.78}$$

The condition being that Eq. (3.76) is satisfied in the selection of Δx and Δτ.

The temperatures $t_2^{\,1}, t_3^{\,1}, t_4^{\,1}, t_2^{\,2}$, ... etc. have been obtained by averaging on the above principle. We may now illustrate this method by solving one practical problem as given in the following example.

Example 3.21

A steel plate (k = 15 W/mK), 30 cm thick, is required to be heat-treated in a furnace at 3000 C, where it is suddenly placed when its temperature is 100 C. The unit surface conductance may be assumed as 120 W/m²K. Determine the time required to reach 700 C at the centre (α = 0.00445 m²/h).

Solution :

Let Δx = 2.5 cm. Thus there will be 12 layers each having a thickness of 2.5 cm. The surface resistance may be represented by an extension of the plate thickness by ;

$$\frac{k}{h} = \frac{15}{120} \times 100 = 12.5 \text{ cm}$$

With these values the graphical solution of the problem is given by Fig. 3.22, which shows the layers of the plate between centre line and surface. The following steps have been taken to draw the graphical plots :

(a) Join the 3000 C point at outer film with 100 C at plane 1. The intersection point t_s represents the temperature at surface of the plane at τ = 0.

(b) Similarly points t_1, t_2, t_3, t_4, t_5 and t_6 are obtained.

(c) The same cycle of plots is repeated by joining 3000 C point with t_1 and the point of intersection with t_2 and so on.

(d) It will be observed from Fig. 3.22 that the number of Δτ which have elapsed, till the temperature at the centre of plane (at plane 6) is 700 C, is 28. The integers written on the individual lines indicate this number of Δτ.

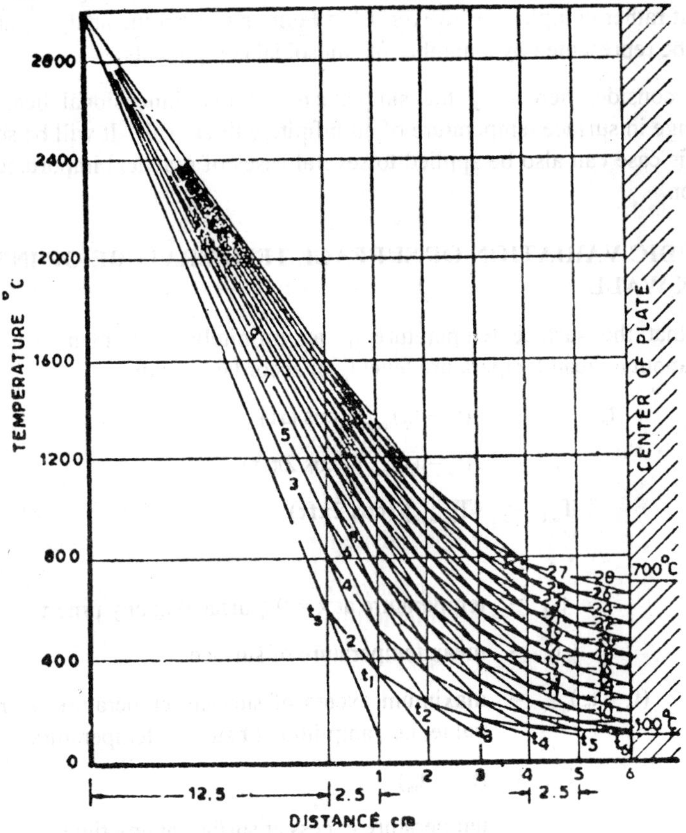

Fig. 3.22 Example 3.21

Thus the central line of plate will reach 700 C after 28 Δτ hrs.

Now from Eq. (3.76);

$$\frac{(\Delta x)^2}{2\alpha(\Delta \tau)} = 1$$

or

$$\Delta\tau = \frac{(\Delta x^2)}{2\alpha} = \frac{(2.5/100)^2}{2 \times 0.00445}$$

$$= 0.007 \text{ hrs}$$

$$\text{Hence total time} = (0.007 \times 60) \times 28 = 12 \text{ min}$$ Ans.

3.' ' PERIODIC HEAT FLOW

Both in nature as well as in engineering applications *periodic changes* of temperature are frequently encountered. For example, the surface temperature of the earth undergoes a periodic variation every day, on account of the daily cycle of the sun. Similarly, in the engineering field, periodic temperature changes occur in internal combustion engines, regenerators of blast furnaces, etc. In general, these periodic changes are not simply sinusoidal but rather complex. However using Fourier's harmonic analysis, any periodic changes can be represented by a number of sinusoidal components.

We shall consider here only the simple case of one-dimensional heat flow with periodic change in surface temperature of an infinitely thick wall. It will be seen that the results for this case can also be applied to several cases of practical importance with fair approximation.

3.15 PERIODIC VARIATION OF SURFACE TEMPERATURE OF INFINITELY THICK WALL

Suppose that the surface temperature of an infinitely thick plane wall oscillates periodically about a mean temperature level t_m in accordance with ;

$$t_{o,\tau} - t_m = \{(t_o - t_m)_{max}\} (\cos \omega\tau)$$

$$= \{(t_o - t_m)_{max}\} (\cos 2\pi f\tau) \quad (3.79)$$

or

$$T_{o,\tau} = \{T_{o,\,max}\} (\cos 2\pi f\tau) \quad (3.80)$$

where

$t_{o,\tau}$ = temperature at x = 0 (surface) at any time τ

t_m = mean temperature of surface

$(t_o - t_m)_{max}$ = maximum excess of surface temperature over the mean value, i.e., amplitude of surface temperature variation

$T_{o,\tau}$ = $(t_{o,\tau} - t_m)$

= temperature excess at surface at any time τ

$T_{o,max}$ = $(t_o - t_m)_{max}$

= maximum temperature excess at surface

ω = $2\pi f$

f = frequency of temperature wave, ie, number of complete changes per unit time

Our problem is to find the temperature at any depth x at any time τ, that is, $t_{x,\tau}$. Since t_m is a constant (the value of which is known) this problem can be equally well solved by finding $T_{x,\tau}$. The governing differential equation is ;

$$\frac{\partial t}{\partial \tau} = \alpha \frac{\partial^2 t}{\partial x^2} \quad (3.2)$$

or

$$\frac{\partial^2 (t - t_m)}{\partial \tau} = \alpha \frac{\partial^2 (t - t_m)}{\partial x^2} \quad \text{(since } t_m \text{ is constant)}$$

or

$$\frac{\partial T}{\partial \tau} = \alpha \frac{\partial^2 T}{\partial x^2} \quad (3.81)$$

The boundary conditions are

$$(i)\ \left.\begin{array}{l} at\ \tau = 0 \\ x = 0 \end{array}\right\}; \qquad T_{0,\tau} = T_{0,max}$$

$$(ii)\ \left.\begin{array}{l} at\ \tau = \tau \\ x = 0 \end{array}\right\}; \qquad T_{0,\tau} = (T_{0,max})(\cos \omega\tau)$$

(a) Let us assume that the solution of Eq. (3.81) is of the from ;

$$T_{x,\tau} = \{T_{0,max}\}\ e^{-mx} \cos(\omega\tau - mx) \tag{3.82}$$

It can be easily seen that Eq. (3.82) satisfies the boundary conditions. However, to verify whether Eq. (3.82) satisfies Eq. (3.81), let us differentiate Eq. (3.82) partially with respect to τ and x. From Eq. (3.82),

$$\frac{\partial T}{\partial \tau} = (T_{0,max})\ e^{-mx} (-\omega) \sin(\omega\tau - mx)$$

$$= -\omega\ (T_{0,max})\ e^{-mx} \sin(\omega\tau - mx) \tag{3.83}$$

$$\frac{\partial T}{\partial x} = -m\ (T_{0,max})\ e^{-mx} \cos(\omega\tau - mx) + m\ (T_{0,max})\ e^{-mx} \sin(\omega\tau - mx)$$

and

$$\frac{\partial^2 T}{\partial x^2} = m^2\ (T_{0,max})\ e^{-mx} \cos(\omega\tau - mx) - m^2\ (T_{0,max})\ e^{-mx} \times \sin(\omega\tau - mx) - m^2\ (T_{0,max})\ e^{-mx} \sin(\omega\tau - mx) - m^2\ (T_{0,max})\ e^{-mx} \cos(\omega\tau - mx)$$

$$= -2m^2\ (T_{0,max})\ e^{-mx} \sin(\omega\tau - mx) \tag{3.84}$$

Substituting Eqs. (3.83) and (3.84) in Eq. (3.81)

$$-\omega\ (T_{0,max})\ e^{-mx} \sin(\omega\tau - mx) = -2\alpha m^2\ \{T_{0,max}\}\ e^{-mx} \times \sin(\omega\tau - mx)$$

or

$$\omega = 2\alpha\, m^2 \tag{3.85}$$

Thus Eq. (3.82) satisfies Eq. (3.81) provided that ;

$$m^2 = \frac{\omega}{2\alpha}$$

or

$$m = +\sqrt{\frac{\omega}{2\alpha}} \tag{3.86}$$

Note that $m = -\sqrt{(\omega/2\alpha)}$ is rejected because this will make the amplitude $T_{x,max}$ at depth x, infinite at large values of x. This is evidently not possible. Hence, the complete solution of Eq. (3.81) is ;

$$T_{x,\tau} = (T_{0,max})\ e^{-\sqrt{(\omega/2\alpha)}\,x} \cos\left(\omega\tau - \sqrt{\frac{\omega}{2\alpha}}\,x\right) \tag{3.87}$$

The temperature variation with time at depth x, given by Eq. (3.87) is shown in Fig. 3.23, while Fig. 3.24 shows the temperature distribution as function of depth at $\omega\tau = 0$ and at $\omega\tau = \pi$.

Several important conclusions can be drawn from Eq. (3.87);

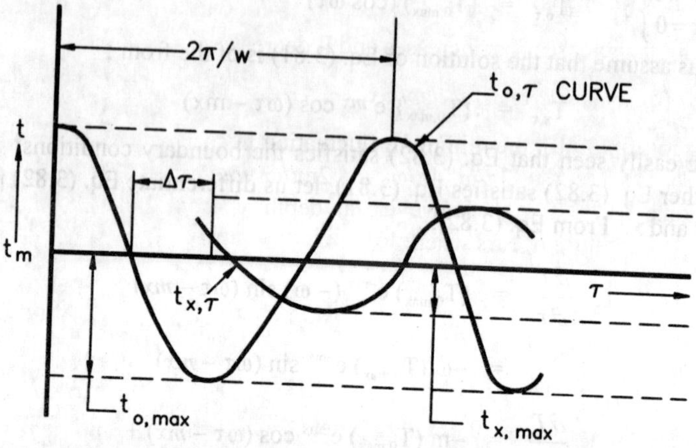

Fig. 3.23 Amplitude of the Temperature Wave

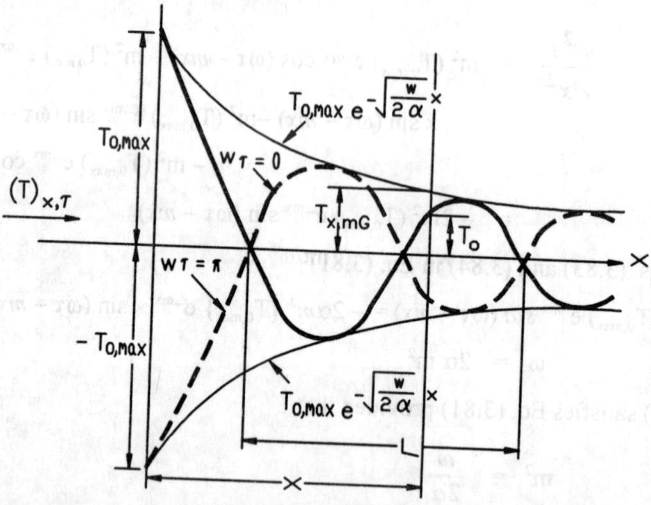

Fig. 3.24 Temperature Distribution in Periodic Heat Flow

(i) Amplitude at depth x

The maximum temperature excess or amplitude at depth x, is obtained by putting

$$\cos\left(\omega\tau - \sqrt{\frac{\omega}{2\alpha}}\,x\right) = 1.$$

Thus the amplitude at depth x is ;

$$T_{x,max} = (T_{o,max})\, e^{-\sqrt{(\omega/2\alpha)}\,x} \tag{3.88}$$

Equation (3.88) shows that amplitude of temperature variation decreases exponentially with increasing depth as illustrated by Fig. 3.24. In other words, the maximum fluctuation in temperature is smaller at distance x than at the surface. Hence the amplitude may become negligible at a relatively small depth. Then it makes no difference whether the wall is thicker than this depth, and Eq. (3.87) can be used for such a relatively thin wall.

Note that $T_{x,max}$ increases if α increases and it decreases if ω increases.

(ii) Time lag or phase difference at depth x

According to Eq. (3.80), the minimum fluctuation in surface temperature occurs when $\cos(\omega\tau) = 1$ or when $\tau = 0, 2\pi/\omega, 4\pi/\omega, \ldots, 2n\pi/\omega$ where $n = 0, 1, 2, 3, \ldots$ etc. Under these conditions $T_{o,\tau} = T_{o,max}$. However, at depth x, from Eq. (3.87) the maximum fluctuation in temperature occurs when ;

$$\cos\left(\omega\tau - \sqrt{\frac{\omega}{2\alpha}}\,x\right) = 1$$

or

$$\omega\tau - \sqrt{\frac{\omega}{2\alpha}}\,x = 0, 2\pi, 4\pi, \ldots, \text{etc.}$$

$$= 2n\pi$$

or

$$\tau = \sqrt{\frac{1}{2\alpha\omega}}\,x + \frac{2n\pi}{\omega} \qquad (3.89)$$

If

$$n = 0; \quad \tau = \sqrt{\frac{1}{2\alpha\omega}}\,x$$

Thus, while the temperature excess at the surface is maximum when $\tau = 2n\pi/\omega$ it is maximum at depth x at a *later* time when $\tau = (2n\pi/\omega) + \sqrt{(1/2\alpha\omega)}\,x$. The delay or time lag, $\Delta\tau$, therefore is ;

$$\Delta\tau = \frac{2n\pi}{\omega} + \sqrt{\frac{1}{2\alpha\omega}}\,x - \frac{2n\pi}{\omega}$$

or

$$\Delta\tau = \sqrt{\frac{1}{2\alpha\omega}}\,x \qquad (3.90)$$

Note that the time lag increases with increasing x, and decreases with increasing α and ω. It is because of this time lag that the inside surface of the walls of a building attain the maximum temperature at about 6 p.m. whereas the outside surface exposed to the sun is hottest at about 1 p.m.

(iii) Form and frequency of wave at depth x

Equation (3.88) shows that the form of the wave does not change with depth; it remains a cosine wave. Also, from Eq. (3.89), it is seen that the time interval between two successive maxima is $(2\pi/\omega)$ which is the same as at the surface. Thus the periodic time $(2\pi/\omega)$ and frequency of the wave $(\omega/2\pi)$ do not change with depth.

(iv) Wave length and velocity of wave

Recall that the wavelength of a wave is the distance between two adjacent crests. Now, at x = 0 the maximum occurs at $\tau = 0$. The next maximum at x = 0 occurs when $\tau = \tau_c = 2\pi/\omega$ (Fig. 3.23). But at time equal τ_c a maximum also occurs at depth x given by ;

$$\cos\left(\omega\tau_c - \sqrt{\frac{\omega}{2\alpha}}\,x\right) = 1$$

or
$$\omega\tau_c - \sqrt{\frac{\omega}{2\alpha}}\,x = 0$$

or
$$x = \omega\tau_c \div \sqrt{\frac{\omega}{2\alpha}}$$

$$= 2\pi\sqrt{\frac{2\alpha}{\omega}}$$

The distance x given by the last equation is therefore the *wavelength* l of the temperature wave.

$$l = 2\pi\sqrt{\frac{2\alpha}{\omega}} \tag{3.91}$$

Since the velocity of a wave is the product of frequency and wavelength, the velocity V of the temperature wave is given by ;

$$V = fl = \frac{l}{\tau_c} = \frac{l\omega}{2\pi}$$

$$= 2\pi\sqrt{\frac{2\alpha}{\omega}}\left(\frac{\omega}{2\pi}\right)$$

or
$$V = \sqrt{2\alpha\omega} \tag{3.92}$$

Note that the velocity with which the temperature wave penetrates the surface depends both on ω and α. The distance travelled by the wave during the time interval τ_1 to τ_2 is given by $V(\tau_2 - \tau_1) = (\tau_2 - \tau_1)\sqrt{2\alpha\omega}$.

Fig. 3.25 shows the temperature distribution in the wall at two instants τ_1 and τ_2 where $\tau_2 > \tau_1$. By these curves the moving of the temperature wave into the wall can be seen. The displacement of the wave in time $(\tau_2 - \tau_1)$ is also shown. The upper and lower boundary curves in Fig. 3.25 represent $(T_{x,max}) = (T_{0,max})\, e^{-mx}$ and $-(T_{x,max}) = -(T_{o,max})\, e^{-mx}$ respectively. It is evident that the temperature distribution for each time, τ = constant, will touch the $(T_{x,max})$ curve when the *cosine* in Eq. (3.87) is equal to 1 and will touch the $-(T_{x,max})$ curve when it equals −1. All other points on the τ = const. curves will be between these two limiting lines. Hence the upper and lower boundary curves form *envelopes* of the system of curves τ = constant.

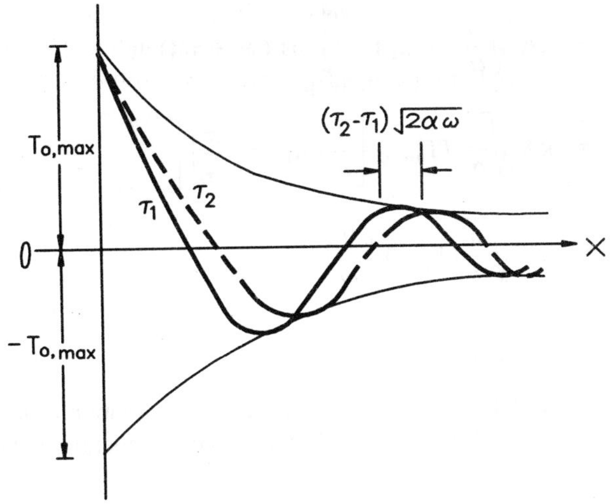

Fig. 3.25 Displacement Curve of Temperature Wave

(v) Heat flow rate and energy storage

Besides the temperature distribution, we are also interested in the rate of heat flow into or out of the plate at any instant and the energy stored in the wall every half cycle. Since the rate at which heat enters the wall is ;

$$q = -kA\left(\frac{\partial t}{\partial x}\right)_{x=0} = -kA\left(\frac{\partial T}{\partial x}\right)_{x=0}$$

we get from Eq. (3.82)

$$q = -kAm\,(T_{0,max})\,e^{-mx}\,[\sin(\omega\tau - mx) - \cos(\omega\tau - mx)]_{x=0}$$

$$= -kAm\,(T_{0,max})\,(\cos\omega\tau - \sin\omega\tau)$$

$$= kA\sqrt{\frac{\omega}{2\alpha}}\,(T_{0,max})\,(\cos\omega\tau - \sin\omega\tau)$$

or
$$q = kA\sqrt{\frac{\omega}{\alpha}}\,(T_{0,max})\left(\frac{\cos\omega\tau}{\sqrt{2}} - \frac{\sin\omega\tau}{\sqrt{2}}\right)$$

so
$$q = kA\sqrt{\frac{\omega}{\alpha}}\,(T_{0,max})\cos\left(\omega\tau + \frac{\pi}{4}\right) \tag{3.93}$$

Note that q is positive in the limits $(\omega\tau + \pi/4) = -\pi/2$ to $+\pi/2$ and negative in the limits $(\omega\tau + \pi/4) = \pi/2$ to $-\pi/2$. In other words, q is positive between the limits ;

$$\tau = \frac{1}{\omega}\left[-\frac{\pi}{2} - \frac{\pi}{4}\right] = -\frac{3\pi}{4\omega}$$

and
$$\tau = \frac{1}{\omega}\left[\frac{\pi}{2} - \frac{\pi}{4}\right] = +\frac{\pi}{4\omega}$$

Integrating Eq. (3.93) between the above limits gives the heat energy stored, Q, in half a cycle.

$$Q = kA\sqrt{\frac{\omega}{\alpha}}\,(T_{0,max})\int_{-3\pi/4\omega}^{\pi/4\omega}\cos(\omega\tau + \pi/4)\,d\tau$$

$$= kA\sqrt{\frac{\omega}{\alpha}}\,(T_{0,max})\left[\frac{1}{\omega}\sin\left(\omega\tau + \frac{\pi}{4}\right)\right]_{3\pi/4\omega}^{\pi/4\omega}$$

$$= kA\sqrt{\frac{\omega}{\alpha}}\,(T_{0,max})\left(\frac{2}{\omega}\right)$$

or

$$Q = kA\frac{2}{\sqrt{\omega\alpha}}(T_{0,max})$$

(b) Note that we had considered in Eq. (3.82) a cosine variation of temperature with time and obtained result as Eq. (3.87). If the surface temperature varies as a sine function, i.e.,

$$T_{0,\tau} = (T_{0,max})\sin(\omega\tau) \tag{3.95}$$

it can be shown, by following an exactly similar procedure that ;

$$T_{x,\tau} = (T_{0,max})\,e^{-\sqrt{(\omega/2\alpha)}\,x}\sin\left(\omega\tau - \sqrt{\frac{\omega}{2\alpha}}\,x\right) \tag{3.96}$$

Moreover, the time lag $\Delta\tau$, the amplitude at depth x, $T_{x,max}$, the wavelength l, the velocity V, and the heat storage in half-cycle Q, are all given by the same equations which were derived above for the cosine variation. Only in Eq. (3.94) for q we have to replace *cos* $(\omega\tau - \pi/4)$ by *sin* $(\omega\tau + \pi/4)$.

Example 3.22

The temperature of the earth's surface in a given locality varies as a sine wave function of time from –20 C to 20 C every 24 hours. If k = 0.63 W/m K, C_p = 1.88 kJ/kg K and ρ = 1600 kg/m³, find;

(a) Amplitude of temperature variation at a depth of 0.3 m.

(b) Time lag of the temperature wave at a depth of 0.3 m.

(c) Temperature at a depth of 0.3 m, 5 hours after the surface temperature reaches a minimum value.

(d) Heat flowing into 1 m² of surface every half-cycle.

Solution :

$$\omega = 2\pi f$$

$$f = \frac{1}{24}$$

so

$$\omega = 2\pi \times \frac{1}{24} = 0.262 \text{ rad/hr}$$

$$t_m = \frac{-20+20}{2} = 0\ C$$

$$T_{o,max} = 20 - 0 = 20\ C$$

$$\alpha = \frac{k}{\rho C_p} = \frac{0.63 \times 3600/1000}{1600 \times 1.88} = \frac{1}{1600}\ m^2/hr$$

(a) From Eq. (3.88),

$$T_{x,max} = (T_{o,max})\ e^{-\sqrt{(\omega/2\alpha)}\,x}$$

$$= (20)\,e^{-\sqrt{(0.262 \times 1600/2)}\,(0.3)}$$

$$= (20)\ e^{-4.35}$$

$$= 0.256\ C$$ Ans.

(b) From Eq. (3.90)

$$\Delta\tau = \sqrt{\frac{1}{2\alpha\omega}}\,x$$

$$= \sqrt{\frac{1600}{2 \times 0.262}}\,(0.3)$$

$$= 16.6\ hr$$ Ans.

(c) From Eq. (3.95), the surface temperature is minimum when *sin* $(\omega\tau) = -1$. This occurs for the first time when $(\omega\tau) = 3\pi/2$ or when;

$$\tau = \frac{3\pi}{2\omega} = \frac{3\pi}{2(2\pi/24)} = 18\ hr$$

Five hours later ; $\tau = 18 + 5 = 23$. At this time,

$$T_{x,\tau} = (T_{o,max})\ e^{-\sqrt{(\omega/2\alpha)}\,x}\ \sin\left[\omega\tau - \sqrt{\frac{\omega\alpha}{2}}\,x\right]$$

$$= (20)\ e^{-4.35}\ \sin\,[0.262 \times 23 - 4.35]$$

$$= 0.256\ \sin 95.5°$$

$$= 0.256 \times 0.995 = 0.25\ C$$

or $$t_{x,\tau} = T_{x,\tau} + t_m$$

$$= 0.25\ C$$ Ans.

(d) From Eq. (3.94).

$$\frac{Q}{A} = \frac{2}{\sqrt{\alpha\omega}}\,k\,(T_{o,max})$$

$$= \frac{2}{\sqrt{0.262/1600}} \times \frac{0.63 \times 3600}{1000} \times 20$$

$$= 5{,}900\ kJ$$ Ans.

3.16 MORE ILLUSTRATIVE EXAMPLES

Example 3.23

A copper cylinder, 4 cm diameter and 25 cm long, initially at 60°C is immersed in a bath at 0°C. The unit surface conductance between the metal and the bath is 540 W/m²K. Determine (i) the temperature of the cylinder after 1.5 minutes and (ii) the total amount of heat transferred to the cylinder during this time. For copper, k = 83 W/mK, ρ = 8666 kg/m³, C_p = 0.41 kJ/kg K.

Solution :

(i) Let us first calculate Bi to decide whether the internal thermal resistance may be neglected

$$\text{Bi} = \frac{hV}{kA}$$

$$V = \frac{\pi}{4}(0.04)^2(0.25) = 3.14 \times 10^{-4} \text{ m}^3$$

$$A = \pi(0.04)(0.25) + (2)(\pi/4)(0.04)^2 = 0.033912 \text{ m}^2$$

so
$$\text{Bi} = (540 \times 3.14 \times 10^{-4})/(83 \times 0.033912) = 0.0602$$

Since Bi < 0.1, the internal thermal resistance may be neglected.

Now,
$$\frac{t - t_a}{t_i - t_a} = e^{-(hA/c\rho V)\tau} \qquad (3.5)$$

$$\frac{hA\tau}{c\rho V} = \frac{(540)(0.033912)(1.5 \times 60)}{410 \times 8666 \times 3.14 \times 10^{-4}} = 1.4773$$

so
$$\frac{t - t_a}{t_i - t_a} = e^{-1.4773} = 0.228$$

Thus
$$\frac{t - 0}{60 - 0} = 0.228$$

or
$$t = 13.68°\text{C}$$
Ans.

(ii) The total amount of heat transferred is ;

$$Q = c\rho V(t_i - t_a)(1 - exp[-(h A/c\rho V)\tau]) \qquad (3.13)$$

$$= (410)(8666)(3.14 \times 10^{-4})(60 - 0) \times (1 - exp[-1.4773])$$

$$= 51677.4 \text{ W}$$
Ans.

Example 3.24

A plastic sheet 25.4 mm thick, initially at 18°C, is placed between two heated steel plates which are maintained at 138°C. Calculate the time required for the midplane of the plastic to reach 126°C. The thermal conductivity and thermal diffusivity of the plastic sheet are 0.1593 W/mK and 2.6942 × 10⁻⁴ m²/hr, respectively.

Solution :

We may assume perfect contact between the plastic sheet and heated plates and neglect surface resistance.

Now, $$\frac{t_o - t_s}{t_i - t_s} = \frac{126-138}{18-138} = \frac{12}{120} = 0.1$$

From curve 1 Fig. 3.4,

$$\frac{\alpha\tau}{S^2} = 1.05; \text{ since } S = (25.4/2) \times 10^{-3} \text{ m}$$

$$\tau = \frac{(1.05)(S^2)}{\alpha} = \frac{(1.05)[(25.4/2)\times 10^{-3}]^2}{(2.6942\times 10^{-4})}$$

$$= 0.629 \text{ hr}$$ Ans.

Example 3.25

An egg (5 cm diameter) is initially at 5°C. It is immersed in boiling water at 100°C for 15 minutes. Calculate the temperature at the egg centre at the end of this period. The heat transfer coefficient from the water to the egg is 5670 W/m²K. The physical properties of an egg may be assumed to be as follows : $\rho = 990$ kg/m³ ; $\alpha = 1.55 \times 10^{-7}$ m²/s, k = 0.64 W/mK.

Solution :

This is a problem of unsteady state conduction with finite surface and internal resistance. Hence we may use Fig. 3.9.

$$\frac{1}{Bi} = \frac{k}{hS} = \frac{0.64}{5670\times(5/2)\times 10^{-2}} = 0.0045$$

$$\alpha\tau/S^2 = (1.55 \times 10^{-7})(15 \times 60)/[2.5^2 (10^{-4})] = 0.223,$$

From Fig. 3.9, for (1/Bi) $\approx$ 0 and $(\alpha\tau/S^2) = 0.223$,

$$\frac{T_o}{T_i} = \frac{t_o - t_g}{t_i - t_g} = 0.22$$

or, $$\frac{t_o - 100}{5 - 100} = 0.22$$

Thus, $t_o = 79.1$°C Ans.

Since the surface resistance is negligible, we may also use curve 6 of Fig. 3.4. This gives the same result.

Example 3.26

An apple, taken out from a refrigerator, has a temperature of 5°C. Determine the time it would take the centre of the apple to reach 20°C under the following conditions : Apple diameter = 10 cm, apple density = 980 kg/m³, specific heat of apple = 2.5 kJ/kg K, apple thermal conductivity = 0.45 W/mK. Ambient air temperature = 25°C. Convective heat transfer coefficient = 5 W/m²K.

Solution :

Again, this is a problem of unsteady state conduction with finite surface and internal resistance.

$$\frac{1}{Bi} = \frac{k}{hS} = \frac{0.45}{(5)\times(0.1/2)} = 1.8$$

$$\frac{T_o}{T_i} = \frac{t_o - t_g}{t_i - t_g} = \frac{20-25}{5-25} = 0.25$$

From Fig. 3.9 for the above values,

$$\frac{\alpha\tau}{S^2} = 1.04$$

$$\tau = (1.04)\frac{S^2}{\alpha} = \frac{(1.04)\times(0.05)^2}{(0.45)/(980\times 2500)} = 14{,}155.6 \text{ s}$$

$$= 3.932 \text{ hr}$$ Ans.

PROBLEMS

3.1 The distribution of the temperature t across a large concrete slab 30 m thick is given by : $t = 20 + 15x - 2x^2$. Compute (a) the heat entering and leaving each face (b) the heat gained or lost by the slab in unit time and (c) the temperature change per unit time at the centre of the slab. ρ = 2180 kg/m³, C_p = 0.84 kJ/kg C, k = 0.7 W/m K.

[Ans. (a) 262 kJ/hr m² (b) 300 kJ/hr m² (c) 0.0054 C/hr]

3.2 A 2.5 cm copper plate is heated to a temperature of 1000 C prior to cold work in ambient air at 25 C. If the minimum cold working temperature is 915 C calculate the time during which the cold working should be completed.

3.3 Determine the temperature response of a 0.9 mm dia. copper wire at 150 C suddenly immersed in (a) water (h = 70 W/m² K) at 35 C (b) air (h = 9) at 35 C. Plot $t - \tau$ curve for both types of cooling.

3.4 The heat transfer coefficient of a 22 gauge chromel-alumel thermocouple in a gas stream is 570 W/m²K. What is the time constant in seconds (C_p = 0.42 kJ/kg K, ρ = 8600 kg/m³).

[Ans. 1.13 sec]

3.5 Determine the amplitude and time lag of a 22 gauge thermocouple in a gas stream whose temperature is given by $t_g = 27 - 6.5\cos(\pi\tau)$ where τ is in hours. The mass flow rate is 10.7 kg/m² sec. Also assume time constant of thermocouple as 1.3 sec. [Ans. –6C, 15.2 C, 1.27 sec]

3.6 A rubber sheet 1.25 cm thick is to be cured at 144°C for 50 minutes. If the sheet is initially at 33°C and heat is applied from both surfaces, find the time required for the centre to reach 143°C. For rubber $\alpha = 0.00029\ m^2/hr$.

[Ans. : 20 min]

3.7 A 5 cm diameter 36 m long cylinder is to be water - quenched from 870 C to 21 C. Compute the time required in minutes for the centre to reach 540 C, assuming that the surface temperature changes instantly from 870 C to 21 C.

[Ans : 7.4 min]

3.8 A rubber cube of 1.25 cm side is to be cured at 144°C. If it is initially at 33°C and if heat is applied from all sides, find the time required for the centre of the cube to reach 143°C. $\alpha = 0.0026\ m^2/hr$

[Ans : 7 min]

3.9 A 5 cm diameter, 5 m long cylinder is initially at 1020°C. It is quenched in water at 20°C. After how much time will the centre reach 520°C. Neglect surface resistance to heat flow. $\alpha = 0.001\ m^2/hr$.

3.10 A long steel rod 4 cm in diameter, initially at 50°C, is put into a furnace at 500°C. Determine the time required for the centre to reach 400°C if $\alpha = 0.04\ m^2/hr$. Neglect surface resistance.

3.11 What will be the temperature at the centre of a steel cylinder of length and diameter both equal to 50 cm, initially at 500°C, 1 hour after its surface is suddenly cooled to 0°C. $\alpha = 0.025\ m^2/hr$.

3.12 Calculate the temperature which the centre of a steel sphere 0.18 m diameter will have attained 5 min after the sphere has been plunged into a heat treating bath at 38 C. Initial temperature of sphere is 600 C, $\rho = 7800\ kg/m^3$, $C_p = 0.0552$ kJ/kg C, k = 37.2 W/mK.

[Ans : 77.3 C]

3.13 Tomato juice filled in cans 25 cm diameter and 25 cm long is heated in a steam oven at 100°C. The initial temperature of the juice is 50°C. What time is required to ensure that all the juice is heated to the temperature of 90°C. Assume that the thermal diffusivity of the juice is $7 \times 10^{-4}\ m^2/hr$ and that the inner surface of the cans is at steam temperature.

3.14 A 10 cm diameter sphere ($\alpha = 0.003\ m^2/hr$) initially at 100°C is plunged into water at 40°C. After how much time will the centre reach 50°. Neglect surface resistance.

3.15 Find the centre temperature of a steel sphere 20 cm diameter, initially at 300°C, 3 minutes after it is plunged into a bath at 40°C. $\alpha = 0.04\ m^2/hr$. Neglect surface resistance.

3.16 Estimate the temperature at the centre of a can of vegetables after 40 min in steam at 100°C. The can is 12 cm long, 8 cm diameter, and initially at 20°C. The thermal diffusivity of vegetables is $0.0005\ m^2/hr$.

3.17 Compute the depth in a wall where the temperature amplitude is 10 % of that at the surface. The surface temperature varies as a cosine function with a period of 24 hr, $\alpha = 0.001\ m^2/hr$.

3.18 If the ground near the earth's surface is at 5°C and during the night the air temperature drops to −13.7°C how far will the temperature of 0°C penetrate into the ground in 6 hr. (h = 4.9 kJ/m^2 K, α = 0.00186 m^2/hr, k = 0.298 W/mK)

[Ans : 10.8 cm]

3.19 The wall of a large furnace is made up of refractory material 21 cm thick, having k = 1.21 W/mK and a diffusivity of 0.0021 m^2/hr. A thermocouple, 7.5 cm from the inside surface indicated a temperature of 680 C 7 hr after the firing up period. If the gas temperature on the inner surface is 1100 C and the initial wall temperature 18.5 C, what would be the temperature at the thermocouple position according to theory. Assume that the inner layer of the refractory assumes the temperature of the hot gases immediately after the firing up period begins.

[Ans : 732 C]

3.20 The outer surface temperature of a thick wall originally at a uniform temperature of −30°C suddenly changes to 60°C.

Calculate :

(i) Temperature at a depth of 20 cm, 4 hr after the change.

(ii) The rate of heat flow per m^2 of a plane located at 30 cm from the surface 4 hr after change.

(iii) The total amount of heat entering the wall in 10 hr per m^2 of surface.

3.21 One face of a brick wall is suddenly raised to 200°C. If the initial temperature of the wall was 20°C, calculate the time required for the temperature at a depth of 10 cm from the surface to reach 100°C. α = 0.0015 m^2/hr.

3.22 Reinforced concrete initially at 20°C is to be poured into a foundation the bottom of which is at −10°C. The thickness of concrete is to be 1 m and its thermal diffusivity is 0.0025 m^2/hr. Calculate the distance the freezing point will advance into concrete before the concrete sets. The initial and final sets occur after 1 hr and 10 hr of pouring.

3.23 Calculate the temperature of the earth 0.5 m below a large gas conduit 10 days after hot gases start flowing into the conduit. The initial temperature of the soil is 20°C and the gas temperature is 420°C. For soil α = 0.005m^2/hr.

3.24 In a heating installation hot water pipes are laid under the floor. If the floor is in intimate contact with soil (k = 0.865 W/mK, α = 0.0014 m^2/hr) initially at a uniform temperature of 20°C lower than that of pipe calculate the rate of heat loss to the ground per m^2 of floor area 100 hr after the start of heating. Also calculate the total loss for this time. What is the total heat loss between 25 hr and 100 hr after start of heating ? [Ans : 93.3 kJ/hr; 11,800 kJ; 9330 kJ]

3.25 A wall 60 cm thick, initially at a uniform temperature of 74°C is exposed on both sides to hot gases at 1070°C. If h = 10.44 W/m^2K, α = 0.0172 m^2/hr, compute the following for τ = 27 hours :

(a) Fourier number.
(b) Temperature at surface and at centre.
(c) Heat absorbed by 1 m^2 of wall.

[Ans : (a) Fo = 5.2; (b) t_s = 700°C, t = 660°C]

3.26 Calculate the Biot and Fourier numbers for a long 15 cm diameter cylinder (k = 0.81 W/mK, C_p = 1.46 kJ/kg°C, ρ = 2200 kg/m³) initially at 0 C which has been heated for 4 hr in an oven at 300°C. (h = 25 W/m²K).

3.27 A wall 0.6 m thick, initially at a uniform temperature of 21 C is exposed on both sides to hot gases at 580 C. Assume that the heat transfer coefficient is constant and equal to 0.1 W/m²K and for the wall k = 15.6 W/mK, ρ = 2580 kg/m³ and C_p = 1.46 kJ/kg C. After 27 hr of heating, compute (a) Fo, the Fourier modulus (b) the temperature at the surface and at the centre plane ; (c) the temperature distribution through the wall; (d) the heat absorbed by 1 m² of the wall.

[Ans : (a) 5.19 (b) 256.4 C at centre]

3.28 A cast iron sphere 15 cm in diameter at a uniform temperature of 370 C is quenched in a water bath which is at a temperature of 38 C. Determine the value of the film coefficient *h* between the sphere and the water if at the end of 5.6 min the temperature at the centre of sphere has dropped to 155 C.

[Ans : Bi = 0.35]

3.29 What length of time is required to heat a 2.5 cm diameter 5 cm long copper (k = 380 W/mK) cylinder from 60°C to 200°C in a furnace whose temperature is 600°C. The value of h at the cylinder surface is 57 W/m²K and α = 0.186 m²/hr.

[Ans : 3.3 min]

3.30 The wall of a rocket combustion chamber is 0.64 cm thick steel (k = 34.8 W/m K, α = 0.0155 m²/hr). With a flame side surface coefficient of 5700 W/m²K, how long may this chamber operate with a wall temperature below 1100°C when with an initial uniform wall temperature of 30°C the combustion gases at 2500°C begin suddenly to flow immediately after firing. The outer wall surface may be assumed insulated because of the low value of *h* at this surface.

[Ans : 2.3 min]

3.31 A disk of plate glass 20 cm diameter and 2.5 cm thick is to be annealed. The disk is originally at 24°C. How long must it remain in an oven at 410°C so that every part of the disk reaches 399°C. h = 14.6 W/m²K, α = 0.00121 m²/hr.

[Ans : 2.3 hr]

3.32 A brass cylinder at 600°C, 5 cm diameter and 5 cm long, is immersed in a water bath at 20°C where h = 1950 W/mK. Determine the temperature at the centre of the cylinder 1 min after it is put in the bath. For brass, k = 87 W/mK, Cp = 0.04 kJ/kg °C, and ρ = 7500 kg/m³.

3.33 A large tank of aluminum (k = 218 W/mK, C_p = 0.84 kJ/kg°C, ρ = 2560 kg/m³) 7.5 cm thick is filled with petrol at 18°C. The outside of the tank is suddenly exposed to hot gases at 538°C. The petrol in contact with the wall will ignite when the wall temperature reaches 204°C. Find the time that will elapse before ignition occurs. The value of h for the outer wall surface and hot gases is 34.8 W/m²K. The value of h for inside surface and petrol is 23.2 W/m²K. Neglect the internal resistance of the wall.

[Ans : 53 min]

3.34 If the temperature of the earth's surface varies sinusoidally from 5°C to 25°C over a 24 hr period, compute :

(a) The temperature at a depth of 20 cm, 18 hr after the surface temperature started to increase.

(b) The time lag of the temperature wave at a depth of 20 cm.

(c) The total heat flowing into surface per m^2 during 12 hr after the surface temperature started to increase.

Given k = 2.0 W/mK; $\alpha = 0.0018\ m^2/hr$.

3.35 The surface temperature of a furnace wall varies periodically as a sine function from 750°C to 850°C. The period is 12 hr. If $\alpha = 0.003\ m^2/hr$ and k = 1.5 W/mK, determine :

(i) Amplitude of temperature change at the surface.

(ii) Time lag of the wave at a depth of 10 cm.

(iii) The surface temperature, 18 hr after the surface temperature begins to exceed the mean value.

(iv) The temperature at a depth of 10 cm, 8 hr after the surface temperature starts to exceed the mean value.

3.36 A long copper rod is insulated throughout its length and one end is alternately heated and cooled through the range 0 to 100°C every 30 minutes. Determine the velocity and wavelength of the temperature wave. $\alpha = 1.14\ cm^2/sec$.

[Ans : 0.089 cm/sec : 161 cm]

3.37 A cold wave lasting for 2 weeks (period = 4 weeks) causes a 20°C fall of a surface temperature. What will be the temperature fall at a depth of 100 cm below the surface if $\alpha = 0.0058\ cm^2/sec$. What is the time lag of the minimum temperature at this depth as compared to the surface ?

[Ans : 4.5°C, 6.7 days.]

3.38 The temperature at the surface of a semi-infinite solid varies according to the equation $t = 27 + 27 (\cos \omega\tau)$ °C where $\omega = \pi/10$ radians per sec. The mean temperature of the solid is maintained at 27 C. The time is considered as zero when the surface temperature is 54 C. Calculate the temperature at a point inside the solid and 5 mm from the surface at the time $\tau = 3$ hr. The thermal conductivity of the solid k = 23 W/mK, the density $\rho = 897\ kg/m^3$ and the specific heat $C_p = 0.42$ kJ/kg C.

[Ans :51.5 C]

3.39 The daily range in temperature at the earth's surface in a given locality is from 23 C to 12 C. Determine (a) the amplitude of temperature oscillation at a depth of 0.30 m, (b) the time lag of the temperature wave at a depth of 0.30 m, and (c) the temperature at a depth of 0.30 m, 5 hr after the surface temperature reaches the minimum temperature. Assume k = 0.52 W/mK; C_p = 1.98 kJ/kg K, and $\rho = 1600\ kg/m^3$.

[Ans : (a) 0.066 C (b) 16.9 hr (c) 0.065 C]

CHAPTER
4

Heat Transfer by Convection

4.1 THE NATURE OF HEAT CONVECTION

One of the major problems encountered in engineering practice is the calculation of heat exchange between a *solid surface* and a *fluid* in contact with it. If the fluid is at *rest*, the problem reduces to that of conduction only. Such a situation, however, is difficult to visualise. Even if we consider water being heated in a vessel, it will be found that it is *not at rest* because of density gradients being developed in the whole body of water. The water in the bottom is hotter and therefore lighter than that at the top, thus it rises up replacing the colder water which flows down. Therefore even in this case the water is set in motion due to heating. The heat exchange by *convection*, strictly speaking, refers to that occuring between a *solid surface* and the *fluid flowing* along it.

It was mentioned in Section 1.4 that convection is not fully a heat transfer mechanism (which will depend on the mere existence of temperature differential) because it depends for its operation on the *motion of fluid* acting as a carrier of energy. The fluid motion along the solid surface may be motivated by any of the two processes given below :

(a) It may be caused by *density difference* produced by *temperature gradient*. Heat exchange in such a situation is known as *free* or *natural convection*.

(b) It may be caused by some external agency such as a *pump or a blower*. The motion of *fluid* in such cases will be faster than that in *free motion* and therefore will result in increased rate of heat transfer. This type of heat transfer is known as *forced convection*.

It is obvious that the transfer of energy in a convection process occurs mechanically and depends upon the motion of fluid particles. Thus in addition to the law of heat conduction and geometry of the system, the heat exchange by convection will also depend upon the laws of *fluid dynamics*.

The Newton's heat convection equation ; $q = h\ A\ \Delta t$ is not as simple as it appears. The equation is actually a definition of the *convective heat transfer coefficient* or *unit thermal convective conductance*, h, rather than representation of the heat convection law. The determination of the *local value* h_x or *average value* h, of this coefficient forms the basis of the computation of heat transfer by convection.

4.2 THE NUSSELT NUMBER

As mentioned earlier, the resistance to heat flow by *convection* is mainly concentrated in a *thin layer* immediately adjacent to the wall surface. Heat transfer occurs due to heat conduction and energy transport by moving fluid within this layer. Hence, the value of heat transfer coefficient, h is determined essentially by considering the thickness and the characteristics of this thin *boundary layer*. The boundary layer analysis will be discussed in later sections.

In Fig. 4.1, is shown a hot wall surface at temperature t_w, over which a fluid (thermal conductivity, k) is flowing with a velocity V_∞. Let the temperature of fluid be t_∞. The temperature gradient in the fluid is shown to be confined in a thin layer in the vicinity of the surface known as *thermal boundary layer* (shown magnified in this figure) of thickness δ_t. Also shown in the figure is the hypothetical layer of fluid of thickness, δ obtained by the dashed line drawn tangent to the actual temperature curve. This layer is assumed completely stagnant, will offer same thermal resistance to heat flow (by conduction only) as that offered by actual boundary layer.

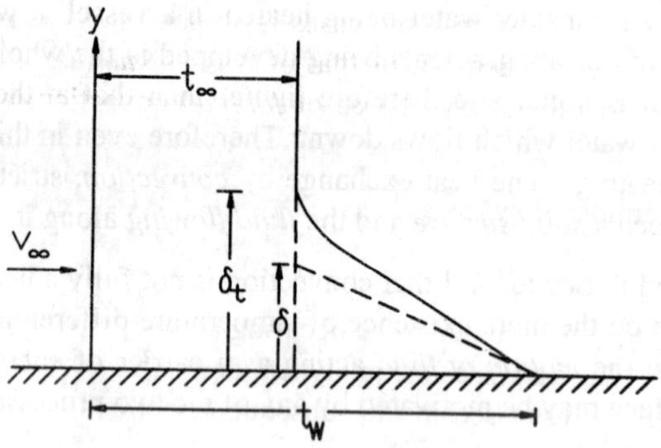

Fig. 4.1 Temperature Variation between Wall and Fluid Confined in a Thin Layer

The rate of heat flow from surface to fluid is given by ;

$$q = -kA\left(\frac{\partial t}{\partial y}\right)_{y=0} = hA(t_w - t_\infty) \tag{4.1}$$

This equation may be written as ;

$$h = \frac{-(\partial t/\partial y)_{y=0}}{(t_w - t_\infty)/L} \times \left(\frac{k}{L}\right) \tag{4.2}$$

or

$$\frac{hL}{k} = \left\{\frac{\partial\left(\dfrac{t_w - t}{t_w - t_\infty}\right)}{\partial(y/L)}\right\}_{y=0} \tag{4.3}$$

(where L is a significant length)

It will be noticed that Eq. (4.3) represents an equality of two dimensionless quantities. The term on the left hand side is known as Nusselt number, given by $Nu = hL/k$ and the term on the right hand side is the dimensionless *slope* of the fluid *temperature distribution curve* at the surface. In practice, the *Nusselt number* is a convenient measure of the convective heat transfer coefficient.

In the hypothetical stagnant layer of thickness δ, the rate of heat transfer is given by ;

$$q/A = k\,(t_w - t_\infty)/\delta = h\,(t_w - t_\infty)$$

$$\text{or} \qquad h = k/\delta \tag{4.4}$$

It is therefore noticed that the thinner the boundary layer the greater the convective coefficient. Hence attempt should be made to reduce the boundary layer thickness to promote heat transfer. This may be accomplished by increasing the velocity and thereby turbulence of the fluid.

4.3 DETERMINATION OF NUSSELT NUMBER

The mechanism of fluid flow in some cases is so complicated that the evaluation of convective heat transfer coefficient, h, or that of *Nusselt number*, Nu, is quite difficult. For practical requirements the study of *free* or *forced* convection may be made for the following cases :

(i) Convection in laminar flow

(ii) Convection in turbulent flow

The flow of fluid may be assumed to be occuring (i) over a flat plate, (ii) within a tube or (iii) outside a tube. The details of the analytical determination of Nu, for a simple case of flat plate only will be discussed in the subsequent sections. However, the results of other cases will be given for reference.

The methods available for the calculation of the value of h or Nu may be broadly classified into three categories :

(i) Dimensional Analysis

(ii) Solution of Boundary Layer Equations.

(iii) Analogy between Heat and Momentum transport.

Unfortunately none of these methods is complete in itself and the exact value of h is obtained with the assistance of experimental data. However a basic knowledge of each of the methods is essential for the engineer engaged in problems connected with heat exchange.

4.4 FORCED CONVECTION

The most extensively used and important heat exchange processes are those of heating and cooling of fluids inside pipes and conduits. As discussed in Section 4.1, the processes will involve heat transfer by *forced convection* as the fluid flowing in the tubes and pipes will be doing so under the action of some pump. The design and analysis of all such heat exchangers require the knowledge of heat transfer coefficient h between the wall surface and the fluid. We shall now discuss in the following sections the determination of h by the various methods mentioned in Section 4.3.

First of all the dimensional analysis approach will be used to determine the nature of correlations involved. The same will then be extended to form empirical equations based on experimental data. Later the boundary layer equations for momentum and energy will be developed and will be solved for laminar flow only. The solution of these equations for turbulent flow is considered out of the scope of this text.

An analogical study will then be made between heat and momentum transfers and relations giving the value of heat transfer coefficient h will be developed.

It will be noticed that the forced convection heat transfer may be correlated by an equation of the form :

$$Nu = \phi(Re)\,\psi(Pr) \tag{4.5}$$

where Nu = Nusselt number ; $\dfrac{hL}{k}$

Re = Reynolds number ; $\dfrac{VD\rho}{\mu}$

Pr = Prandtl number ; $\dfrac{C_p\mu}{k}$

(Note that Pr is a property parameter)

4.5 DIMENSIONAL ANALYSIS APPLIED TO FORCED CONVECTION

In the words of Langhaar, *dimensional analysis* is a method by which we deduce information about a phenomenon from the single premise that the phenomenon can be described by a *dimensionally correct equation* among certain variables. The generality of the method is both its weakness and strength. With little effort, a *partial solution* to nearly all problmes is obtained. On the other hand a complete solution is not obtained nor is the inner mechanism of a phenomenon revealed by dimensional reasoning alone.

In the method of determining the solutions of problems by dimensional analysis we have to seek the assistance of experimental results. Dimensional analysis provides us with only a set of dimensionless groups, involving the different variables, related together by an *unknown function*. The form of this function has to be determined experimentally. However, dimensional analysis reduces the large number of original variables to only a few dimensionless groups. Hence the number of experiments required to find a relation between them is very much reduced.

This method is particularly important in the study of *convection* as it is easily used for determining convective heat transfer coefficient from empirical equations obtained by correlating experimental data with the aid of dimensional analysis.

The combination of *variables into dimensionless groups* is governed by *Buckingham π-theorem*. This theorem states that in a physical system involving n variables having m *fundamental dimensions* the number of independent dimensionless groups which may be obtained by combination of those variables is $(n - m)$.

For instance let $A_1, A_2, A_3, \ldots, A_n$ be the variables affecting a particular problem, then we may write ;

$$F(A_1, A_2, A_3, \ldots, A_n) = 0 \tag{4.6}$$

If the number of fundamental dimensions is m ; then a relationship of the form ;

$$\phi\,(\pi_1, \pi_2, \pi_3, \ldots\ldots, \pi_{n-m}) = 0 \tag{4.7}$$

will hold, where π_1, π_2, π_3, etc. are the dimensionless groups of the form ;

$$\pi = f\,(A_1^{x}, A_2^{y}, A_3^{z}, \ldots\ldots, \ldots\ldots) \tag{4.8}$$

Nusselt developed the basic equations for forced and free convection making use of the method of dimensional analysis in 1906 and 1915 respectively. We shall now discuss the application of this method to *forced convection.*

On the basis of experience we premise that the forced convection heat transfer coefficient, h is a function of the variables listed in Table 4.1. The dimensional formulae of these variables are also given in the table in terms of the *fundamental dimensions* of mass [M], length [L], time [θ] and temperature [T]. Thus we may write,

$$h = f\,(D, k, \rho, C_p, \mu, V) \tag{4.9}$$

or

$$F(h, D, k, \rho, C_p, \mu, V) = 0 \tag{4.10}$$

Applying dimensional analysis, Eq. (4.8) may be written as ;

$$\pi = (h)^a\,(D)^b\,(k)^c\,(\rho)^d\,(C_p)^e\,(\mu)^f\,(V)^g \tag{4.11}$$

Table 4.1 Variables affecting forced convection heat transfer

No.	Variable	Symbol	Dimensional Formula
1.	Fluid velocity	V	L/θ
2.	Fluid density	ρ	M/L^3
3.	Thermal conductivity of fluid	k	$ML/T\theta^3$
4.	Specific heat of fluid	C_p	$L^2/T\theta^2$
5.	Fluid viscosity	μ	M/Lθ
6.	Characteristic linear dimension of heat transfer surface	D	L

Since the dimensions of h are $[M/T\theta^3]$ we get by writing the dimensions of all variables :

$$0 = \left[\frac{M}{T\theta^3}\right]^a [L]^b \left[\frac{ML}{T\theta^3}\right]^c \left[\frac{M}{L^3}\right]^d \left[\frac{L^2}{T\theta^2}\right]^e \left[\frac{M}{L\theta}\right]^f \left[\frac{L}{\theta}\right]^g \tag{4.12}$$

Equating dimensions on the two sides of Eq. (4.12), we obtain,

For M ; $\quad a + c + d + f = 0 \quad$ (i)

L ; $\quad b + c - 3d + 2e - f + g = 0 \quad$ (ii)

θ ; $\quad -3a - 3c - 2e - f - g = 0 \quad$ (iii)

T ; $\quad -a - c - e = 0 \quad$ (iv)

Thus we have four equations between seven exponents a, b, c, d, e, f, g. With the help of these equations we can express any four exponents in terms of the remaining three. Let us express d, e, f, g in terms of a, b, c.

from Eq. (iv) ;

$$e = -a - c \tag{A}$$

adding Eqs. (ii) and (iii)

$$-2f - 3d - 3c - 3a + b + c = 0$$

from Eq. (i), $d = -a - c - f$

therefore,

$$-2f + 3a + 3c + 3f - 3c - 3a + b + c = 0$$

or $$f + b + c = 0$$

or $$f = -b - c \tag{B}$$

further ; $$d = b - a \tag{C}$$

from Eqs. (iii), (A) and (B),

$$-g = 3a + 3c + 2e + f$$

$$= 3a + 3c + 2(-a - c) - b - c$$

$$= a - b$$

or $$g = b - a \tag{D}$$

Since there are seven variables in the dimensional equation (4.12) and there are four fundamental dimensions we should expect (7 – 4) = 3 independent dimensionless numbers or π –terms. The π –terms, π_1, π_2, π_3, as shown in Table 4.2 can be obtained by assigning arbitrary values to the exponents a, b, c.

Table 4.2 Development of Non-Dimensional Parameters

h	D	k	ρ	Cp	μ	V	π-term	Name assigned to π-term
a	b	c	d	e	f	g		
0	1	0	1	0	–1	1	$\pi_1 = (VD\rho/\mu)$	Reynolds number, Re
0	0	–1	0	1	1	0	$\pi_2 = (\mu C_p/k)$	Prandtl number, Pr
1	0	0	–1	–1	0	–1	$\pi_3 = (h/\rho C_p V)$	Stanton number, St

(Header rows: "Variables" spans h–V; "Exponent" spans a–g.)

The parameters can be rewritten as a new dimensionless parameter given by ;

$$\pi_4 = \pi_3 \times (\pi_1 \times \pi_2) = \frac{hD}{k}$$

This is defined as Nusselt numbers

Thus, $$F(\pi_1, \pi_2, \pi_4) = 0$$

or $$F\left\{\left(\frac{VD\rho}{\mu}\right), \left(\frac{C_p\mu}{k}\right), \left(\frac{hD}{k}\right)\right\} = 0 \tag{4.13}$$

or
$$\frac{hD}{k} = \phi\left(\frac{VD\rho}{\mu}\right)\psi\left(\frac{\mu C_p}{k}\right)$$

$$Nu = \phi\,(Re)\,\psi\,(Pr) \tag{4.5}$$

Here, two more dimensionless groups are defined for future use :

$$St = \text{Stanton number} = \frac{h}{\rho c_p V}$$

$$Gt = \text{Graetz number} = \frac{\rho A V c_p}{kL}$$

4.6 EXPERIMENTAL RESULTS IN FORCED CONVECTION

Equation (4.5) represents a generalized correlation for forced convection heat transfer process. The form of the functions, ϕ (Re) and ψ (Pr) will depend upon the various factors affecting the fluid flow. On the basis of theoretical analysis or experimental results, correlations may be specified for different conditions in which heat is exchanged by convection. Some of the correlations used for determining the heat transfer coefficient for forced convection are discussed in succeeding Sections.

4.7 CORRELATIONS FOR FORCED CONVECTION IN LAMINAR FLOW

(a) Laminar flow over a flat plate

(i) The local values of heat transfer coefficient h and the Nusselt number Nu for this case are given by the theoretically obtained correlation ;

$$h_x = 0.332\,k\,(Pr)^{1/3}\left(\frac{V_\infty}{\nu x}\right)^{1/2} \tag{4.14}$$

or
$$Nu_x = \frac{h_x x}{k} = 0.332\,(Pr)^{1/3}\,(\mathrm{Re_x})^{1/2} \tag{4.15}$$

(ii) The average values of h and Nu are then obtained by ;

$$h = \frac{1}{L}\int_0^L \mathrm{h_x}\,dx = \frac{1}{L}\,(0.332)\,k\,(Pr)^{1/3}\int_0^L (V_\infty/\nu x)^{1/2}\,dx$$

or
$$h = 0.664\,\frac{k}{L}\,(Pr)^{1/3}\left(\frac{V_\infty L}{\nu}\right)^{1/2} \tag{4.16}$$

or
$$Nu = \frac{hL}{k} = 0.664\,(Pr)^{1/3}\,(R_{eL})^{1/2} \tag{4.17}$$

In Eqs. (4.14) to (4.17) the fluid properties (ρ, μ, C_p, k) should be evaluated at the film temperature, ie, the average of the temperature of the fluid and the temperature of the surface of the plate.

(iii) The results given by Eqs. (4.14) to (4.17) are valid for all fluids except liquid metals (Pr << 1). For the case of liquid metals the two correlations which reproduce experimental data well are as follows :

1. Sparrow and Gregg obtained the relation ;

$$\frac{Nu_x}{\sqrt{Re_x P_r}} = \frac{0.564}{1+0.9\sqrt{Pr}} \tag{4.18}$$

$$(\text{for } 0.006 < Pr < 0.03)$$

2. Eckert obtained the relation ;

$$Nu_x = \frac{\sqrt{Re_x\, Pr}}{1.55\sqrt{Pr+3.09\sqrt{0.372-0.15\,Pr}}} \tag{4.19}$$

$$(\text{for } Pr << 1)$$

(b) Laminar flow inside tubes

In the case of flow inside tubes, both velocity and temperature generally vary from point to point at a given cross-section. For example, in laminar flow, the velocity is zero at the tube wall and it is maximum at the center line (r = 0). If the tube radius is R, the mean velocity V_m may be obtained by dividing the volume flow rate by the flow area :

so

$$V_m = \left[\int_0^R V(2\pi r)\,dr\right] \div \left(\int_0^R 2\pi r\,dr\right)$$

$$= \left[\int_0^R V(2\pi r)\,dr\right] \div (\pi R^2) \tag{4.20}$$

where V = velocity at radius $r = r$

The mean temperature of the fluid, also called *bulk temperature* or *cup-mixing temperature*, is defined as the temperature which any quantity of the fluid, passing the cross-sectional area of the pipe in a time interval, would assume if it were collected and perfectly mixed in a cup.

Thus the *mean temperature*, t_m, is by definition ;

$$t_m = \frac{\text{Enthalpy flow rate at a given cross-section}}{C_p \times \text{Mass flow rate at the given cross-section}}$$

$$t_m = \left[\int_0^R V\rho C_p t(2\pi r)\,dr\right] \div \left[\int_0^R C_p V\rho(2\pi r)\,dr\right] \tag{4.21}$$

If C_p = constant,

$$t_m = \left[\int_0^R (Vtr)\,dr\right] \div \left[\int_0^R (Vr)\,dr\right]$$

or $$t_m = \frac{2}{R^2 V_m} \int_0^R (Vtr)\, dr \tag{4.22}$$

The heat transfer coefficient, h, for flow in tubes is defined in terms of the above mean (or bulk) temperature of the fluid as below,

$$h = (q/A) / (t_w - t_m) \tag{4.23}$$

where t_w = tube wall temperature

(i) Sellars, Tribus and Klein observed that the local heat transfer coefficient at the beginning of the tube is infinitely large. It decreases rapidly in the direction of flow approaching a constant value as the temperature distribution becomes fully established. They derived from theoretical considerations the following equations for fully developed flow :

1. If the heat flux (q/A) is constant or the wall temperature increase is linear

$$Nu = hD/k = 4.36 \tag{4.24}$$

2. If the wall temperature t_w is uniform

$$Nu = hD/k = 3.66 \tag{4.25}$$

(ii) Later Hausen developed an empirical equation for average Nusselt number for fully developed flow with uniform wall temperature as :

$$Nu = \frac{hD}{k} = 3.66 + \frac{0.0668(\mathrm{D/L})(Re\,Pr)}{1+0.04\{(\mathrm{D/L})\,Re\,Pr\}^{2/3}} \tag{4.26}$$

(iii) Sieder and Tate have proposed the following empirical relation :

$$\frac{hD}{k_b}\left(\frac{\mu_w}{\mu_b}\right)^{0.14} = 1.86\left(\frac{4}{\pi}\right)^{1/3}\left(\frac{W C_{pb}}{k_b L}\right)^{1/3}$$

where $W = \rho A V$ and $A = (\pi/4)\, D^2$

$$= 1.86\left[\left(\frac{VD\rho}{\mu_b}\right)\left(\frac{\mu C_p}{k}\right)_b\left(\frac{D}{L}\right)\right]^{1/3} \tag{4.27}$$

or $$Nu = 1.86\,(Re\,Pr\,D/L)^{1/3}\,(\mu_b/\mu_w)^{0.14} \tag{4.28}$$

where μ_b = fluid viscosity at bulk temperature of fluid.

μ_w = fluid viscosity at tube wall temperature.

In Eq. (4.27) the average heat transfer coefficient is based on the arithmetic mean of the inlet and outlet temperature differences. This equation is valid for, $(Re\,Pr)\,(D/L) > 10$

(iv) For the case of liquid metals (low Pr) the following equations may be used :

$$hD/k = 8.0 \text{ (for uniform } q/A) \tag{4.29}$$

$$hD/k = 5.75 \text{ (for uniform } t_w) \tag{4.30}$$

Example 4.1

Water at 60 C enters a tube 2.5 cm diameter and 3.0 m long at a mean flow velocity of 2 cm/s. Calculate the exit water temperature, T_{b2} if the wall temperature is constant at 80 C.

Solution :

The Reynolds number at the inlet bulk temperature is first determined. The properties of water at 60 C are :

$$\rho = 985 \text{ kg/m}^3$$

$$C_p = 4.18 \text{ kJ/kg K}$$

$$k = 0.651 \text{ W/mK}$$

$$Pr = 3.02$$

$$\mu = 4.71 \times 10^{-4} \text{ kg/ms}$$

now
$$Re = \frac{(985)(0.02)(0.025)}{4.71\times10^{-4}} = 1045$$

so the flow is laminar.

$$Re\ Pr\ (D/L) = \frac{(1045)(3.02)(0.025)}{3}$$

$$= 26.3 > 10$$

So Eq. (4.27) is applicable. We do not yet know the mean bulk temperature, T_b to evaluate properties so we first make the calculation on the basis of 60 C, determine an exit bulk temperature, and then make a second iteration to obtain a more precise value. When inlet and outlet conditions are designated with subscripts 1 and 2, respectively, the energy balance becomes

$$q = h\,\pi\,DL\left(T_w - \frac{T_{b1}+T_{b2}}{2}\right) = mC_p\,(T_{b2} - T_{b1}) \qquad \text{(a)}$$

At the wall temperature of 80°C we have

$$\mu_w = 3.55 \times 10^{-4} \text{ kg/m.s}$$

$$Nu = (1.86)\,(26.3)^{1/3}\left(\frac{4.71}{3.55}\right)^{0.14} = 5.75 \qquad (4.28)$$

or
$$h = \frac{k\,Nu}{D} = \frac{(0.651)(5.75)}{0.025} = 149.8 \text{ W/mK}$$

The mass flow rate is

$$m = \rho\,\frac{\pi D^2}{4}\,V = \frac{(985)\,\pi\,(0.025)^2\,(0.02)}{4}$$

$$= 9.67 \times 10^{-3} \text{ kg/s}$$

Putting the value for h into Eq. (a) along with m and $T_{b1} = 60°C$ and $T_w = 80°C$ gives :

$$q = (149.8)\,\pi\,(0.025)\,(3.0)\left(80 - \frac{T_{b2} + 60}{2}\right)$$

$$= (9.67 \times 10^{-3})\,(4180)\,(T_{b2} - 60) \qquad \text{(b)}$$

This equation can be solved to give

$$T_{b2} = 72 \text{ C}$$

We should therefore evaluate properties at

$$T_{b\text{-mean}} = \frac{72 + 60}{2} = 66 \text{ C}$$

The new values at 66 C are,

$$\rho = 982 \text{ kg/m}^3, \; C_p = 4185 \text{ J/kg°C}$$

$$\mu = 4.36 \times 10^{-4} \text{ kg/ms}$$

$$k = 0.656 \text{ W/mK}$$

and $Pr = 2.78$

therefore $$Re = \frac{(0.02)(0.025)(982)}{4.36 \times 10^{-4}} = 1126$$

now, $$Re\,Pr\,(D/L) = \frac{(1126)(2.78)(0.025)}{3} = 26.1$$

$$Nu = (1.86)\,(26.1)^{1/3}\left(\frac{4.36}{3.55}\right)^{0.14} = 5.7 \qquad (4.28)$$

$$h = \frac{(0.656)(5.7)}{0.025} = 149.6 \text{ W/m}^2\text{K}$$

Putting this value of h back into Eq. (a) we obtain

$$T_{b2} = 71.9 \text{ C} \qquad \text{Ans.}$$

This checks, and no further iteration is required.

(c) Laminar flow through ducts of other geometry

(i) The values of Nu for flow through parallel plates and triangular ducts are listed in Table 4.3 ; the values for rectangular ducts are given by Fig. 4.2. Here D_e is the *equivalent* diameter or hydraulic diameter.

Table 4.3 Nusselt number for fully developed laminar flow in ducts

Geometry	Velocity distribution	Condition at wall	$h\,D_e/k$
Parallel plates	Parabolic	q/A uniform	8.23
Parallel plates	Parabolic	t_w uniform	7.60
Triangular duct	Parabolic	q/A uniform	3.00
Triangular duct	Parabolic	t_w uniform	2.35

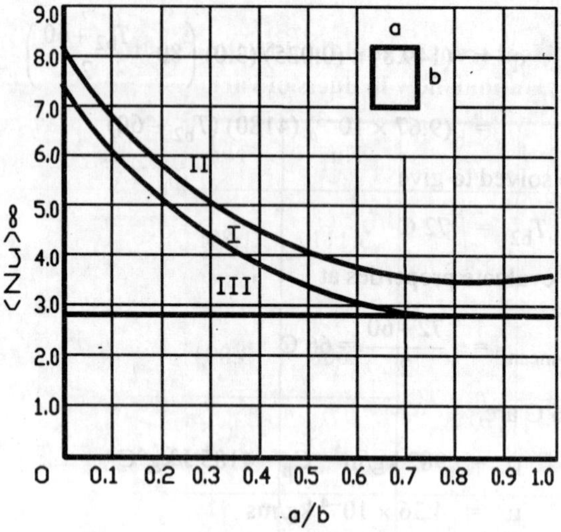

Fig. 4.2 Values of Nu for fully Developed Laminar Flow in Rectangular Ducts, Curve (I): Uniform wall Temperature, (II) : Uniform Heat flux per unit length with peripheral wall Temperature and (III) : Uniform Heat flux Axially and Peripherally.

(ii) Shah and London have presented the heat transfer and fluid friction information for fully developed laminar flow in ducts of various cross section as given in Table 4.4. Here,

Nu_1 = average Nusselt number for uniform heat flux in flow direction and uniform wall temperature at particular flow cross section

Nu_2 = average Nusselt number for uniform heat flux both in flow direction and around periphery.

Nu_3 = average Nusselt number for uniform wall temperature.

f.Re = product of friction factor and Reynolds number.

Example 4.2

A flat plate heated to 80 C is cooled by an air stream at 20 C flowing @ 10 m/sec. Calculate the value of local heat transfer coefficient at x = 30 cm.

Solution :

$$t_w = 80\text{ C} ; t_g = 20\text{ C}$$

The average temperature is t = 50 C. From the properties for air at this temperature we get ;

$$k = 2.83 \times 10^{-2} \text{ W/mK}$$

$$\nu = 17.95 \times 10^{-6} \text{ m}^2/\text{s}$$

$$\text{Pr} = 0.698$$

$$\text{Re}_x = \frac{10 \times 30/100}{17.95 \times 10^{-6}} = 165{,}000$$

Table 4.4 Heat transfer and fluid friction for fully developed laminar flow in ducts of various cross sections

Geometry ($L/D_h > 100$)	$\frac{b}{a}$	Nu_1	Nu_2	Nu_3	$f.Re$
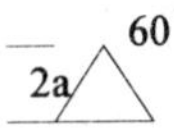 60° 2a 2a	$\sqrt{\frac{3}{2}}$	3.111	1.892	2.47	13.333
	1	3.608	3.091	2.976	14.227
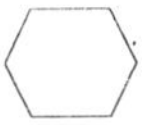		4.002	3.862	3.34	15.054
2b 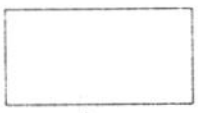 2a	$\frac{1}{2}$	4.123	3.017	3.391	15.548
		4.364	4.364	3.657	16.000
2b  2a	$\frac{1}{4}$	5.099	4.35	3.66	18.700
2b 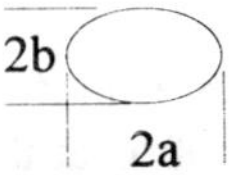 2a	0.9	5.331	2.930	4.439	18.233
2b  2a	$\frac{1}{8}$	6.490	2.904	5.597	20.585
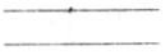	0	8.235	8.235	7.541	24.000
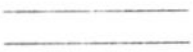 Insulated	0	5.385	-	4.861	24.000

At this Reynolds number the boundary layer may be laminar to turbulent.

(a) If the flow is laminar, Eq. (4.15) is applicable and we get at x = 30 cm.

$$Nu_x = 0.332 \times (0.698)^{1/3} \times (165{,}000)^{1/2}$$
$$= 123$$

Thus the heat transfer coefficient :

$$h_x = 123 \times \frac{k}{x} = 123 \times \frac{2.83 \times 10^{-2}}{0.3}$$
$$= 10.10 \text{ W/m}^2\text{K} \qquad \text{Ans.}$$

(b) If the flow is turbulent, Eq. (4.31) of Section 4.8 is applicable and we get ;

$$St_x = \frac{Nu_x}{Re_x\,Pr} = \frac{0.0296(165{,}000)^{-0.2}}{1+\{0.87(1.5\times0.698^{-1/6})(165000)^{0.1}(-0.302)\}}$$

$$= 0.0031$$

Prandtl suggested that the boundary layer may be assumed to start from the leading edge (even if it is not actually so).

then ;
$$h_x = Nu_x\frac{k}{x} = (0.0031\times Re_x\times Pr)\frac{k}{x}$$

$$= 0.0031\times165{,}000\times(0.698)\times\frac{2.83\times10^{-2}}{0.3}$$

$$= 33.8\ \text{W/m}^2\text{K}$$ **Ans.**

If we use Eq. (4.36) for turbulent flow we get ;

$$Nu_x = 0.0296\,(Re_x)^{0.8}\,(Pr)^{1/3}$$

$$= 0.0296\,(165{,}000)^{0.8}\,(0.698)^{1/3} = 390$$

or
$$h_x = Nu_x\left(\frac{k}{x}\right) = \frac{390\times0.0283}{0.3} = 36.4\ \text{W/m}^2\text{K}$$ **Ans.**

It will be noticed that h_x is much higher for turbulent flow than that for laminar flow.

Example 4.3

A fluid flows through a 10 cm ID pipe. Assume that the velocity is uniform over the cross-section of the pipe but the temperature varies linearly from 100°C at the pipe wall to 0°C at the center line. Calculate the heat transfer coefficient based on the mean fluid temperature if the heat flow rate from the pipe wall is 11,600 W/m².

Solution :

From Eq. (4.21), taking $\rho\,C_p$ = constant ;

$$t_m = \left\{\int_0^R tV(2\pi r)\,dr\right\} \div \left\{\int_0^R V(2\pi r)\,dr\right\}$$

Here, Vr = constant

$$t = 0+\frac{(100-0)}{(0.05-0)}r = 2000\,r$$

so
$$t_m = \left\{\int_0^R 2000\,r\,(r\,dr)\right\} \div \int_0^R r\,dr$$

$$= 2000\times\frac{2}{3}R = 2000\times\frac{2}{3}\times0.05$$

or
$$t_m = 66.67°C$$

so $$h = \frac{(q/A)}{(t_w - t_m)}$$

$$= \frac{11{,}600}{(100-66.67)}$$

$$= 350 \text{ W/m}^2\text{K}$$ Ans.

Example 4.4

1000 kg/hr of cheese at 15°C is pumped through a tube 7.5 cm in diameter. After passing through an unheated length of about 50 diameters, it passes through a 1.2 m length of tube maintained at 90°C. Calculate the heat transfer coefficient and the mean temperature of cheese leaving the heated section. For cheese the following values may be used :

$$\rho = 1100 \text{ kg/m}^3$$

$$\mu = 86{,}400 \text{ kg/hr-m}$$

$$C_p = 2.85 \text{ kJ/kg K}$$

$$k = 0.43 \text{ W/mK} = 1.55 \text{ kJ/hr m K}$$

Solution :

$$V = \left(\frac{1000}{\rho}\right)\frac{1}{A} = \frac{1000}{1100} \times \frac{4}{(\pi)(7.5)^2 \times 10^{-4}}$$

$$= 207 \text{ m/hr}$$

$$Re = \frac{VD\rho}{\mu} = \frac{(207)(0.075)(1100)}{86{,}400}$$

$$= 0.198$$

$$Pr = \frac{\mu C_p}{k} = \frac{86{,}400 \times 2.85}{1.55} = 158{,}500$$

From Eq. (4.26)

$$Nu = 3.66 + \frac{0.0668(D/L)(Re\,Pr)}{1 + 0.04\{(D/L)\,Re\,Pr\}^{2/3}}$$

$$= 3.66 + \frac{0.0668(0.075/1.2)(0.198 \times 158{,}500)}{1 + 0.04\left[\frac{0.075}{1.2} \times 0.198 \times 158{,}500\right]^{2/3}}$$

$$= 3.66 + \frac{131}{1 + 0.04(1970)^{2/3}}$$

$$= 3.66 + 17.9$$

$$= 21.56$$

or $$h = Nu\left(\frac{k}{D}\right) = 21.56 \times \frac{0.43}{0.075}$$

$$= 123.6 \text{ W/m}^2\text{K} = 443 \text{ kJ/hr m}^2\text{K} \qquad \text{Ans.}$$

Let t_1 and t_2 denote the temperature of cheese at entrance and exit of the heated section. Then, mean temperature of cheese is given by ;

$$t_m = \left(\frac{t_1 + t_2}{2}\right) = \left(\frac{15 + t_2}{2}\right)$$

But, $$q = h A (t_w - t_m) = W C_p (t_2 - t_1)$$

or $$(443)(\pi \times 0.075 \times 1.2)\left(90 - \frac{15 + t_2}{2}\right) = 1000 \times 2.85 (t_2 - 15)$$

$$125\left(82.5 - \frac{t_2}{2}\right) = 2850 (t_2 - 15)$$

or $$t_2 = 18.3°\text{C} \qquad \text{Ans.}$$

Example 4.5

A tube 4 m long is held at 100°C by steam jacketing. A fluid flows through the tube at the rate 175 kg/hr. The tube diameter is 20 mm. The mean temperature (or bulk temperature) of the fluid is 30°C. The property values of the fluid are given below :

Temperature	30°C	65°C	100°C
C_p (kJ/kg-K)	1.92	2.02	2.09
k (kJ/hr-m-K)	0.54	0.50	0.46
ρ (kg/m^3)	875	850	825
ν (m^2/sec)	12×10^{-6}	5.1×10^{-6}	2.6×10^{-6}

Solution :

$$t_b = 30°\text{C} \quad \text{and} \quad t_w = 100°\text{C}$$

$$t_f = \frac{t_b + t_w}{2} = \frac{30 + 100}{2} = 65°\text{C}$$

Let us first evaluate Pr and Re by taking the property values at t_f.

$$Pr_f = \left(\frac{\mu C_p}{k}\right)_f = \frac{(5.1 \times 10^{-6} \times 3600 \times 850)(2.02)}{0.50} = 62$$

$$Re_f = \left(\frac{V D}{\nu_f}\right)$$

or $$V = \frac{W}{\rho A} = \frac{175}{850 \times \frac{\pi}{4}(0.02)^2} = 655 \text{ m/hr}$$

$$Re_f = \frac{655 \times 0.02}{5.1 \times 10^{-6} \times 3600} = 715$$

Hence the flow is laminar. Applying Eq. (4.26) ;

$$\text{Nu} = 3.66 + \frac{0.0668(0.02/4)(715)(62)}{1+0.04\left\{\left(\frac{0.02}{4}\right)(715)(62)\right\}^{2/3}}$$

$$= 3.66 + 5.9 = 9.56$$

Hence $$h = 9.56 \times \frac{0.50}{0.02} = 238 \text{ kJ/hr-m}^2\text{-C}$$

$$= 66.3 \text{ W/m}^2\text{K}$$ Ans.

Alternatively, we may use Eq. (4.27) of Sieder and Tate ;

$$\frac{hD}{k_b} = 1.86\left(\frac{\mu_b}{\mu_w}\right)^{0.14}\left[\frac{VD\rho}{\mu_b}\left(\frac{\mu C_p}{k}\right)_b \frac{D}{L}\right]^{1/3}$$

$$\left(\frac{\mu_b}{\mu_w}\right)^{0.14} = \left(\frac{875\times 12\times 10^{-6}}{825\times 2.6\times 10^{-6}}\right)^{0.14} = 1.25$$

$$\frac{VD\rho}{\mu_b} = \frac{W}{\rho A}\times\frac{D}{\mu_b}\times\rho = \frac{175}{(\pi/4)(0.02)\times 12\times 10^{-6}\times 875\times 3600}$$

$$= 295$$

$$\left(\frac{\mu C_p}{k}\right)_b = \frac{(875\times 12\times 10^{-6}\times 3600)(1.92)}{0.54} = 134$$

$$\frac{D}{L} = \frac{0.02}{4} = 0.005$$

or $$\frac{hD}{k_b} = 1.86\,(1.25)\,[295\times 134\times 0.005]^{1/3}$$

$$= (1.86)\,(1.25)\,(198)^{1/3}$$

$$= 13.6$$

Hence $$h = \frac{13.6\times 0.54}{0.02} = 370 \text{ kJ/hrm}^2\text{K}$$

$$= 103 \text{ W/m}^2\text{K}$$ Ans.

4.8 CORRELATIONS FOR FORCED CONVECTION IN TURBULENT FLOW

(A) Turbulent flow over flat plates

(i) For the case of flow over flat plates, laminar boundary layer exists in the neighbourhood of the leading edge. It is only after a certain distance from the leading edge, when $Re_x > 5\times 10^5$ that the flow becomes turbulent. The general equation giving the local heat transfer coefficient is :

$$(St)_x = \frac{0.0296(Re_x)^{-0.2}}{1+\{0.87(1.5Pr^{-1/6})(Re_x)^{0.1}(Pr-1)\}} \tag{4.31}$$

where *St* is called the *Stanton number*. It was defined in Table 4.2 as ratio of the Nusselt number to the product of Reynolds and Prandtl numbers.

$$St = \frac{h}{\rho V C_p} = \frac{Nu}{Re\, Pr} \tag{4.32}$$

$$St_x = \frac{h_x}{\rho V C_p} = \frac{Nu_x}{Re_x\, Pr} \tag{4.33}$$

(ii) Sometimes another dimensionless number *j* is used to express the heat transfer correlations for turbulent flow. This is called *Colburn's j-factor* and is defined as ;

$$j = \left(\frac{h}{\rho V C_p}\right)\left(\frac{\mu C_p}{k}\right)^{2/3} = (St)\,(Pr)^{2/3} \tag{4.34}$$

(iii) A simplified form of Eq. (4.31) for turbulent flow over a flat plate is ;

$$j_x = \left(\frac{h_x}{C_p \rho V}\right)(Pr)^{2/3} = 0.0296\left(\frac{\nu}{V_\infty x}\right)^{0.2} \tag{4.35}$$

or

$$j_x = (St)_x\,(Pr)^{2/3} = \frac{0.0296}{(Re)_x^{\;0.2}} \tag{4.35a}$$

or

$$\frac{h_x\, x}{k} = Nu_x = 0.0296\,(Re_x)^{0.8}\,(Pr)^{1/3} \tag{4.36}$$

From Eq. (4.35) the average value of *h* between x = 0 and x = L is given by ;

$$j = \left(\frac{h}{C_p \rho V_\infty}\right)(Pr)^{2/3} = 0.037\left(\frac{\nu}{V_\infty L}\right)^{0.2} \tag{4.37}$$

or

$$j = (St)\,(Pr)^{2/3} = \frac{0.037}{(Re_L)^{0.2}} \tag{4.37a}$$

or

$$\frac{hL}{k} = Nu = 0.037\,(Re_L)^{0.8}\,(Pr)^{1/3} \tag{4.38}$$

In Eqs. (4.31) and (4.35) to (4.38) the fluid properties, C_p, μ, and *k* are evaluated at the film temperature.

(iv) Over a flat plate a laminar boundary layer exists over the forward portion between $x = 0$ and $x = L_{tr}$, and a turbulent boundary layer exists beyond L_{tr}. With this simple model the average value of *h* over a plate of length $L > L_{tr}$ is obtained by combining Eqs. (4.16) and (4.36) and by assuming that the turbulent boundary layer extrapolates to δ = 0 at x = 0.

$$h = \frac{1}{L}\left[\int_0^{L_{tr}} (h_x)_{lam}\, dx + \int_{L_{tr}}^{L} (h_x)_{turb}\, dx\right]$$

$$= \frac{1}{L}\left[\int_0^{L_{tr}}(h_x)_{\text{lam}}\,dx + \int_0^{L}(h_x)_{\text{turb}}\,dx - \int_0^{L_{tr}}(h_x)_{turb}\,dx\right]$$

$$= [(1/L)\,k\,(Pr)^{1/3}] \times [0.664\,(Re_{tr})^{0.5} + 0.037\,(Re_L)^{0.8} - 0.037\,(Re_{tr})^{0.8}]$$

or $$\frac{hL}{k} = 0.037\,(Pr)^{1/3}\,[(Re_L)^{0.8} + 17.9\,(Re_{tr})^{0.5} - (Re_{tr})^{0.8}] \qquad (4.39)$$

If $Re_{tr} = 3.2 \times 10^5$, Eq. (4.39) yields ;

$$Nu = \frac{hL}{k} = 0.037\,(Pr)^{1/3}\,[(Re_L)^{0.8} - 15{,}500] \qquad (4.40)$$

(b) Turbulent flow inside tubes

(i) The relation similar to Eq. (4.26) for forced convection in turbulent flow inside tubes is ;

$$St = \frac{0.0384\,(\text{Re})^{-0.25}}{1+(1.5Pr^{-1/6})(Re)^{-1.8}\,(Pr-1)} \qquad (4.41)$$

(ii) For fully developed flow in circular tubes, there are six widely accepted correlations of the data as given below :

1. Mc Adams has proposed a general correlation exactly on the lines of Eq. (4.5) as follows ;

$$\frac{hD}{k_b} = 0.023\left(\frac{V\,D\rho}{\mu_b}\right)^{0.8}\left(\frac{\mu C_p}{k}\right)^{0.4} \qquad (4.42)$$

or $$Nu = 0.023\,(Re)^{0.8}\,(Pr)^{0.4} \qquad (4.42a)$$

Here all the fluid properties are evaluated at the mean bulk temperature of the fluid.

2. Colburn suggested a relation in terms of his *j-factor* :

$$j = \left(\frac{h}{C_{pb}\,\rho V}\right)\left(\frac{\mu C_p}{k}\right)_f^{2/3} = \frac{0.023}{(V\,D\rho/\mu_f)^{0.2}} \qquad (4.43)$$

or $$St = 0.023\,(Re)^{-0.2}\,(Pr)^{-2/3} \qquad (4.43a)$$

Here all the fluid properties, except C_p in the computation of Stanton number, are evaluated at the film temperature $t_f = (t_w + t_b)/2$. In the Stanton number the value of C_p at fluid bulk temperature is used.

3. Sieder and Tate proposed the equation ;

$$\left(\frac{h}{C_{pb}\,\rho V}\right)\left(\frac{\mu C_p}{k}\right)_b^{2/3}\left(\frac{\mu_w}{\mu_b}\right)^{0.14} = \frac{0.027}{(V\,D\rho/\mu_b)^{0.2}} \qquad (4.44)$$

or $$St = 0.027\,(Re)^{-0.2}\,(Pr)^{-2/3}\left(\frac{\mu_w}{\mu_b}\right)^{-0.14} \qquad (4.44a)$$

Here all the fluid properties, except μ_w in viscosity ratio term are evaluated at the bulk temperature t_b. μ_w is evaluated at the wall temperature t_w.

Eqs. (4.41) to (4.44) may be applied for the following range of variables :

$$Pr : \text{from } 0.5 \text{ to } 120, \; Re : \text{from } 2300 \text{ to } 10^7, \text{ and } (L/D) > 50$$

4. Dittus and Boelter recommend for fully developed turbulent flow in smooth tubes :

$$Nu = 0.023\,(Re)^{0.8}\,(Pr)^n \tag{4.45}$$

where $n = 0.4$ for heating, $T_w > T_b$

$n = 0.3$ for cooling, $T_w < T_b$

All properties in this correlation are evaluated at the bulk temperature t_b. Eq. (4.45) has been verified experimentally to within ± 25 percent for uniform wall temperature as well as uniform heat flux conditions within the following ranges of parameters ;

$$0.7 < Pr < 160, \; (L/D) > 60 \text{ and } Re > 6000$$

Since this correlation does not take into account variations in physical properties due to temperature gradient at a given cross-section, it should only be used for moderate temperature differences $(T_b - T_w)$.

5. Petukhov developed more complex equations as below :

$$Nu = \frac{(f/8)\,Re\,Pr}{1.07 + 12.7(f/8)^{0.5}\,(Pr^{2/3} - 1)}\left(\frac{\mu_b}{\mu_w}\right)^n \tag{4.46}$$

where $n = 0.11$ for $T_w > T_b$

$n = 0.25$ for $T_w < T_b$

$n = 0$ for constant heat flux or for gases.

All properties are evaluated at $T_f = (T_w + T_b)/2$ except for μ_b and μ_w. The friction factor is obtained by ;

$$1/f = (1.82 \log_{10} Re - 1.64)^2 \tag{4.47}$$

Equation (4.46) is applicable for the following ranges :

$$Pr = 0.5 \text{ to } 2000, \; Re = 10^4 \text{ to } 5 \times 10^6 \text{ and } (\mu_b/\mu_w) = 0 \text{ to } 40$$

Another version of the above is the Petukhov -Popov equation :

$$Nu = \frac{(f/8)\,Re\,Pr}{K_1 + K_2\,(f/8)^{1/2}\,(Pr^{2/3} - 1)} \tag{4.48}$$

where $f = (1.82 \log_{10} Re - 1.64)^{-2}$ (4.47)

$K_1 = 1 + 3.4 f$

$K_2 = 11.7 + (1.8/Pr^{1/3})$

6. The Sleicher-Rouse equation is :

$$Nu = 5 + 0.015\,Re_f^{\,a}\,Pr_w^{\,b}$$

where $$a = 0.88 - \frac{0.24}{4 + Pr_w} \qquad (4.49)$$

$$b = \frac{1}{3} + 0.5\, e^{-0.6\, Pr_w}$$

The equation is valid for $0.1 < Pr_w < 10^5$ and $10^4 < Re < 10^6$.

(iii) For the case of liquid metals the following equation agrees well with experimental results :

$$Nu = 6.7 + 0.041\, (Re\, Pr)^{0.793}\, (e)^{41.8Pr} \qquad (4.50)$$

for $0.005 < Pr < 0.05$ and $Re > 10{,}000$

In terms of the Peclet number, Pe defined as $Pe = Re\, Pr$, Eq. (4.50) may be written as;

$$Nu = 6.7 + 0.0041\, (Pe)^{0.793}\, e^{41.8Pr} \qquad (4.51)$$

While using Eq. (4.50) the fluid properties may be evaluated at t_w because the temperature difference $(t_w - t_b)$ will not be large on account of the very high magnitudes of h. If desired, the effect of property variation may be taken into account by the equation ;

$$\frac{hD}{k_b} = Nu_w \left(\frac{\alpha_w}{\alpha_b}\right) \qquad (4.52)$$

where (Nu_w) is Nusselt number evaluated by Eq. (4.50) taking the fluid properties at t_w.

For design purpose, any of the following six equations may be used for liquid metals with reasonable accuracy ;

1. $Nu = 7 + 0.025\, Pe^{0.8}$ (4.53)
2. for uniform heat flux :

   $Nu = 0.625\, Pe^{0.4}$ (4.54)
3. for uniform heat flux ; $Re\, Pr > 100$; $L/D > 30$; and $3.6 \times 10^3 < Re < 9.05 \times 10^5$:

   $Nu = 4.82 + 0.0185\, (Re\, Pr)^{0.827}$ (4.55)
4. for constant surface temperature ; $Re\, Pr > 100$; $L/D > 30$:

   $Nu = 5.0 + 0.025\, (Re\, Pr)^{0.8}$ (4.56)
5. for uniform heat flux ; $Re\, Pr < 100$:

   $Nu = 3.0\, Re^{0.0833}$ (4.57)

   In Eqs. (4.53) to (4.57) the properties are evaluated at bulk temperature.
6. Heat transfer from a sphere to liquid sodium for $3.56 \times 10^4 < Re < 1.525 \times 10^5$:

   $Nu = 2 + 0.386\, (Re\, Pr)^{0.5}$ (4.58)

(iv) For fully developed turbulent flow of air in a helical coil, the following relation may be used for the ratio of h_c (in the coil) to h (in straight pipe)

$$\frac{h_c}{h} = 1 + 3.5\, (D/D_{HE}) \qquad (4.59)$$

where D_{HE} = helix diameter.

(c) Turbulent flow in non-circular conduits

For flow through the annular space between concentric pipes, Wiegand has proposed an equation for the heat transfer coefficient on the outer wall of the inner pipe as below :

$$\frac{h D_e}{k_b} = 0.023 \left(\frac{V D_e \rho}{\mu_b}\right)^{0.8} \left(\frac{\mu C_p}{k}\right)_b^{0.4} \left(\frac{D_2}{D_1}\right)^{0.45} \tag{4.60}$$

where D_1 = outside diameter of inner pipe

D_2 = inside diameter of outer pipe

D_e = hydraulic diameter.

For computing h at the inner wall of the outer pipe any of the equations for flow through circular tubes may be used with substitution of D_e for D. Even for the outer wall of the inner pipes these equations are quite satisfactory if we use D_e for D.

For flow through rectangular or triangular ducts reasonably accurate value of h may be obtained by multiplying the RHS of Eqs. (4.42) to (4.46) by 0.75 and by substitution of D_e for D in these equations.

Example 4.6

A flat plate 30 cm × 30 cm at 90°C is placed in an air stream of 60 m/sec. The temperature of air is 0°C. The critical *Re* where transition from laminar to turbulent flow occurs is 3.2×10^5 ; calculate rate of heat transfer from the upper surface.

Solution :

$$t_w = 90\text{ C}; \; t_a = 0\text{ C}$$

Hence average temperature is 45 C. From the property tables for air we get at 45 C:

$$k = 2.8 \times 10^{-2} \text{ W/mK}$$

$$\nu = 17.45 \times 10^{-6} \text{ m}^2\text{/sec}$$

$$Pr = 0.698$$

$$Re_{tr} = 320{,}000 = \frac{L_{tr} V}{\nu}$$

$$L_{tr} = \frac{320{,}000 \times 17.45 \times 10^{-6}}{60}$$

$$= 0.093 \text{ m} = 9.3 \text{ cm}$$

(a) Laminar flow exists up to 9.3 cm. Applying Eq. (4.17) to this length ;

$$Nu = 0.664 \times (0.698)^{1/3} \times (320{,}000)^{1/2}$$

$$= 304$$

Thus the average heat transfer coefficient ;

$$h = 304 \times \frac{2.8\times10^{-2}}{0.093} = 93 \text{ W/m}^2\text{K}$$

(b) Turbulent flow exists between x = 9.3 to x = 30 cm. From Eq. (4.36), for turbulent flow,

$$h_x = \frac{Nu_x\, k}{x} = 0.0296\,\frac{k}{x}\,(Re_x)^{0.8}\,(Pr)^{1/3}$$

The average value of *h* for the turbulent portion is given by ;

$$h = \frac{1}{L-L_{tr}}\int_{L_{tr}}^{L} h_x\, dx$$

or

$$h = \frac{1}{(0.3-0.093)}\left[0.0296(k)(Pr)^{1/3}\left(\frac{V}{\nu}\right)^{0.8} \times \int_{0.098}^{0.8}\left(\frac{1}{(x)^{0.2}}\right)dx\right]$$

$$= \frac{1}{0.207}\left\{0.0296(0.028)(0.698)^{1/3}\left(\frac{60}{17.45\times10^{-6}}\right)^{0.8}\left[\frac{0.3^{0.8}-0.093^{0.8}}{0.8}\right]\right\}$$

$$= \frac{1}{0.207}\left[0.0296\times0.028\times0.89\times1.68\times10^{5}\times\frac{(0.3804-0.15)}{0.8}\right]$$

$$= 177 \text{ W/m}^2\text{K}$$

Hence the heat transfer rate is given by,

$$q = [(0.3 \times x_{tr} \times 93) + \{0.3\,(0.3 - x_{tr}) \times 177\}] \times 90$$

$$= [(0.3 \times 0.093 \times 93) + \{0.3 \times 0.207 \times 177\}] \times 90$$

$$= 1220 \text{ W}$$

Ans.

Aliter

Since the value of Re_{tr} in this problem is the same as that used in deriving Eq. (4.40), we can use Eq. (4.40) directly :

$$Nu = \frac{hL}{k} = 0.037\,(Pr)^{1/3}\,[Re^{0.8} - 15{,}500]$$

$$= 0.037\,(0.698)^{1/3}\left[\left(\frac{60\times0.3}{17.45\times10^{-6}}\right)^{0.8} - 15{,}500\right]$$

$$= 0.037\,(0.89)\,(65{,}000 - 15{,}500)$$

$$= 1625$$

so

$$h = 1625 \times \frac{k}{L} = \frac{1625\times0.028}{0.3}$$

or $\quad h = 151\ W/m^2K$

$$q = 151 \times 0.3 \times 0.3 \times 90$$

or $\quad q = 1222\ W$ Ans.

Example 4.7

Calculate the heat transfer coefficient for water at 60°C flowing through a 0.625 cm diameter tube with a velocity of 0.9 m/sec. The tube wall temperature is 40 C.

Solution :

$$t_f = \frac{60+40}{2} = 50°C$$

At 50°C, for water,

$$\mu = 2.167\ kg/m\ hr$$

$$k = 0.63\ W/mK = 2.27\ kJ/m\ hr\ K$$

Taking $\quad C_p = 4.187\ kJ/kg\ K$ and $\rho = 1000\ kg/m^3$

$$Pr = \frac{\mu C_p}{k} = \frac{2.167 \times 4.187}{2.27} = 4.0$$

$$Re = \frac{VD\rho}{\mu} = \frac{0.9 \times (0.625 \times 10^{-2}) \times 1000}{2.167/3600}$$

$$= 9{,}350$$

From Eq. (4.43)

$$h = \frac{0.023 C_p \rho V}{(Pr)^{2/3}(Re)^{0.2}} = \frac{0.023(4.187)(1000)(0.9)(3600)}{(4)^{2/3}(9350)^{0.2}}$$

$$= 20{,}000\ kJ/hr\ m\ C$$

$$= 5{,}500\ W/m^2K$$

Ans.

Example 4.8

Water flows in a 50 mm diameter tube at a velocity of 80 cm/sec. The length of tube is 3 m. Determine the heat transfer coefficient if t_w = 70°C and the bulk temperature of water is 30°C.

Solution :

The property values for water at t_b = 30°C and t_f = (30 + 70)/2 = 50°C are :

Temperature	30°C	50°C
ρ (kg/m^3)	1000.0	988.1
Cp (kJ/kgK)	4.187	4.187
k (W/mK)	0.63	0.648
ν (m^2/sec)	9.2×10^{-7}	5.56×10^{-7}
Pr	6.13	3.54

Let us first calculate Re to find out whether flow is laminar or turbulent.

$$\text{Re} = \frac{VD}{\nu} = \frac{0.8 \times 0.05}{9.2 \times 10^{-7}} = 4.35 \times 10^4$$

Hence the flow is turbulent. Applying McAdams equation (4.42) we have ;

$$\frac{hD}{k_b} = 0.023 \left(\frac{VD\rho}{\mu_b}\right)^{0.8} \left(\frac{\mu C_p}{k}\right)_b^{0.4}$$

$$= 0.023\,(4.35 \times 10^4)^{0.8}\,(6.13)^{0.4}$$

$$= 0.023\,(5160)\,(2.07)$$

$$= 246$$

or $$h = \frac{246 \times 0.63}{0.05} = 3080 \text{ W/m}^2\text{K}$$ Ans.

Aliter

We may also find the value of *h* by applying the Colburn Eq. (4.43)

$$\left(\frac{h}{C_p \rho V}\right)\left(\frac{\mu C_p}{k}\right)_f^{2/3} = \frac{0.023}{(VD\rho/\mu_f)^{0.2}}$$

$$\left(\frac{\mu C_p}{k}\right)_f = \text{Pr}_f = 3.54$$

$$\text{Re} = \left(\frac{VD\rho}{\mu_f}\right) = \frac{0.8 \times 0.05}{5.56 \times 10^{-7}} = 7.2 \times 10^4$$

so $$\left(\frac{h}{4.187 \times 988.1 \times 0.8 \times 3600}\right)(3.54)^{2/3} = \frac{0.023}{(7.2 \times 10^4)^{0.2}}$$

$$h = \frac{0.023 \times 988.1 \times 0.8 \times 3600 \times 4.187}{(3.54)^{2/3}\,(7.2 \times 10^4)^{0.2}}$$

$$= \frac{18.25 \times 3600 \times 4.187}{2.325 \times 9.4}$$

$$= 12{,}560 \text{ kJ/hr m}^2\text{K} = 3500 \text{ W/m}^2\text{K}$$ Ans.

Example 4.9

In a heat exchanger water flows through a long 2.2 cm internal diameter copper tube at a bulk velocity of 2 m/sec and is heated by steam condensing at 150°C on the outside of the tube. The water enters at 15°C and leaves at 60°C. Find the heat transfer coefficient for water.

Solution :

Mean bulk temperature $$t_b = \frac{15+60}{2} = 37.5°\text{C}.$$

The physical properties of water at 37.5°C are :

$$\rho = 990 \text{ kg/m}^3$$

$$C_p = 4.16 \text{ kJ/kg K}$$

$$\mu = 0.00069 \text{ kg/m sec}$$

$$k = 0.63 \text{ W/mK} = 2.28 \text{ kJ/hr m K}$$

Hence,
$$Re = \frac{2 \times (2.2/100) \times 990}{0.00069}$$

$$= 63{,}200$$

$$Pr = \frac{(0.00069 \times 3600)(4.16)}{2.28}$$

$$= 4.58$$

Because the flow is turbulent and the tube is quite long we may calculate the heat transfer coefficient from any of the equations (4.42) to (4.46)

(a) Using McAdams equation (4.42)

$$h = \frac{(0.023)(0.63)}{(2.2/100)}(63{,}200)^{0.8}(4.58)^{0.4}$$

$$= 8400 \text{ W/m}^2\text{K}$$

Ans.

(b) To use the Colburn equation (4.43) we should know the film temperature. This, in turn, requires a knowledge of the wall temperature on the water side. If we assume that the average coefficient for condensation is of the same magnitude as the heat transfer coefficient for water, and neglect the temperature drop over the copper wall, the wall temperature will be (150 + 37.5)/2 = 93.75°C. Hence, the water side film temperature will be :

$$t_f = \frac{t_b + t_w}{2} = \frac{37.5 + 93.75}{2} = 65.6°\text{C}$$

At 65.6°C, the properties of water are :

$$\rho = 980 \text{ kg/m}^3$$

$$k = 0.67 \text{ W/mK} = 2.404 \text{ kJ/hr m K}$$

$$C_p = 4.187 \text{ kJ/kg K}$$

$$\mu = 0.000438 \text{ kg/m sec}$$

Hence,
$$Pr_f = \left(\frac{\mu C_p}{k}\right)_f = \frac{(0.000438 \times 3600)(4.187)}{2.404}$$

$$= 2.74$$

and
$$Re = \frac{VD\rho}{\mu_f} = \frac{2 \times 0.022 \times 980}{0.000438}$$

$$= 98{,}600$$

From Eq. (4.43)

$$\frac{h}{(C_p)_b \rho V} = \frac{0.023}{(Pr_f)^{2/3}(Re)^{0.2}}$$

$$= \frac{0.023}{(2.74)^{2/3}(98{,}600)^{0.2}} = 1.18 \times 10^{-3}$$

or

$$h = (1.18 \times 10^{-3})(4.187)(980)(2 \times 3600)$$

$$= 34{,}800 \text{ kJ/hr m}^2 \text{ K}$$

$$= 9700 \text{ W/m}^2\text{K}$$

Ans.

(c) According to Sieder and Tate equation (4.44),

$$\frac{h}{\rho V C_p} = 0.027\,(Pr)^{-2/3}\,(Re)^{-0.2}\left(\frac{\mu_w}{\mu_b}\right)^{-0.14}$$

Here Pr and Re are evaluated at the bulk temperature of the fluid as in the case of McAdams equation.

Hence, $Re = 63{,}200$

$Pr = 4.58$

At wall temperature of 93.75°C, the value of $\mu_w = 0.0003$ kg/m sec

$$\text{Hence, } h = (0.027)(990)(2 \times 3600)(4.187)(4.58)^{-2/3} \times (63{,}200)^{-0.2}\left(\frac{0.0003}{0.00069}\right)^{-0.14}$$

$$= 36{,}039 \text{ kJ/hr m}^2\text{W K} = 10{,}025 \text{ W/m}^2\text{K}$$

Ans.

(d) If we use the Dittus and Boelter Eq. (4.45), the result will be the same as that obtained from the McAdams Eq. (4.42). Ans.

(e) The Petukhov equation (4.46) will give,

$$Nu = \frac{(f/8)\,Re\,Pr}{1.07 + 12.7(f/8)^{0.5}(Pr^{2/3} - 1)}\left(\frac{\mu_b}{\mu_w}\right)^n$$

Since $T_w > T_b$, $n = 0.11$

Also, at film temperature Re = 98600 :

$Pr = 2.74$,

whereas μ_b (at 37.5 C) $= 0.00069$ kg/m s

μ_w (at 93.75 C) $= 0.000297$ kg/m s

The value of f may be computed from Eq. (4.47) or obtained from Fig.4.3.

$$f = (1.82 \log_{10} Re - 1.64)^{-2}$$

or

$$f = [1.82 \log_{10} 98{,}600 - 1.64]^{-2}$$

$$= 0.018$$

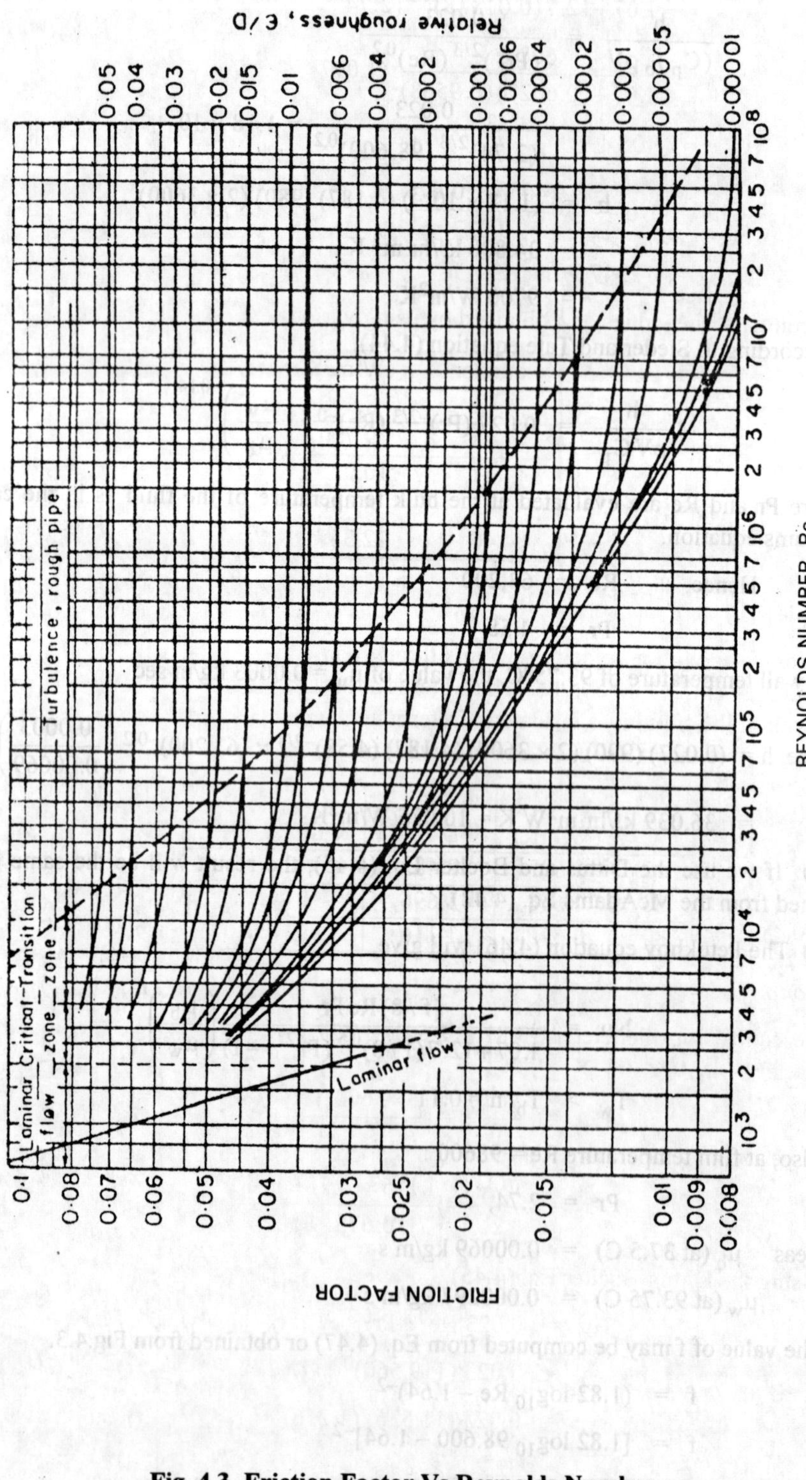

Fig. 4.3 Friction Factor Vs Reynolds Number

$$\text{Nu} = \frac{(0.018/8)(98600)(2.74)}{1.07+12.7(0.018/8)^{0.5}(2.74^{2/3}-1)}\left(\frac{0.00069}{0.000297}\right)^{0.11}$$

$$= \frac{608.6}{1.07+(0.6027)(0.9568)}(1.097)$$

$$= 405.4$$

$$h = \text{Nu}\frac{k}{D} = \frac{(405.4)(0.67)}{0.022}$$

$$= 12347 \text{ W/m}^2\text{K}$$ Ans.

To compare, the values of h obtained by using various equations are as below :

Equation	h (W/m²K)
4.42	8,400
4.43	9,700
4.44	10,025
4.45	8,400
4.46	12,347

Example 4.10

Water flows with a velocity of 3 m/s through an annulus between a 3.8 cm ID outer tube and 2.54 cm OD inner tube. The water is at 80°C and the inner wall is at 40°C. The outer wall of the annulus is insulated. Determine the value of the heat transfer coefficient by using different correlations. The following properties of water may be used ;

T (°C)	k (W/mK)	μ (kg/m hr)	ρ (kg/m³)	C_p (kJ/kgK)
40	0.623	2.485	992	4.18
60	0.658	1.696	981	4.18
80	0.675	1.116	973	4.18

Solution :

The equivalent diameter is (3.81 – 2.54) = 1.27 cm. So at the bulk temperature of 80°C,

$$\text{Re} = \frac{VD\rho}{\mu} = \frac{(3\times3600)(0.0127)(973)}{1.116} = 119{,}560$$

$$\text{Pr} = \frac{\mu C_p}{k} = \frac{1.116\times(4.18\times1000)}{0.675\times3600} = 1.92$$

(a) Using the Dittus-Boelter Eq. (4.45)

$$\text{Nu} = 0.023\ \text{Re}^{0.8}\ \text{Pr}^{0.3}$$

$$= 0.023\ (119{,}560)^{0.8}\ (1.92)^{0.3}$$

$$= 0.023\ (11{,}536)\ (1.216)$$

$$= 323$$

So $h = (Nu)k/D = (323 \times 0.675)/0.0127$

$= 17,167 \text{ W/m}^2\text{K}$ Ans.

(b) The Sieder-Tate Eq. (4.44a) gives

$$Nu = 0.027\ Re^{0.8}\ Pr^{1/3} \left(\frac{\mu_b}{\mu_w}\right)^{-0.14}$$

$$= 0.027\ (119,560)^{0.8}\ (1.92)^{1/3} \left(\frac{1.116}{2.485}\right)^{0.14}$$

$$= 0.027\ (11,536)\ (1.24)\ (0.894)$$

$$= 345$$

So $h = 18,336 \text{ W/m}^2\text{K}$ Ans.

(c) The Petukhov-Popov Eq. (4.48) gives

$$f = (1.82 \log_{10} Re - 1.64)^{-2} = (9.241 - 1.64)^{-2}$$

$$= 0.0173$$

$$K_1 = 1 + 3.4\ f = 1.0588$$

$$K_2 = 11.7 + 1.8/Pr^{1/3} = 11.7 + (1.8/1.24) = 13.15$$

$$Nu = \frac{(f/8)(Re)(Pr)}{K_1 + K_2 (f/8)^{1/2} (Pr^{2/3} - 1)}$$

$$= \frac{(0.0173/8) \times (119560) \times (1.92)}{1.0588 + 13.15(0.0173/8)^{1/2}\ (1.92^{2/3} - 1)}$$

$$= \frac{496.4}{1.0588 + 0.6115 \times 0.5451}$$

$$= 357$$

so, $h = 18,974 \text{ W/m}^2\text{K}$ Ans.

(d) The Sleicher-Rouse Eq. (4.49) gives :

$$Nu = 5 + 0.015\ Re_f^{\,a}\ Pr_w^{\,b}$$

$$a = 0.88 - \frac{0.24}{4 + Pr_w}$$

$$b = \frac{1}{3} + \frac{0.5}{e^{0.65 Pr_w}}$$

At 40°C,

$$Pr_w = (2.485)\ (4.18 \times 1000)/(0.623 \times 3600) = 4.63$$

so $a = 0.88 - \dfrac{0.24}{4 + 4.63} = 0.852$

$$b = \frac{1}{3} + \frac{0.5}{e^{0.6 \times 4.63}} = 0.364$$

At 60°C, $$Re_f = \frac{(3 \times 3600)(0.0127)(981)}{1.696} = 79{,}320$$

so $$Nu = 5 + 0.015\,(79{,}320)^{0.852}\,(4.63)^{0.384}$$

$$= 5 + 0.015\,(14.937)\,(1.748) = 397$$

hence $$h = \frac{397 \times 0.658}{0.0127} = 20{,}569 \text{ W/m}^2\text{K}$$ Ans.

Example 4.11

Calculate the heat transfer coefficient for mercury flowing through a 1.5 cm ID tube at the rate of 30,000 kg/hr. The bulk temperature of mercury may be taken as 40°C. For mercury the following values may be used at 40°C :

$$k = 8.7 \text{ W/mK} = 31.4 \text{ kJ/hr m K}$$

$$C_p = 0.138 \text{ kJ/kg K}$$

$$\mu = 5.4 \text{ kg/m hr}$$

Solution :

$$Re = \frac{VD\rho}{\mu} = \frac{GD}{\mu}$$

$$G = \frac{W}{A} = \frac{30{,}000}{(\pi/4)(1.5 \times 10^{-2})^2} = 1.63 \times 10^8 \text{ kg/hr m}^2$$

$$Re = \frac{1.63 \times 10^8 \times 1.5 \times 10^{-2}}{5.4} = 4.53 \times 10^5$$

$$Pr = \frac{\mu C_p}{k} = \frac{5.4 \times 0.138}{31.4} = 0.0238$$

$$Pe = (Re \times Pr) = 4.53 \times 10^5 \times 0.0238$$

$$= 10{,}750$$

since Re > 10,000 and Pr is in the range 0.005 to 0.05 we may use Eq. (4.50) ;

$$Nu = 6.7 + 0.0041\,(Pe)^{0.793}\,e^{41.8Pr}$$

$$= 6.7 + 0.0041\,(10{,}750)^{0.793}\,e^{41.8 \times 0.0238}$$

$$= 24.1$$

or $$h = \frac{24.1 \times 8.7}{1.5 \times 10^{-2}} = 14{,}000 \text{ W/m}^2\text{ K}$$ Ans.

If we use Eq. (4.53) we get ;

$$Nu = 7 + 0.025\,Pe^{0.8}$$

$$= 7 + 0.025\,(10{,}750)^{0.8}$$

$$= 7 + 0.025\,(1680) = 49$$

or $$h = \frac{49 \times 8.7}{1.5 \times 10^{-2}} = 28{,}400 \text{ W/m}^2\text{K}$$ Ans.

If we use Eq. (4.50), the result is ;

$$Nu = 0.625\,(Pe)^{0.4}$$

$$= 0.625\,(10{,}750)^{0.4}$$

$$= 0.625\,(41) = 25.6$$

So $$h = \frac{25.6 \times 8.7}{1.5 \times 10^{-2}} = 14{,}900 \text{ W/m}^2\text{K}$$ Ans.

Example 4.12

In a double pipe heat exchanger, water flows through the inner pipe maintaining its outer surface at a mean temperature of 60°C. Steam at 1.4 bars flows through the annular space between the two pipes at the rate of 110 kg/hr. The temperature of steam at entrance to the exchanger is 205°C. Using the Colburn equation determine the exit temperature of steam if the pipes are 2.5 m long. The outer diameter of the inner pipe is 3.35 cm and the inner diameter of the outer pipe which is insulated is 5.25 cm. For steam, use $C_p = 2.0$ kJ/kg K, $\rho = 0.688$ kg/m^3 and k and μ values at 1 atm pressure.

Solution :

The area of flow A_f is ;

$$A_f = \frac{\pi}{4}\,(D_2^2 - D_1^2) = \frac{\pi}{4}\,(5.25^2 - 3.35^2)$$

$$= 12.9 \text{ cm}^2 = 12.9 \times 10^{-4} \text{ m}^2$$

$$G = \frac{W}{A_f} = \frac{110}{12.9 \times 10^{-4}} = 8.53 \times 10^4 \text{ kg/hr m}^2$$

$$D_e = D_2 - D_1 = 5.25 - 3.35 = 1.9 \text{ cm or } 0.019 \text{ m}$$

Let us assume (as a first approximation) that the steam leaves at 145°C, i.e., it undergoes a temperature drop of 60°C. Then

$$t_b = \frac{205 + 145}{2} = 175 \text{ C}$$

$$t_w = 60°\text{C}$$

$$t_f = \frac{t_b + t_w}{2} = \frac{175 + 60}{2} = 117.5 \text{ C}$$

At 117.5 C, 1 atm, the properties of steam are :

$$\mu = 0.045 \text{ kg/hr m}$$

$$k = 0.089 \text{ kJ/hr m K} = 0.0252 \text{ W/mK}$$

Hence,

$$Pr_f = \left(\frac{\mu C_p}{k}\right)_f = \frac{0.045 \times 2.0}{0.089} = 1.0$$

$$Re = \frac{G D_e}{\mu} = \frac{8.53 \times 10^4 \times 0.019}{0.045}$$

$$= 3.6 \times 10^4$$

From Eq. (4.43),

$$\frac{h}{GC_{pb}} = \frac{0.023}{Re^{0.2}\, Pr_f^{\ 2/3}} = \frac{0.023}{(36000)^{0.2}\,(1)^{2/3}}$$

or $$\frac{h}{8.53 \times 10^4 \times 2.0} = \frac{0.023}{8.12}$$

or $$h = 485 \text{ kJ/hr m}^2 \text{ K}$$

Now, if we denote the inlet and exit temperature of steam by t_1 and t_2, a heat balance gives ;

$$q = W\, C_p\, (t_1 - t_2) = h\, A\, \Delta t_m$$

where Δt_m = mean temperature difference between steam and pipe surface.

Hence,

$$110 \times 2.0\, (205 - t_2) = (485)\left(\pi \times \frac{3.35}{100} \times 2.5\right)\Delta t_m$$

or $$(205 - t_2) = 0.58\, \Delta t_m \qquad \text{(a)}$$

It will be shown in Chapter 7 that Δt_m is given by ;

$$\Delta t_m = \frac{\Delta t_2 - \Delta t_1}{\log_e\left(\dfrac{\Delta t_2}{\Delta t_1}\right)}$$

where Δt_2 and Δt_1 are the temperature differences at exit and entrance, respectively.

Hence, $$\Delta t_m = \frac{(t_2 - 60) - (t_1 - 60)}{\log_e\left(\dfrac{t_2 - 60}{t_1 - 60}\right)}$$

$$\Delta t_m = \frac{205 - t_2}{\log_e\left(\dfrac{145}{t_2 - 60}\right)} \qquad \text{(b)}$$

From Eqs. (a) and (b),

$$(205 - t_2) = 0.58\, \frac{205 - t_2}{\log_e\left(\dfrac{145}{t_2 - 60}\right)}$$

or $$\log_e\left(\frac{145}{t_2-60}\right) = 0.58$$

$$\frac{145}{t_2-60} = 1.786$$

or $$t_2 = 141.3°C$$

Note that this value of t_2 agrees fairly well with our assumed value of 145°C. This means that our values of t_b and t_f were close enough and the property values of k and μ were taken at the correct temperature. Hence second trial is not required.

Thus,

Outlet temperature of steam = 141.3°C Ans.

Example 4.13

205 kg per hour of air are cooled from 100 C to 30 C by flowing through a 3.5 cm ID pipe coil bent into a helix of 0.6 m diameter. Calculate the value of the air-side heat transfer coefficient if the properties of air at 65°C are :

$$k = 0.0298 \text{ W/mK}$$

$$\mu = 0.003 \text{ kg/hr m}$$

$$Pr = 0.7$$

Solution :

Mass velocity of air, $$G = \frac{205\times 4}{\pi\times(3.5)^2\,10^{-4}}$$

$$= 2.14\times 10^4 \text{ kg/m}^2\text{hr}$$

From Eq. (4.42)

$$\frac{hD}{k} = 0.023\left(\frac{DG}{\mu}\right)^{0.8}\times\left(\frac{\mu C_p}{k}\right)^{0.4}$$

$$= 0.023\left[\left(\frac{3.5\times 10^{-2}\times 21.4\times 10^4}{0.003}\right)^{0.8}(0.7)^{0.4}\right] = 200$$

or $$h = \frac{200\times 0.0298}{0.035}$$

$$= 170 \text{ W/m}^2\text{K}$$

From Eq. (4.59) ;

$$\frac{h_e}{h} = 1+3.5\,\frac{D}{D_{HE}}$$

$$= 1+3.5\left(\frac{3.5}{60}\right)$$

$$= 1.204$$

or $$h_{coil} = (1.204)(170)$$

$$= 205.0 \text{ W/m}^2\text{K}$$ Ans.

4.9 CORRELATIONS FOR FORCED CONVECTION IN LAMINAR OR TURBULENT FLOW OVER SUBMERGED BODIES

A few of the important experimental correlations for laminar and turbulent flow normal to the outside surface of cylinders, spheres, tube banks, etc., are summarized in Tables 4.5 and 4.6, supported by Table 4.7. By normal flow is meant that the direction of flow is at right angles to the axis of the cylinder or tube. For flow parallel to the outside of tubes the value of h may be determined by using the equations for flow inside tubes provided that the equivalent diameter D_e is used in place of D.

Table 4.5 Mean heat transfer coefficient *h* for flow over cylinders, spheres, etc.,

Fluid	*Flow geometry*	*Correlation*	*Eq. No.*	*Range of applications*
Air water and hydro-carbon-oils	Flow normal to single cylinder, ie, at right angles to axis of cylinder	$\frac{hD}{k_f}(Pr_f)^{-0.3} = 0.35 + 0.56\left(\frac{DG}{\mu_f}\right)^{0.52}$	(4.61)	$1 < \frac{DG}{\mu_f} < 1000$
Air	Flow normal to single cylinder	$\frac{hD}{k_f}(Pr_f)^{-0.3} = 0.26\left(\frac{DG}{\mu_f}\right)^{0.60}$	(4.62)	$1000 < \frac{DG}{\mu_f} < 50{,}000$
Gases	Flow normal to single sphere	$\frac{hD}{k_f} = 0.37\left(\frac{DG}{\mu_f}\right)^{0.6}(Pr_f)^{1/3}$	(4.63)	$10 < \frac{DG}{\mu_f} < 70{,}000$
Air	Flow normal to spheres	$(hD/k_f) = 2.0 + 0.6\,(Pr_f)^{1/3}\,(Re_f)^{0.5}$	(4.64)	-
Water and oil	Flow normal to single sphere	$\frac{hD}{k_f}(Pr_f)^{-0.3} = 0.97 + 0.68\left(\frac{DG}{\mu_f}\right)^{0.5}$	(4.65)	$1 < \frac{DG}{\mu_f} < 1000$
	Flow through packed bed of spheres	$\frac{hD_p}{k} = 0.8\left(\frac{D_p G_0}{\mu}\right)^{0.7}(Pr)^{1/3}$	(4.66)	7 < D (mm) < 12.5
Air	Flow across short fins consisting of strips or pins	$\frac{h}{(C_p)_b G_{max}}\left[\frac{\mu C_p}{k}\right]_f^{2/3} = \left[\frac{z_p G_{max}}{\mu_f}\right]^{-0.5}$	(4.67)	$260 < \left[\frac{z_p G_{max}}{\mu_f}\right] < 12{,}000$

h - mean heat transfer coefficient
D - outside diameter
D_p - mean particle diameter - (6) (total particle surface) / (volume of particles)
G_0 - mass flow rate per unit cross-sectional area of empty bed
G_{max} - maximum mass velocity
Z_p - 2 (z + y) where z - width of fin and y - thickness of fin.

Table 4.6 : Mean heat transfer coefficient h for flow over tube banks

Fluid	Flow geometry	Correlation	Eq. No.	Range of application
Gases	Flow normal to tube banks 10 or more rows deep (staggered tubes)	$\frac{hD}{k_f} = 0.33 \left(\frac{DG_{max}}{\mu_f}\right)^{0.6} (Pr_f)^{1/3}$	(4.68)	$2000 < \frac{DG_{max}}{\mu_f} < 32,000$
	Flow normal to tube banks 10 or more rows deep (in-line tubes)	$\frac{hD}{k_f} = 0.26 \left(\frac{DG_{max}}{\mu_f}\right)^{0.6} (Pr_f)^{1/3}$ (If the number of rows *N* is less than ten multiply h from Eqs. (4.68) or (4.69) by ratio given in Table 4.8)	(4.69)	$2000 < \frac{DG_{max}}{\mu_f} < 32,000$
Air	Flow normal to banks of tubes 10 rows deep (staggered or in-line)	$\frac{hD}{k_f} = b \left(\frac{DG_{max}}{\mu_f}\right)^{n} (Pr_f)^{1/3}$ (Value of b and n are given in Table 4.7. If N is different from 10 multiply h from Eq. (4.70) by ratio given in Table 4.8)	(4.70)	$2000 < \frac{DG_{max}}{\mu_f} < 40,000$

h - mean heat transfer coefficient
D - outside diameter of tubes
G_{max} - maximum mass velocity

Table 4.7 Values of b and n for Eq. (4.70)

$x_L = \frac{S_L}{D}$	$x_T = 1.25$		$x_T = 1.5$		$x_T = 2$		$x_T = 3$	
	b	n	b	n	b	n	b	n
Staggered								
0.600	-	-	-	-	-	-	0.236	0.636
0.900	-	-	-	-	0.495	0.571	0.445	0.581
1.000	-	-	0.552	0.558	-	-	-	-
1.125	-	-	-	-	0.531	0.565	0.575	0.560
1.250	0.575	0.556	0.561	0.554	0.576	0.556	0.579	0.562
1.500	0.501	0.568	0.511	0.562	0.502	0.568	0.542	0.568
2.000	0.448	0.572	0.462	0.568	0.535	0.556	0.498	0.570
3.000	0.344	0.592	0.395	0.580	0.488	0.562	0.467	0.574
In line								
1.250	0.386	0.592	0.305	0.608	0.111	0.704	0.0703	0.752
1.500	0.407	0.586	0.278	0.620	0.112	0.702	0.0753	0.744
2.000	0.464	0.570	0.332	0.602	0.254	0.632	0.220	0.648
3.000	0.322	0.601	0.396	0.584	0.415	0.581	0.317	0.608

The equivalent diameter of a tube bundle can be determined as follows. Let us draw a plane at right angles to the axes of the tubes. The area of this plane which is bounded by the four planes containing the centre lines of the tubes in two adjacent longitudinal and two adjacent transverse rows minus the cross-sectional area of one tube, is called the free cross-sectional area of tube bundle. The hatched areas in Figs. 4.4 (a) and 4.4 (b) represent the free cross-sectional areas for in-line and staggered tubes. In both cases this area comes out to be $(S_L\, S_T) - \pi D^2/4$. The sum of the arcs formed by the intersection of the outside surfaces of the tubes and the free cross-sectional area is called *wetted perimeter*. It can be easily seen that the wetted perimeter is (πD) in both Figs. 4.4 (a) and 4.4 (b). The equivalent diameter is then :

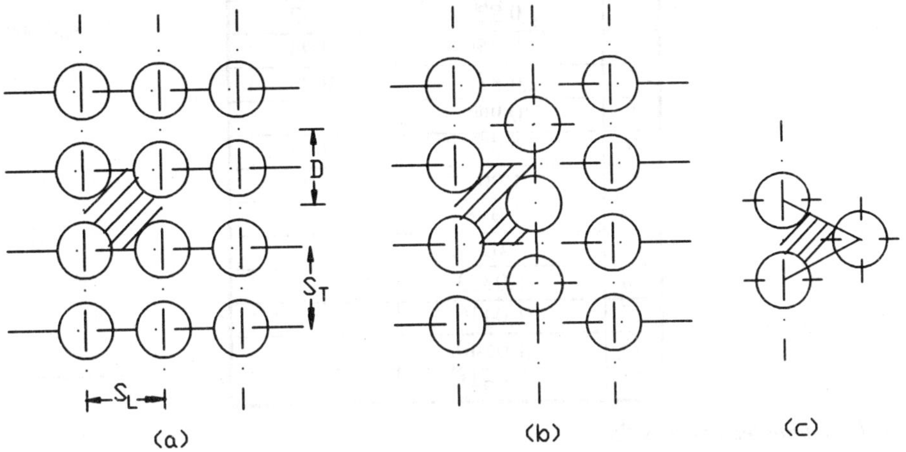

Fig. 4.4 In-line and Staggered Tubes

$$D_e = \frac{4 \times (\text{free cross-sectional area})}{\text{wetted perimeter}}$$

or
$$D_e = \frac{4(S_L\, S_T - \pi D^2/4)}{\pi D} \tag{4.71}$$

where S_L = longitudinal spacing or pitch between centres of tubes

S_T = Transverse spacing or pitch between centres of tubes

x_L = S_L/D, $x_T = S_T/D$, D = outside diameter

Some authors compute the free cross-sectional area for staggered tubes by joining the centres of every three tubes, as shown in Fig. 4.4 (c). In this case the free cross-sectional area is $(\frac{1}{2}\, S_L\, S_T - \pi D^2/8)$, and the wetted perimeter is $\pi D/2$. This gives ;

$$D_e = \frac{4(\frac{1}{2} S_L\, S_T - \pi D^2/8)}{(\pi D/2)}$$

or
$$D_e = \frac{4(S_L\, S_T - \pi D^2/4)}{\pi D} \tag{4.72}$$

Note that both the equations (4.71) and (4.72) give the same result if S_L and S_T are fixed.

Some specific cases of flow over submerged bodies are further discussed below.

Table 4.8 Ratio of h for N rows deep to that for 10 rows deep

N	Staggered tubes	In-line tubes
1	0.68	0.64
2	0.75	0.80
3	0.83	0.87
4	0.89	0.90
5	0.92	0.92
6	0.95	0.94
7	0.97	0.96
8	0.98	0.98
9	0.99	0.99
10	1.00	1.00
12	1.018	-
15	1.0215	-
18	1.0323	-
25	1.0430	-
35	1.0540	-
72	1.0645	-

(a) Tube bundles in cross flow

The determination of the convective heat transfer coefficient between a bank of tubes and a fluid flowing at right angles to the tubes is very important for the design of many types of heat exchangers. The heat transfer in flow over tubes depends largely on the flow pattern and turbulence which in turn depend on the velocity of the fluid and the size and arrangement of tubes. Several correlations for computing the value of *h* are given in Table 4.6.

Correlations have been developed by Zukauskas which predict tube bank heat transfer within 6 percent for 10 or more rows and are valid for $0.7 < Pr < 500$.

The correlations are of the form

$$Nu = \frac{hD}{k} = C Re^m \, Pr^{0.36} \, (Pr/Pr_w)^{0.25} \tag{4.73}$$

where D is tube diameter, Re is based on D, subscript *w* denotes tube wall and the properties are evaluated at bulk fluid temperature except Pr_w which is evaluated at tube wall temperature. The correlations are summarized in Table 4.9.

(b) Cylinders and Spheres

(i) For flow of gases and liquids over cylinders and spheres, the average heat transfer coefficient may be calculated by ;

$$\frac{hD}{k_f} = C\left(\frac{V_\infty D}{\nu_f}\right)^n Pr_f^{1/3} \tag{4.82}$$

where the properties are evaluated at the film temperature. The values of the constants *C* and *n* are given below :

Re_f	C	n
0.4 - 4	0.989	0.330
4 - 40	0.911	0.385
40 - 4000	0.683	0.466
4000 - 40,000	0.193	0.618
40,000 - 400,000	0.0266	0.805

Table 4.9 Zukauskas Correlations for heat transfer over tube banks (Ref. Fig. 4.4)

Correlation	*Eq. No.*	*Range of Re*	*Tube arrangement*
$Nu = 0.8\ Re^{0.4}\ Pr^{0.36}\ (Pr/Pr_w)^{0.25}$	(4.74)	10 – 100	In-line
$Nu = 0.9\ Re^{0.4}\ Pr^{0.36}\ (Pr/Pr_w)^{0.25}$	(4.75)	10 – 100	Staggered
$Nu = 0.27\ Re^{0.63}\ Pr^{0.36}\ (Pr/Pr_w)^{0.25}$	(4.76)	$10^3 - 2 \times 10^5$	In-line $S_T/S_L \geq 0.7$
$Nu = 0.35\ (S_T/S_L)^{0.2}\ Re^{0.6}\ Pr^{0.36}\ (Pr/Pr_w)^{0.25}$	(4.77)	$10^3 - 2 \times 10^5$	Staggered $S_T/S_L < 2$
$Nu = 0.40\ Re^{0.6}\ Pr^{0.36}\ (Pr/Pr_w)^{0.25}$	(4.78)	$10^3 - 2 \times 10^5$	Staggered $S_T/S_L \geq 2$
$Nu = 0.021\ Re^{0.84}\ Pr^{0.36}\ (Pr/Pr_w)^{0.25}$	(4.79)	$> 2 \times 10^5$	In-line
$Nu = 0.022\ Re^{0.84}\ Pr^{0.36}\ (Pr/Pr_w)^{0.25}$	(4.80)	$> 2 \times 10^5$	Staggered $Pr > 1$
$Nu = 0.019\ Re^{0.84}$	(4.81)	$> 2 \times 10^5$	Staggered $Pr = 0.7$

(ii) A correlation for cylinder in cross flow of liquids and gases has been given by Zukauskas

$$Nu = \frac{hD}{k} = C\left(\frac{V_\infty D}{\nu}\right)^m Pr^n \left(\frac{Pr}{Pr_w}\right)^{0.25} \tag{4.83}$$

where all fluid properties are evaluated at bulk fluid temperature except for Pr_w which is evaluated at tube wall temperature.

For $Pr < 10$, $n = 0.36$ and for $Pr > 10$, $n = 0.37$. Values of C and m are as follows :

Re	C	m
1 – 40	0.75	0.4
40 – 10^3	0.51	0.5
$10^4 - 2 \times 10^5$	0.26	0.6
$2 \times 10^5 - 10^6$	0.076	0.7

(iii) The heat transfer coefficient from liquids to cylinders in cross flow may be better calculated by ;

$$Nu_f = (0.35 + 0.56\ Re_f^{0.5})\ Pr_f^{0.3} \tag{4.84}$$

This relation may be used for $0.1 < Re_f < 10^5$ if excessive free stream turbulence is not present.

(iv) Eckert and Drake recommend the following relations for heat transfer from tubes in cross flow :

$$Nu = (0.43 + 0.50\, Re^{0.5})\, Pr^{0.38} \left(\frac{Pr_f}{Pr_w}\right)^{0.25} ;\ 1 < Re \times 10^3 \tag{4.85}$$

$$Nu = 0.25\, Re^{0.6}\, Pr^{0.38} \left(\frac{Pr_f}{Pr_w}\right)^{0.25} ;\ 10^3 < Re < 2 \times 10^5 \tag{4.86}$$

For gases the fluid properties are evaluated at the film temperature and the Prandtl number ratio may be dropped. For liquids the properties are evaluated at the free stream temperature.

(v) Churchill and Bernstein recommend the following relation which is applicable over a very wide range of experimental data :

$$Nu = 0.3 + \frac{0.62\, Re^{1/2}\, Pr^{1/3}}{[1+(0.4/Pr)^{2/3}]^{3/4}} [1 + (Re/282{,}000)^{5/8}]^{4/5} \tag{4.87}$$

for $10^2 < Re < 10^7 ;\ Pe > 0.2$ (4.87)

In the range of Reynolds number between 20,000 and 400,000 and $Pr > 0.2$ the following equation is better than Eq. (4.87)

$$Nu = 0.3 + \frac{0.62\, Re^{1/2}\, Pr^{1/3}}{[1+(0.4/Pr)^{2/3}]^{1/4}} [1 + (Re/282{,}000)^{1/2}] \tag{4.88}$$

Equations (4.87) and (4.88) were proposed on the basis of data for air, water and liquid sodium. Properties in these equations are evaluated at the film temperature.

(vi) For cylinders with noncircular cross section in gases Jakob recommends

$$Nu_p = B\, Re_b^{\,n} \tag{4.89}$$

The values of B and n are given in Table 4.10

(vii) Several correlations for flow of gases and liquids past spheres are given in Table 4.5. Achenbach has obtained relations applicable over a still wider range of Reynolds numbers for air (Pr = 0.71).

$$Nu = 2\,(0.25 + 3 \times 10^{-4}\, Re^{1.6})^{1/2} ;$$

$$100 < Re < 3 \times 10^5 \tag{4.90}$$

$$Nu = 430 + a\, Re + b\, Re^2 + c\, Re^3 ;$$

$$3 \times 10^5 < Re < 5 \times 10^6 \tag{4.91}$$

where $a = 5 \times 10^{-3};\ b = 0.25 \times 10^{-9};\ c = -\,3.1 \times 10^{-17};$

(viii) Vliet and Leppert recommend the following equation for heat transfer from spheres to oil and water for Re = 1 to 200,000

$$Nu\, Pr^{0.3} \left(\frac{\mu_w}{\mu}\right)^{0.25} = 1.2 + 0.53\, Re^{0.54} \tag{4.92}$$

All properties are evaluated at free stream temperature except μ_w which is computed at the surface temperature of the sphere.

Table 4.10 Values of B and N for equation (4.89)

Flow direction and profile	Re_D			
	From	To	n	b
→ 	5,000	100,000	0.588	0.222
→ 	2,500	15,000	0.612	0.224
→ 	2,500	7,500	0.624	0.261
→ 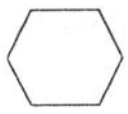	5,000	100,000	0.638	0.138
→ 	5,000	19,500	0.638	0.144
→ 	5,000	100,000	0.675	0.092
→ 	2,500	8,000	0.699	0.160
→ 	4,000	15,000	0.731	0.205
→ 	19,500	100,000	0.782	0.035
→ 	3,000	15,000	0.804	0.085

(ix) Whitaker recommends a single equation for gases and liquids flowing past spheres which is valid for *Re* from 3.5 to 80,000 and *Pr* from 0.7 to 380.

$$Nu = 2 + (0.4\,Re^{1/2} + 0.06\,Re^{2/3})\,Pr^{0.4}\left(\frac{\mu}{\mu_w}\right)^{0.25} \qquad (4.93)$$

The properties are evaluated at the free stream temperature.

(c) Bluff objects

The heat transfer coefficient in the separated wake regions behind a flat plate (width L) placed perpendicular to the flow and a half-round cylinder (diameter, D) may be calculated by the following equations for $Re = 1$ to 4×10^5.

Normal flat plate :

$$Nu = \frac{hL}{k} = 0.20\,Re_L^{2/3} \qquad (4.94)$$

Half-round cylinder with flat rear surface :

$$Nu = \frac{hD}{k} = 0.16\,Re_D^{2/3} \qquad (4.95)$$

The heat transfer coefficients from square plates to air at various angles to the free stream may be computed by

$$\left(\frac{h}{C_p \rho V_\infty}\right) Pr^{2/3} = 0.930\, Re_L^{-1/2} \tag{4.96}$$

for Re_L from 2×10^4 to 10^5. L is the length of the edge of the plate.

Example 4.14

In a cooling tower the average diameter of the water drops from the water spray is 1.5 mm. The relative velocity between the water drops and the air current is 0.9 m/sec. The air and water temperatures may be taken to be 15°C and 80°C respectively. Find the convective film coefficient.

Solution :

$$t = \frac{15+80}{2} = 47.5°C$$

At 47.5°C, the properties of air at 1 atm are :

$$\rho = 1 \text{ kg/m}^3$$

$$C_p = 1.0 \text{ kJ/kg K}$$

$$k = 0.1 \text{ kJ/hr m K} = 0.0278 \text{ W/mK}$$

$$\mu = 0.0708 \text{ kg/hr m}$$

Hence,

$$Pr_f = \left(\frac{\mu C_p}{k}\right)_f = \frac{0.0708 \times 1.0}{0.1}$$

$$= 0.708$$

$$Re_f = \frac{VD\rho}{\mu_f}$$

$$= \frac{(0.9 \times 3600)(1.5 \times 10^{-3})(1)}{0.0708}$$

$$= 69$$

From Eq. (4.63)

$$\frac{hD}{k_f} = 0.37\left(\frac{DG}{\mu_f}\right)^{0.6} (Pr_f)^{1/3}$$

$$= 0.37\,(69)^{0.6}\,(0.708)^{1/3}$$

$$= 0.37 \times 12.7 \times 0.893$$

$$= 4.2$$

or $$h = \frac{4.2 \times 0.0278}{1.5 \times 10^{-3}} = 78 \text{ W/m}^2\text{K}$$ **Ans.**

From Eq. (4.64)

$$\frac{hD}{k_f} = 2 + 0.6\,(Pr_f)^{1/3}\,(Re_f)^{0.5}$$

$$= 2 + 0.6\,(0.893)\,(69)^{0.5}$$

$$= 2 + 4.45 = 6.45$$

or $$h = \frac{6.45 \times 0.0278}{1.5 \times 10^{-3}}$$

$$= 120\ W/m^2K$$ Ans.

Example 4.15

Air at 25°C flows with a velocity of 50 m/s normal to a wire of diameter 0.0394 mm. The surface temperature of the wire is maintained at 50°C. Calculate the heat transfer coefficient.

Solution :

$$T_f = (25 + 50)/2 = 37.5°C$$

At 37.5°C the properties of air are :

$$k = 0.027\ W/mK$$

$$\nu = 17.66 \times 10^{-6}\ m^2/s$$

$$Pr = 0.706$$

Now, $$Re = \frac{V_\infty D}{\nu} = \frac{(50)(0.0394 \times 10^{-3})}{17.66 \times 10^{-6}} = 112$$

$$Pe = Re.\ Pr = (112)\,(0.706) = 78.8$$

Using Eq. (4.82) with $C = 0.683$ and $n = 0.466$

$$Nu = (0.683)\,(112)^{0.466}\,(0.706)^{1.3} = 5.47$$

$$h = Nu\left(\frac{k}{D}\right) = (5.47)\,\frac{0.027}{0.0394 \times 10^{-3}}$$

$$= 3755\ W/m^2K$$ Ans.

If we use Eq. (4.87)

$$Nu = 0.3 + \frac{(0.62)(112)^{0.5}(0.705)^{1/3}}{\left[1+\left(\frac{0.4}{0.705}\right)^{2/3}\right]^{3/4}}\left[1+\left(\frac{112}{282{,}000}\right)^{5/8}\right]^{4/5} \qquad (4.88)$$

$$= 4.27$$

$$h = \frac{(4.27)(0.027)}{0.0394 \times 10^{-3}} = 2926\ W/m^2K$$ Ans.

Example 4.16

Air is being heated by blowing it over a bank of 5 rows of tubes. The tubes are 2.5 cm OD and 1.2 m long. The tubes are spaced with their centres at the corners of equilateral triangles ; the centre to centre distance is 2 tube diameters. There are 5 tubes per row. There is space of 1.25 cm between the outside tubes and the duct. The surface temperature of the tubes may be assumed to be 108.5°C. If 14.5 m³/min of air is heated from 15°C to 40°C, calculate the heat transfer coefficient for air.

Solution :

$$t_b = \frac{15+40}{2} = 27.5°C$$

$$t_f = \frac{t_b + t_w}{2} = \frac{27.5+108.5}{2} = 68°C$$

For air at 68°C ;

$$k = 0.1 \text{ kJ/hr m K} = 0.028 \text{ W/mK}$$

$$\mu = 2.04 \times 10^{-5} \text{ kg/m sec}$$

$$C_p = 1.0 \text{ kJ/kg K}$$

$$Pr = 0.72$$

The density is based on air at 15°C. This value at 15°C is 1.216 kg/m³.

The minimum free cross-sectional area for flow is computed as follows :

Flow area between duct and outer tubes,

$$= 2\left(1.2 \times \frac{1.25}{100}\right) = 0.03 \text{ m}^2$$

Flow area between the 5 tubes of a row,

$$= 4\left(1.2 \times \frac{2.5}{100}\right) = 0.12 \text{ m}^2$$

Total flow area $A_f = 0.15 \text{ m}^2$

Hence, the maximum mass velocity is ;

$$G_{max} = \frac{W}{A_f} = \frac{14.5 \times 1.216}{0.15} \times \frac{1}{60}$$

$$= 1.96 \text{ kg/m}^2 \text{ sec}$$

or

$$\frac{DG_{max}}{\mu_f} = \frac{2.5}{100} \times 1.96 \times \frac{1}{2.04 \times 10^{-5}}$$

$$= 2400$$

Let us use Eq. (4.70) ;

$$\frac{hD}{k_f} = b\left(\frac{DG_{max}}{\mu_f}\right)^n (Pr_f)^{1/3}$$

From Table 4.7, for $x_L = x_T = 2$, and staggered arrangement, b = 0.535, n = 0.556

so $$\frac{hD}{k_f} = 0.535\,(2400)^{0.556}\,(0.72)^{1/3} = 0.535 \times 75 \times 0.9$$

$$= 36.1$$

or $$h = \frac{36.1 \times 0.028}{0.025} = 40.4 \text{ W/m}^2\text{K}$$

Since the number of rows N is less than 10 we should multiply the above value of h by the ratio from Table 4.8. This ratio is 0.92.

or $$h = 40.4 \times 0.92$$

$$= 37.2 \text{ W/m}^2\text{ K}$$ Ans.

Aliter

We may also use Eq. (4.68) ;

$$\frac{hD}{k_f} = 0.33\left(\frac{DG_{max}}{\mu_f}\right)^{0.6}(Pr_f)^{1/3}$$

$$Pr_f = \left(\frac{\mu C_p}{k}\right)_f = \frac{2.04 \times 10^{-5} \times 3600 \times 1.0}{0.1}$$

$$= 0.72$$

so $$\frac{hD}{k_f} = 0.33\ (2400)^{0.6}\,(0.72)^{1/3}$$

$$= 31.8$$

or $$h = \frac{31.8 \times 0.028}{0.025} = 36.3 \text{ W/m}^2\text{K}$$

Multiplying by the ratio 0.92 from Table 4.8 for N = 5,

$$h = 36.3 \times 0.92$$

$$= 33.6 \text{ W/m}^2\text{K}$$ Ans.

Aliter

Let us use one of the equations given by Zukauskas. At the bulk temperature of 27.5° C.

$$\rho = 1.175 \text{ kg/m}^3;\ k = 0.0265 \text{ W/mK},\ \mu = 0.06579\ \frac{\text{kg}}{\text{hr m}}$$

$$Pr = 0.7015$$

At the surface temperature of the tube $Pr_w = 0.687$.

$$G_{max} = \frac{W}{A_f} = \frac{14.5 \times 1.175}{0.15 \times 60} = 1.8931\ \frac{\text{kg}}{\text{m}^2\text{s}}$$

$$\frac{DG_{max}}{\mu} = \frac{0.025 \times 1.8931}{0.06579/3600} = 2590$$

Since $S_T/S_L = 2/2 = 1$, we may use Eq. (4.77).

$$Nu = 0.35 \left(\frac{S_T}{S_L}\right)^{0.2} (Re)^{0.6} (Pr)^{0.36} \left(\frac{Pr}{Pr_w}\right)^{0.25}$$

$$= (0.35)\,(1)^{0.2}\,(2590)^{0.6}\,(0.7015)^{0.36}\,(0.7015/0.687)^{0.25}$$

$$= (0.35)\,(1)\,(111.7)\,(0.88)\,(1.0052)$$

$$= 34.6$$

$$\therefore \quad h = 34.6 \times \frac{0.0265}{0.025} = 36.66 \text{ W/m}^2\text{K}$$ Ans.

Example 4.17

In a shell and tube heat exchanger brine flows inside the tubes and water flows outside the tubes and essentially parallel to them. There is only one pass of water. The water enters at 13°C and leaves at 7°C. The flow rate is 150 kg/min. The shell is 12.7 cm inside diameter and there are 19 tubes, each 1.9 cm OD spaced at the apexes of 2.5 cm equilateral triangles.

(a) Determine the water-side heat transfer coefficient.

(b) If the shell is made of 20 cm ID pipe and if it is provided with transverse baffles spaced 7.5 cm apart so that the water flows essentially normal to the tubes, find the value of h.

Solution :

$$t_b = \frac{13+7}{2} = 10 \text{ C}$$

At 10 C the properties of water are,

$$\rho = 998.4 \text{ kg/m}^3$$

$$C_p = 4.187 \text{ kJ/kg K}$$

$$k = 0.58 \text{ W/m K} = 2.09 \text{ kJ/hr m K}$$

$$\mu = 4.75 \text{ kg/hr m}$$

$$Pr = 9.5$$

Flow area for water,

$$= \frac{\pi}{4}(12.7)^2 - (19)\frac{\pi}{4}\,1.9^2$$

$$= 126.5 - 54 = 72.5 \text{ cm}^2$$

or

$$V = \left(\frac{150}{998.4} \times \frac{1}{60}\right) \frac{1}{72.5 \times 10^{-4}} = 0.346 \text{ m/sec}$$

From Eq. (4.71), the equivalent diameter for the tube bundle is :

$$D_e = \frac{4(S_L S_T - \pi D^2/4)}{\pi D}$$

$$S_L = 2.5 \text{ cm}$$

$$S_T = 2.5 \sin 60 = 2.16 \text{ cm}$$

$$D = 1.9 \text{ cm}$$

so
$$D_e = \frac{4\left(2.5 \times 2.16 - \frac{\pi}{4} \times 1.9^2\right)}{\pi \times 1.9} = 1.72 \text{ cm}$$

$$Re = \frac{V D_e \rho}{\mu} = \frac{(0.346 \times 3600)(1.72 \times 10^{-2})(998.4)}{4.75} = 4400$$

From Eq. (4.42)

$$\frac{h D_e}{k} = 0.023 (Re)^{0.8} (Pr)^{0.4}$$

$$= 0.023 (4400)^{0.8} (9.5)^{0.4}$$

$$= 0.023 \times 815 \times 2.46$$

$$= 46$$

or
$$h = \frac{46 \times 0.58}{1.72 \times 10^{-2}} = 1560 \text{ W/m}^2\text{K}$$ Ans.

(b) In this case, there will be 8 tubes across any diameter of the shell because the tube spacing is 2.5 cm and shell diameter is 20 cm. Also, on account of the transverse baffles, the water will make several passes over the tubes. The flow area for water will be,

$$A_f = (20 \times 7.5) - (8 \times 1.9 \times 7.5)$$

$$= 150 - 114 = 36 \text{ cm}^2$$

or
$$G = \frac{150}{60} \times \frac{1}{36 \times 10^{-4}} = 695 \text{ kg/m}^2\text{sec}$$

$$\frac{DG}{\mu} = \frac{(1.9 \times 10^{-2})(695 \times 3600)}{4.75} = 10{,}000$$

From Eq. (4.68),

$$\frac{hD}{k_f} = 0.33 \left(\frac{DG_{max}}{\mu_f}\right)^{0.6} (Pr_f)^{1/3}$$

$$= 0.33 (10000)^{0.6} (9.5)^{1/3}$$

$$= 0.33 \times 252 \times 2.12$$

$$= 177$$

$$\text{or} \qquad h = \frac{177 \times 0.58}{1.9 \times 10^{-2}}$$

$$= 5410 \text{ W/m}^2\text{K} \qquad \text{Ans.}$$

4.10 HYDRODYNAMIC AND THERMAL BOUNDARY LAYERS OVER A FLAT PLATE

Another method of evaluating the heat transfer coefficient during convection, as proposed in Section 4.3, is the solution of *Boundary Layer Equations*. We shall now discuss the boundary layer, its equations, and how they can be solved.

Consider a very thin flat plate parallel to an unbounded flow stream of velocity V_∞. The flow upstream of the plate is *one-dimensional* i.e., $V_\infty = V_x$ and $V_y = 0$. When the fluid meets the leading edge of the plate, friction causes it to adhere to the surface and a *laminar boundary layer* is developed as shown in Fig. 4.5. At any distance x from the leading edge the x component of fluid velocity V_x increases from $V_x = 0$ at the plate surface to $V_x = V_\infty$ at the edge of the boundary layer. The velocity distribution in the boundary layer is also shown in the figure.

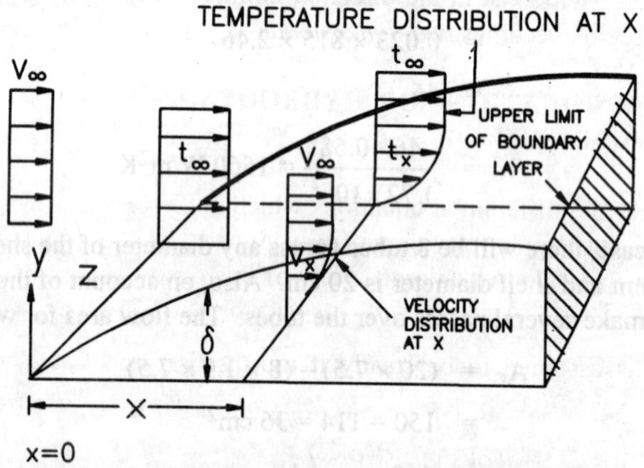

Fig. 4.5 Boundary Layer over a Flat Plate

If the approaching free stream temperature is t_∞, a *thermal boundary layer* will also exist along with the *velocity* or *hydrodynamic boundary layer*. The relationship between the two boundary layers will be discussed later. For the moment suffice it to say that at any distance x the temperature of the fluid will vary from $t = t_s$ at plate surface to $t = t_\infty$ at the edge of the thermal boundary layer.

Considering the hydrodynamic boundary layer it is clear that as the boundary layer thickness increases with x, more and more fluid is included in the boundary layer and is retarded by friction. As a result the average velocity of the fluid in x direction is decreased. In order to satisfy the continuity of flow the initially parallel stream lines of the oncoming stream must diverge away from the plate so that the flow area between the stream lines is increased. This is shown in Fig. 4.6. On account of this divergence of

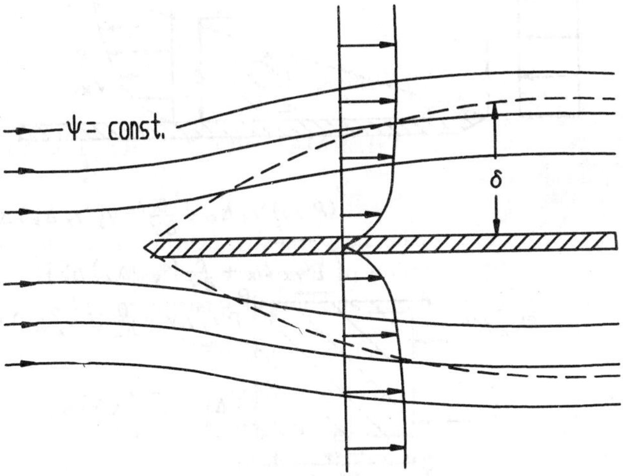

Fig. 4.6 Stream Lines for flow over a Flat Plate

stream lines the velocity vectors will no longer be parallel to the x-direction and a small velocity component V_y will exist in the direction normal to the place. Thus the flow in the boundary layer region becomes *two-dimensional.*

4.11 MOMENTUM EQUATION FOR HYDRODYNAMIC BOUNDARY LAYER OVER A FLAT PLATE

Let us now derive the differential equations which apply to the boundary layer. For this purpose we consider a control volume ($\Delta x \times \Delta y \times 1$) located in the boundary layer region, as shown in Fig. 4.7. An enlarged view of this control volume is also shown in the figure.

Since the flow is assumed to be steady it must satisfy the *steady flow continuity equation*

$$\frac{\partial(\rho V_x)}{\partial x} + \frac{\partial(\rho V_y)}{\partial y} = 0 \tag{4.98}$$

or

$$\frac{\partial G_x}{\partial x} + \frac{\partial G_y}{\partial y} = 0$$

(where G indicates mass velocity, ρV)

To obtain the momentum equation let us make a momentum balance in the x direction:

$$\left\{\begin{matrix}\text{Rate of x} - \text{momentum}\\ \text{leaving control volume}\end{matrix}\right\} - \left\{\begin{matrix}\text{Rate of x} - \text{momentum}\\ \text{entering control volume}\end{matrix}\right\} +$$

$$\left\{\begin{matrix}\text{Rate of accumulation}\\ \text{of x} - \text{momentum}\\ \text{within control volume}\end{matrix}\right\} = \left\{\begin{matrix}\text{Summation of external}\\ \text{forces in x} - \text{direction}\end{matrix}\right\} \tag{4.99}$$

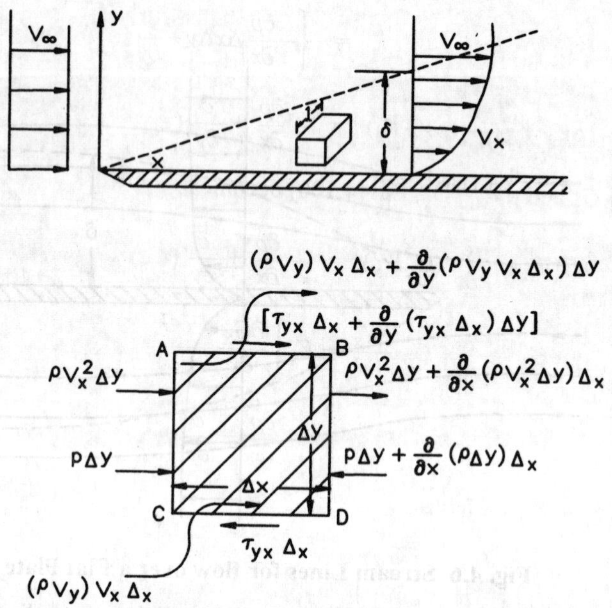

Fig. 4.7 Derivation of Boundary Layer Momentum Equation (4.107)

(i) Rate of x-momentum entering control volume due to flow across faces AC and CD

$$= \rho V_x^2 \Delta y + (\rho V_y) V_x \Delta x$$

(ii) Rate of x-momentum leaving control volume due to flow across faces *BD* and *AB*,

$$= \rho V_x^2 \Delta y + \frac{\partial}{\partial x}(\rho V_x^2 \Delta y)\Delta x + (\rho V_y) V_x \Delta x + \frac{\partial}{\partial y}(\rho V_y V_x \Delta x)\Delta y$$

$$= \rho V_x^2 \Delta y + \frac{\partial}{\partial x}(\rho V_x^2)\Delta x \Delta y + \rho V_y V_x \Delta x + \frac{\partial}{\partial y}(\rho V_y V_x)\Delta x \Delta y$$

(iii) Rate of accumulation of x-momentum within control volume is zero because the flow is *steady*.

(iv) The external forces acting on the control volume in the x-direction consist of the normal pressure force and the viscous force. Hence, summation of external forces on control volume in x-direction,

$$= p\,\Delta y - \left\{p\,\Delta y + \frac{\partial}{\partial x}(p\,\Delta y)\Delta x\right\} - \tau_{yx}\Delta x + \left\{\tau_{yx}\Delta x + \frac{\partial}{\partial y}(\tau_{yx}\Delta x)\Delta y\right\}$$

$$= -\frac{\partial p}{\partial x}\Delta x \Delta y + \frac{\partial}{\partial y}(\tau_{yx})\Delta x \Delta y$$

Hence Eq. (4.99) gives on substitution of the above four relations ;

$$\left\{\rho V_x^2 \Delta y + \frac{\partial}{\partial x}(\rho V_x^2)\Delta x \Delta y + \rho V_y V_x \Delta x + \frac{\partial}{\partial y}(\rho V_y V_x)\Delta x \Delta y\right\}$$

$$- \{\rho V_x^2 \Delta y - \rho V_y V_x \Delta x\} + \{0\}$$

$$= \left\{-\frac{\partial p}{\partial x}\Delta x\Delta y+\frac{\partial}{\partial y}(\tau_{yx})\Delta x\,\Delta y\right\}$$

or
$$\frac{\partial}{\partial x}(\rho V_x^2)+\frac{\partial}{\partial y}(\rho V_y V_x) = -\frac{\partial p}{\partial x}+\frac{\partial}{\partial y}(\tau_{yx}) \quad (4.100)$$

Since $\rho V_x = G_x$ and $\rho V_y = G_y$, Eq. (4.100) becomes :

$$\frac{\partial}{\partial x}(G_x V_x)+\frac{\partial}{\partial y}(G_y V_x) = -\frac{\partial p}{\partial x}+\frac{\partial}{\partial y}(\tau_{yx})$$

or
$$G_x\frac{\partial V_x}{\partial x}+G_y\frac{\partial V_x}{\partial y}+V_x\left(\frac{\partial G_x}{\partial x}+\frac{\partial G_y}{\partial y}\right)$$
$$= -\frac{\partial p}{\partial x}+\frac{\partial}{\partial y}(\tau_{yx})$$

But, from Eq. (4.98) we have :

$$Vx\left(\frac{\partial G_x}{\partial x}+\frac{\partial G_y}{\partial y}\right) = 0$$

hence,
$$G_x\frac{\partial V_x}{\partial x}+G_y\frac{\partial V_x}{\partial y} = -\frac{\partial p}{\partial x}+\frac{\partial}{\partial y}(\tau_{yx}) \quad (4.101)$$

In order to simplify Eq. (4.101) further, let us make the following assumptions for the boundary layer :

(a) The x component of velocity is much larger than the y-component.

$$V_x >> V_y \quad (4.102)$$

(b) The variation in V_x with y is much larger than the variation of V_x with x or the variation of V_y with x or y,

$$\left.\begin{array}{l}\frac{\partial V_x}{\partial y}>>\frac{\partial V_x}{\partial x}\\ \frac{\partial V_x}{\partial y}>>\frac{\partial V_y}{\partial x}\\ \frac{\partial V_x}{\partial y}>>\frac{\partial V_y}{\partial y}\end{array}\right\} \quad (4.103)$$

(c) The variation of p with y is negligible

$$\partial p/\partial y \approx 0 \quad (4.104)$$

(d) It also follows from Eq. (4.104)

$$\partial p/\partial x \approx 0$$

Now,
$$\tau_{yx} = \mu\left[\frac{\partial V_y}{\partial x}+\frac{\partial V_x}{\partial y}\right]=\mu\frac{\partial V_x}{\partial y}$$

{assumption (b)}

Also, $$\frac{\partial p}{\partial x} = 0$$

Hence Eq. (4.101) yields ;

$$G_x \frac{\partial V_x}{\partial x} + G_y \frac{\partial V_x}{\partial y} = \frac{\partial}{\partial y}\left(\mu \frac{\partial V_x}{\partial y}\right) \quad (4.105)$$

or $$G_x \frac{\partial V_x}{\partial x} + G_y \frac{\partial V_x}{\partial y} = \mu \frac{\partial^2 V_x}{\partial y^2} \quad (4.106)$$

$$\rho V_x \frac{\partial V_x}{\partial x} + \rho V_y \frac{\partial V_x}{\partial y} = \mu \frac{\partial^2 V_x}{\partial y^2}$$

or $$V_x \frac{\partial V_x}{\partial x} + V_y \frac{\partial V_x}{\partial y} = \nu \frac{\partial^2 V_x}{\partial y^2} \quad (4.107)$$

Equation (4.107) is the *momentum equation* for the hydrodynamic boundary layer. The boundary conditions for this partial differential equation are ;

(i) $V_x = V_\infty$ at $x = 0$

(ii) $V_x = V_\infty$ at $y = \infty$

(iii) $V_y = 0$ at $y = 0$

(iv) $V_x = 0$ at $y = 0$

4.12 SOLUTION OF MOMENTUM EQUATION OF BOUNDARY LAYER OVER A FLAT PLATE

In order to solve Eq. (4.107) for the case of flow over a flat plate let us define a stream function ϕ such that :

$$\frac{\partial \phi}{\partial y} = V_x \quad (4.108)$$

$$-\frac{\partial \phi}{\partial x} = V_y \quad (4.109)$$

Note that the above definition of stream function satisfies the continuity equation (4.98). The next step is to transform the partial differential equation (4.107) into an ordinary differential equation by defining a new variable β (a function of x and y) as follows :

Let $$\beta = \frac{y}{\sqrt{x}}\sqrt{\frac{V_\infty}{\nu}} \quad (4.110)$$

$$\phi = \sqrt{V_\infty \nu x}\, f(\beta) \quad (4.111)$$

Now, $$V_x = \frac{\partial \phi}{\partial y} = \frac{\partial \phi}{\partial \beta}\frac{\partial \beta}{\partial y} = \sqrt{V_\infty \nu x}\,\frac{df}{d\beta}\,\frac{1}{\sqrt{x}}\sqrt{\frac{V_\infty}{\nu}}$$

{f is abbreviated for $f(\beta)$ of Eq. (4.111)}

or
$$V_x = V_\infty \frac{df}{d\beta} \tag{4.112}$$

or
$$\frac{\partial V_x}{\partial x} = V_\infty \frac{\partial}{\partial x}\left(\frac{\partial f}{\partial \beta}\right) = V_\infty \frac{\partial}{d\beta}\left(\frac{df}{d\beta}\right)\frac{\partial \beta}{\partial x}$$

$$= V_\infty \frac{d^2 f}{d\beta^2}\left(\frac{-y}{2x^{3/2}}\sqrt{\frac{V_\infty}{\nu}}\right)$$

or
$$\frac{\partial V_x}{\partial x} = -\frac{(V_\infty)^{3/2}\, y d^2 f}{2x\sqrt{x\nu}\, d\beta^2} \tag{4.113}$$

and
$$\frac{\partial V_x}{\partial y} = V_\infty \frac{\partial}{\partial y}\left(\frac{df}{d\beta}\right)$$

$$= V_\infty \frac{\partial}{\partial y}\left(\frac{df}{d\beta}\right)\frac{\partial \beta}{\partial y}$$

or
$$\frac{\partial V_x}{\partial y} = V_\infty \frac{d^2 f}{d\beta^2}\sqrt{\frac{V_\infty}{\nu x}} \tag{4.114}$$

similarly,
$$\frac{\partial^2 V_x}{\partial y^2} = \frac{V_\infty{}^2}{\nu x}\frac{d^3 f}{d\beta^3} \tag{4.115}$$

Again,
$$V_y = -\frac{\partial \phi}{\partial x}$$

but
$$\frac{\partial \phi}{\partial x} = \sqrt{V_\infty \nu}\;\frac{1}{2\sqrt{x}} f + \sqrt{V_\infty \nu x}\;\frac{\partial}{\partial x} f$$

$$= \frac{\sqrt{V_\infty \nu}}{2\sqrt{x}} f + \sqrt{V_\infty \nu x}\;\frac{df}{d\beta}\frac{\partial \beta}{\partial x}$$

$$= \frac{\sqrt{V_\infty \nu}}{2\sqrt{x}} f + \sqrt{V_\infty \nu x}\;\frac{df}{d\beta}\left[\frac{-y}{2x^{3/2}}\sqrt{\frac{V_\infty}{\nu}}\right]$$

or
$$\frac{\partial \phi}{\partial x} = \frac{\sqrt{V_\infty \nu}}{2\sqrt{x}} f - \frac{y}{2x} V_\infty \frac{df}{d\beta}$$

or
$$V_y = \frac{y}{2x} V_\infty \frac{df}{d\beta} - \frac{1}{2}\frac{\sqrt{V_\infty r}}{x} f \tag{4.116}$$

Substituting Eqs. (4.113), (4.114), (4.115) and (4.116) in Eq. (4.107) we get ;

$$\left\{-V_\infty \frac{df}{d\beta}\frac{V_\infty{}^{3/2}\, y}{2x\sqrt{x\nu}}\frac{d^2 f}{d\beta^2}\right\} + \left\{\frac{y}{2x} V_\infty \frac{df}{d\beta} - \frac{1}{2}\sqrt{\frac{V_\infty \nu}{x}} f\right\} \times$$

$$\left\{V_\infty \frac{d^2 f}{d\beta^2}\sqrt{\frac{V_\infty}{x\nu}}\right\} = r\left(\frac{V_\infty^2\, d^3 f}{\nu x\, d\beta^3}\right)$$

Simplification of the above equation results in the following ;

$$\frac{d^3 f}{d\beta^3} + \frac{f}{2}\frac{d^2 f}{d\beta^2} = 0 \tag{4.117}$$

or

$$2f''' + ff'' = 0 \tag{4.118}$$

Thus the partial differential equation (4.107) for the boundary layer velocity distribution is transformed into an ordinary differential equation of the third order. The boundary conditions for Eq. (4.118) can be found out as below.

Since $\beta = 0$ at $y = 0$ and $\beta = \infty$ at $y = \infty$ it follows from Eqs. (4.112) and (4.116) that ;

(i) At $y = \infty,\ V_x = V_\infty \dfrac{df}{d\beta} = V_\infty$

or at

$$\beta = \infty, \frac{df}{d\beta} = 1 \tag{4.119}$$

(ii) At $y = 0,\ V_y = \left\{\dfrac{y}{2x} V_\infty \dfrac{df}{d\beta} - \dfrac{1}{2}\sqrt{\dfrac{V_\infty \nu}{x}}\, f\right\} = 0$

or at

$$\beta = 0,\ f(\beta) = 0 \tag{4.120}$$

(iii) At $y = 0, V_x = V_\infty \dfrac{df}{d\beta} = 0$

or at

$$\beta = 0, \frac{df}{d\beta} = 0 \tag{4.121}$$

To summarize, the momentum equation (4.107) for boundary layer reduces to ;

$$2f''' + ff'' = 0 \tag{4.118}$$

where the boundary conditions are ;

$$f(\beta) = 0 \text{ at } \beta = 0$$

$$f'(\beta) = 0 \text{ at } \beta = 0$$

$$f'(\beta) = 1 \text{ at } \beta = \infty$$

The general solution of Eq. (4.118) cannot be obtained and it is necessary to solve it by *numerical methods* or by a series expansion. A solution was first presented by Blasius in 1908. Additional solutions have since been obtained. The solution is of the form :

$$f(\beta) = \frac{\alpha\beta^2}{2!} - \frac{1}{2}\frac{\alpha^2\beta^5}{5!} + \frac{11}{4}\frac{\alpha^3\beta^8}{8!} - \frac{375}{8}\frac{\alpha^4\beta^{11}}{11!} + \ldots\ldots$$

(where $\alpha = 0.332$) (4.122)

By using Eq. (4.122) the velocity distribution can be represented by a single curve with dimensionless coordinates (V_x/V_∞) and β, as shown in Fig. 4.8. It enables us to obtain two important results :

(i) Thickness of hydrodynamic boundary layer, δ.

(ii) Skin friction coefficient, C_f.

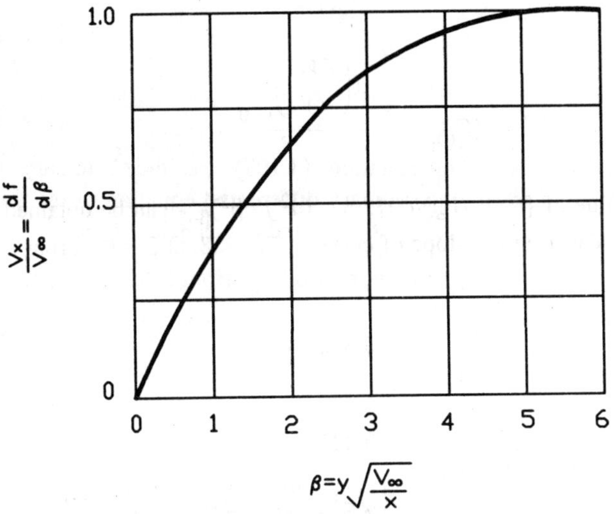

Fig. 4.8 Solution of Boundary Layer Equation (4.107)

Recall that at the edge of the boundary layer $V_x = 0.99 \times V_\infty$. Hence the thickness δ of the boundary layer is given by that value of y at which $V_x/V_\infty = 0.99$. From Fig. 4.8 we note that $V_x/V_\infty = 0.99$ is at $\beta = 5.0$. Hence the value of β at the edge of boundary layer ($y = \delta$) is given by ;

$$\beta = \frac{y}{\sqrt{x}}\sqrt{\frac{V_\infty}{\nu}} = 5.0$$

or
$$\frac{\delta}{\sqrt{x}}\sqrt{\frac{V_\infty}{\nu}} = 5.0$$

or
$$\delta = 5.0\sqrt{\frac{\nu x}{V_\infty}}$$

or
$$\frac{\delta}{x} = 5.0\sqrt{\frac{\nu}{V_\infty x}}$$

or
$$\frac{\delta}{x} = \frac{5.0}{\sqrt{Re_x}} \tag{4.123}$$

where
$$Re_x = \frac{V_\infty x}{\nu}$$

It represents the *local Reynolds number* at distance x from the leading edge of the plate. From Eq. (4.123) the *boundary layer thickness* δ can be evaluated at any value of x.

The *skin friction coefficient*, C_f is defined as the ratio of the shear stress at the plate to the dynamic head, $\frac{1}{2}\rho V_\infty^2$. Thus at any value of x the local value of the skin friction coefficient $(C_f)_x$ is given by ;

$$(C_f)_x = \frac{(\tau_{yx})_{y=0}}{\frac{1}{2}\rho V_\infty^2} \tag{4.124}$$

or
$$(C_f)_x = \frac{\mu\left(\frac{\partial V_x}{\partial y}\right)_{y=0}}{\frac{1}{2}\rho V_\infty^2} \tag{4.125}$$

Now, the value of $\{\partial (V_x/V_\infty)/\partial\beta\}_{y=0} = (d^2 f/d\beta^2)_{\beta=0}$ can be obtained either from Eq. (4.122) or by measuring the slope of curve in Fig. 4.8. at $\beta = 0$. From either method we get ;

$$\left\{\frac{\partial(V_x/V_\infty)}{\partial\beta}\right\}_{\beta=0} = 0.332 \tag{4.126}$$

or
$$\frac{1}{V_\infty}\left\{\frac{\partial V_x}{\partial\beta}\right\}_{\beta=0} = 0.332$$

But
$$\left(\frac{\partial V_x}{\partial y}\right)_{y=0} = \left\{\left(\frac{\partial V_x}{\partial\beta}\right)_{y=0}\right\} \times \left(\frac{\partial\beta}{\partial y}\right)$$

or
$$\left(\frac{\partial V_x}{\partial y}\right)_{y=0} = 0.332\, V_\infty \sqrt{\frac{V_\infty}{x\nu}}$$

hence
$$(C_f)_x = \frac{0.332\,\mu V_\infty \sqrt{\frac{V_\infty}{\nu x}}}{\frac{1}{2}\rho V_\infty^2} = 0.664\sqrt{\frac{\nu}{V_\infty x}}$$

or
$$(C_f)_x = \frac{0.664}{\sqrt{(Re)_x}} \tag{4.127}$$

The average value of the skin friction coefficient, C_f can be obtained by integrating $(C_f)_x$ from $x = 0$ to $x = L$ (where L is the length of the plate) and then dividing the result by L.

$$C_f = \frac{1}{L}\int_0^L (C_f)_x\, dx$$

$$= \int_{x=0}^{x=L} \frac{0.664\, dx}{\sqrt{(V_\infty x)/\nu}}$$

$$= \frac{1.328}{\sqrt{V_\infty L/\nu}}$$

so
$$C_f = \frac{1.328}{\sqrt{(Re)_L}} \tag{4.128}$$

Knowing C_f the average shear stress on the plate can be calculated.

4.13 ENERGY EQUATION FOR THERMAL BOUNDARY LAYER OVER A FLAT PLATE

The energy equation for the thermal boundary layer is obtained by following a procedure similar to that given above for the momentum equation. We consider a control volume ($\Delta x \times \Delta y \times 1$) located in the boundary layer region as shown in Fig. 4.9. An enlarged view of this control volume is also shown in the figure. We now make a first law energy balance for control volume as follows :

$$\left\{\begin{matrix}\text{Rate of heat transfer}\\ \text{across control surface}\end{matrix}\right\} - \left\{\begin{matrix}\text{Rate of work transfer}\\ \text{across control surface}\end{matrix}\right\} = \left\{\begin{matrix}\text{Rate of outflow of}\\ \text{energy with mass}\\ \text{from control volume}\end{matrix}\right\} -$$

$$\left\{\begin{matrix}\text{Rate of inflow of}\\ \text{energy with mass}\\ \text{into control volume}\end{matrix}\right\} + \left\{\begin{matrix}\text{Rate of accumulation}\\ \text{of energy within}\\ \text{control volume}\end{matrix}\right\} \qquad (4.129)$$

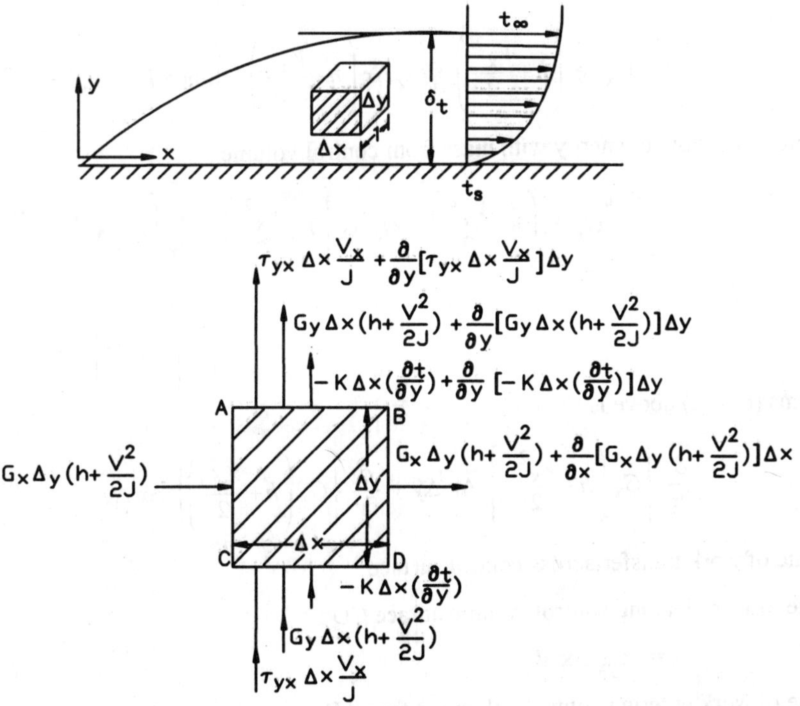

Fig. 4.9 Derivation of Boundary Layer Energy Equation (4.133)

The various terms in Eq. (4.134) can be evaluated as follows :

(i) Rate of heat transfer across control surface :

(a) Rate of heat conduction across face CD ;

$$= \; -k \, \Delta x \, \frac{\partial t}{\partial y}$$

(b) Rate of heat conduction across face AB ;

$$= -k\,\Delta x\,\frac{\partial t}{\partial y} + \frac{\partial}{\partial y}\left(-k\,\Delta x\,\frac{\partial t}{\partial y}\right)\Delta y$$

So, net rate of heat transfer to control volume ;

$$= -k\,\Delta x\,\frac{\partial t}{\partial y} - \left\{-k\,\Delta x\,\frac{\partial t}{\partial y} - k\,\frac{\partial^2 t}{\partial y^2}\,\Delta x\,\Delta y\right\}$$

$$= +k\,\frac{\partial^2 t}{\partial y^2}\,(\Delta x\,\Delta y)$$

(ii) Rate of energy transfer with mass into control volume :

(a) Rate of inflow of energy with mass into the control volume ;

$$= G_x\,\Delta y\left(h + \frac{V^2}{2}\right) + G_y\,\Delta x\left(h + \frac{V^2}{2}\right)$$

$$= G_x\,\Delta y\left(h + \frac{V_x^{\,2}}{2}\right) + G_y\,\Delta x\left(h + \frac{V_x^{\,2}}{2}\right) \quad \text{(Since } V^2 = V_x^{\,2} + V_y^{\,2} \approx V_x^{\,2}\text{)}$$

(b) Rate of outflow of energy with mass from control volume :

$$= \left\{G_x\,\Delta y\left(h + \frac{V_x^{\,2}}{2}\right) + G_y\,\Delta x\left(h + \frac{V_x^{\,2}}{2}\right) + \frac{\partial}{\partial x}\left[G_x\,\Delta y\left(h + \frac{V_x^{\,2}}{2}\right)\Delta x\right]\right.$$

$$\left. + \frac{\partial}{\partial y}\left[G_y\,\Delta x\left(h + \frac{V_x^{\,2}}{2}\right)\right]\Delta y\right\}$$

So, items (b) - (a) above is :

$$\frac{\partial}{\partial x}\left\{G_x\left(h + \frac{V_x^{\,2}}{2}\right)\right\}\Delta x\,\Delta y + \frac{\partial}{\partial y}\left\{G_y\left(h + \frac{V_x^{\,2}}{2}\right)\right\}\Delta x\,\Delta y$$

(iii) Rate of work transfer across control surface

(a) Rate of work leaving control volume at face CD ;

$$= \tau_{yx}\,\Delta x\,V_x$$

(b) Rate of work entering control volume at face AB ;

$$= \tau_{yx}\,\Delta x\,V_x + \frac{\partial}{\partial y}\,(\tau_{yx}\,\Delta x\,V_x)\,\Delta y$$

Net rate of work leaving control volume

$$= \tau_{yx}\,\Delta x\,V_x - \left\{\tau_{yx}\,\Delta x\,V_x + \frac{\partial}{\partial y}\,(\tau_{yx}\,\Delta x\,V_x)\,\Delta y\right\}$$

$$= -\frac{\partial}{\partial y}\,(\tau_{yx}\,V_x)\,\Delta x\,\Delta y$$

(iv) Rate of accumulation of energy within control volume equals *zero* because of steady state.

Now, substitution of the above four relations in Eq. (4.129) gives ;

$$\left[+k \frac{\partial^2 t}{\partial y^2} \Delta x\, \Delta y + \frac{\partial}{\partial y}(\tau_{yx} V_x) \Delta x\, \Delta y \right]$$

$$= \frac{\partial}{\partial x} \left\{ G_x \left(h + \frac{V_x^2}{2} \right) \right\} \Delta x\, \Delta y + \frac{\partial}{\partial y} \left\{ G_y \left(h + \frac{V_x^2}{2} \right) \right\} \Delta x\, \Delta y + 0$$

or $$+k \frac{\partial^2 t}{\partial y^2} + \frac{\partial}{\partial y}(\tau_{yx} V_x) = \frac{\partial}{\partial x} \left\{ G_x \left(h + \frac{V_x^2}{2} \right) \right\} + \frac{\partial}{\partial y} \left\{ G_y \left(h + \frac{V_x^2}{2} \right) \right\} \quad (4.130)$$

Now,

(i) $$\frac{\partial}{\partial y}(\tau_{yx} V_x) = \frac{\partial}{\partial y}\left(\mu V_x \frac{\partial V_x}{\partial y} \right) = \mu V_x \frac{\partial^2 V_x}{\partial y^2} + \mu \left(\frac{\partial V_x}{\partial y} \right)^2$$

(ii) $$\frac{\partial}{\partial x} \left\{ G_x \left(h + \frac{V_x^2}{2} \right) \right\} = G_x \frac{\partial}{\partial x} \left(h + \frac{V_x^2}{2} \right) + \left(h + \frac{V_x^2}{2} \right) \frac{\partial G_x}{\partial x}$$

(iii) $$\frac{\partial}{\partial y} \left\{ G_y \left(h + \frac{V_x^2}{2} \right) \right\} = G_y \frac{\partial}{\partial y} \left(h + \frac{V_x^2}{2} \right) + \left(h + \frac{V_x^2}{2} \right) \frac{\partial G_y}{\partial y}$$

Hence, Eq. (4.130) becomes,

$$k \frac{\partial^2 t}{\partial y^2} + \mu V_x \frac{\partial^2 V_x}{\partial y^2} + \mu \left(\frac{\partial V_x}{\partial y} \right)^2 = G_x \frac{\partial}{\partial x} \left(h + \frac{V_x^2}{2} \right) +$$

$$G_y \frac{\partial}{\partial y} \left(h + \frac{V_x^2}{2} \right) + \left(h + \frac{V_x^2}{2} \right) \left(\frac{\partial G_x}{\partial x} + \frac{\partial G_y}{\partial y} \right)$$

or $$k \frac{\partial^2 t}{\partial y^2} + \mu V_x \frac{\partial^2 V_x}{\partial y^2} + \mu \left(\frac{\partial V_x}{\partial y} \right)^2 = G_x \frac{\partial}{\partial x} \left(h + \frac{V_x^2}{2} \right) + G_y \frac{\partial}{\partial y} \left(h + \frac{V_x^2}{2} \right) \quad (4.131)$$

$$\left\{ \text{Since} \left(\frac{\partial G_x}{\partial x} + \frac{\partial G_y}{\partial y} \right) = 0 \right\}$$

Again $$\frac{\partial}{\partial x} \left(h + \frac{V_x^2}{2} \right) = \frac{\partial h}{\partial x} + V_x \frac{\partial V_x}{\partial x}$$

$$= C_p \frac{\partial t}{\partial x} + \frac{V_x\, \partial V_x}{\partial x}$$

$$\frac{\partial}{\partial y} \left(h + \frac{V_x^2}{2} \right) = \frac{\partial h}{\partial y} + V_x \frac{\partial V_x}{\partial y}$$

$$= C_p \frac{\partial t}{\partial y} + \frac{V_x\, \partial V_x}{\partial y}$$

Hence the RHS of Eq. (4.131) becomes,

$$G_x C_p \frac{\partial t}{\partial x} + G_x V_x \frac{\partial V_x}{\partial x} + G_y C_p \frac{\partial t}{\partial y} + G_y V_x \frac{\partial V_x}{\partial y}$$

$$= G_x C_p \frac{\partial t}{\partial x} + G_y C_p \frac{\partial t}{\partial y} + V_x \left(G_x \frac{\partial V_x}{\partial x} + G_y \frac{\partial V_x}{\partial y} \right)$$

$$= G_x C_p \frac{\partial t}{\partial x} + G_y C_p \frac{\partial t}{\partial y} + V_x \left(\mu \frac{\partial^2 V_x}{\partial y^2} \right)$$

[use has been made of Eq. (4.106)]

Thus Eq. (4.131) now takes the form :

$$k \frac{\partial^2 t}{\partial y^2} + \mu V_x \frac{\partial^2 V_x}{\partial y^2} + \mu \left(\frac{\partial V_x}{\partial y} \right)^2 = G_x C_p \frac{\partial t}{\partial x} + G_y C_p \frac{\partial t}{\partial y} + V_x \left(\mu \frac{\partial^2 V_x}{\partial y^2} \right)$$

or $$G_x C_p \frac{\partial t}{\partial x} + G_y C_p \frac{\partial t}{\partial y} = k \frac{\partial^2 t}{\partial y^2} + \mu \left(\frac{\partial V_x}{\partial y} \right)^2$$

or $$\rho V_x C_p \frac{\partial t}{\partial x} + \rho V_y C_p \frac{\partial t}{\partial y} = k \frac{\partial^2 t}{\partial y^2} + \mu \left(\frac{\partial V_x}{\partial y} \right)^2$$

or $$V_x \frac{\partial t}{\partial x} + V_y \frac{\partial t}{\partial y} = \alpha \frac{\partial^2 t}{\partial y^2} + \frac{\mu}{\rho C_p} \left(\frac{\partial V_x}{\partial y} \right)^2 \qquad (4.132)$$

Eq. (4.132) is the energy equation for boundary layer. It turns out that the term $\left(\frac{\mu}{\rho C_p} \right) \left(\frac{\partial V_x}{\partial y} \right)^2$ may be neglected in comparision to $\alpha\, \partial^2 t/\partial y^2$. Hence we finally get ;

$$V_x \frac{\partial t}{\partial x} + V_y \frac{\partial t}{\partial y} = \alpha \frac{\partial^2 t}{\partial y^2} \qquad (4.133)$$

The temperature boundary conditions are

$$t = t_w \text{ at } y = 0 \; ; \; t = t_\infty \text{ at } y = \infty$$

If we compare Eq. (4.133) with Eq. (4.107) it is at once apparent that if $\alpha = \nu$ ie, thermal and momentum diffusivities are equal, then the two equations become exactly similar. Hence the velocity and temperature distribution must be similar to each other.

4.14 SOLUTION OF ENERGY EQUATION FOR BOUNDARY LAYER OVER A FLAT PLATE

Eq. (4.133) can be transformed into an ordinary differential equation by using the variable β and the stream function ϕ defined earlier. In addition let us define another variable θ as given below :

$$\theta = \frac{t - t_w}{t_\infty - t_w} = f(\beta)$$

Now $$V_x = V_\infty \frac{df}{d\beta} \tag{4.112}$$

$$V_y = \frac{y}{2x} V_\infty \frac{df}{d\beta} - \frac{1}{2}\sqrt{\frac{V_\infty \nu}{x}}\, f(\beta) \tag{4.116}$$

Also,

$$\frac{\partial t}{\partial x} = (t_\infty - t_w)\frac{\partial \theta}{\partial x} = (t_\infty - t_w)\frac{\partial \theta}{\partial \beta}\frac{\partial \beta}{\partial x}$$

or $$\frac{\partial t}{\partial x} = (t_\infty - t_w)\left\{\frac{-y}{2x^{3/2}}\sqrt{\frac{V_\infty}{\nu}}\right\}\frac{d\theta}{d\beta} \tag{4.134}$$

and $$\frac{\partial t}{\partial y} = (t_\infty - t_w)\frac{\partial \theta}{\partial y} = (t_\infty - t_w)\frac{\partial \theta}{\partial \beta}\frac{\partial \beta}{\partial y}$$

so $$\frac{\partial t}{\partial y} = (t_\infty - t_w)\sqrt{\frac{V_\infty}{\nu x}}\frac{d\theta}{d\beta} \tag{4.135}$$

and $$\frac{\partial^2 t}{\partial y^2} = (t_\infty - t_w)\frac{V_\infty}{\nu x}\frac{d^2\theta}{d\beta^2} \tag{4.136}$$

Substituting Eqs. (4.112), (4.116), (4.134), (4.135) and (4.136) in Eq. (4.133) we get ;

$$\left(V_\infty \frac{df}{d\beta}\right)(t_\infty - t_w)\left\{\frac{-y}{2x^{3/2}}\sqrt{\frac{V_\infty}{\nu}}\right\}\frac{d\theta}{d\beta} + \left\{\frac{y}{2x}V_\infty\frac{df}{d\beta} - \right.$$

$$\left.\frac{1}{2}\sqrt{\frac{V_\infty \nu}{x}}\, f(\beta)\right\}(t_\infty - t_w)\sqrt{\frac{V_\infty}{\nu x}}\frac{d\theta}{d\beta} = \alpha\left\{(t_\infty - t_w)\frac{V_\infty}{\nu x}\frac{d^2\theta}{d\beta^2}\right\}$$

or $$-\frac{1}{2}\frac{V_\infty}{x} f(\beta)\frac{d\theta}{d\beta} = \alpha\frac{V_\infty}{\nu x}\frac{d^2\theta}{d\beta^2}$$

or $$\frac{\alpha}{\nu}\frac{d^2\theta}{d\beta^2} + \frac{1}{2} f(\beta)\frac{d\theta}{d\beta} = 0$$

or $$\frac{d^2\theta}{d\beta^2} + \frac{1}{2}\frac{\nu}{\alpha} f(\beta)\frac{d\theta}{d\beta} = 0$$

$$\frac{d^2\theta}{d\beta^2} + \frac{1}{2}\Pr f(\beta)\frac{d\theta}{d\beta} = 0 \tag{4.137}$$

The boundary conditions for Eq. (4.137) are ;

$$\theta = 0 \quad \text{at} \quad \beta = 0$$

$$\theta = 1 \quad \text{at} \quad \beta = \infty$$

The solution to Eq. (4.137) was first given by Pohlhausen. Without going into the details of the solution we shall simply state the results ;

$$\frac{d\theta}{d\beta} = \left\{\left(\frac{d\theta}{d\beta}\right)_{\beta=0}\right\} \times exp\left\{-\frac{Pr}{2}\int_0^\beta f\,d\beta\right\}$$

$$\theta\,(\beta) = \left\{\left(\frac{d\theta}{d\beta}\right)_{\beta=0}\right\} \times \int_0^\beta exp\left\{-\frac{Pr}{2}\int_0^\beta f\,d\beta\right\} d\beta$$

We can evaluate $\left(\frac{d\theta}{d\beta}\right)_{\beta=0}$ by letting $\beta \rightarrow \infty$ so that $\theta\,(\beta) = 1$.

Then $$\left(\frac{d\theta}{d\beta}\right)_{\beta=0} = \frac{1}{\int_0^\infty exp\left\{-\frac{Pr}{2}\int f\,d\beta\right\} d\beta} \qquad (4.138)$$

Eq. (4.138) gives the dimensionless slope of the temperature distribution at the surface where $\beta = 0$. It is a function only of Prandtl number and Pohlhausen found that it could be represented by ;

$$\left(\frac{d\theta}{d\beta}\right)_{\beta=0} = 0.332\,Pr^{1/3} \qquad (4.139)$$

The temperature distribution curves for various values of *Pr* are shown in Fig. 4.10 where the coordinates are $\theta\,(\beta)$ and β.

Since $(d\theta/d\beta)_{\beta=0} = 0.332$ for Pr = 1, the temperature distribution curve for Pr = 1 is the same as for the velocity distribution in Fig. 4.8.

Fig. 4.10 enables us to determine the thickness of the thermal boundary layer δ_t and also the value of the local and average heat transfer coefficients h_x and h.

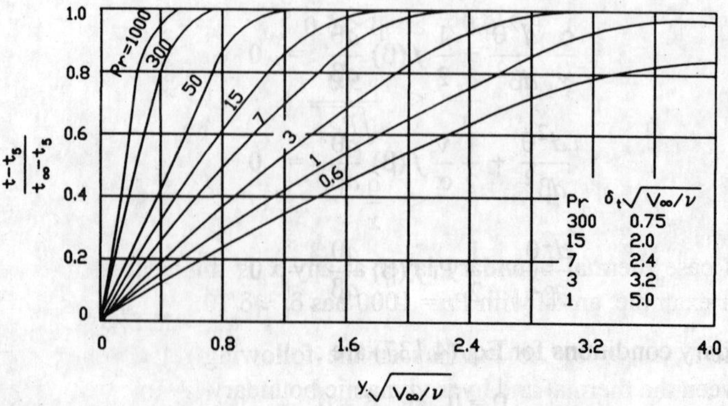

Fig. 4.10 Solution of Boundary Layer Energy Equation (4.133)

4.15 THICKNESS OF THERMAL BOUNDARY LAYER, δ_t

This is given by that value of y at which $(t - t_w) = 0.99 \times (t_\infty - t_w)$ or $\theta = 0.99$. From Fig. 4.10 we find the following ;

(a) Case 1 :

$$Pr = 1$$

$$\theta = 0.99 \text{ at } \beta = 5.0$$

or $$y\sqrt{\frac{V_\infty}{\nu x}} = \beta = \delta_t \sqrt{\frac{V_\infty}{\nu x}} = 5.0$$

or $$\delta_t = 5.0 \sqrt{\frac{\nu x}{V_\infty}}$$

or $$\frac{\delta_t}{x} = \frac{5.0}{\sqrt{Re_x}} = \frac{\delta}{x} \tag{4.140}$$

In this case the thickness of the thermal boundary layer δ_t, at any value of x equals the thickness of the velocity boundary layer δ.

(b) Case 2 :

$$Pr < 1$$

$$\theta = 0.99 \text{ at } \beta > 5.0$$

or $$\frac{\delta_t}{x} > \frac{5.0}{\sqrt{Re_x}}$$

$$> \frac{\delta}{x} \tag{4.141}$$

In this case thermal boundary layer at any x is *thicker* than the velocity boundary layer.

(c) Case 3 :

$$Pr > 1$$

$$\theta = 0.99 \text{ at } \beta < 5.0$$

or $$\frac{\delta_t}{x} < \frac{5.0}{\sqrt{Re_x}}$$

$$< \frac{\delta}{x} \tag{4.142}$$

In this case thermal boundary layer at any x is thinner than the velocity boundary layer. For example, an oil with $Pr = 1000$ has $\delta_t \approx \delta/10$.

In general, according to Pohlhausen the following relationship may be assumed to exist between the thermal and hydrodynamic boundary layers ;

$$\delta_t = \frac{\delta}{Pr^{1/3}} \tag{4.143}$$

4.16 LOCAL AND AVERAGE HEAT TRANSFER COEFFICIENTS IN FORCED CONVECTION

(i) The *rate of heat flow* per unit area at any location x along the plate is given by ;

$$\frac{q_x}{A} = -k\left(\frac{\partial t}{\partial y}\right)_{y=0} = h_x(t_w - t_\infty) \qquad (4.1)$$

or

$$h_x = \frac{-k\left(\frac{\partial t}{\partial y}\right)_{y=0}}{(t_w - t_\infty)} \qquad (4.2)$$

Now,

$$\left(\frac{d\theta}{d\beta}\right)_{y=0} = 0.332\, Pr^{1/3} \qquad (4.139)$$

or

$$\left[\frac{d(t-t_w)/(t_\infty - t_w)}{d\beta}\right]_{y=0} = 0.332\, Pr^{1/3}$$

or

$$\frac{1}{(t_\infty - t_w)}\left[\frac{dt}{d\beta}\right]_{y=0} = 0.332\, Pr^{1/3}$$

or

$$\frac{dt}{d\beta} = 0.332\,(t_\infty - t_w)\, Pr^{1/3}$$

$$\left(\frac{\partial t}{\partial y}\right)_{y=0} = \left\{\left(\frac{dt}{d\beta}\right)_{y=0}\right\}\left(\frac{\partial \beta}{\partial y}\right)$$

or

$$\left(\frac{\partial t}{\partial y}\right)_{y=0} = 0.332\,(t_\infty - t_w)\, Pr^{1/3}\sqrt{\frac{V_\infty}{\nu x}}$$

Eq. (4.2) then yields the value of *local heat transfer* coefficient in forced convection ;

$$h_x = -\frac{(k)(0.332)(t_\infty - t_w)\, Pr^{1/3}}{(t_w - t_\infty)}\sqrt{\frac{V_\infty}{\nu x}}$$

$$= 0.332\, k\, Pr^{1/3}\sqrt{\frac{V_\infty}{\nu x}}$$

or

$$Nu_x = 0.332\sqrt[3]{Pr}\,\sqrt{Re_x} \qquad (4.15)$$

(ii) The total heat transfer from the plate can be found as follows ;

$$q = \int_0^L h_x\, b\,(t_w - t_\infty)\, dx$$

(where b = width of the plate)

$$q = (t_w - t_\infty)\int_0^L 0.332\, k\, b\, Pr^{1/3}\sqrt{\frac{V_\infty}{\nu x}}\, dx$$

or $$q = b\,(t_w - t_\infty)\,(0.664)\,k\,Pr^{1/3}\sqrt{\frac{V_\infty L}{\nu}} \tag{4.144}$$

but $$q = h\,(bL)\,(t_w - t_\infty) \tag{4.145}$$

From Eqs. (4.144) and (4.145), the *average heat transfer coefficient*, **h** is given by ;

$$h = \frac{0.664\,k\,Pr^{1/3}}{L}\sqrt{\frac{V_\infty L}{\nu}}$$

or $$h = \frac{0.664}{L}\,k\,Pr^{1/3}\sqrt{Re_L} \tag{4.17}$$

Example 4.18

Air at 25 C and atmospheric pressure is flowing over a flat plate @ 3 m/sec. If the plate is 1 m wide and the temperature t_w = 75 C, calculate the following at a location 1 m from leading edge.

(a) Hydrodynamic boundary layer thickness, δ.

(b) Local friction coefficient, $(C_f)_x$.

(c) Thermal boundary layer thickness $(\delta)_t$.

(d) Local heat transfer coefficient, h_x.

Solution :

The average temperature of the film ;

$$t_f = \frac{t_w + t_{air}}{2} = \frac{25+75}{2} = 50 \text{ C.}$$

From the property tables of air we get the following values at 50 C ;

$$\rho = 1.093 \text{ kg/m}^3$$

$$C_p = 1 \text{ kJ/kg K}$$

$$\mu = 19.6 \times 10^{-6} \text{ kg/m sec}$$

$$k = 0.10 \text{ kJ/m hr K} = 0.0028 \text{ W/m K}$$

$$Pr = 0.698$$

Thus ; $$(Re)_{x=1} = \frac{(3)(1)(1.093)}{19.6\times10^{-6}} = 1.65 \times 10^5$$

(a) The value of δ is given by Eq. (4.123) ;

$$\delta = \frac{5x}{\sqrt{Re_x}} = \frac{5\times1}{\sqrt{1.65\times10^5}} \text{ m} = 1.23 \text{ cm}$$ Ans.

(b) The value of $(C_f)_x$ is given by Eq. (4.127) ;

$$(C_f)_x = \frac{0.664}{\sqrt{Re_x}} = \frac{0.664}{\sqrt{1.65\times10^5}} = 0.001635$$ Ans.

(c) The value of $(\delta)_t$ is given by Eq. (4.143) ;

$$(\delta)_t = \delta/Pr^{1/3}$$

$$= 1.23/(0.698)^{1/3} = 1.39 \text{ cm}$$ Ans.

(d) The value of h_x is given by Eq. (4.14)

$$h_x = 0.332 \frac{k}{x} (Re_x)^{1/2} (Pr)^{1/3}$$

$$= 0.332 \times \frac{0.0028}{1} (1.65 \times 10^5)^{1/2} (0.698)^{1/3}$$

$$= 3.24 \text{ W/m}^2\text{K}$$ Ans.

4.17 INTEGRAL METHOD FOR SOLVING BOUNDARY LAYER PROBLEMS

In this approximate method devised by Von Karmon an attempt is made to satisfy the continuity, momentum and energy equations for the entire thickness (rather than at every point) of the boundary layer.

(A) Velocity boundary layer

Let us consider the steady flow of an incompressible fluid over a flat plate (Fig. 4.11a) having unit width perpendicular to the plane of the paper, and focus attention on the control volume ABCD.

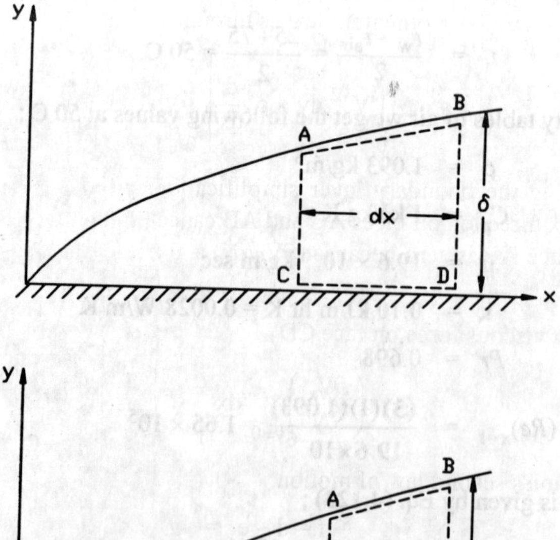

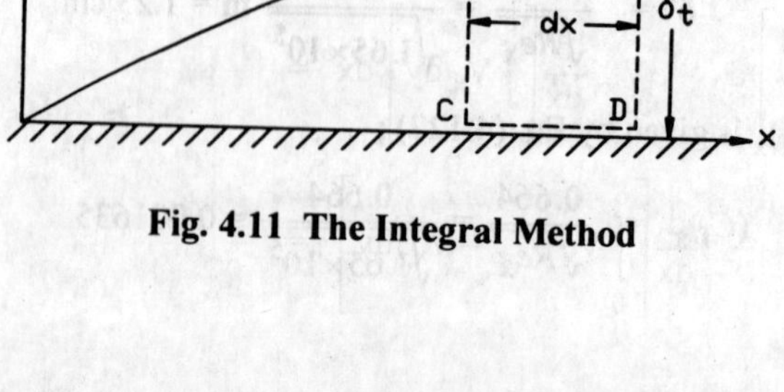

Fig. 4.11 The Integral Method

(a) Rate at which mass enters at face AC

$$= \rho \int_0^\delta V_x \, dy$$

(b) Rate at which mass leaves at face BD

$$= \rho \int_0^\delta V_x \, dy + \frac{d}{dx}\left[\rho \int_0^\delta V_x \, dy\right] dx$$

(c) It is apparent that the continuity equation can be satisfied only if the rate at which mass flows in across face AB

$$= \rho \frac{d}{dx}\left[\int_0^\delta V_x \, dy\right] dx$$

(d) Rate at which x direction momentum enters through face AC

$$= \rho \int_0^\delta V_x^2 \, dy$$

(e) Rate at which x direction momentum enters through face AB

$$= \rho V_\infty \frac{d}{dx}\left[\int_0^\delta V_x \, dy\right] dx$$

(f) Rate at which x direction momentum leaves through face BD

$$= \rho \int_0^\delta V_x^2 \, dy + \rho \frac{d}{dx}\left[\int_0^\delta V_x^2 \, dy\right] dx$$

Now, considering the boundary layer simplification, dp/dx = 0, it is clear that the pressure forces in x direction on faces AC and AB cancel that on face BD. Moreover, the viscous force on face AB is zero because $(\partial V_x/\partial y) = 0$ at the outer edge of the boundary layer.

(g) The x direction viscous force on face CD

$$= -\mu \left(\frac{\partial V_x}{\partial y}\right)_{y=0} dx$$

Applying Newton's second law of motion

$$f - d - e = g$$

$$\rho \frac{d}{dx}\left[\int_0^\delta V_x^2 \, dy\right] dx - \rho V_\infty \frac{d}{dx}\left[\int_0^\delta V_x \, dy\right] dx = -\mu \left(\frac{\partial V_x}{\partial y}\right)_{y=0} dx$$

or

$$\frac{d}{dx}\left[\int_0^\delta V_x (V_\infty - V_x) dy\right] = \frac{\mu}{\rho}\left(\frac{\partial V_x}{\partial y}\right)_{y=0} \qquad (4.146)$$

Pohlhausen suggested a velocity distribution of the form

$$V_x = a_1 + a_2 y + a_3 y^2 + a_4 y^3$$

where the constants a_1 to a_4 may be evaluated from the conditions :

$$y = 0, V_x = 0$$

$$y = 0, \frac{\partial^2 V_x}{\partial y^2} = 0$$

(From Eq. 4.107 by putting $V_x = V_y = 0$ at $y = 0$)

$$y = \delta, V_x = V_\infty$$

$$y = \delta, \frac{\partial V_x}{\partial y} = 0$$

We get :

$$a_1 = 0, a_3 = 0, a_2 = 3 V_\infty/2\delta, a_4 = - V_\infty/2\delta^3$$

Hence,

$$V_x = V_\infty \left[\frac{3}{2}\left(\frac{y}{\delta}\right) - \frac{1}{2}\left(\frac{y}{\delta}\right)^3\right] \quad (4.147)$$

Substitution of Eq. (4.147) in Eq. (4.146) gives

$$\frac{d}{dx}\left[\int_0^\delta V_\infty^2 \left\{\frac{3}{2}\left(\frac{y}{\delta}\right) - \frac{1}{2}\left(\frac{y}{\delta}\right)^3\right\}\left\{1 - \frac{3}{2}\left(\frac{y}{\delta}\right) + \frac{1}{2}\left(\frac{y}{\delta}\right)^3\right\} dy\right] = \frac{3\mu V_\infty}{2\rho\delta} \quad (4.148)$$

Integration yields

$$\frac{d}{dx}\left[\frac{39}{280}\delta\right] = \frac{3\mu V_\infty}{2\rho\delta}$$

or

$$\delta \, d\delta = \frac{140}{13}\frac{\nu}{V_\infty} dx \quad (4.149)$$

Integrating Eq. (4.149) from $\delta = 0$ to $\delta = \delta$ and $x = o$ to $x = x$

$$\frac{\delta^2}{2} = \frac{140}{13}\frac{\nu}{V_\infty}$$

or

$$\delta = \sqrt{\frac{280\nu}{13 V_\infty}} \quad (4.150)$$

and

$$\frac{\delta}{x} = \frac{4.64}{\sqrt{Re_x}} \quad (4.151)$$

The local and average skin friction coefficients are then given by

$$C_{fx} = 0.646/(Re_x)^{0.5} \quad (4.152)$$

$$C_f = 1.292/(Re_L)^{0.5} \quad (4.153)$$

Comparison of Eqs. (4.151), (4.152) and (4.153) with the corresponding equations (4.123), (4.127) and (4.128) shows that the agreement between the integral method and the exact method is very good.

(B) Thermal Boundary Layer

An energy balance for the steady state control volume of Fig. 4.11b gives

{Rate of enthalpy flow out of control volume} – {Rate of enthalpy flow into control volume}
= {Rate of heat transfer to control volume} (4.154)

Rate of enthalpy flow into control volume across face AC

$$= \rho C_p \int_0^{\delta_t} V_x t \, dy \tag{i}$$

Rate of enthalpy flow into control volume across face AB (see para C above)

$$= \rho C_p t_\infty \left[\frac{d}{dx} \left\{ \int_0^{\delta_t} V_x \, dy \right\} dx \right] \tag{ii}$$

Rate of enthalpy flow out of control volume across face BD

$$= \rho C_p \int_0^{\delta_t} V_x t \, dy + \rho C_p \frac{d}{dx} \left[\int_0^{\delta_t} V_x t \, dy \right] dx \tag{iii}$$

Now, the rate of heat transfer to the control volume across face AB is zero because $(\partial t / \partial y)_{y=\delta_t} = 0$. Moreover, considering the boundary layer simplification that the temperature gradient in the x direction is negligible compared to that in the y direction, the rate at which heat is conducted into the control volume across face AC is equal to the rate at which it is conducted out of control volume across face BD. Hence we have to consider only the rate at which heat is transferred into control volume by conduction across face CD. This rate is

$$= -k \left(\frac{\partial t}{\partial y} \right)_{y=0} dx \tag{iv}$$

Substituting (i) to (iv) in Eq. (4.154),

$$\frac{d}{dx} \left[\int_0^{\delta_t} V_x (t_\infty - t) \, dy \right] dx = \alpha \left(\frac{\partial t}{\partial y} \right)_{y=0} \tag{4.155}$$

This is the integral equation of the thermal boundary layer. Assume that the temperature profile is given by

$$t = A_1 + A_2 y + A_3 y^2 + A_4 y^3$$

where the constants A_1 to A_4 may be evaluated from the conditions :

$$y = 0, \text{ to } t = t_w$$

$$y = 0, \frac{\partial^2 t}{\partial y^2} = 0$$

(From Eq. 4.133 by putting $V_x = V_y = 0$ at $y = 0$)

$$y = \delta_t, t = t_\infty$$

$$y = \delta_t, \frac{\partial t}{\partial y} = 0$$

Then : $A_1 = t_w$, $A_3 = 0$, $A_2 = 3 - (3/2)(t_w - t_\infty)$, $A_4 = (t_w - t_\infty)/2\,\delta_t^3$. Therefore,

$$t = t_w = -\frac{3}{2\delta_t}(t_w - t_\infty)\,y + \frac{1}{2\delta_t^{\,3}}(t_w - t_\infty)\,y^3$$

or

$$\frac{t - t_\infty}{t_w - t_\infty} = 1 - \frac{3}{2}\frac{y}{\delta_t} + \frac{1}{2}\left(\frac{y}{\delta_t}\right)^3 \quad (4.156)$$

Substituting Eqs. (4.156) and (4.147) in Eq. (4.155),

$$\frac{d}{dx}\left[\int_0^{\delta_t} V_\infty \left\{\frac{3}{2}\left(\frac{y}{\delta}\right) - \frac{1}{2}\left(\frac{y}{\delta}\right)^3\right\}(t_w - t_\infty)\left\{1 - \frac{3}{2}\left(\frac{y}{\delta_t}\right) + \frac{1}{2}\left(\frac{y}{\delta_t}\right)^3\right\}dy\right] = \frac{3\alpha}{2\,\delta_t}(t_w - t_\infty)$$

$$\frac{3}{20}\frac{d}{dx}\left[\frac{\delta_t^{\,2}}{\delta}\left\{1 - \frac{1}{14}\left(\frac{\delta_t}{\delta}\right)^2\right\}\right] = \frac{3\alpha}{2\,V_\infty\,\delta_t} \quad (4.157)$$

If $\delta_t \le \delta$, as would be the case for $Pr \ge 1$, $(1/14)(\delta_t^2/\delta) << 1$. Hence

$$\delta_t \frac{d}{dx}\frac{\delta_t^{\,2}}{\delta} = \frac{10\alpha}{V_\infty}$$

Putting

$$\delta = \sqrt{\frac{280}{13}\frac{\nu x}{V_\infty}} \quad (4.150)$$

We get

$$\delta_t \frac{d}{dx}\left(\frac{\delta_t^{\,2}}{\sqrt{x}}\right) = \frac{10\alpha}{V_\infty}\sqrt{\frac{280}{13}\frac{\nu}{V_\infty}} \quad (4.158)$$

It can be verified that the solution of Eq. (4.158) is of the form

$$\delta_t = C_1\,x^{1/2} \text{ (where } C_1 \text{ is constant)}$$

Substituting the above in Eq. (4.158),

$$C_1^{\,3} = \left[\frac{20\alpha}{V_\infty}\sqrt{\frac{280}{13}\frac{\nu}{V_\infty}}\right]$$

$$\delta_t = \left[\frac{20\alpha}{V_\infty}\sqrt{\frac{280}{13}\frac{\nu}{V_\infty}}\right]^{1/3} x^{1/2}$$

or

$$\delta_t = \frac{4.52}{Pr^{1/3}}\sqrt{\frac{\nu x}{V_\infty}} \quad (4.159)$$

The ratio (δ_t/δ) is then,

$$\frac{\delta_t}{\delta} = \left(\frac{13}{14\,Pr}\right)^{1/3} = \frac{1}{1.026\,Pr^{1/3}} \tag{4.160}$$

The local heat transfer coefficient,

$$h_x = \frac{-k\left(\frac{\partial t}{\partial y}\right)_{y=0}}{t_w - t_\infty} = \frac{3k}{2\delta_t}$$

or

$$h_x = 0.332\,k\,Pr^{1/3}\,(V_\infty/\nu_x)^{1/2} \tag{4.161}$$

$$Nu_x = 0.332\,Pr^{1/3}\,Re_x^{\ 1/2} \tag{4.162}$$

$$Nu_L = 0.664\,Pr^{1/3}\,Re_L^{\ 1/2} \tag{4.163}$$

Note that Eq. (4.162) is the same as that given by the exact solution, Eq. (4.15).

4.18 STANTON NUMBER

An interesting point is observed by making a simultaneous study of Eqs. (4.15) and (4.127). If we divide Eq. (4.15) by dimensionless product $(Re_x\,Pr^{1/3})$ we get ;

$$\left(\frac{Nu_x}{Re_x\,Pr^{1/3}}\right) = \left(\frac{0.332}{Re_x^{\ 1/2}}\right)$$

or

$$\left(\frac{Nu_x}{Re_x\,Pr}\right) Pr^{2/3} = \frac{1}{2}\left(\frac{0.664}{Re_x^{\ 1/2}}\right)$$

The term in the bracket on the right hand side is given by Eq. (4.127) as $(C_f)_x$. Also, we define the term in the bracket on the left hand side as *Stanton number*, $St_x = (Nu_x/Re_x\,Pr)$.

$$\text{Thus} \qquad St_x\,Pr^{2/3} = \left(\frac{(C_f)_x}{2}\right) j_x \tag{4.164}$$

This relationship was proposed by Colburn and illustrates the inter-relationship between the two processes while defining a new parameter *Stanton number*, which was also obtained as one of the dimensionless groups π_3 (Table 4.2).

4.19 SIMILARITY RELATIONS IN FORCED CONVECTION

Nusselt employed the method of similarity to heat transfer problems, as Reynolds did for fluid flow problems. Just as a significant parameter *Reynolds number* is found to characterize the pattern of fluid flow, we shall see that some other parameters play a significant role in heat transfer by forced convection.

Let there be two systems which are geometrically, hydrodynamically and thermally similar. Let them be denoted by suffix 1 and 2. Then we may write that :

$$x_2 = X x_1\,;\,y_2 = X y_1\,;\,l_2 = X l_1$$

$$V x_2 = V\,Vx_1\,;\,V y_2 = V\,Vy_1$$

$$\rho_2 = A\rho_1$$
$$\mu_2 = B\mu_1$$
$$C_{p2} = C\,C_{p1}$$
$$k_2 = D\,k_1$$
$$(t_w - t_\infty)_2 = E\,(t_w - t_\infty)_1$$
$$h_2 = F\,h_1$$
$$g_2 = G\,g_1$$
$$\beta_2 = H\beta_1.$$ (It will be used in a Section 4.27)

We will now apply the similarity considerations to the three basic equations (4.107), (4.133) and (4.1).

(i) Now, Eq. (4.107) may be written for systems 1 and 2 as

$$\rho_1 V_{x1}\frac{\partial V_{x1}}{\partial x_1} + \rho_1 V_{y1}\frac{\partial V_{x1}}{\partial y_1} = \mu_1\frac{\partial^2 V_{x1}}{\partial {y_1}^2} \qquad \text{(a)}$$

$$\rho_2 V_{x2}\frac{\partial V_{x2}}{\partial x_2} + \rho_2 V_{y2}\frac{\partial V_{x2}}{\partial y_2} = \mu_2\frac{\partial^2 V_{x2}}{\partial {y_2}^2} \qquad \text{(b)}$$

Transforming Eq. (b) with the aid of identities given above,

$$\frac{AV^2}{X}\left(\rho_1 V_{x1}\frac{\partial V_{x1}}{\partial x_1} + \rho_1 V_{y1}\frac{\partial V_{x1}}{\partial y_1}\right) = \frac{BV}{X^2}\mu_1\frac{\partial^2 V_{x1}}{\partial {y_1}^2} \qquad \text{(c)}$$

Comparing Eqs. (a) and (c) we get :

$$\frac{AV}{X} = \frac{B}{X^2} \text{ (for hydrodynamic similarity)} \qquad (4.165)$$

This equation requires that :

$$\left(\frac{\rho_2}{\rho_1}\right)\left(\frac{V_{x2}}{V_{x1}}\right)\left(\frac{x_2}{x_1}\right) = \left(\frac{\mu_2}{\mu_1}\right)$$

or

$$\frac{x_1 V_{x1}\rho_1}{\mu_1} = \frac{x_2 V_{x2}\rho_2}{\mu_2} \qquad (4.166)$$

$$Re_1 = Re_2$$

We are already familiar with this dimensionless quantity and recognise it as *Reynolds number*.

(ii) Similarly, Eq. (4.133) may be written for system 2 as follows :

$$V_{x2}\frac{\partial t_2}{\partial x_2} + V_{y2}\frac{\partial t_2}{\partial y_2} = \alpha_2\frac{\partial^2 t_2}{\partial {y_2}^2}$$

Transforming this equation in terms of system 1 we have ;

$$\frac{VE}{X}\left(V_{x1}\frac{\partial t_1}{\partial x_1}+V_{y1}\frac{\partial t_1}{\partial y_1}\right) = \frac{DE}{ACX^2}\frac{k_1}{\rho_1 C_{p1}}\frac{\partial^2 t_1}{\partial {y_1}^2}$$

In order that thermal similarity exists ;

$$\frac{VE}{X} = \frac{DE}{ACX^2}$$

or

$$VACX = D \tag{4.167}$$

Dividing Eq. (4.167) by Eq. (4.165) we get ;

$CB = D$, which requires that ;

$$\left(\frac{C_{p2}}{C_{p1}}\right)\left(\frac{\mu_2}{\mu_1}\right) = \frac{k_2}{k_1}$$

or

$$\frac{\mu_1 C_{p1}}{k_1} = \frac{\mu_2 C_{p2}}{k_2} \text{ or } Pr_1 = Pr_2 \tag{4.168}$$

This ratio is recognized as *Prandtl number*, which is only a property parameter and signifies the characteristics of the fluid.

Further Eq. (4.167) requires that ;

$$\left(\frac{V_{x2}}{V_{x1}}\right)\left(\frac{\rho_2}{\rho_1}\right)\left(\frac{C_{p2}}{C_{p1}}\right)\left(\frac{x_2}{x_1}\right) = \left(\frac{k_2}{k_1}\right)$$

or

$$\frac{V_{x1}\,x_1}{(k_1/\rho_1 C_{p1})} = \frac{V_{x2}\,x_2}{(k_2/\rho_2 C_{p2})} \text{ or } Pe_1 = Pe_2 \tag{4.159}$$

The ratio is called Peclet number.

(iii) Finally, Eq. (4.1) may be written for system 2 as ;

$$\left\{-k_2\left(\frac{\partial t_2}{\partial y_2}\right)_{y_2=0}\right\} = h_2\,(t_w - t_\infty)_2$$

or

$$-\frac{DE}{X}\,k_1\left(\frac{\partial t_1}{\partial y_1}\right)_{y_1=0} = FE\,h_1\,(t_w - t_\infty)_1$$

Thus a comparison of this equation with Eq. (4.1) written for system 1, leads to ;

$$\frac{DE}{X} = FE$$

or

$$D = FX \tag{4.170}$$

or

$$\left(\frac{k_2}{k_1}\right) = \left(\frac{x_2}{x_1}\right)\left(\frac{h_2}{h_1}\right)$$

or

$$\frac{h_1 x_1}{k_1} = \frac{h_2 x_2}{k_2} \text{ or } Nu_1 = Nu_2 \tag{4.4}$$

This dimensionless group is recognized as *Nusselt number* as discussed in Section 4.2. To sum up, we obtain the dimensionless parameters *Re*, *Pr* and *Nu* which are significant in forced convection.

4.20 MOMENTUM AND HEAT EXCHANGE IN TURBULENT FLOW (EDDY VISCOSITY AND EDDY THERMAL DIFFUSIVITY)

(a) Nature of Turbulence

It is well known that in laminar flow the layers of fluid slide over each other with no *macroscopic* mixing (the mixing is only on the molecular scale) and the velocity in macroscopic steady flow is constant at any point. In turbulent flow, on the other hand, there is *visible macroscopic* mixing by eddy motion between the layers, and even in over-all steady flow the velocity at a point fluctuates about some mean value. In laminar flow exchanges of momentum and heat occur on a molecular scale. This is to say that faster moving and high temperature molecules cross into slower moving and low temperature molecules and vice-versa. In turbulent flow, this exchange of momentum and heat on the molecular scale is present ; but in addition to this there is momentum and heat exchange on *macroscopic* scale on account of eddy mixing. This is to say that faster moving particles (of macroscopic size) cross into slower-moving fluid and vice-versa causing momentum transfer ; similarly heat exchange is caused by the transport of high temperature fluid and vice-versa. Note that this exchange occurs between the moving fluid particles in a direction at right angles to the main stream. It is thus evident that both momentum and heat exchange would be considerably more in turbulent flow than in laminar flow. It is for this reason that friction loss is usually much higher in turbulent flow than in laminar flow.

(b) Mean and Fluctuating Velocities

As stated earlier, in turbulent flow the velocity at any point varies with time in a chaotic manner both in magnitude and direction. Therefore, strictly speaking, there is no such thing as steady state in turbulent flow. However, we can define a mean velocity in the x direction by ;

$$\overline{V}_x = \frac{1}{\tau}\int_0^\tau V_x \, d\tau \tag{4.171}$$

where V_x denotes the instantaneous velocity in x direction and is a function of time. Since the frequency of the velocity fluctuation is large τ need be only a few seconds. Similar expressions can be written for other velocity components and pressure. If all these mean quantities are constant over successive time intervals τ we may characterize the turbulent flow to be at steady state (strictly speaking, at steady state with respect to the mean flow).

The instantaneous values of the variables may be expressed as the sum of a mean value and a fluctuating value

$$V_x = \overline{V}_x + V'_x \tag{4.172a}$$

$$V_y = \overline{V}_y + V'_y \tag{4.172b}$$

$$V_z = \overline{V}_z + V'_z \tag{4.172c}$$

$$p = \bar{p} + p' \tag{4.172d}$$

$$t = \bar{t} + t' \tag{4.172e}$$

where V'_x, V'_y, V'_z = x, y, z components of fluctuating velocity

t' = fluctuating temperature

p' = fluctuating pressure

It is evident from the above discussion that

$$\overline{V'_x} = \frac{1}{\tau}\int_0^\tau V'_x \, d\tau = 0$$

$$\overline{V'_y} = 0, \ \overline{V'_z} = 0, \ \overline{p'} = 0, \ \overline{t'} = 0$$

If the flow is *one-dimensional* the *mean* values of the *y* and *z* components of the velocity will be zero ;

$$\overline{V_y} = \overline{V_z} = 0$$

so that in this case

$$V_x = \overline{V_x} + V'_x \tag{4.172a}$$

$$V_y = V'_y \tag{4.173}$$

$$V_z = V'_z \tag{4.174}$$

A typical plot of V_x versus τ is shown in Fig. 4.12.

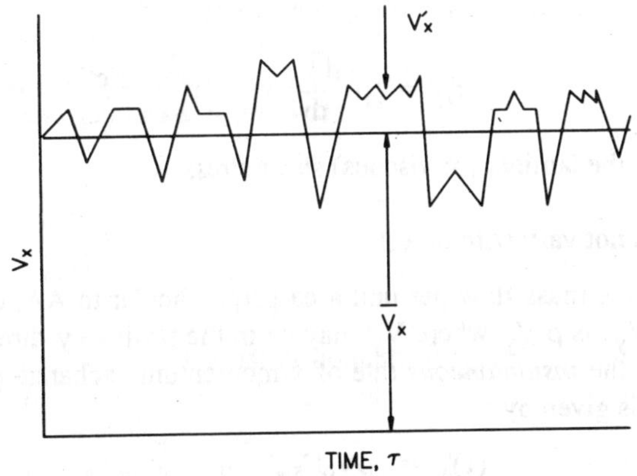

Fig. 4.12 Fluctuations of Velocity in Turbulent Flow

(c) Eddy Viscosity and Eddy Thermal Diffusivity

In order to express quantitatively the momentum and heat exchange in turbulent flow let us consider an arbitrary plane A-A, parallel to the surface in a one-dimensional turbulent flow stream (Fig. 4.13). Let V_x, V_y and t denote respectively the instantaneous *x* and *y* velocities, and temperature, of the fluid at plane A-A. Then from (4.172), (4.173), and (4.172e),

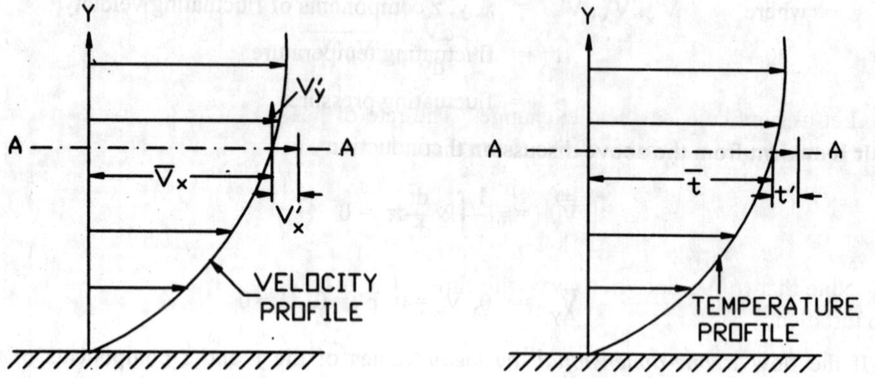

Fig. 4.13 Velocity and Temperature Profiles in Turbulent Flow

$$V_x = \bar{V}_x + V_x' \tag{4.172a}$$

$$V_y = V'_y \tag{4.173}$$

$$t = \bar{t} + t' \tag{4.172e}$$

The rate of x-momentum exchange per unit area of plane AA; due to molecular exchange (or due to viscosity of the fluid) is given by ;

$$(\tau)_l = \mu \frac{d\,\bar{V}_x}{dy} \tag{4.175}$$

where $(\tau)_l$ denotes the laminar (or viscous) shear stress.

Note that $(\tau)_l$ does not vary with time.

At any instant the mass flow per unit area perpendicular to AA, due to y-component of velocity $V_y = V_y'$, is $\rho\, V_y'$ where V_y' may be in the positive y-direction or negative y-direction. Hence, the *instantaneous* rate of x-momentum exchange per unit area of AA due to turbulence is given by ;

$$(\tau)'_t = \rho V_y' V_x'$$

where $(\tau)'_t$ denotes the *turbulent shear stress*. Note that $(\tau)_t'$ varies from instant to instant. The mean turbulent shear stress may then be expressed by

$$(\tau)_t = \rho\, \overline{V_y'}\, \overline{V_x'} \tag{4.176}$$

Note that although the time average of $\rho\, V_y'$ (ie. $\rho\, \overline{V_y'}$) will be zero (since $\overline{V_y'} = 0$) the time average of $\rho\, V_y'\, V_x'$ (ie. $\rho \overline{V_y' V_x'}$) will not be zero.

$$\tau = (\tau)_l + (\tau)_t$$

$$= \mu \frac{d\bar{V}_x}{dy} + \rho \overline{V_x'V_y'}$$

or
$$\frac{\tau}{\rho} = \nu \frac{d\bar{V}_x}{dy} + \overline{V_x'V_y'} \qquad (4.177)$$

Let us now consider heat exchange. The rate of heat exchange per unit area of AA due to molecular transport (due to thermal conductivity of the fluid) is given by ;

$$\left(\frac{q}{A}\right)_l = -k \frac{d\bar{t}}{dy} \qquad (4.178)$$

Note that $(q/A)_l$ does not vary with time. The instantaneous rate of heat transfer due to turbulence per unit area of AA is given by ;

$$\left(\frac{q}{A}\right)_t' = \rho V_y' C_p t'$$

Also, note that $(q/A)_t'$ varies from instant to instant. The mean rate of turbulent heat transfer may be expressed by ;

$$\left(\frac{q}{A}\right)_t = \rho C_p \overline{V_y't'} \qquad (4.179)$$

From (4.178) and (4.179), the total rate of heat transfer per unit area AA will be ;

$$\frac{q}{A} = -k \frac{d\bar{t}}{dy} + \rho C_p \overline{V_y't'}$$

or
$$\frac{q}{A} = -\rho C_p \left[\alpha \frac{d\bar{t}}{dy} - \overline{V_y't'}\right] \qquad (4.180)$$

Unfortunately Eqs. (4.167) and (4.180) are not much use because we do not know the values of V_y', V_x', t' etc. For this reason the turbulent momentum exchange is considered equivalent to an increase in the viscosity effect so that the effective viscosity is much greater than the actual and varies with the degree of turbulence. For this purpose we define an *eddy viscosity* or *eddy momentum diffusity* e_m, given by :

$$\overline{V_x'V_y'} = \varepsilon_M \frac{dV}{dy} \qquad (4.181)$$

where ε_M is similar to kinematic viscosity ν.

Similarly, the turbulent heat exchange may be considered equivalent to an increase in the thermal diffusivity of the fluid so that the effective thermal diffusivity is much greater than the actual and varies with the degree of turbulence. For this purpose we define an eddy thermal diffusivity ε_H by ;

$$\overline{V_y't'} = -\varepsilon_H \frac{d\bar{t}}{dy} \qquad (4.182)$$

Substitution of Eqs. (4.181) and (4.182) into Eqs. (4.177) and (4.180) gives ;

$$\frac{\tau}{\rho} = (\nu + \varepsilon_M)\frac{d\overline{V}_x}{dy} \tag{4.183}$$

$$\frac{q}{A} = -\rho C_p (\alpha + \varepsilon_H)\frac{d\bar{t}}{dy} \tag{4.184}$$

Values of ε_M and ε_H for a particular case may be obtained from measurements of the mean velocity and temperature distributions. From a physical stand point ε_M and ε_H should be closely related. As a first approximation they have been taken to be equal ; $\varepsilon_M/\varepsilon_H = 1$. The ratio $\varepsilon_M/\varepsilon_H$ is called the *turbulent* Prandtl number, analogus to the *laminar* Prandtl number (ν/α).

4.21 ANALOGY BETWEEN HEAT AND MOMENTUM TRANSFERS, REYNOLDS ANALOGY

Reynolds established a relationship between friction force and heat transfer. This relation is called *Renolds analogy*. Subsequent investigators (Prandtl, Colburn, Von Karman, Martinelli and others) have modified the relation established by Reynolds.

(a) Statement of Reynolds Analogy

Reynolds developed a relation between heat transfer and fluid friction applicable to fluids with Pr = 1. His analogy may be expressed by any of the following equivalent equations :

$$St = \frac{h}{\rho V_\infty C_p} = \frac{C_f}{2}; \; Pr = 1 \tag{4.185}$$

$$\frac{(q/A)}{\tau} = -C_p \frac{dt}{dV_x}; \; Pr = 1 \tag{4.186}$$

(b) Derivation of Reynolds Analogy

We shall now demonstrate Reynolds analogy by considering laminar as well as turbulent flow over a flat plate. For the case of laminar flow Eqs. (4.73) and (4.99) give ;

$$V_x \frac{\partial V_x}{\partial x} + V_y \frac{\partial V_x}{\partial y} = \nu \frac{\partial^2 V_x}{\partial y^2} \tag{4.107}$$

$$V_x \frac{\partial t}{\partial x} + V_y \frac{\partial t}{\partial y} = \alpha \frac{\partial^2 t}{\partial y^2} \tag{4.133}$$

(i) As already pointed out in Section 4.13 and 4.14, if Pr = 1, the velocity and temperature distributions in the boundary layer would be similar and the velocity distribution curve (for Pr =1) in Fig. 4.8 would coincide with the temperature distribution curve (for Pr = 1) in Fig. 4.10. Hence we can write ;

$$\left\{\frac{\partial\left(\dfrac{t-t_w}{t_\infty - t_w}\right)}{\partial y}\right\}_{y=0} = \left\{\frac{\partial\left(\dfrac{V_x}{V_\infty}\right)}{\partial y}\right\}_{y=0}$$

or
$$\frac{1}{(t_\infty - t_w)}\left\{\frac{\partial(t-t_w)}{\partial y}\right\}_{y=0} = \frac{1}{V_\infty}\left(\frac{\partial V_x}{\partial y}\right)_{y=0} \qquad \text{(a)}$$

Since Pr = 1 or $k = \mu C_p$, we can multiply the LHS of Eq. (a) by k and RHS by μC_p. This gives;

$$\frac{1}{(t_w - t_\infty)}\left\{(-k)\frac{\partial(t-t_w)}{\partial y}\right\}_{y=0} = \frac{C_p}{V_\infty}\mu\left(\frac{\partial V_x}{\partial y}\right)_{y=0}$$

$$\frac{1}{(t_\infty - t_w)}\left(\frac{q}{A}\right) = \frac{C_p}{V_\infty}\tau = \frac{C_p}{V_\infty}(C_f)\left(\frac{1}{2}\rho V_\infty^2\right)$$

(since $C_f = \tau/\frac{1}{2}\rho V_\infty^2$)

or
$$h = \rho V_\infty C_p\left(\frac{C_f}{2}\right)$$

or
$$\frac{h}{\rho V_\infty C_p} = \frac{C_f}{2} \qquad (4.185)$$

It can be easily shown that the above relation also holds for local values of *h* and C_f, ie.,

$$\frac{h_x}{\rho V_\infty C_p} = \frac{(C_f)_x}{2}$$

(ii) Now, let us consider turbulent flow. For this case Eqs. (4.183) and (4.184) give :

$$\frac{\tau}{\rho} = (\nu + \varepsilon_M)\frac{d\bar{V}_x}{dy} \qquad (4.183)$$

$$\frac{q}{A} = -\rho C_p(\alpha + \varepsilon_H)\frac{d\bar{t}}{dy} \qquad (4.184)$$

If $\nu = \alpha$ and $\varepsilon_M = \varepsilon_H$, ie, both laminar as well as turbulent Prandtl numbers are equal to unity,

$$\nu + \varepsilon_M = \alpha + \varepsilon_H$$

In this case, dividing Eq. (4.184) by (4.183) gives

$$\frac{(q/A)}{\tau} = -C_p\frac{d\bar{t}}{d\bar{V}_x} = -C_p\frac{dt}{dV_x} \qquad (4.186)$$

where t and V_x in Eq. (4.186) denote the mean values.

The equivalence of Eqs. (4.185) and (4.186) can be readily shown as follows. We may write Eq. (4.186) as ;

$$\frac{(q/A)_w}{\tau_w C_p}dV_x = -dt$$

where the suffix *w* refers to the wall surface, ie, y = 0

Integrating Eq. (4.186) over the boundary layer thickness :

$$\frac{(q/A)_w}{\tau_w C_p} \int_0^{V_\infty} dV_x = -\int_{t_w}^{t_\infty} dt$$

or
$$\frac{(q/A)_w}{\tau_w C_p} (V_\infty) = (t_w - t_\infty) \tag{b}$$

Since, by definition ;

$$h_x = \frac{(q/A)_w}{(t_w - t_\infty)}$$

$$\tau_w = (C_f)_x \frac{\rho V_\infty^2}{2}$$

Eq. (b) may be written as ;

$$\frac{h_x}{\rho V_\infty C_p} = \frac{(C_f)_x}{2}$$

or
$$\left(\frac{1}{\rho V_\infty C_p}\right)\left\{\frac{1}{L}\int_0^L h_x\, dx\right\} = \left(\frac{1}{2}\right)\left\{\frac{1}{L}\int_0^L (C_f)_x\, dx\right\}$$

or
$$\frac{h}{\rho V_\infty C_p} = \frac{C_f}{2} \tag{4.185}$$

(c) General Form of Reynolds Analogy

The general form of Reynolds analogy, applicable to all fluids ($Pr = 1$ or $Pr \neq 1$) may be expressed by ;

$$j = (St)(Pr)^{2/3} = \frac{C_f}{2} \tag{4.187}$$

Equation (4.187) may be deduced by using Eqs. (4.128) and (4.17) for flow over a flat plate ;

$$C_f = \frac{1.328}{\sqrt{Re_L}}$$

$$h = 0.664 \frac{k}{L} Pr^{1/3} \sqrt{Re_L}$$

Hence,

$$\left(\frac{h}{\rho V_\infty C_p}\right)(Pr)^{2/3} = \frac{0.664}{\rho V_\infty C_p} \frac{k}{L} \frac{\mu C_p}{k} \sqrt{\frac{V_\infty L \rho}{\mu}}$$

$$= \frac{0.664}{\sqrt{V_\infty L\rho/\mu}} = \frac{C_f}{2}$$

Thus
$$(St)\, Pr^{2/3} = \frac{C_f}{2} \tag{4.187}$$

4.22 FREE CONVECTION

As mentioned in Section 4.1 *free convection* heat transfer occurs due to the motion of fluid by *density difference* only. Any hot body placed in a fluid at a higher or lower temperature than the body, will have transfer of heat by *free convection.*

Due to the temperature difference between the body and the surrounding fluid, heat flows between the two, and causes density change in the fluid in the vicinity of body surface. The fluid with higher density flows downwards while that with lower density rises up. This motion of fluid is responsible for *heat convection.* Since there is no external agency causing the motion of fluid, and motion is caused *naturally* due to gravity alone, we call the phenomenon as *natural* or *free convection.* Heat exchange from steam radiators in heating of buildings or from hot vessels of the kitchen are examples of free convection.

Gravity, however, is not the only body force which produces free convection. In high speed rotating machines, there may be centrifugal forces developing within, which may also cause free convection currents. We shall, however, not take such factors into account but discuss only the gravitational free convection.

Figure 4.14 shows the velocity and temperature distribution in a fluid near a heated flat plate placed vertically in air. The velocity reaches a maximum value at a distance of about 2.5 mm from the surface. At a distance of about 25 mm the velocity again becomes zero. The temperature profile, however, has a continuously decreasing tendency as the distance from the surface increases.

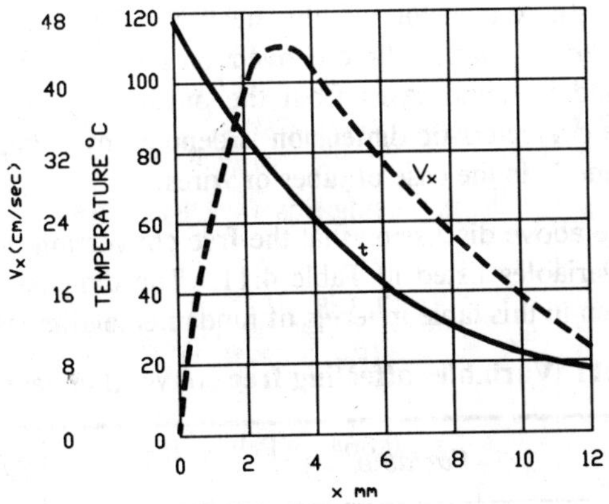

Fig. 4.14 Velocity and Temperature Variation in Free Convection

As will be discussed later the physical interpretation of the Nusselt number applies to free convection also. Hence attention will be directed again towards the determination of *Nu* or the heat transfer coefficient *h*. It may be pointed out that the value of *h* in this case is relatively low as compared to that of forced convection.

We will first deal with the *dimensional analysis* method to find a correlation for the heat transfer coefficient. The solution of boundary layer equation by the integral method will next be discussed. Finally empirical equations for different cases which have been obtained on the basis of experimental data will be presented.

It will be noticed that a general correlation of the form of Eq. (4.5) for the case of free convection, is of the type ;

$$Nu = \phi(\text{Gr})\,\psi(\text{Pr}) \tag{4.188}$$

where Gr = A new dimensionless parameter, replacing Re of forced convection. It is known as *Grashof number* ;

$$= \frac{L^3 \rho^2 g \beta \Delta t}{\mu^2} \tag{4.189}$$

4.23 DIMENSIONAL ANALYSIS APPLIED TO FREE CONVECTION

It is reasonable to assume that the *free convection heat transfer coefficient, h,* would depend on the same variables as the forced convection heat transfer coefficient. Thus h should be a function of V, ρ, k, μ, C_p and D. Since the circulation of the fluid is not caused by means of an external agency but it takes place due to the difference in density between the various layers of fluid on account of temperature difference between them, the fluid velocity V itself depends on the following variables ;

(a) Temperature difference between the surface and bulk fluid, Δt.

(b) Coefficient of volume expansion of the fluid, β

(c) Acceleration due to gravity, g.

Hence the variable V should be replaced by the variables Δt, β and g.

Another point of difference should be noted. In forced convection the fluid generally flows inside or outside of tubes or in the annular space between two tubes. For this reason the characteristic dimension of the heat transfer surface is denoted by D, signifying an *equivalent diameter*. Free convection heat transfer generally takes place from plates, walls and in some cases from the outside surfaces of tubes, wires and cylinders. Hence the characteristic dimension is denoted by L (signifying length) in the case of flat surfaces and D in the case of tubes or wires.

It follows from the above discussion that the free convection heat transfer coefficient will depend on the variables listed in Table 4.11. The dimensional formulae of these variables are also given in this table in terms of fundamental dimensions M, L, T, and θ.

Table 4.11 Variables affecting free convection heat transfer

No.	*Variable*	*Symbol*	*Dimensional Formula*
1.	Fluid density	ρ	M/L^3
2.	Thermal conductivity of fluid	k	$ML/T\theta^3$
3.	Specific heat	Cp	$L^2/T\theta^2$
4.	Fluid viscosity	μ	$M/L\theta$
5.	Characteristic linear dimension of heat transfer surface	L	L
6.	Temperature difference between surface and bulk fluid	Δt	T
7.	Coefficient of volume expansion of fluid	β	1/T
8.	Acceleration due to gravity	g	L/θ^2

Thus we may write,

$$F\,(h, L, \rho, C_p, k, \Delta t, g, \beta, \mu) = 0$$

From dimensional analysis,

$$\pi = (h)^a (L)^b (\rho)^c (C_p)^d (k)^e (\Delta t)^f (g)^i (\beta)^j (\mu)^n$$

Since the dimensional formula for h is $\left[\dfrac{M}{T\theta^3}\right]$ we obtain,

$$0 = \left[\frac{M}{T\theta^3}\right]^a [L]^b \left[\frac{M}{L^3}\right]^c \left[\frac{L^2}{T\theta^2}\right]^d \left[\frac{ML}{T\theta^3}\right]^e [T]^f \left[\frac{L}{\theta^2}\right]^i \left[\frac{1}{T}\right]^j \left[\frac{M}{L\theta}\right]^n \qquad (4.190)$$

Equating dimensions on the two sides of Eq. (4.190) ;

$$M: \quad a + c + e + n = 0 \qquad (i)$$

$$L: b - 3c + 2d + e + i - n = 0 \qquad (ii)$$

$$\theta: -3a - 2d - 3e - 2i - n = 0 \qquad (iii)$$

$$T: \quad -a - d - e + f - j = 0 \qquad (iv)$$

Since we have only four equations between nine unknown exponents, let us express any four exponents in terms of the remaining five. Expressing f, i, n and d in terms of a, b, c, e and j.

From Eq. (i)

$$n = -a - c - e \qquad (A)$$

From Eqs. (ii) and (A),

$$b - 3c + 2d + e + i - (-a - c - e) = 0$$

or

$$b - 2c + 2d + 2e + a + i = 0 \qquad (v)$$

From Eqs. (iii) and (A),

$$-3a - 2d - 3e - 2i - (-a - c - e) = 0$$

or

$$-2a - 2d - 2e - 2i + c = 0 \qquad (vi)$$

Adding Eqs. (v) and (vi),

$$b - c - a - i = 0$$

or

$$i = b - c - a \qquad (B)$$

Substituting Eq. (B) in Eq. (v),

$$b - 2c + 2d + 2e + a + (b - c - a) = 0$$

or

$$2d = 3c - 2b - 2e$$

or

$$d = \frac{3}{2}c - b - e \qquad (C)$$

from Eqs. (iv) and (C),

$$-a - \left[\frac{3}{2}c - b - e\right] - e + f - j = 0$$

or
$$f = a - b + j + \tfrac{3}{2} c \qquad \text{(D)}$$

We can now proceed to find the π-terms by assigning arbitrary values to the five exponents *a*, *b*, *c*, *e*, and *j* as shown in Table 4.12.

From Table 4.12 it will be observed that a simple value of unity is assigned to each of the above exponents (except c = 2) while others are assigned zero. The parameter π_2 is recognised as the reciprocal of Prandtl number. A recombination of other parameters leads us to a few more recognised dimensionless numbers

Table 4.12 : Development of non-dimensional parameters

Variables									π-terms
h	*L*	ρ	*k*	β	C_p	*g*	μ	Δ*t*	
Exponents									
a	*b*	*c*	*e*	*j*	*d*	*i*	*n*	*f*	
0	0	0	0	1	0	0	0	1	$\pi_1 = \beta \Delta t$
0	0	0	1	0	−1	0	−1	0	$\pi_2 = (\mu C_p/k)^{-1}$
0	1	0	0	0	−1	1	0	−1	$\pi_3 = (gL/C_p \Delta t)$
1	0	0	0	0	0	−1	−1	1	$\pi_4 = (h \Delta t/g\mu)$
0	0	2	0	0	3	−2	−2	3	$\pi_5 = \dfrac{\rho^2 C_p^3 \Delta t^3}{g^2 \mu^2}$

(i) A new parameter π_6 may be written as :

$$\pi_6 = \frac{\pi_3 \pi_4}{\pi_2}$$

or
$$\pi_6 = \frac{gL}{C_p \Delta t} \frac{h \Delta t}{g\mu} \frac{C_p \mu}{k}$$

$$= \frac{hL}{k}$$

It is recognised as Nusselt number.

(ii) Again, another parameter π_7 may be written as :

$$\pi_7 = \pi_5 (\pi_1) (\pi_3)^3$$

$$= \frac{\rho^2 C_p^3 \Delta t^3}{g^2 \mu^2} \left(\frac{g^3 L^3}{C_p^3 \Delta t^3} \right) \beta \Delta t$$

$$= \frac{\rho^2 g L^3 \beta \Delta t}{\mu^2}$$

This *dimensionless* number is known as *Grashof number.*

(iii) It will be noted that the five π-terms of Table 4.12 are thus covered by three dimensionless groups π_2, π_6 and π_7

Thus
$$F(\pi_2, \pi_6, \pi_7) = 0$$

or $$F\left\{\left(\frac{hL}{k}\right),\left(\frac{C_p\mu}{k}\right),\left(\frac{L^3\rho^2 g\beta\Delta t}{\mu^2}\right)\right\} = 0$$

or $$\frac{hL}{k} = \phi\left(\frac{L^3\rho^2 g\beta\Delta t}{\mu^2}\right)\psi\left(\frac{C_p\mu}{k}\right) \tag{4.191}$$

or $$Nu = \phi(Gr)\,\psi(Pr) \tag{4.188}$$

Aliter

Note that equation (4.188) for free convection can also be obtained from Eq. (4.13) by the following reasoning. As already pointed out, the fluid velocity V in free convection depends on the force due to the buoyancy of the fluid. This force arises because of the difference in temperature between the surface and the main body of the fluid. When the temperature difference between the surface and the main body of the fluid is Δt the average temperature of the fluid near the surface is $(\Delta t/2)$ higher or lower than the surrounding fluid.

Now, $$\beta = (1/v)(\Delta v/\Delta t)$$

Hence change in specific volume of fluid due to temperature difference $\Delta t/2$ is given by ; $\Delta v = (\beta v\, \Delta t)/2 = (va - v)$

Now, $$\Delta\rho = \rho_a - \rho = \frac{1}{v_s} - \frac{1}{v} = \frac{v - v_s}{v_s\, v} \approx \frac{v - v_s}{v^2} = -\frac{\Delta v}{v^2}$$

or $$\Delta\rho = -\frac{1}{v^2}\,\beta v\,\frac{\Delta t}{2}$$

so $$\Delta\rho = -\frac{\beta\Delta t}{v\times 2} = -\beta\rho\,\frac{\Delta t}{2}$$

The *buoyant force* per unit volume of fluid is then given by,

$$\text{Buoyant force} = (\rho - \rho_s)g$$
$$= -\Delta\rho\, g$$
$$= \beta\,\rho\, g\,\frac{\Delta t}{2}$$

The work done by this force *per unit volume of fluid* when it acts through a height of surface L equals,

$$\text{Work done} = L\times\beta\,\rho\, g\,\frac{\Delta t}{2}$$

If all this work is converted into kinetic energy of motion in producing a velocity V, then ;

$$\frac{\rho V^2}{2} = L\,\beta\,\rho\, g\,\frac{\Delta t}{2}$$

or $$V^2 = L\,\beta\, g\,\Delta t$$

Let us now substitute this value of V in Eq. (4.13) ;

$$\frac{hD}{k} = \phi\left(\frac{VD\rho}{\mu}\right)\psi\left(\frac{\mu C_p}{k}\right) \tag{4.13}$$

or

$$\frac{hL}{k} = \phi\left(\frac{VL\rho}{\mu}\right)\psi\left(\frac{\mu C_p}{k}\right)$$

$$= \phi\left(\frac{V^2 L^2 \rho^2}{\mu^2}\right)\psi\left(\frac{\mu C_p}{k}\right)$$

$$= \phi\left(\frac{L\beta g \Delta t L^2 \rho^2}{\mu^2}\right)\psi\left(\frac{\mu C_p}{k}\right)$$

or

$$\frac{hL}{k} = \phi\left(\frac{L^3 \rho^2 g \beta \Delta t}{\mu^2}\right)\psi\left(\frac{\mu C_p}{k}\right) \tag{4.191}$$

or

$$Nu = \phi(Gr)\,\psi(Pr) \tag{4.188}$$

It would be confirmed from the above that the *Grashof number* may be considered to be a particular form of the Reynolds number, that is,. a number pertaining to fluid flow due to buoyancy force caused by temperature difference.

4.24 INTEGRAL METHOD APPLIED TO FREE CONVECTION FROM A VERTICAL PLATE

Consider the vertical flat plate of length L and unit width perpendicular to the plane of the paper (Fig. 4.15). If such a plate is heated, the fluid in its immediate neighbourhood experiences a temperature increase due to heat transfer from the plate and starts to flow upward. As a result, both hydrodynamic and thermal boundary layers begin to form with zero thickness at the bottom edge of the plate and with increasing thickness in the upward direction. Both layers remain laminar up to certain distance from the bottom, after which they become turbulent.

Within the boundary layer, the temperature decreases from $t = t_w$ at the plate surface to $t = t_\infty$ at $y = \delta_t$, where δ_t is the thickness of the thermal boundary layer and t_∞ the uniform temperature of the fluid outside the heated region. However, unlike the case of forced flow of the fluid over the plate, the velocity V_x is zero at the surface ($y = 0$), reaches a maximum value somewhere within the boundary layer and becomes zero again at the edge of the hydrodynamic boundary layer (Fig. 4.15).

For simplicity let us assume that both boundary layers have the same thickness δ. Let us consider the control volume ABCD of length dx, height δ and unit width perpendicular to the plane of the paper (Fig. 4.15). From Newton's second law of motion

$$\left\{\begin{matrix}\text{Rate at which x direction}\\ \text{momentum leaves the}\\ \text{control volume}\end{matrix}\right\} - \left\{\begin{matrix}\text{Rate at which x direction}\\ \text{momentum enters the}\\ \text{control volume}\end{matrix}\right\} = \left\{\begin{matrix}\text{Sum of external forecs}\\ \text{in x direction acting on}\\ \text{the control volume}\end{matrix}\right\} \tag{i}$$

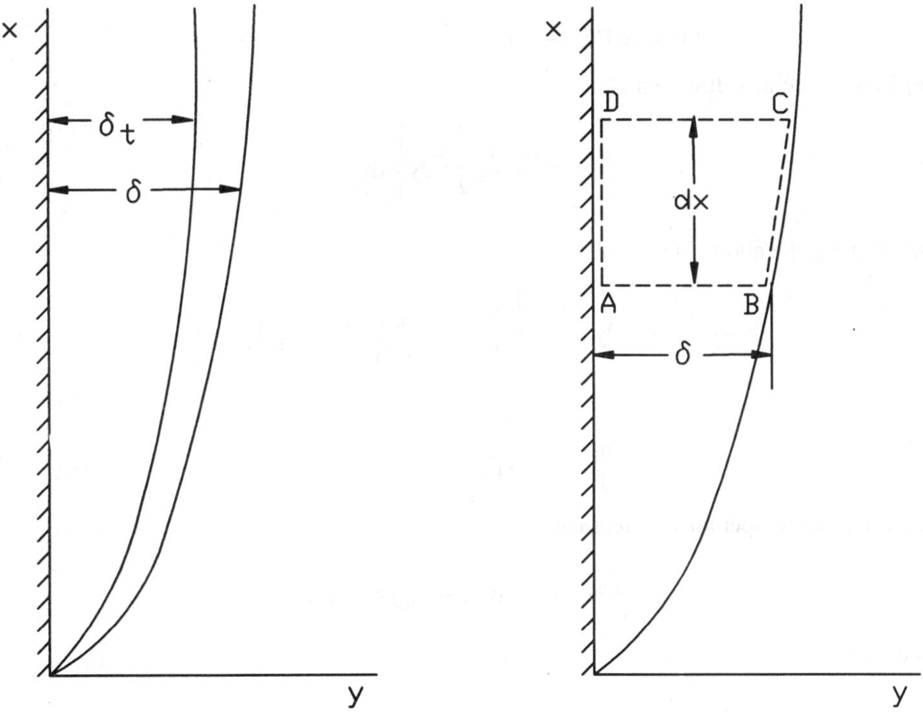

Fig. 4.15 Velocity and Temperature Boundary Layers in Free Convection from Vertical Plate

(a) Rate at which x direction momentum enters control volume across face AB

$$= \rho \int_0^{\delta} V_x^2 \, dy$$

(b) Rate at which x direction momentum leaves control volume across face CD

$$= \rho \left[\int_0^{\delta} V_x^2 \, dy + \frac{d}{dx} \left\{ \int_0^{\delta} V_x^2 \, dy \right\} dx \right]$$

(c) Pressure force in x direction

$$\text{On face AB} = \rho\,\delta$$

$$\text{On face BC} = \left(p + \frac{dp}{2} \right) d\,\delta$$

$$\text{On face CD} = -\left(p + \frac{dp}{2} \right) (\delta + d\delta)$$

(d) Viscous force in x direction

$$\text{On face AD} = -\mu \left(\frac{\partial V_x}{\partial y} \right)_{y=0} dx$$

$$\text{On face BC} = 0$$

(e) Body force in x direction

$$= -\left\{ g \int_0^{\delta} \rho \, dy \right\} dx$$

Substituting the above in (i)

$$\rho \frac{d}{dx} \left[\int_0^{\delta} V_x^2 \, dy \right] = \left[\frac{dp}{dx} \delta - \mu \left(\frac{\partial V_x}{\partial y} \right)_{y=0} - g \int_0^{\delta} \rho \, dy \right]$$

Now,

$$\frac{dp}{dx} = -\rho_\infty g$$

and for small temperature difference

$$\frac{\rho - \rho_\infty}{\rho} = -\beta (T - T_\infty) = -\beta (t - t_\infty)$$

Therefore,

$$\frac{d}{dx} \left[\int_0^{\delta} V_x^2 \, dy \right] = g\beta \left[\int_0^{\delta} (t - t_\infty) dy \right] - \frac{\mu}{\rho} \left(\frac{\partial V_x}{\partial y} \right)_{y=0} \qquad (4.192)$$

We also have the energy equation (4.155) which gives (on putting $\delta_t = \delta$)

$$\frac{d}{dx} \left[\int_0^{\delta} V_x (t - t_\infty) dy \right] = -\alpha \left(\frac{\partial t}{\partial y} \right)_{y=0} \qquad (4.193)$$

To solve Eqs. (4.192) and (4.193), we assume that the temperature profile may be approximated by a parabola as

$$\frac{t - t_\infty}{t_w - t_\infty} = \left(1 - \frac{y}{\delta} \right)^2 \qquad (4.194)$$

and the velocity profile is given by

$$V_x = V_{x1} \frac{y}{\delta} \left(1 - \frac{y}{\delta} \right)^2 \qquad (4.195)$$

where V_{x1} is an arbitrary function of x. Note that Eq. (4.195) satisfies the conditions that, $V_x = 0$ at $y = 0$ and $y = \delta$ and that $(\partial V_x / \partial y) = 0$ at $y = \delta$.

Introducing the velocity and temperature profiles,

$$\int_0^{\delta} V_x^2 \, dy = \frac{V_{x1}^2 \, \delta}{105}$$

$$\int_0^\delta (t-t_\infty)\, dy = \frac{(t_w - t_\infty)\delta}{3}$$

$$\int_0^\delta V_x\,(t - t_\infty)\, dy = \frac{1}{30}\, Vx1\,(t_w - t_\infty)\,\delta$$

So, the boundary layer equations become

$$\frac{1}{105}\,\frac{d}{dx}\,(V_{x1}{}^2\,\delta) = \left[\frac{1}{3}\, g\,\beta\,(t_w - t_\infty)\,\delta - \frac{\nu V_{x1}}{\delta}\right]$$

$$\frac{1}{30}\,(t_w - t_\infty)\,\frac{d}{dx}\,(V_{x1}\,\delta) = \frac{2\alpha}{\delta}\,(t_w - t_\infty)$$

To solve these equations let us try

$$V_{x1} = C_1\, x^m$$

$$\delta = C_2\, x^n$$

The above equations satisfy the conditions that at x = 0, both V_{x1} and δ equal zero. Introducing these expressions we get

$$\frac{2m+n}{105}\, C_1{}^2\, C_2\, x^{2m+n-1} = \left[g\beta(t_w - t_\infty)\frac{C_2\, x^n}{C_3} - \frac{C_1}{C_2}\,\nu\, x^{m-n}\right]$$

and

$$\frac{m+n}{30}\, C_1\, C_2\, x^{m+n-1} = \frac{2\alpha}{C_2}\, x^{-n}$$

Since these equations must be valid for any value of x it is evident that

$$2m + n - 1 = n = m - n$$

$$m + n - 1 = -n$$

This gives,

$$m = \frac{1}{2},\ n = \frac{1}{4}$$

Thus, we obtain

$$\frac{C_1^2\, C_2}{84} = g\,\beta\,(t_w - t_\infty)\,\frac{C_2}{3} - \frac{C_1}{C_2}\,\nu$$

$$\frac{C_1\, C_2}{40} = \frac{2\alpha}{C_2}$$

Hence

$$C_1 = 5.17\,\nu\left(0.952 + \frac{\nu}{\alpha}\right)^{-1/2}\left[\frac{g\beta(t_w - t_\infty)}{\nu^2}\right]^{1/2}$$

$$C_2 = 3.93\left(0.952 + \frac{\nu}{\alpha}\right)^{1/4}\left[\frac{g\beta(t_w - t_\infty)}{\nu^2}\right]^{-1/4}\left(\frac{\nu}{\alpha}\right)^{-1/2}$$

$$V_{x1} = 5.11\,\nu\,(0.952 + Pr)^{-1/2}\left[\frac{g\beta(t_w - t_\infty)}{\nu^2}\right]^{1/2} x^{1/2}$$

$$\delta = 3.93\,(0.952 + Pr)^{1/4}\,Pr^{-1/2}\left[\frac{g\beta(t_w - t_\infty)}{\nu^2}\right]^{-1/4} x^{1/4} \qquad (4.196)$$

$$\frac{\delta}{x} = 3.93\,(0.952 + Pr)^{1/4}\,(Pr)^{-1/2}\,(Gr_x)^{-1/4} \qquad (4.197)$$

The local heat transfer coefficient

$$h_x = \frac{-k\left(\dfrac{\partial t}{\partial y}\right)_{y=0}}{t_w - t_\infty} = \frac{2k}{\delta}$$

Therefore,
$$Nu_x = \frac{h_x\,x}{k} = \frac{2x}{\delta}$$

$$= \frac{2}{3.93(0.952+Pr)^{1/4}(Pr)^{-1/2}(Gr_x)^{-1/4}}$$

$$Nu_x = 0.508\,(0.952 + Pr)^{-1/4}\,(Pr)^{1/2}\,(Gr_x)^{1/4} \qquad (4.198)$$

It can be shown that for a plate of length L, the average heat transfer coefficient may be determined by

$$Nu_L = 0.677\,(0.952 + Pr)^{-1/4}\,(Pr)^{1/2}\,(Gr_L)^{1/4} \qquad (4.199)$$

4.25 EMPIRICAL EQUATIONS FOR FREE CONVECTION BASED ON EXPERIMENTAL RESULTS

Equation (4.188) represents a generalised correlation for free convection. The form of the functions of Gr and Pr may be determined for different cases with the help of experimental results. Some of the important relationships are given below.

(a) Laminar Flow Over Vertical Plates and Cylinders (Gr. Pr ≈ 10^4 to 10^8)

(i) For free convection from vertical plates Eckert derived the following relations theoretically (Sec. 4.24) for the local Nusselt number Nu_x and the average Nusselt number Nu for a plate of height L :

$$Nu_x = 0.508\left(\frac{Pr}{0.952+Pr}\right)^{1/4}(Gr_x\,Pr)^{1/4} \qquad (4.198)$$

$$Nu = 0.677\left(\frac{Pr}{0.952+Pr}\right)^{1/4}(Gr\,Pr)^{1/4} \qquad (4.199)$$

These equations successfully correlate data for air and short vertical plates. The characteristic dimension is taken to be the height (L) of the plate in Eq. (4.199).

(ii) However, Eckert and Jackson recommend ;

$$Nu_f = 0.555\,\{(Gr)_f\,(Pr)_f\}^{1/4} \qquad (4.200)$$

(for $Gr_f\,Pr_f = 10^5$ to 10^8. The suffix *f* indicates average film temperature)

(iii) McAdams recommends the following equation for vertical plates and vertical cylinders (characteristic dimension, height L) :

$$\frac{hL}{k_f} = 0.59\left[\left(\frac{L^3\rho_f^2 g\beta_f\,\Delta t}{\mu_f^{\,2}}\right)\left(\frac{\mu C_p}{k}\right)_f\right]^{1/4} \tag{4.201}$$

or $$(Nu)_f = 0.59\,\{(Gr)_f\,(Pr_f)\}^{1/4}$$

(for $Gr_f\ Pr_f$ from 10^4 to 10^8) (4.201a)

(iv) Churchill and Chu have proposed the following correlation

$$Nu = 0.68 + \frac{0.67\,Ra^{1/4}}{[1+(0.492/Pr)^{9/16}]^{4/9}} \tag{4.202}$$

$$(10^{-1} < Ra < 10^9)$$

where Ra = (Gr) (Pr) is called the Rayleigh number. Properties are evaluated at the mean film temperature.

(v) For a vertical plate, or a plate tilted at an angle θ from the vertical with the heated surface facing down, Fujii and Imura recommend ;

$$Nu_L = 0.56\,(Gr_L\,Pr\cos\theta)^{1/4} \tag{4.203}$$

for $10^5 < Gr_L\,Pr\cos\theta < 10^{11}$

and $0 \le \theta \le 89°$

(b) Laminar Flow Over Horizontal Cylinders (Gr Pr = 10^4 to 10^8)

(i) The data on heat transfer from horizontal pipes to air and to water are represented by :

$$\frac{hD}{k_f} = 0.53\left\{\left(\frac{D^3\rho_f^{\,2} g\beta_f\,\Delta t}{\mu_f^{\,2}}\right)\left(\frac{\mu C_p}{k}\right)_f\right\}^{1/4} \tag{4.204}$$

or $$(Nu)_f = 0.53\,\{(Gr)_f\,(Pr)_f\}^{1/4} \tag{4.204a}$$

(ii) For laminar flow over cylinders larger than wires in diameter, the data for all fluids, metallic or non-metallic are well represented by ;

$$(Nu)_f = 0.53\left\{\left(\frac{Pr}{Pr+0.952}\right)_f (Gr.Pr)_f\right\}^{1/4} \tag{4.205}$$

(iii) The data for liquid metals is represented by ;

$$(Nu)_f = 0.53\,[(Gr)_f\,(Pr)_f^{\,2}]^{1/4} \tag{4.206}$$

Note that for ordinary non-metallic liquids with high Pr, 0.952 is small in comparison to Pr, and Eq. (4.205) reduces to Eq. (4.204a).

(iv) Churchill and Chu proposed the following correlation for horizontal cylinders :

$$Nu = 0.36 + \frac{0.518\,Ra^{1/4}}{[1+(0.559/Pr)^{9/16}]^{4/9}} \quad (10^{-6} < Ra < 10^9) \tag{4.207}$$

(c) Laminar Flow Over Horizontal Plates

Fischenden and Saunders have correlated data from horizontal square plates exposed to air. They recommend ;

(i) For horizontal heated plates facing up or cooled plates facing down ;

$$(Nu)_f = 0.54\,[(Gr)_f\,(Pr)_f]^{1/4} \quad (4.208)$$

$$(\text{for } Gr_f\,Pr_f = 10^5 \text{ to } 2 \times 10^7)$$

(ii) For horizontal heated plates facing down or cooled plates facing up ;

$$(Nu)_f = 0.27\,[(Gr)_f\,(Pr)_f]^{1/4} \quad (4.209)$$

$$(\text{for } Gr_f\,Pr_f = 3 \times 10^5 \text{ to } 10^{10})$$

In Eqs.(4.208) and (4.209) the characteristic dimension (to compute Gr and Nu) is taken to be the side of the plate.

(iii) For a 2-dimensional horizontal plate of length L, Fujii and Imura recommend ;

Heated surface facing down,

$$Nu_l = 0.58\,(Gr_l\,Pr)^{1/5} \quad (4.210)$$

$$(\text{for } 10^6 < Gr_L\,Pr < 10^{11})$$

Heated surface facing up,

$$Nu_l = 0.16\,(Gr_L\,Pr)^{1/3} \quad (4.211)$$

$$(\text{for } 7 \times 10^6 < Gr_L\,Pr < 2 \times 10^8)$$

and

$$Nu_L = 0.13\,(Gr_L\,Pr)^{1/3} \quad (4.211a)$$

$$(\text{for } 5 \times 10^8 < Gr_L\,Pr)$$

(iv) For circular plates of diameter d with heated surface facing down Kadambi and Drake suggest ;

$$Nu_d = 0.82\,(Gr_d\,Pr)^{1/5}\,Pr^{0.034} \quad (4.212)$$

(v) For a cooled circular plate facing down in a liquid metal we may use ;

$$Nu_d = 0.26\,(Gr_d\,Pr^2)^{0.35} \quad (4.213)$$

(d) Turbulent Flow Over Vertical Plates and Cylinders (Gr. Pr $\approx 10^9$ to 10^{12})

(i) For this case McAdams recommends ;

$$\frac{hL}{k_f} = 0.13\left[\left(\frac{L^3\,\rho_f^{\,2}\,g\beta_f\,\Delta t}{\mu_f^{\,2}}\right)\left(\frac{\mu\,C_p}{k}\right)_f\right]^{1/3} \quad (4.214)$$

or

$$(Nu)_f = 0.13\,\{(Gr)_f\,(Pr)_f\}^{1/3} \quad (4.214a)$$

$$(\text{for } Gr_f\,Pr_f = 10^9 \text{ to } 10^{12})$$

The equation may also be applied to liquids and gases being heated inside vertical tubes at low velocities.

(ii) Churchill and Chu recommend,

$$\text{Nu} = \left\{0.825 + \frac{0.387\,\text{Ra}^{1/6}}{[1+(0.492/\text{Pr})^{9/16}]^{8/27}}\right\}^2 \quad (4.215)$$

for the entire range of Rayleigh numbers up to 10^{12}.

(iii) Eckert and Jackson suggest the equation ;

$$\text{Nu} = 0.021\,\{(\text{Gr})_f(\text{Pr})_f\}^{0.4} \quad (4.216)$$

$$(\text{for } \text{Gr}_f\,\text{Pr}_f = 10^{10} \text{ to } 10^{12})$$

(e) Turbulent Flow Over Horizontal Cylinders (Gr. Pr = 10^9 to 10^{12})

(i) For horizontal cylinders (characteristic dimension-diameter) Fischenden and Saunders recommend for the flow of gases ;

$$(\text{Nu})_f = 0.10\,\{(\text{Gr})_f(\text{Pr})_f\}^{1/2} \quad (4.217)$$

$$(\text{for } \text{Gr}_f\,\text{Pr}_f > 10^8)$$

(ii) Churchill and Chu recommend,

$$\text{Nu} = \left\{0.60 + \frac{0.387\,\text{Ra}^{1/6}}{[1+(0.559/\text{Pr})^{9/16}]^{8/27}}\right\}^2 \quad (10^9 < \text{Ra} < 10^{12}) \quad (4.218)$$

(f) Turbulent Flow over Horizontal Plates

For horizontal heated plates facing up or cooled plates facing down in air or gas Fischenden and Saunders suggest ;

$$(\text{Nu})_f = [0.14\,(\text{Gr})_f(\text{Pr})_f]^{1/3} \quad (4.219)$$

$$(\text{for } \text{Gr}_f\,\text{Pr}_f = 2\times10^7 \text{ to } 3\times10^{10})$$

(g) Flow Over Spheres and Rectangular Blocks

For spheres and rectangular blocks, the equation for *horizontal* cylinders may be used, provided that the characteristic dimension L is computed by

$$\frac{1}{L} = \frac{1}{L_v} + \frac{1}{L_h} \quad (4.220)$$

where L_v = vertical dimension,

L_h = horizontal dimension.

For a sphere, the characteristic dimension from Eq. (4.220) comes out to be (D/2).

(h) Flow Inside Vertical or Horizontal Enclosed Air Spaces

(i) In the case of enclosed air spaces, the Grashof and Nusselt numbers are computed by using the thickness δ of the air space (ie, the clearance between the two enclosing surfaces) as the characteristic dimension.

Thus, $$(\text{Nu})_f = \frac{h\delta}{k_f} \quad (4.221)$$

or $$(Gr)_f = \left(\frac{\delta^3 \rho_f^2 g \beta_f \Delta t}{\mu_f^2}\right) \quad (4.222)$$

where δ = thickness of air space or clearance between the two surfaces.

(ii) For vertical air space Jakob suggests the equations ;

$$(Nu)_f = 1 \; ; \text{ for } (Gr)_f < 2 \times 10^3 \quad (4.223a)$$

$$(Nu)_f = 0.2 \left(\frac{L}{\delta}\right)^{-1/9} \{(Gr)_f (Pr)_f\}^{1/4} \quad (4.223b)$$

$$\{(\text{for } (Gr_f) = 2.0 \times 10^4 \text{ to } 2 \times 10^5)$$

$$(Nu)_f = 0.07 \left(\frac{L}{\delta}\right)^{-1/9} \{(Gr)_f (Pr)_f\}^{1/3} \quad (4.223c)$$

$$(\text{for } (Gr_f) = 2.1 \times 10^5 \text{ to } 1.1 \times 10^7)$$

(iii) For vertical air spaces Graaf and Held suggest ;

$$(Nu)_f = 1 \; ; \text{ for } (Gr)_f < 7 \times 10^7 \quad (4.224a)$$

$$(Nu)_f = 0.0384 \, (Gr_f)^{0.37} \quad (4.224b)$$

$$(\text{for } (Gr_f) = 10^4 \text{ to } 8 \times 10^4)$$

$$(Nu)_f = 0.0317 \, (Gr)_f^{0.37} \; ; \quad (4.224c)$$

$$(\text{for } (Gr)_f > 2 \times 10^5)$$

(iv) For horizontal air spaces Jakob suggests ;

$$(Nu)_f = 1 \; ; \text{ for } (Gr)_f < 10^3 \quad (4.225a)$$

$$(Nu)_f = 0.21 \, \{(Gr)_f (Pr)_f\}^{1/4} \; ; \quad (4.225b)$$

$$(\text{for } (Gr)_f = 10^4 \text{ to } 3.2 \times 10^5)$$

$$(Nu)_f = 0.075 \, \{(Gr)_f (Pr)_f\}^{1/3} \; ; \quad (4.225c)$$

$$(\text{for } (Gr)_f = 3.2 \times 10^5 \text{ to } 10^7)$$

(v) For a vertical fluid space of height L and thickness δ the correlation of Berkovsky and Polevikov may be used :

$$(Nu)_f = 0.22 \left(\frac{L}{\delta}\right)^{-1/4} \left(\frac{Pr}{0.2 + Pr} Ra_\delta\right)^{0.28} \quad (4.226)$$

$$\left(\text{for } 2 < \frac{L}{\delta} < 10 \; ; \; Pr < 10 \; ; \; Ra_\delta < 10^{10}\right)$$

$$Nu_\delta = 0.18 \left(\frac{Pr}{0.2 + Pr} Ra_\delta\right)^{0.29} \quad (4.227)$$

$$\left(\text{for } 1 < \frac{L}{\delta} < 2 \; ; \; 10^{-3} < Pr < 10^5 \; ; \; 10^3 < \frac{Ra_\delta \, Pr}{0.2 + Pr}\right)$$

where Ra, the Rayleigh number, is the product of Grashof and Prandtl numbers (Gr. Pr) and $Nu_\delta = h\delta/k$.

(vi) Data are lacking for aspect ratios $L/\delta < 1$. However some findings are as follows.

$$\text{As} \quad Ra_\delta \to \infty, \; Nu_\delta \to \left(\frac{L}{\delta}\right) Ra_\delta^{1/4} \tag{4.228}$$

(for $L/\delta = 0.01$ and 0.02)

$$Nu_f = 0.014 \; Ra_\delta^{0.38} \tag{4.229}$$

(for $L/\delta = 0.0625$ and $2 \times 10^8 < Ra_\delta < 2 \times 10^9$)

$$Nu_\delta = 0.748 \; Ra_\delta^{0.226} \tag{4.230}$$

(for $L/\delta = 0.5$; $2 \times 10^{10} < Ra_\delta < 10^{11}$ and $3.0 \le Pr \le 4.3$)

(vii) For horizontal air spaces Graaf and Held suggest ;

$$(Nu)_f = 1 \text{ for } (Gr)_f < 2 \times 10^3 \tag{4.231a}$$

$$(Nu)_f = 0.0507 \, (Gr)_f^{0.1} ; \tag{4.231b}$$

(for $(Gr)_f = 2 \times 10^3$ to 5×10^4)

$$(Nu)_f = 3.8 \text{ for } (Gr)_f = 5 \times 10^4 \text{ to } 2 \times 10^5 \tag{4.231c}$$

$$(Nu)_f = 0.0426 \, (Gr)_f^{0.37} \text{ for } (Gr)_f > 2 \times 10^5 \tag{4.231d}$$

(viii) In a horizontal fluid layer with heating from above, heat transfer is by conduction only. Heating from below results in conduction heat transfer only if $Ra_\delta <$ 1700 where the length scale is the spacing enclosing the layer.

Hollands et. al. proposed the following equation for horizontal air layers heated from below over a very wide range of Rayleigh numbers :

$$Nu_\delta = 1 + 1.44 \left[1 - \frac{1708}{Ra_\delta}\right] + \left[\left(\frac{Ra_\delta}{5830}\right)^{1/3} - 1\right] \tag{4.232}$$

If the quantity inside any of the square brackets is negative it is to be taken as zero. The equation is valid for $Ra_\delta = 1700$ to 10^8. For water, in the range $Ra_\delta = 1700$ to 3.5×10^9,

$$Nu_\delta = 1 + 1.44 \left[1 - \frac{1708}{Ra_\delta}\right] + \left[\left(\frac{Ra_\delta}{5830}\right)^{1/3} - 1\right] + 2.0 \left[\frac{Ra_\delta^{1/3}}{140}\right]^{[1 - \ln(Ra_\delta^{1/3}/140)]} \tag{4.233}$$

(i) Flow Inside Cylindrical and Spherical Cavities

(i) For natural convection inside spherical cavities of diameter D the relation given by Eqs. (4.234) may be used with the following values of C and n :

$$\frac{hD}{k} = C \, [Gr_d \, Pr]^n \tag{4.234}$$

where

$Gr_d\,Pr$	C	n
$10^4 - 10^9$	0.59	1/4
$10^9 - 10^{12}$	0.13	1/3

(ii) For the gap between concentric spheres of diameters D_o and D_i, Raithby and Hollands suggest ;

$$\frac{k_{eff}}{k} = 0.74\left[\frac{b^{1/4}}{D_o\,D_i\,(D_i^{-7/5}+D_o^{-7/5})^{5/4}}\right]\times Ra_b^{1/4}\left(\frac{Pr}{0.861+Pr}\right)^{1/4} \tag{4.235}$$

$$\left(\text{for } 0.70 \le Pr \le 4200 \text{ and } 10 \le \left[\frac{b^{1/4}}{(D_o\,D_i)(D_i^{-7/5}+D_o^{-7/5})^5}\right] Ra_b \le 10^7\right)$$

where $b = (D_o - D_i)/2$ and Ra_b is based on the temperature difference across the gap.

(iii) Similarly, for concentric cylinders, the same authors suggest ;

$$\frac{k_{eff}}{k} = 0.386\left[\frac{\ln(D_o/D_i)}{b^{3/4}\left(\dfrac{1}{D_i^{3/5}}+\dfrac{1}{D_o^{3/5}}\right)^{5/4}}\right]\times\left(\frac{Pr}{0.861+Pr}\right)^{1/4} Ra_b^{1/4} \tag{4.236}$$

$$\left(\text{for } 0.70 \le Pr \le 6000 \text{ and } 10 \le \left[\frac{\ln(D_o/D_i)}{b^{3/4}\left(\dfrac{1}{D_i^{3/5}}+\dfrac{1}{D_o^{3/5}}\right)^{5/4}}\right]^4 Ra_b \le 10^7\right)$$

The effective thermal conductivity k_{eff} is the thermal conductivity that a motionless fluid, with conductivity k, in the gap must have so as to transfer the same amount of heat as the moving fluid. Knowing *k*, the conduction equation may be used to compute the heat transfer rate.

Example 4.19

A vertical plate is heated from one side and maintained at 96 C. On the other side is air at atmospheric pressure and 20 C. Calculate the local heat transfer coefficient at a distance 20 cm from lower edge, and average value over the 20 cm length.

Solution :

$$t_w = 96\text{ C} ;\ t_{air} = 20\text{ C}$$

$$\text{hence} \quad t_f = \frac{96+20}{2} = 58\text{ C}$$

At 58 C the properties of air from tables are as given below ;

$$\nu = 18.77 \times 10^{-6} \text{ m}^2/\text{sec}$$

$$k = 2.98 \times 10^{-2} \text{ W/mK}$$

$$\beta = \frac{1}{273+58} = \frac{1}{331} \text{ K}^{-1}$$

hence $$Gr_x = \frac{g\beta x^3 \rho^2 \Delta t}{\mu^2} = \frac{(9.806)(0.2)^3(96-20)}{331\times(18.77\times10^{-6})^2}$$

$$= 51 \times 10^6$$

Now, using Eq. (4.198) we have

$$Nu_x = \frac{h_x x}{k} = 0.508\left(\frac{Pr}{0.952+Pr}\right)^{1/4} (Gr_x \times Pr)^{1/4}$$

or $$h_x = \frac{2.98\times10^{-2}}{0.2}\left\{0.508\sqrt[4]{0.694}\,(1.646)^{-1/4}\,(51\times10^6)^{1/4}\right\}$$

$$= 5.6 \text{ W/m}^2\text{K}$$

The average value of heat transfer coefficient is ;

$$h = \frac{1}{L}\int_0^L h_x \, dx = \frac{4}{3}(h_L)$$

(since x = L = 0.2 m)

$$= \frac{4}{3} \times 5.6 = 7.45 \text{ W/m}^2\text{K}$$ Ans.

The same answer can be obtained by using Eq. (4.199) directly.

Example 4.20

Determine the rate of heat loss from a 100 mm diameter steam pipe placed horizontally in ambient air at 30 C. The length of pipe is 4 m and wall temperature t_w = 170 C.

Solution :

The property values of air at t_f = 100 C are ;

$$k = 0.0322 \text{ W/mK}$$

$$\nu = 23.31 \times 10^{-6} \text{ m}^2/\text{sec}$$

$$Pr_f = 0.69 \text{ and } \beta = \frac{1}{373} \text{ K}^{-1}$$

hence $$Gr = \frac{(9.806)(0.1)^3(170-30)}{373\times(23.31\times10^{-6})^2}$$

$$= 6.94 \times 10^6$$

Using Eq. (4.204) we get ;

$$Nu = 0.53\,(6.94 \times 10^6 \times 0.69)^{1/4}$$
$$= 25.2$$

or $$h = \frac{N_u k}{D} = \frac{25.2 \times 0.0322}{0.1} = 8.1 \text{ W/m}^2\text{K}$$

The heat loss $= h \times A \times \Delta t = 8.1 \times (\pi \times 0.1 \times 4) \times 140$

$= 1420$ W **Ans.**

Example 4.21

A horizontal cylinder 2.5 cm diameter and 0.6 m long is suspended in water at 20 C. Calculate the rate of heat transfer if the cylinder surface is at 55 C. What would be the rate of heat transfer if the cylinder is suspended in light oil at 20 C. The following properties (at 37.5 C) may be used ;

		Water	*Oil*
ρ	(kg/m^3)	992	905
μ	(kg/hr m)	2.47	82.0
k	(W/mK)	0.622	0.133
β	(K^{-1})	3.96×10^{-4}	7.2×10^{-4}
Pr		4.64	324

Solution :

Now, $$Gr_f = \frac{D^3 \rho^2 g\, \beta \Delta t}{\mu^2}$$

$$= \frac{(0.025)^3 (992^2)(9.81 \times 3600^2)(3.96 \times 10^{-4})(55-20)}{(2.47)^2}$$

$$= 4.45 \times 10^6$$

From Eq. (4.204)

$$Nu_f = 0.53\,[Gr_f\, Pr_f]^{1/4}$$
$$= 0.53\,(4.45 \times 10^6 \times 4.64)^{1/4}$$
$$= 0.53 \times 67.3$$
$$= 35.7$$

or $$h = \frac{35.7 \times 0.622}{0.025} = 890 \text{ W/m}^2\text{K}$$

or $$q = h\, A\, \Delta t$$
$$= (890)\,(\pi \times 0.025 \times 0.6)\,(35)$$
$$= 1470 \text{ W}$$ **Ans.**

In the case of oil,

$$Gr_f = \frac{(0.025)^3 (905)^2 (9.81 \times 3600^2)(7.2 \times 10^{-4})(55-20)}{(82)^2}$$

$$= 6100$$

So $$(Nu)_f = 0.53\,(6100 \times 324)^{1/4}$$
$$= 0.53 \times 36.8 = 19.5$$

or $$h = \frac{19.5 \times 0.133}{0.025}$$
$$= 104 \text{ W/m}^2\text{K}$$

hence $$q = (104)\,\{\pi \times 0.025 \times 0.6\,(35)\}$$
$$= 171 \text{ W}$$ Ans.

Example 4.22

The parallel outer and inner walls of a building are 4 m high and 5 m long. The walls are 10 cm apart. The outer surface of the inner wall is at 25 C and the inner surface of the outer wall is at 5 C.

(a) Calculate the total heat loss per hr.

(b) If the air space is divided in half by a sheet of aluminum foil, 0.025 mm thick parallel to the walls, what would be the heat loss per hr.

Solution :

It would be appropriate to evaluate the properties of air at the mean temperature of the two surfaces, ie. at (5 + 25)/2 = 15 C. At 15 C ;

$$\rho = 1.225 \text{ kg/m}^3$$
$$\mu = 0.065 \text{ kg/hr m}$$
$$C_p = 1.0 \text{ kJ/kg K}$$
$$k = 0.025 \text{ W/m}^2\text{K} = 0.082 \text{ kJ/hr m K}$$
$$\beta = 1/288 \text{ K}^{-1}$$

Now $$Pr_f = \frac{0.065 \times 1.0}{0.089}$$
$$= 0.72$$

$$Gr_f = \frac{\delta^3 \rho^2 g\, \beta\, \Delta t}{\mu^2}$$
$$= (0.1)^3\,(1.225)^2\,(9.81)\,(3600)^2 \left(\frac{1}{288}\right) \frac{(25-5)}{(0.065)^2}$$
$$= 3.15 \times 10^6$$

(a) From (4.196 c),

$$(Nu)_f = 1 \text{ ; for } (Gr)_f < 2 \times 10^3 \qquad (4.223a)$$

$$\frac{h\delta}{k_f} = 0.07 \left(\frac{L}{\delta}\right)^{-1/9} (Gr_f\, Pr_f)^{1/3}$$

$$= 0.07\left(\frac{4}{0.1}\right)^{-1/9} [(3.15 \times 10^6)(0.72)]^{1/3}$$

$$= 0.07 \times \frac{1}{1.505} \times 131.3$$

$$= 6.18$$

$$h = \frac{6.18 \times 0.0252}{0.1} = 1.56 \text{ W/m}^2\text{K}$$

$$q = h\,A\,\Delta t$$

$$= 1.56\,(4 \times 5)\,(20) = 624 \text{ W}$$ **Ans.**

If we use Eq. (4.224)

$$Nu_f = 0.0317\,(Gr_f)^{0.37} = 0.0317\,(3.15 \times 10^6)^{0.37}$$

$$= 0.0317 \times 256 = 8.1$$

so

$$h = \frac{8.1 \times 0.0252}{0.1} = 2.03$$

and

$$q = 2.03 \times 400 = 812 \text{ W}$$ **Ans.**

(b) In this case there are 2 air spaces, each 5 cm thick. Since the temperature difference between the two surfaces is only 20 C, we may assume the same properties of air in the two air spaces and evaluate these properties at 15 C. Thus the resistance of each air space to heat flow will be the same and the temperature of the aluminum foil will be midway between the two surfaces, i.e, 15 C. Then for each air space ;

$$\delta = 0.05 \text{ m}$$

$$\Delta t = 10 \text{ C}$$

so

$$Gr_f = \frac{\delta^3 \rho^2 g \beta \Delta t}{\mu^2}$$

$$= \frac{(0.05)^2 (1.225)^2 (9.81 \times 3600 \times 3600)(10)}{(0.065)^2 (288)}$$

$$= 1.97 \times 10^5$$

From (4.223b) ;

$$Nu_f = 0.2\left(\frac{L}{\delta}\right)^{-1/9} \{(Gr)_f (Pr)_f\}^{1/4}$$

$$= 0.2\left(\frac{4}{0.05}\right)^{-1/9} \{1.97 \times 10^5 \times 0.72\}^{1/4}$$

$$= 0.2 \times \frac{1}{1.626} \times 19.4$$

$$= 2.39$$

or

$$h = \frac{2.39 \times 0.0252}{0.05} = 1.2 \text{ W/m}^2\text{K}$$

and
$$q = h A \Delta t$$
$$= 1.2 \times 20 \times 10$$
$$= 240 \text{ W} \qquad \text{Ans.}$$

In this case use of Eq. (4.224c)gives ;
$$Nu_f = 0.0317\,(1.97 \times 10^5)^{0.37}$$
$$= 2.89$$

So
$$h = 1.45 \text{ W/m}^2\text{K}$$

Thus
$$q = 290 \text{ W} \qquad \text{Ans.}$$

Example 4.23

A circular pan of water 150 mm in diameter and 80 mm deep is placed on a burner. The bottom of the pan is maintained at 100 C and the top surface of water is initially at 20 C. Calculate the initial rate of heat transfer from the burner to the water.

Solution :

Taking the properties of water at 60 C,

$\beta = 5.18 \times 10^{-4}\ \text{K}^{-1}$; Pr = 3.02 : K = 0.657 W/mK and $\nu = 0.478 \times 10^{-6}\ \text{m}^2/\text{s}$

So
$$Ra_\delta = \frac{g\beta\,\Delta t\,\delta^3\,Pr}{\nu^2} = \frac{(9.8)(5.18\times10^{-4})(100-20)(0.08)^3\,(3.02)}{(0.478\times10^{-6})^2}$$
$$= 2.75 \times 10^9$$

From Eq. (4.233) ;

$$Nu_\delta = 1 + 1.44\left[1 - \frac{1708}{2.75\times10^9}\right] + \left[\left(\frac{2.75\times10^9}{5830}\right)^{1/3} - 1\right] +$$
$$2.0\left[\frac{(2.75\times10^9)^{1/3}}{140}\right]^{[1-\ln\{(2.75\times10^9)^{1/3}/140\}]}.$$
$$= 1 + 1.44 + 76.8 + 0.1 = 79.3$$
$$= 79.3$$

$$h = Nu_\delta\,\frac{k}{\delta} = \frac{(79.3)(0.657)}{0.08}$$
$$= 651 \text{ W/m}^2\text{K} \qquad \text{Ans.}$$

$$q = (651)\left(\frac{\pi\times0.15^2}{4}\right)(100-20)$$
$$= 920 \text{ W} \qquad \text{Ans.}$$

4.26 SIMILARITY RELATIONS IN FREE CONVECTION

In Section 4.13, *similarity* considerations were applied to forced convection. The dimensionless parameters Re, Nu, and Pr were obtained as a result. The same treatment is now extended to free convection.

It will be noted that the application of this technique to Eqs. (4.99) and (4.1) results in Pr and Nu respectively as shown in Section 4.18. We should, however, expect that Gr will be obtained instead of Re, when Eq. (4.107) is used. The mathematical analysis involved is given below.

Since gravity effects are playing significant part in fluid motion, Eq. (4.107) should be modified accordingly and may be written for system 2 as :

$$\rho_2 V x_2 \frac{\partial V_{x2}}{\partial x_2} + \rho_2 V y_2 \frac{\partial V_{x2}}{\partial y_2} = \rho_2 g_2 \beta_2 (t_w - t_\infty)_2 + \mu_2 \frac{\partial^2 V_{x2}}{\partial {y_2}^2}$$

Transforming this equation for system 1, we have ;

$$\frac{AV^2}{X}\left(\rho_1 V_{x1} \frac{\partial V_{x1}}{\partial x_1} + \rho_1 V_{y1} \frac{\partial V_{x1}}{\partial y_1}\right)$$

$$= \text{EAGH}\,\{\rho_1 \beta_1 g_1 (t_w - t_\infty)_1\} + \frac{BV}{X^2}\left(\mu_1 \frac{\partial^2 V_{x1}}{\partial {y_1}^2}\right)$$

For hydrodynamic similarity

(i) $$\frac{BV}{X^2} = \text{EAGH}$$

(ii) $$\frac{AV^2}{X} = \frac{BV}{X^2} \quad \text{or } AVX = B$$

The second identity has already been found as equal to Re in Eq. (4.166). Considering the first identity we have ;

$$BV = \text{EAGHX}^2$$

or

$$\left(\frac{\mu_2}{\mu_1}\right)\left(\frac{V_{x2}}{V_{x1}}\right) = \left\{\frac{(t_w - t_\infty)_2}{(t_w - t_\infty)_1}\right\}\left(\frac{\rho_2}{\rho_1}\right)^2\left(\frac{g_2}{g_1}\right)\left(\frac{\beta_2}{\beta_1}\right)\left(\frac{x_2}{x_1}\right)^2$$

$$\frac{{\rho_1}^2 g_1 \beta_1 \Delta t_1 x_1^2}{\mu_1 V_{x1}} = \frac{{\rho_2}^2 g_2 \beta_2 \Delta t_2 {x_2}^2}{\mu_2 V_{x2}} \tag{4.237}$$

Multiplying Eq. (4.237) by Eq. (4.166) to eliminate V_x we get ;

$$\left(\frac{{\rho_1}^2 g_1 \beta_1 \Delta t_1 {x_1}^3}{{\mu_1}^2}\right) = \left(\frac{{\rho_2}^2 g_2 \beta_2 \Delta t_2 {x_2}^3}{{\mu_2}^2}\right) \tag{4.238}$$

or $$Gr_1 = Gr_2$$

This ratio is the Grashof number as already explained in Eq. (4.189). From similarity considerations, therefore, it is evident that Grashof number plays a significant part in free convection.

4.27 HEAT TRANSFER IN HIGH SPEED FLOW

The processes in a high speed boundary layer are not adiabatic. However, it is usual to relate them to adiabatic processes. The conversion of kinetic energy to enthalpy is then described by

$$i_o = i_\infty + \frac{V_\infty^{\ 2}}{2} \qquad (4.239)$$

where i_o is the stagnation enthalpy and i_∞ is the free stream enthalpy of the gas.

For an ideal gas Eq. (4.239) gives

$$T_o = T_\infty + \frac{V_\infty^2}{2C_p} \qquad (4.240)$$

or, in terms of the Mach number M,

$$\frac{T_o}{T_\infty} = 1 + \frac{\gamma - 1}{2} M_\infty^{\ 2} \qquad (4.241)$$

where $M_\infty = \dfrac{V_\infty}{a_\infty}$ = free stream Mach number

$a = (\gamma\, g_c\, R\, T_\infty)^{1/2}$ = sonic velocity at free stream conditions.

In a real boundary layer the fluid is not brought to rest reversibly and there is some dissipation of kinetic energy due to viscous shearing. For this reason a *recovery factor*, *r* is defined as

$$r = \frac{T_{aw} - T_\infty}{T_o - T_\infty} \qquad (4.242)$$

where T_{aw} is the actual adiabatic wall temperature ie, temperature of the gas at wall surface (see Fig. 4.16).

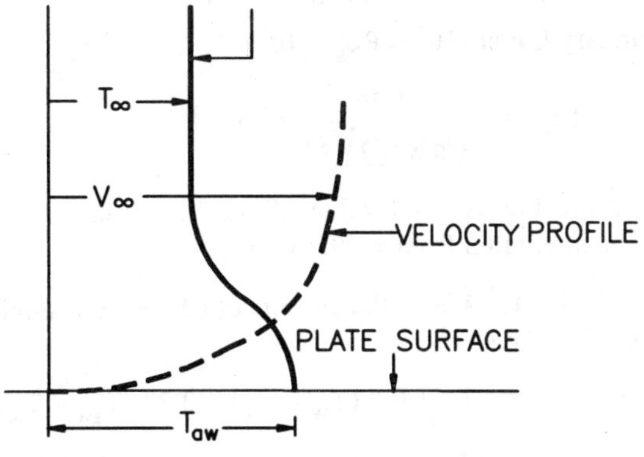

Fig. 4.16 (a) Velocity and Temperature Distribution in High Speed Flow on an Insulated Plate

Experiment show that r may be calculated by the following equations :

Laminar flow : $$r = Pr^{1/2} \tag{4.243}$$

Turbulent flow : $$r = Pr^{1/3} \tag{4.244}$$

The rate of heat transfer may be calculated by :

$$q = hA\,(T_w - T_{aw}) \tag{4.245}$$

using an average value of heat transfer coefficient, h.

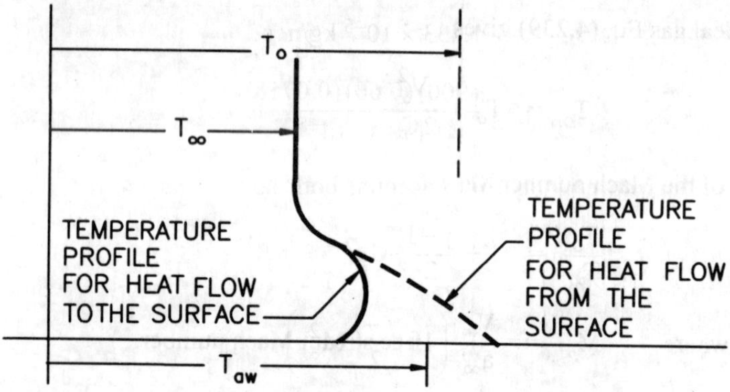

Fig. 4.16 (b) Temperature Profiles in High Speed Boundary Layer for Heating and Cooling

The local value of the heat transfer coefficient h_x may be obtained from :

(i) Laminar Boundary Layer ($Re_x < 10^5$)

$$St_x = \left(\frac{h_x}{C_p\,\rho V_\infty}\right) = 0.332\,(Re_x)^{-1/2}\,(Pr)^{-2/3} \tag{4.246}$$

(ii) Turbulent Boundary Layer ($10^5 < Re_x < 10^7$)

$$St_x = 0.0288\,(Re_x)^{-1/5}\,(Pr)^{-2/3} \tag{4.247}$$

(iii) Turbulent Boundary Layer ($10^7 < Re_x < 10^9$)

$$St_x = \frac{2.46}{(ln\,Re_x)^{2.584}}\,(Pr)^{-2/3} \tag{4.248}$$

In order to determine the average value of the heat transfer coefficient the above expressions have to be integrated from $x = 0$ to $x = L$.

Note that in Eqs. (4.246) to (4.248) the properties of the fluid should be evaluated at a reference temperature T_{ref}, given by :

$$T_{ref} = T_\infty + 0.5\,(T_w - T_\infty) + 0.22\,(T_{aw} - T_\infty) \tag{4.249}$$

The above procedure is used because in high speed flow the temperature gradients in a boundary layer are large and therefore the variation in physical properties of the fluid is substantial.

Example 4.24

Air at 0.05 atm and - 40 C flows with a velocity of 900 m/s over a flat plate 60 cm long and 50 cm wide. How much cooling would be required to maintain the plate temperature at 40 C.

Solution :

Let us first compute the Reynolds number on the basis of free stream conditions.

$$\rho_\infty = p_\infty/RT = \frac{(1.0132\times10^5)(0.05)}{(287)(233)} = 0.0758 \text{ kg/m}^3$$

$$\mu_\infty = 1.433 \times 10^{-5} \text{ kg/m s}$$

$$Re_{L,\infty} = \frac{(900)(0.60)(0.0758)}{1.433\times10^{-5}} = 2.856 \times 10^6$$

We have, therefore to take into account both laminar and turbulent boundary layer heat transfer.

(i) Laminar Heat Transfer

$$T_o = T_\infty\left(1+\frac{\gamma-1}{2}M_\infty^2\right) \qquad (4.240)$$

$$a_\infty = \sqrt{\gamma RT_\infty} = \sqrt{1.4\times287\times233} = 306 \text{ m/s}$$

$$M_\infty = \frac{V_\infty}{a_\infty} = \frac{900}{306} = 2.94$$

$$T_0 = (233)(1 + 0.2 \times 2.94^2) = 636 \text{ K}$$

As a first approximation let Pr = 0.7

From (Eq. 4.243), $r = (Pr)^{1/2} = (0.7)^{1/2} = 0.837$

Now, $$\frac{T_{aw}-T_\infty}{T_o-T_\infty} = 0.837 \qquad (4.242)$$

or $$T_{aw} = 233 + (636 - 233)(0.837)$$
$$= 570 \text{ K}$$

also $$T_{ref} = 233 + 0.5(40 + 40) + 0.22(570 - 233) \qquad (4.249)$$
$$= 347.1 \text{ K}$$

The value of Pr at 347 K is 0.697 which checks with the assumption made.

At the reference temperature 347.1 K,

$$\rho = \frac{(1.0132\times10^5)(0.05)}{(287)(347.1)} = 0.0508 \text{ kg/m}^3$$

$$\mu = 2.07 \times 10^{-5} \text{ kg/m s}$$

$$k = 0.03 \text{ W/mK}$$

$$C_p = 1.009 \text{ kJ/kg K}$$

If $$(Re)_{crit} = 5 \times 10^5$$

$$= \frac{V_\infty \rho x_{crit}}{\mu}$$

then $$x_{crit} = \frac{(5\times10^5)(2.07\times10^{-5})}{(900)(0.0508)} = 0.226 \text{ m}$$

Integrating Eq. (4.246) from x = 0 to x = 0.226 we get the average value of Nu as ;

$$Nu = \frac{h x_{crit}}{k} = (0.664)\,(Re_{crit})^{1/2}\,Pr^{1/3}$$

$$= (0.664)\,(5 \times 10^5)^{1/2}\,(0.697)^{1/3}$$

$$= 416.3$$

so $$h = \frac{(416.3)(0.03)}{0.226} = 55.26 \text{ W/m}^2\text{K}$$

and $$q = h A (T_w - T_{aw})$$

$$= (55.26)\,(0.226 \times 0.5)\,(313 - 570)$$

$$= -1604.8 \text{ W}$$

(ii) Turbulent Heat Transfer

Assuming $$Pr = 0.7,$$

$$r = (Pr)^{1/3} = (0.7)^{1/3} = 0.888$$

$$\frac{T_{aw} - T_\infty}{T_o - T_\infty} = 0.888$$

$$T_{aw} = 233 + (636 - 233)\,(0.88)$$

$$= 587.6 \text{ K}$$

$$T_{ref} = 233 + (0.5)\,(40 + 40) + 0.22\,(587.6 - 233)$$

$$= 351 \text{ K}$$

(Pr = 0.695 which checks)

Now, $$\rho = \frac{(1.0132\times10^5)(0.05)}{287\times351} = 0.0502 \text{ kg/m}^3$$

$$\mu = 2.09 \times 10^{-5} \text{ kg/m s}$$

$$k = 0.0302 \text{ W/mK}$$

$$C_p = 1.009 \text{ kJ/kg K}$$

So $$St_x\, Pr^{2/3} = 0.0296\, Re_x^{-1/5} \qquad (4.222)$$

or $$h_x = Pr^{-2/3}\,\rho\, V_\infty\, C_p\,(0.0296)\left(\frac{V_\infty \rho x}{\mu}\right)^{-1/5}$$

$$= (0.695)^{-2/3}\,(0.0502 \times 900 \times 1009) \times \left(\frac{900 \times 0.0502 \times x}{2.09 \times 10^{-5}}\right)^{-1/5} \times 0.0296$$

$$= 93.7\,x^{-1/5}$$

and $$h = \frac{\int_{0.226}^{0.6} h_x\,d_x}{(0.6-0.226)} = \frac{42.2}{0.374}$$

$$= 112.8\ \text{W/m}^2\text{K}$$

or $$q = (112.8)\,(0.6 - 0.226)\,(0.5)\,(313 - 587.6)$$

$$= -5792.3\ \text{W}$$

(iii) Total cooling required is therefore,

$$q = 1604.8 + 5792.3 = 7397.1\ \text{W}$$ Ans.

4.28 ILLUSTRATIVE EXAMPLES

Example 4.25

A 40 cm × 40 cm steel plate, 1 cm thick is kept vertically in still air at 30 C. Determine the rate of heat transfer from the plate to air if the surface temperature of the plate is 290 C.

Solution :

$$\text{Mean film temperature} = \frac{290+30}{2} = 160\ \text{C}$$

From Table B1, $k = 0.03638$ W/mK; $\nu = 30.08 \times 10^{-6}$ m²/s ; $Pr = 0.682$

$$Ra_L = Gr_L\,Pr = (9.81)\,\frac{1}{(273+160)}\,\frac{(0.4)^3\,(290-30)}{(30.08)^2 \times 10^{-12}} \times 0.682$$

$$= 2.8416 \times 10^8$$

(i) Let us use the correlations proposed by Churchill and Chu :

$$Nu_L = 0.68 + \frac{0.670\,Ra_L^{\ 0.25}}{[1+(0.492/Pr)^{9/16}]^{4/9}} \qquad (4.202)$$

$$(10^1 < Ra_L < 10^9)$$

$$Nu_L = \left\{0.825 + \frac{0.387\,Ra_L^{\ 1/6}}{[1+(0.492/Pr)^{9/16}]^{8/27}}\right\}^2 \qquad (4.215)$$

$$(10^9 < Ra_L < 10^{12})$$

Since $Ra_L < 10^9$, Eq. 4.202 gives,

$$Nu_L = 67.146$$

or $$h = \frac{67.146 \times 0.03638}{0.4} = 6.107\ \text{W/m}^2\text{K}$$

Heat transfer rate $q = hA\,\Delta t$

$= (6.107)(2 \times 0.4 \times 0.4)(290 - 30) = 508.1$ W Ans.

(ii) If we use McAdams equation, (4.201a) we obtain,

$$Nu_L = 0.59\,(Ra_L)^{0.25}$$

$$= 0.59\,(2.8416 \times 10^8)^{0.25} = 76.6$$

This value of Nu_L gives $h = 6.967$ W/m²K and $q = 579.6$ W. Ans.

If we use Eq. (4.185)

$$Nu_L = 83.14$$

$$q = 629.1 \text{ W}$$ Ans.

Example 4.26

The outer surface temperature of a 4 cm OD pipe carrying steam is 150 C. Compute the rate of heat transfer per m length to ambient air at 30 C. What would be the rate of heat transfer per unit length if the surface of the pipe is covered with rain water at 30 C. The emissivity of the pipe is 0.9. The physical properties of air and water at 90 C are as follows :

Air : $k = 0.03127$ W/m K ; $\nu = 22.10 \times 10^{-6}$ m²/s ;

$Pr = 0.69$

Water : $k = 0.680$ W/mK ; $\nu = 0.326 \times 10^{-6}$ m²/s ;

$Pr = 1.95$; $\beta = 0.72 \times 10^{-3}$ K⁻¹

Solution :

We can use Eq. (4.204 a) to determine the value of Nu. However, the following correlations recommended by Churchill and Chu for free convection from a horizontal cylinder are believed to yield better results :

$$Nu = 0.36 + \frac{0.518\,Ra^{1/4}}{[1+(0.559/Pr)^{9/16}]^{4/9}} \quad (10^{-6} < Ra < 10^9) \qquad (4.207)$$

$$Nu = \left\{0.60 + \frac{0.387\,Ra^{1/6}}{[1+(0.559/Pr)^{9/16}]^{8/27}}\right\}^2 \quad (10^9 < Ra < 10^{12}) \qquad (4.218)$$

where $Ra = Gr\,Pr$

For air, at mean film temperature of (150 + 30)/2 = 90 C,

$$Ra = Gr\,Pr = \left(\frac{g\beta L^3}{\nu^2}\Delta t\right)(Pr)$$

$$= 9.81 \times \frac{1}{(273+90)} \times \frac{(0.04)^3(150-30)(0.69)}{(22.1)^2 \times 10^{-12}} = 29.32 \times 10^4$$

For water, similarly at 90 C,

$$Ra = (9.81)\frac{(0.72\times10^{-3})(0.04)^3(120)(1.95)}{(0.326)^2\times10^{-12}}$$

$$= 9.953\times10^8$$

Hence, for air, using Eq. 4.207,

$$Nu = 0.36+\frac{(0.518)(29.32\times10^4)^{1/4}}{[1+(0.559/0.69)^{9/16}]^{4/9}} = 9.45$$

and for water $Nu = 0.36+\dfrac{(0.518)(9.953\times10^8)^{1/4}}{[1+(0.559/1.95)^{9/16}]^{4/9}} = 77.3$

We then obtain :

$$h_{air} = Nu\frac{k}{D} = \frac{9.45\times0.03127}{0.04} = 7.39\ W/m^2K$$

$$h_{water} = Nu\frac{k}{D} = \frac{77.3\times0.68}{0.04} = 1314\ W/m^2K$$

The convective heat transfer rate per m length is then,

$$q_{air} = h\,A\,\Delta t = (7.39)\,(\pi\times0.04\times1)\,(150-30)$$

$$= 111.4\ W$$

The rate of radiant heat transfer will be ;

$$q_{rad} = A\,(F_g)_{12}\,\sigma\,(T_1^4-T_2^4) = (\pi\times0.04\times1)\,(0.9)\times5.67\times10^{-8}\times(423^4-303^4)$$

$$= 151\ W \quad \text{since } (F_g)_{12} = \varepsilon$$

For air,

$$q_{total} = 111.4+151 = 262.4\ W \qquad \text{Ans.}$$

Now,

$$q_{water} = (1314)\,(\pi\times0.04\times1)\,(120) = 19{,}808\ W \qquad \text{Ans.}$$

In the case of water there would be hardly any radiant heat transfer.

Example 4.27

Determine the rate of heat transfer by free convection between two plates 1.8 m long and 1m wide, having a spacing of 2 cm between them. The plates are inclined, the larger side making an angle of 30 C to the horizontal. The lower plate is at 80 C and the upper plate at 60 C.

Solution :

We can apply Eq. (4.232), for natural convection in a horizontal air space. However, Ra_δ may be replaced by $Ra_\delta \cos\alpha$ where α is the angle of inclination to the horizontal.

Then, $$Nu_\delta = 1+1.44\left[1-\frac{1708}{Ra_\delta\cos\alpha}\right]+\left[\left(\frac{Ra_\delta\cos\alpha}{5830}\right)^{1/3}-1\right]$$

The mean temperature is (80 + 60)/2 = 70 C. Hence the properties of air are : Pr = 0.694 ; $\nu = 20.02 \times 10^{-6}$ m²/s ; k = 0.02964 W/mK

$$Ra_\delta = (9.81)\frac{1}{(273+70)}\frac{(0.02)^3(0.694)}{(20.02)^2 \times 10^{-12}} = 7923.7$$

$$Ra_\delta \cos\alpha = (7923.7)(\cos 30°) = 6862.1$$

$$Nu_\delta = 1 + 1.44\left[1 - \frac{1708}{6862.1}\right] + \left[\left(\frac{6862.1}{5830}\right)^{1/3} - 1\right]$$

$$= 2.137$$

Thus $h = (Nu_\delta)(k/\delta) = (2.137)(0.02964/0.02)$

$$= 3.167 \text{ W/m}^2\text{K}$$

or $q = hA\,\Delta t = (3.167)(1.8 \times 1)(80 - 60) = 114$ W **Ans.**

Example 4.28

The heat transfer coefficient for a gas flowing over a thin flat plate 3 m long 0.3 m wide varies with distance from the leading edge according to

$$h(x) = 10\,x^{-1/4} \text{ W/m}^2\text{K}$$

Calculate (a) the average heat transfer coefficient, (b) the rate of heat transfer between the plate and the gas if the plate is at 170 C and the gas is at 30 C, and (c) the local heat flux 2 m from the leading edge.

Solution :

(a)
$$h = \frac{1}{L}\int_0^L h(x)\,dx = \frac{1}{L}\int_0^L 10\,x^{-1/4}\,dx$$

$$= \frac{40}{3L}L^{3/4} = \frac{40}{3}\frac{(3)^{3/4}}{3} = 10.131 \text{ W/m}^2\text{K}$$ **Ans.**

(b)
$$q = hA\,(t_p - t_g)$$

$$= (10.131)(3 \times 0.3)(170 - 30) = 1276.5 \text{ W}$$ **Ans.**

(c)
$$\frac{q_x}{A} = h_x\,(t_p - t_g)$$

$$h_x = 10 \times 2^{-1/4} = 8.409 \text{ W/m}^2$$

$$\frac{q_x}{A} = (8.409)(170 - 130) = 1177.26 \text{ W/m}^2$$ **Ans.**

Example 4.29

A fluid flows through a tube of radius R. Heat is supplied at the tube surface at a constant rate q_w per unit area and time. Derive a differential equation for the temperature of the fluid. The velocity profile is given by

$$V_z - V_{z,max}\left(1 - \frac{r^2}{R^2}\right)$$

Solution :

For steady one-dimensional flow of fluid having constant properties, let us make an energy balance on a control volume consisting of an annular ring as shown in Fig. 4.17

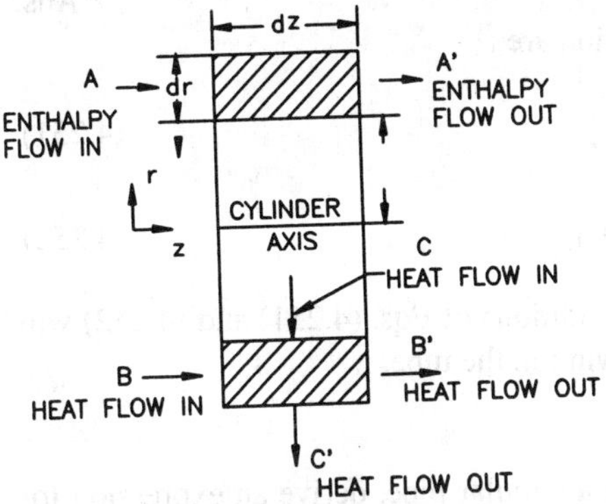

A	$\rho\, 2\pi r\, dr\, V_z\, C_p\, T$
A′	$\rho\, 2\pi\, r\, dr\, V_z\, C_p\, T + \frac{\partial}{\partial z}(\rho\, 2\pi r\, dr\, V_z\, C_p T)\, dz$
B	$-\,k\,(2\pi r\, dr)\frac{\partial T}{\partial z}$
B′	$-\,k\,(2\pi r\, dr)\frac{\partial T}{\partial z} \;-\, k\,(2\pi r\, dr)\frac{\partial^2 T}{\partial z^2}\, dz$
C	$-\,k\,(2\pi r\, dz)\frac{\partial T}{\partial r}$
C′	$-\,k\,(2\pi r\, dz)\frac{\partial T}{\partial r} + \frac{\partial}{\partial r}\,[-\,k\,(2\pi r\, dz)\frac{\partial T}{\partial r}\,]\, dr$

Fig. 4.17 Example 4.29

Net rate of enthalpy flow into control volume,

$$= \; -\,\rho\,(2\pi\, r\, dr)\, C_p \frac{\partial}{\partial z}(V_z\, T)\, dz \qquad (a)$$

Net rate of heat flow into control volume in r direction,

$$= \; k\,(2\,\pi\, dz)\, \frac{\partial}{\partial r}\left(r\frac{dT}{dr}\right) dr \qquad (b)$$

Net rate of heat flow into control volume in z direction,

$$= \; k\,(2\,\pi\, r\, dr)\, \frac{\partial^2 T}{dz^2}\, dz \qquad (c)$$

Under steady state,

$$(a) + (b) + (c) = 0 \qquad (d)$$

Substituting the concerned expressions in (d) and dividing by (2π r dr dz), we get

$$\rho\, C_p \frac{\partial}{\partial z}(V_z\, T) \;=\; \frac{k}{r}\frac{\partial}{\partial r}\left(r\frac{dT}{dr}\right) + k\,\frac{\partial^2 T}{\partial z^2} \qquad (e)$$

Neglecting heat conduction in the *z* direction (axial direction) and noting V_z is a function of r alone we obtain,

$$\rho\, C_p\, V_z \frac{\partial T}{\partial z} \;=\; \frac{k}{r}\frac{\partial}{\partial r}\left(r\frac{\partial T}{\partial r}\right) \qquad (f)$$

Substituting for the velocity profile, we get,

$$\rho\, C_p\, V_{z,max}\left(1-\frac{r^2}{R^2}\right)\frac{\partial T}{\partial z} = \frac{k}{r}\frac{\partial}{\partial r}\left(r\frac{\partial T}{\partial r}\right)$$

or
$$V_{z,max}\left(1-\frac{r^2}{R^2}\right)\frac{\partial T}{\partial z} = \frac{\alpha}{r}\frac{\partial}{\partial r}\left(r\frac{\partial T}{\partial r}\right) \quad (4.250)$$

Ans.

The boundary conditions for the above equation are :

At $$r = 0, \frac{\partial T}{\partial r} = 0 \quad (4.251)$$

At $$r = R, -k\frac{\partial T}{\partial r} = q_w \quad (4.252)$$

The solution of Eq. (4.250) with boundary conditions of Eqs. (4.251) and (4.252) will give the temperature profile within the fluid flowing in the tube.

Example 4.30

For the case of laminar flow of fluid through a circular pipe, derive an expression for the difference $(T_w - T_b)$ between the pipe wall temperature and fluid bulk temperature at any location z where the flow is fully developed both hydrodynamically and thermally. Assume (a) the heat flux at the wall is constant (b) the wall temperature is constant.

Solution :

An energy balance for an elementary length dz of the pipe gives,

$$\rho\, A\, V\, C_p\, dT_b = (2\pi\, R\, dz)q_w$$

(V is the average velocity of the fluid, and q_w the heat flux at the wall). Since $A = \pi R^2$,

$$\frac{dT_b}{dz} = \frac{2\,q_w}{\rho\, V\, C_p\, R} = \frac{2\alpha\, q_w}{k\, V R} \quad (4.253)$$

(a) Case 1 : q_w = constant

$$\frac{dT_b}{dz} = \text{constant}$$

(i) Thus we see that the bulk temperature, T_b, of the fluid varies linearly with z. If the pipe length is L and the fluid bulk temperatures at inlet and outlet are T_{bi} and T_{bo}, respectively,

$$T_{bo} - T_{bi} = (2\alpha\, q_w\, L)/(k\, V\, R) \quad (4.254)$$

A fully developed velocity profile means that the profile does not change with z, ie, for any radial location,

$$\frac{d(V_z / V_{z,max})}{dz} = 0 \quad (4.255)$$

The entrance length for fully developed laminar flow is generally given by ;

$$\frac{Z_e}{D} = 0.05\, Re = 0.05\,\frac{VD}{\nu} \quad (4.256)$$

(V is the average velocity and D the tube diameter)

For a fully developed temperature profile the condition is that, at any radial location, where the temperature of the fluid is T,

$$\frac{d[(T_w - T)/(T_w - T_b)]}{dz} = 0 \tag{4.257}$$

T_w is the wall temperature and T_b the fluid bulk temperature given by Eq. (4.21),

$$T_b = \left[\int_0^R V\rho C_p\, t(2\pi r)\,dr\right] \div \left[\int_0^R V\rho C_p (2\pi r)\,dr\right]$$

The entry length for the flow to be fully developed thermally is given by ;

$$\frac{Z_t}{D} = 0.05\ \mathrm{Re\ Pr} \tag{4.258}$$

Now, we define a local heat transfer coefficient h_z and apply Newton's law of cooling.

$$q_w = h_z (T_w - T_b)$$

If $(T_w - T_b)$ varies with z, h_z also varies with z for the case of constant wall flux q_w. This holds in the entrance region. In the thermally developed flow region $(T_w - T_b)$ does not vary with *z* and h_z is constant. Fig. 4.18 shows the variation of T_b, T_w, and h_z with *z* for constant q_w. The development of the temperature profile is also indicated.

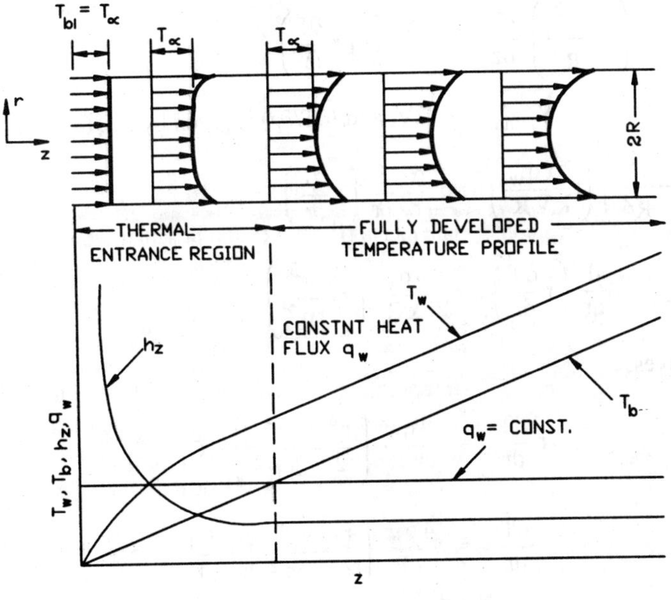

Fig. 4.18 Example 4.30 (a)

To determine the temperature T in the thermally developed region we can apply the condition of Eq. (4.257) to Eq. (4.253) :

$$\frac{\partial}{\partial z}\left[\frac{(T_w - T)}{(T_w - T_b)}\right] = 0$$

or $$-\frac{(T_w - T)}{(T_w - T_b)^2}\left[\frac{\partial T_w}{\partial z} - \frac{\partial T_b}{\partial z}\right] + \frac{1}{(T_w - T_b)} \times \left[\frac{\partial T_w}{\partial z} - \frac{\partial T}{\partial z}\right] = 0$$

or $$\frac{\partial T}{\partial z} = \frac{\partial T_w}{\partial z} - \frac{(T_w - T)}{(T_w - T_b)}\left[\frac{\partial T_w}{\partial z} - \frac{\partial T_b}{\partial z}\right]$$

or $$\frac{\partial T}{\partial z} = \frac{\partial T_w}{\partial z} - \frac{T_w - T}{T_w - T_b}\,\frac{\partial T_w}{\partial z} + \frac{T_w - T}{T_w - T_b}\,\frac{\partial T_b}{\partial z}$$

It may be seen from Fig. 4.18 that T_b and T_w lines are parallel in thermally developed region.

So, $$\frac{\partial T_w}{\partial z} = \frac{\partial T_b}{\partial z}$$

Hence, $$\frac{\partial T}{\partial z} = \frac{\partial T_w}{\partial z}$$

Using Eq. (4.253)

$$\frac{\partial T}{\partial z} = \frac{\partial T_w}{\partial z} = \frac{\partial T_b}{\partial z} = \frac{2\alpha q_w}{k V R} = \text{constant} \qquad (4.259)$$

From Eq. (4.250),

$$V_{z,max}\left(1 - \frac{r^2}{R^2}\right)\frac{\partial T}{\partial z} = \frac{\alpha}{r}\frac{\partial}{\partial r}\left(r\frac{\partial T}{\partial r}\right)$$

Since $V_{z,max} = 2V$, and $(\partial T/\partial z)$ is given by Eq. (4.259)

$$2V\left(1 - \frac{r^2}{R^2}\right)\left(\frac{2\alpha q_w}{k V R}\right) = \frac{\alpha}{r}\frac{\partial}{\partial r}\left(r\frac{\partial T}{\partial r}\right)$$

or $$\frac{d}{dr}\left(r\frac{dT}{dr}\right) = \frac{4q_w}{kR}\left(r - \frac{r^3}{R^2}\right) \qquad (4.260)$$

Integration gives,

$$r\frac{dT}{dr} = \frac{4q_w}{kR}\left(\frac{r^2}{2} - \frac{r^4}{4R^2}\right) + C_1$$

or $$\frac{dT}{dr} = \frac{4q_w}{kR}\left(\frac{r}{2} - \frac{r^3}{4R^2}\right) + \frac{C_1}{r}$$

Integrating again,

$$T = \frac{4q_w}{kR}\left(\frac{r^2}{4} - \frac{r^4}{16R^2}\right) + C_1 \ln r + C_2 \qquad (4.261)$$

The boundary conditions are

At
$$r = 0, \frac{dT}{dr} = 0 \tag{4.262}$$
$$r = R, T = T_w \tag{4.263}$$

On applying these to Eq. (4.261) the first condition gives $C_1 = 0$; the second condition gives,

$$C_2 = T_w - \frac{4q_w}{kR}\frac{3R^2}{16}$$

Then Eq. (4.261) becomes,

$$T = T_w + \frac{q_w R}{k}\left(\frac{r^2}{R^2} - \frac{r^4}{4R^4} - \frac{3}{4}\right) \tag{4.264}$$

Ans.

(ii) In order to determine the bulk temperature of the fluid we can use Eq. (4.21),

$$T_b = \int_0^R [\rho V C_p T (2\pi r) dr] \div \int_0^R [\rho V C_p (2\pi r) dr]$$

where $V = 2V\left(1 - \frac{r^2}{R^2}\right)$ and T is given by Eq. (4.264)

So
$$T_b = \frac{\int_0^R 2V\left(1-\frac{r^2}{R^2}\right)\left[T_w + \frac{q_w R}{k}\left(\frac{r^2}{R^2} - \frac{r^4}{4R^4} - \frac{3}{4}\right)\right] r dr}{\int_0^R 2V\left(1-\frac{r^2}{R^2}\right) r dr}$$

or
$$T_b = T_w - \frac{11}{24}\frac{q_w R}{k}$$

Since
$$R = D/2$$

$$T_w - T_b = \frac{11}{48}\frac{q_w D}{k} \tag{4.265}$$

Ans.

Now, since $q_w = h_z (T_w - T_b)$, we get

$$\frac{q_w}{h_z} = \frac{11}{48}\frac{q_w D}{k} \tag{4.266}$$

or
$$\frac{h_z D}{k} = Nu_z = \frac{48}{11} = 4.364 \tag{4.267}$$

Thus, for fully developed temperature profile, the local Nusselt number is constant and equals 4.364.

Ans.

(b) Case 2 : T_w = const.

$$\frac{dT_b}{dz} = \frac{2\alpha q_w}{kVR} \tag{4.253}$$

Since $$q_w = h_z (T_w - T_b), \tag{4.259}$$

we have, $$\frac{dT_b}{dz} = \frac{2\alpha h_z}{kVR}(T_w - T_b)$$

$$= -\frac{d(T_w - T_b)}{dz} = \frac{2\alpha h_z}{kVR}(T_w - T_b)$$

(since T_w = constant)

or $$\frac{d(T_w - T_b)}{(T_w - T_b)} = -\frac{2\alpha h_z}{kVR}\, dz \tag{4.268}$$

Integrating the above equation from the pipe entrance ($z = 0$, $T_b = T_{bi}$) to any coordinate z where $T_b = T_{bz}$, we get,

$$\int_{T_{bi}}^{T_{bz}} \frac{d(T_w - T_b)}{(T_w - T_b)} = -\frac{2\alpha}{kVR}\int_0^z h_z\, dz$$

or $$\ln \frac{T_w - T_{bz}}{T_w - T_{bi}} = -\frac{2\alpha z}{kVR}\, \bar{h}_z \tag{4.269}$$

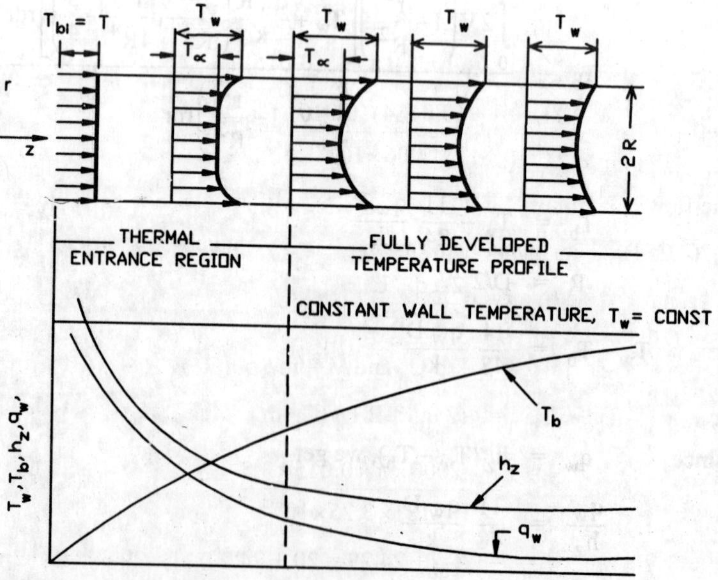

Fig. 4.19 Example 4.30 (b)

Here $\bar{h}_z$ denotes the average value of h_z from z = 0 to z = z and is equal to,

$$\bar{h}_z = \frac{1}{z}\int_0^z h_z\, d_z \tag{4.270}$$

Thus, $$\frac{T_w - T_{bz}}{T_w - T_{bi}} = \exp\left[-\frac{2\alpha z}{kVR}\bar{h}_z\right] \tag{4.271}$$

This equation is valid for flows in both undeveloped and fully developed regions.

The local value of the Nusselt number in the fully developed flow regime for constant wall temperature is found to be

$$Nu_z = \frac{h_z D}{k} = 3.658 \tag{4.272}$$

Fig. 4.19 shows the variations of T_b, h_z and q_w with z for the case of constant wall temperature. The variation in temperature profile is also shown.

Example 4.31

Water flows through a 1.5 cm ID tube at the rate of 3.25×10^{-3} kg/s. The tube is 3.5 m long. Water enters at a bulk temperature of 20 C. Compute the water outlet temperature and tube wall temperature the outlet. The wall heat flux is 2000 W/m^2.

Solution :

Let us first determine the Reynolds number by using the fluid properties at inlet temperature. At 20 C, the values for water are :

$$\rho = 1000 \text{ kg/m}^3 \text{ ; } \nu = 1.006 \times 10^{-6} \text{ m}^2\text{/s}$$

$$C_p = 4181 \text{ J/kg K, Pr} = 7.02$$

$$V = \frac{\dot{m}}{\rho A} = \frac{3.25 \times 10^{-3}}{(1000)(\pi/4)(0.015)^2} = 0.0184 \text{ m/s}$$

$$Re = \frac{VD}{\nu} = \frac{(0.0184)(0.015)}{1.006 \times 10^{-6}} = 274.4$$

Hence the flow is laminar.

From Eq. (4.254),

$$T_{bo} - T_{bi} = \frac{2\alpha q_w L}{kVR}$$

Since $\alpha = k/\rho\, C_p$ and $V = \dot{m}/(\rho\, \pi\, R^2)$

we have, $$T_{bo} - T_{bi} = (2\, q_w\, \pi\, R\, L)/(C_p\, \dot{m})$$

$$= \frac{2000 \times \pi \times 0.015 \times 3.5}{4181 \times 3.25 \times 10^{-3}} = 24.29 \text{ C}$$

or $$T_{bo} = T_{bi} + 24.29 = 20 + 24.29 = 44.29 \text{ C}$$

The only property of the fluid used in the above equation is C_p. If the value of C_p is taken at an average bulk temperature of $(t_{bi} + t_{bo})/2 = 32.15$ C the resulting value of T_{bo} will be about 44.30 C. It hardly differs from the value 44.29 C as found above. However, a second iteration should be performed wherever considered necessary.

Now, the entry length for the temperature profile to be fully developed is given by ;

$$\frac{z_t}{D} = 0.05\,(Re)\,(Pr) \tag{4.258}$$

or
$$z_t = 0.05\,(274.4)\,(7.02)\,(0.015)$$
$$= 1.445 \text{ m}$$

Since the pipe is 3.5 m long i.e. $L > z_t$, therefore the flow is fully developed.

Since
$$T_{wo} - T_{bo} = \frac{11}{48}\frac{q_w D}{k} \tag{4.265}$$
$$= \left(\frac{11}{48}\right)\frac{2000 \times 0.015}{0.616}$$
$$= 11.16 \text{ C}$$

(here k has been taken at a mean bulk temperature of 32 C)

So,
$$T_{wo} = 44.29 + 11.16 = 55.45 \text{ C} \qquad \text{Ans.}$$

Example 4.32

Water, flowing through a 4 cm ID, 8.6 m long pipe, at the rate of 1 kg/s, is heated from 30 C to 90 C. The inner wall of the pipe is maintained at a constant temperature of 120 C. Determine the average value of the heat transfer coefficient. Water has the following properties at 60 C : $\rho = 983.2$ kg/m^3; $\nu = 0.478 \times 10^{-6}$ m^2/s ; k = 0.659 W/mK; Pr = 2.98 ; $C_p = 4178$ J/kg K.

Solution :

The rate of heat transfer q is given by,

$$q = \dot{m}\, C_p\,(T_{bo} - T_{bi})$$
$$q = -\rho A\, V\, C_p\,[(T_w - T_{bo}) - (T_w - T_{bi})] \tag{a}$$

Applying Eq. (4.269), for z= L,

$$V = \frac{2\alpha L \bar{h}_L}{k R} \times \frac{1}{\ln[(T_w - T_{bo})/(T_w - T_{bi})]} \tag{b}$$

Substituting for V from Eq. (b) in Eq. (a),

$$q = \frac{\rho A C_p\, 2\alpha L \bar{h}_L}{k R} \times \frac{[(T_w - T_{bo}) - (T_w - T_{bi})]}{\ln[(T_w - T_{bo})/(T_w - T_{bi})]}$$

Putting A = π R2, α = k/ρ Cp

$$q = 2\pi R L \bar{h}_L \frac{[(T_w - T_{bo}) - (T_w - T_{bi})]}{\ln[(T_w - T_{bo})/(T_w - T_{bi})]}$$

or
$$q = (\pi DL)\,(\ln)\,(\Delta T_m)$$

where
$$\Delta T_m = [(T_w - T_{bo}) - (T_w - T_{bi})] \div \ln\left(\frac{T_w - T_{bo}}{T_w - T_{bi}}\right)$$

Now,
$$q = \dot{m}\, C_p\,(T_{bo} - T_{bi}) = 1 \times 4178 \times (90 - 30) = 250680 \text{ W}$$

so $$\Delta T_m = \frac{(120-90)-(120-30)}{\ln(30/90)} = \frac{60}{\ln 3} = 54.61 \text{ C}$$

or $$\bar{h}_L = \frac{q}{(\pi DL)(\Delta T_m)} = \frac{250680}{\pi \times 0.04 \times 8.6 \times 54.61}$$

$$= 4249.7 \text{ W/m}^2\text{K} \qquad \text{Ans.}$$

If we use McAdam's equation (4.42a) or Dittus and Boelter equation (4.45) we obtain the following result :

$$Nu = 0.023\,(Re)^{0.8}\,(Pr)^{0.4}$$

$$Re = \frac{VD}{\nu} = \frac{\dot{m}(2R)}{\rho(\pi R^2)\nu} = \frac{1 \times 2 \times 10^6}{983.2 \times \pi \times 0.02 \times 0.478} = 67764$$

so $$Nu = 0.023\,(67764)^{0.8}\,(2.98)^{0.4} = 260.75$$

or $$\bar{h}_L = \frac{(Nu)(k)}{D} = \frac{(260.75)(0.659)}{0.04}$$

$$= 4295.9 \text{ W/m}^2\text{K} \qquad \text{Ans.}$$

This value agrees well with the previous value of 4249.7 W/m^2K.

Example 4.33

Calculate the length of 6 cm ID steel tube required to heat 18 kg/s of mercury from 100 C to 200 C. The tube wall temperature is 40 C above the bulk mean temperature of mercury throughout its length. For mercury, the following property values may be used at 150°C : ρ = 13264 kg/m^3, k = 11.49 W/mK, C_p = 136.5 J/kg K, $\nu = 0.0853 \times 10^{-6}$ m^2/s, Pr = 0.0134.

Solution :

Let us first determine the value of Reynolds number to decide whether the flow is laminar or turbulent.

Velocity is given by,

$$V = \frac{\text{volume flow rate}}{\text{flow area}} \; \frac{(\dot{m}/\rho)}{(\pi D^2/4)}$$

$$= \frac{(18/13264)}{(\pi \times 0.06^2)/4} = 0.48 \text{ m/s}$$

so $$Re = \frac{VD}{\nu} = \frac{0.48 \times 0.06}{0.0853 \times 10^{-6}} = 33.76 \times 10^4$$

Hence flow is turbulent (Re > 2000)

Here the heat flux is constant (constant temperature difference between tube wall and fluid). Hence we may use Eq. (4.55) :

$$Nu = 4.82 + 0.0185\,(Re\,Pr)^{0.827}$$

$$= 4.82 + 0.0185\,(33.76 \times 10^4 \times 0.0134)^{0.827} = 24.33$$

so $$h = (Nu)\frac{k}{D} = (24.33)\frac{(11.49)}{0.06} = 4659.2 \text{ W/m}^2\text{ K}$$

The heat transfer rate is.

$$q = \dot{m}\, C_p\,(T_2 - T_1) = h\,A\,\Delta T$$

where $T_1 = 100$ C, $T_2 = 200$ C, $A = \pi$ DL,

$$\Delta T = 40 \text{ C}$$

Hence,

$$18 \times 136.5 \times (200 - 100) = (4659.2)\,(\pi \times 0.06 \times L)\,(40)$$

$$L = 7 \text{ m}$$ Ans.

Note that L/D = 7/0.06 = 116.7 which is much greater than 30 as required for Eq. (4.55).

Example 4.34

The crank case of an automobile engine is approximately 75 cm long, 30 cm wide and 10 cm deep. The surface temperature of the crank case is 70 C. Estimate the rate of heat flow from the crank case to atmospheric air at 10 C at a road speed of 100 km/hr. Assume that the boundary layer is turbulent over the entire surface and for the front and rear surfaces the same average value of h be used as for the bottom and sides.

Solution :

The average value of the heat transfer coefficient h for turbulent flow over a flat plate is given by,

$$\frac{hL}{k} = 0.037\,(Re_L)^{0.8}\,(Pr)^{1/3} \qquad (4.38)$$

The physical properties of air at the film temperature $t_f = (70 + 10)/2 = 40$ C are : k = 0.02755 W/mK, $\nu = 16.96 \times 10^{-6}$ m²/s, and Pr = 0.699.

$$Re_L = \frac{VD}{\nu} = \left(\frac{100 \times 1000}{3600}\right)\left(\frac{0.75}{16.96 \times 10^{-6}}\right) = 1.23 \times 10^6$$

so $$\frac{hL}{k} = 0.037\,(1.23 \times 10^6)^{0.8}\,(0.699)^{1/3} = 2442.5$$

or $$h = \frac{(2442.5)(0.02755)}{0.75} = 89.7 \text{ W/m}^2\text{K}$$

Surface area = area of bottom, sides, front and rear

$$= (0.75 \times 0.3) + (0.75 \times 0.1 \times 2) + (0.3 \times 0.1 \times 2)$$

$$= 0.435 \text{ m}^2$$

Thus heat transfer rate, q

$$= h\,A\,\Delta t$$

$$= (89.7)\,(0.435)\,(70 - 10)$$

$$= 2341.7 \text{ W}$$ Ans.

PROBLEMS

(Note : 1 Nm = 1 Ws = 1 J)

4.1 Calculate Reynolds number for the following data : D = 15 cm ; V = 3 m/sec ; ρ = 14,400 kg/m^3, and μ = 0.0372 kg/m sec.

4.2 Calculate Nusselt number for the following : L = 15 cm ; k = 2.9 W/mK ; h = 45 W/m^2 C.

4.3 Calculate Pradtl number for the following data : Cp = 0.5 kJ/kg C ; k = 2.9 W/m K; μ = 5 poise (gm/cm sec).

4.4 Calculate the value of Stanton number for the data of Problems 4.3 and 4.4 if V = 3 m/sec and ρ = 900 kg/m^3.

4.5 Show that :

$$\text{Nu}\,(\text{Pr})^{-1/3} = \{\text{St}\,(\text{Pr})^{2/3}\}\,\text{Re}$$

4.6 Using dimensional analysis show that the rate of heat transfer q from a body of linear dimension d, submerged in a fluid flowing at velocity V is given by ;

$$\frac{q}{k d\,\Delta t} = C\left(\frac{k}{\rho C_p d V_\infty}\right)^n$$

4.7 During winter season the wind is blowing parallel to the short side of a flat roof at speed of 75 km/hr. The size of roof is 9 m × 20 m, and the temperature of air is – 5 C and that of roof surface is 5 C. Calculate the heat loss through the roof.

4.8 A plate at 90 C is located parallel to an air stream flowing @ 75 m/sec. The temperature of air is 0 C. The plate is 60 cm long and 45 cm wide. The width is along the direction of main stream. Assume the transition Reynolds number where the boundary layer changes from laminar to turbulent as $Re_{tr} = 4 \times 10^5$. Calculate ;

(a) Average heat transfer coefficient in laminar and turbulent regions.

(b) Average friction coefficient in laminar and turbulent regions

4.9 An airplane wing is 1.2 m wide and its surface is at 24 C. The aeroplane is flying at 240 km/hr in air at 0 C and 67 cm Hg pressure. Calculate the heat loss per m length of the wing.

4.10 A flat plate at 90 C is located parallel to a water stream having a free stream velocity of 60 m/sec and at 0 C. The plate is 45 cm long and 60 cm wide. The flow in the boundary layer changes from laminar to turbulent at Re = 4×10^5. Find Nu for the plate and the thickness of thermal boundary layer at a distance of 20 cm from the leading edge.

[Ans. 61,900 ; 0.225 cm]

4.11 A cooling fin 7.5 cm along the leading edge and 45 cm long in the flow direction is located parallel to an air stream. The surface temperature of the fin is 88 C and the air temperature is 10 C. The free stream velocity of air is 27 m/sec. Determine the rate of heat transfer from the fin to air.

[Ans. 163 W]

4.12 The velocity profile for water flowing in a pipe 20 cm in diameter is given by $V = 190\ r^2 - 96\ r$ and the temperature profile is given by $t = 80\ r^2 + 45$ where r is the distance in meters measured from the inside surface. Calculate the average (cup-mixing) temperature of water.

4.13 300 kg/hr of water is to be heated from 30 C to 45 C while flowing through a horizontal 2 cm diameter steel pipe. If the pipe is jacketed with steam condensing at 100 C, what length of pipe is required.

(Hint : The flow will be found to be laminar in pipe, calculate LMTD)

4.14 A heat exchanger is to be designed to cool lubricating oil from 60 C to 40 C. The oil will flow through the tube 5 mm diameter at the rate of 1 m/sec. Calculate the length of the tube required using the properties of oil as ; $\rho = 800$ kg/m^3, k = 0.12 W/m K ; $\mu = 12$ centipoise ; $C_p = 0.45$ kJ/kg C.

4.15 280,000 liters/hr of air is flowing through a 10 cm diameter pipe at 50 C. If the entrance and discharge temperatures are 20 C and 35 C respectively, calculate the heat transfer coefficient.

4.16 Liquid sodium flows through a 10 cm diameter pipe at an average velocity of 3 m/sec. The wall temperature is 275 C and the mean sodium temperature is 200 C. Find the coefficient of heat transfer for the film. Given that for sodium at 237.5 C; $\rho = 900$ kg/m^3 ; $C_p = 1.34$ kJ/kg C ; $\mu = 44.64 \times 10^{-5}$ kg/m sec ; k = 67.5 W/mK.

4.17 Two coaxial tubes 100 cm long form an annular space with the diameters 1.2 cm and 1.0 cm. Ethylene at 100 C is flowing through the space with a velocity of 20 m/sec. Calculate the heat energy picked up by the fluid for $\Delta t = 10$ C. The properties of ethylene are ; $\rho = 0.924 \times 10^{-3}$ gm/cm^3 ; $\mu = 124 \times 10^{-6}$ poise (gm/cm s, $C_p = 0.44$ J/gm C ; $k = 250 \times 10^{-6}$ J/sec cm C.)

4.18 The inside surface of a condenser tube 2 cm in diameter is held at 50 C by water at 35 C flowing through it at an average velocity of 1 m/sec. It is observed that for turbulent flow $\log_e (Nu\ Pr^{-0.4})$ is linear with $\log_e$ (Re). In particular $(Nu\ Pr^{-0.4})$ = 40 and 250 for Re = 10^4 and 10^5 respectively. Calculate the surface coefficient of heat transfer from the inside surface of the tube to the water flowing within.

4.19 A condenser is required to handle 10, 000 kg/hr of steam at a pressure of 0.1 bar. The heating surface is made of horizontal brass tubes 3 cm ID and 1 mm thick. The tube length is 3 m. The cooling water @ 4000,000 kg/hr at 25 C circulates with a velocity of 21 m/sec. How many tubes are required in the condenser.

4.20 Water at an average temperature of 27 C is flowing through a smooth 5 cm ID pipe at velocity of 1 m/sec. If the temperature of inner surface of pipe is 50 C determine (a)unit surface conductance (b) heat flux (c) pressure drop per m ; $f = 0.046/Re^{0.2}$.

[Ans. 4140 W/m^2 K (b) 15,000 W]

4.21 Considering a fluid to have a uniform velocity across a 10 cm ID pipe and at a given location the temperature to vary linearly from 40 C at pipe wall to 20 C at centre find (a) the bulk temperature (b) the heat transfer coefficient (c) the thermal conductivity of fluid if the heat flux is 29,000 W/m^2.

[Ans. (a) 20 C, (b) 1,450 W/m^{2}K, (c) 24.2 W/mK]

4.22 Air flows at a velocity of 0.3 m/sec through the annular space between two concentric tubes. The outer tube is 5 cm ID and the inner tube is 3.125 cm OD. The air enters at 16 C and leaves at 32 C. The temperature of the outside surface of the inner tube is 50 C. Determine the heat transfer coefficient between the sir and the inner tube

[Ans. 3.5 W/m^2K]

4.23 Mercury at mean bulk temperature of 65 C flows through a copper tube 5 cm ID at a velocity of 7.2 m/sec. The temperature at the inside surface of the tube is 80 C. Calculate the rate of heat transfer if the tube is 9 m long.

[Ans. 905×10^3 W]

4.24 Calculate the value of h for water flowing through a 5 cm dia tube at 2.1 m/sec velocity. The mean bulk temperature of water is 15 C and the tube is at 25 C.

[Ans. 5700 W/m^2K]

4.25 Water flows through the annular space between two tubes. The mean bulk temperature of water is 50 C. The larger diameter of the annulus is 10 cm and the smaller diameter is 5 cm. The temperature at the larger surface of the annulus is 5 C and at the smaller surface 15 C. Determine h (a) between the water and the smaller tube (b) between water and the larger tube.

4.26 Calculate the heat transfer coefficient for liquid sodium at 200 C flowing through a 7.5 cm diameter tube maintained at 250 C. The bulk velocity of sodium is 3 m/sec. For liquid sodium at 200 C ;

$$C_p = 1.34 \text{ kJ/kg K}$$
$$\rho = 900 \text{ kg/m}^3$$
$$k = 80 \text{ W/mK}$$
$$\mu = 0.435 \times 10^{-3} \text{ kg/m sec}$$

4.27 In a counter-flow heat exchanger lubricating oil is cooled from 40 C to 30 C by water entering at 15 C. The velocity of oil is 0.6 m/sec and it flows through the inside tube 4.375 cm ID and 0.5 cm OD. If the film coefficient between the outside tube surface and water is 525 W/m^2K, find the flow rate of oil and water for 9 m^2 of inside tube surface. For oil :

$$C_p = 1.93 \text{ kJ/kg K}$$
$$\mu = 81.5 \text{ kg/hr m}$$
$$k = 0.128 \text{ W/mK}$$
$$\rho = 896 \text{ kg/m}^3$$

[Ans. 0.85 kg/sec ; 0.0039 kg/sec]

4.28 In a double pipe heat exchanger, liquid sodium flows through the inner pipe at the rate of 910 kg/hr while mercury flows in the annulus at the rate of 18,200 kg/hr. Sodium enters at 500 C and leaves at 300 C. Mercury enters at 50 C and leaves at 150 C. Determine the length of the exchanger. The dimensions of the pipes are :

	ID	OD
Inner pipe	2.060 cm	2.625 cm
Outer pipe	4.025 cm	4.750 cm

The following average properties for mercury and sodium may be used.

	Mercury	*Sodium*
ρ (m^3/kg)	13,344	832
k (W/mK)	10	72.5
Cp (kJ/kgK)	0.134	1.34
μ (kg/hr m)	4.5	1.08

4.29 Plot a curve showing the increase in laminar boundary layer thickness for a distance of 30 cm, by calculating the value of δ at 5 cm intervals, if the fluid flowing is air at a temperature of 20 C at a velocity of 75 m/sec.

4.30 A steam pipe 8.75 cm OD is exposed to air. Wind at a speed of 50 km per hr blows squarely on the pipe. The surface temperature of the pipe is 96 C and the air temperature is 10 C. Calculate the heat loss per m length of the pipe.

[Ans. 4460 W]

4.31 Water at 15 C flows across a tube bundle at a free stream velocity of 21 m/sec. The outside diameter of the tubes is 8.8 cm. Both longitudinal and transverse spacing of the tubes are 22.5 cm. The tubes are staggered. If the tubes are at 40 C, estimate the heat transfer coefficient.

[Ans. 12.2 W/m^2 K]

4.32 A tube bundle consists of 15 transverse and 12 longitudinal rows of 0.625 cm OD tubes arranged staggered with S_L = 0.78 cm and S_T = 0.94 cm. Air at 21 C flows normal to the axes of the tubes. The tube surfaces are at 94 C. The free stream air velocity is 0.9 m/sec. Find h for the tube bundle.

[Ans. 183 W/m^2 K]

4.33 Calculate Grashof number with the following data :

$d = 25$ mm ; $t_\infty = 150$ C ; $t_w = 100$ C ; $\nu = 2.31 \times 10^{-5}$ m^2/sec ; Pr = 0.69.

4.34 A horizontal cylinder 2.5 cm diameter and 10 cm in length is suspended in still air which is at a temperature of 20 C. Calculate the rate of heat transfer from the cylinder surface if its temperature is 60 C.

[Ans. 290 W]

4.35 If $Nu = 0.53\,[(Pr^2\,Gr)/(0.952 + Pr)]^{1/4}$ show that ;

(a) for Pr > > 1 ; $Nu = 0.53\,(Gr\,Pr)^{1/4}$

(b) for Pr < < 1 (for liquid metals) ; $Nu = 0.53\,(Gr\,Pr^2)^{1/4}$

4.36 A transformer is to be cooled by free convection. The sheet is 30 cm tall and 30 cm × 4 cm in cross section. The surface temperature is 95 C. The ambient air temperature is 40 C. Calculate the heat dissipation.

4.37 An enclosed vertical square layer of air 60 cm × 60 cm × 10 cm thick is heated from below at a rate of 58 W while its upper surface is kept at 25 C by means of a cooling device. Calculate the temperature of bottom surface.

4.38 The vertical air layer of problem 4.37 is heated @ 58 W from one of its large surfaces instead of bottom. The opposite surface is kept at 25 C. Calculate the temperature of the warmer surface.

4.39 Water is heated in a tank by horizontally immersed steam pipes 10 cm OD. The temperature of water is 45 C while that of pipe surface 95 C. The convection currents of one pipe do not affect those of the other. Calculate h.

4.40 Specify a consistent set of units for the group $(\mu^2/k^2\ \rho^2\ V^2)^{1/2}$ to have the reciprocal dimension of h.

4.41 Calculate the heat transfer coefficient at 30 cm from the upper edge of a vertical surface at 102 C. What is the thickness of film at this location.

4.42 What is the rate of heat transfer by free convection from a 20 cm × 25 cm horizontal flat plate facing upward to air at 27 C. The plate is at 150 C.

[Ans. 660 W]

4.43 The temperature of the surface of a horizontal cylinder in still air is 250 C. The air temperature is 50 C. What should be the diameter of the cylinder if the Grashof number is 2×10^8.

4.44 A horizontal cylinder 5 cm in diameter is placed in a fluid for which k = 0.028 W/mK. If the Grashof and Prandtl numbers are 10,000 and 10, respectively, find the heat loss by natural convection per m length of the cylinder. The temperature difference between cylinder surface and the fluid is 70 C.

4.45 The heat transfer coefficient for a wire at 55 C placed in quiet air at 20 C is 25 W/m^2K. If the wire temperature is increased to 110 C what will be the new value of h.

4.46 Compute the free convection heat transfer coefficient from a horizontal wire 1.25 mm diameter at 730 C to nitrogen gas at 30 C. For nitrogen,

$$\mu = 0.025 \text{ kg/hr m}$$

$$C_p = 1.0 \text{ kJ/kg K}$$

$$k = 0.043 \text{ W/m K}$$

4.47 Compare the heat loss per m^2 of outside surface of a pipe 7.5 cm diameter and a wire 0.025 cm diameter both at 100 C placed in still air at 20 C.

CHAPTER

5

Heat Transfer with Change of Phase

5.1 HEAT TRANSFER ACCOMPANIED BY CHANGE OF PHASE

In the preceding chapter the conditions of the system did not envisage any change in the *chemical composition* or that of *phase* of the fluid flowing. There are many industrial processes however, where the transfer of thermal energy is accompanied by *change of phase*. The process of heat transfer in such cases is not merely that of *convection*, but is very much *different* and complex. The phenomena of transfer of heat associated with change of phase will be described in brief in this chapter.

The two important processes which fall in this category are those of *boiling liquids* and *condensing vapors*. These processes, in addition to all the variables associated with convection, depend also upon those variables which govern the *change of phase* ; from liquid to vapor in *boiling* and from vapor to liquid in *condensation*. These additional variables like surface tension of fluid, latent heat and pressure, make the phenomena so complex that it has not been possible to evolve detailed general equations describing the process or their analytical solutions so far. During the last few years, however, considerable progress has been reported in the understanding of the phenomena of evaporation and condensation.

Until recently the processes involving heat transfer with change of phase were found primarily in steam generators and condensers of thermal power plants, evaporators and condensers of refrigerating systems and in chemical industry. But the cooling problem in *nuclear reactors* or *rocket motors* has thrown a challenge to engineers to develop the knowledge of *boiling heat transfer* so as to dissipate heat energy at rates as large as 10^8 W/m^2. The large magnitude of this rate may be realised by comparing it to maximum heat transfer rates in a modern boiler (2×10^5 W/m^2).

5.2 PHENOMENON OF BOILING

Boiling, though a familiar phenomenon, is extremely complicated. In order to understand the heat transfer mechanism during boiling, we shall first discuss the phenomenon of *boiling* as such and the existence of various *regimes of boiling*.

Boiling is the process of vapor formation in a liquid receiving heat. Usually, the temperature of the boiling liquid is assumed to be equal to the *saturation temperature*, t_s, corresponding to the boiling pressure. However, experience shows that the liquid is at a slightly higher temperature than t_s.

If water is allowed to boil over a horizontal surface, the temperature profile may be represented by Fig. 5.1. It is observed that the bulk of the water is at a fairly uniform temperature (slightly higher than t_s = 100 C). In a layer of 2.5 mm thick over the heated surface, however, there is a steep gradient up to the surface temperature of 109.1 C, when the heating is done at the rate of 43,000 kJ/hr m^2.

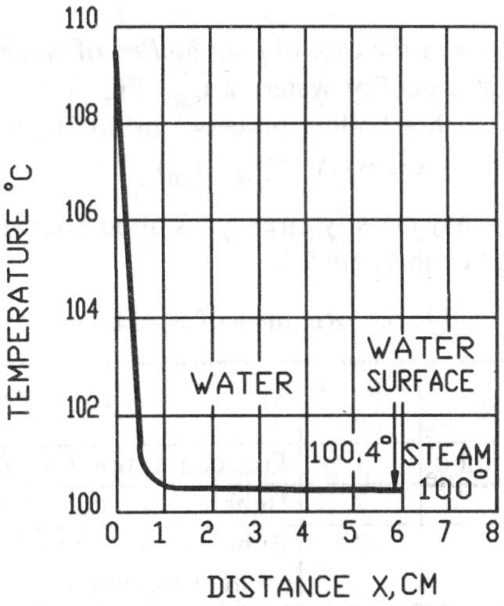

Fig. 5.1 Boiling of Water Over Heated Surface

In general, the boiling process which finds maximum engineering application is the one in which the heating surface is submerged beneath the free surface of the liquid in a container, as discussed above. This type is classified as *pool boiling*. The movement of liquid in this case is caused only by free convection or by stirring action of the bubbles.

If the boiling occurs on the walls of a tube through which liquid flows with considerable velocity, then this velocity influences the liquid motion thereby causing the bubble growth and separation as well. Such boiling is called *forced convection boiling*.

In ordinary *pool boiling*, the liquid is originally well below the saturation (evaporation) temperature. Some time after heat addition through the heating surface, vapor bubbles appear on the surface. They, after having grown up, separate from the surface and move into the bulk of the liquid. If the rate of heat flow is sufficiently large, boiling in the vicinity of the surface may take place even when the bulk temperature is below the boiling point. In such a case the bubbles of vapor, after travelling a short path, vanish. obviously they condense in the bulk of the liquid, which is cooler than their own temperature at boiling point. This type of boiling is called *local boiling* or *subcooled boiling*.

It is only after the liquid bulk has been heated sufficiently that the vapor bubbles will travel all the way to the liquid surface and start a flow of vapor into the space above the surface. Then, the actual evaporation process occurs. Such boiling is called *boiling with net evaporation*.

The most extensively investigated boiling process is the *pool boiling with net evaporation*. This process will be discussed in detail. The discussion will be followed by the presentation of empirical or semi-empirical correlations of the data on boiling heat transfer.

5.3 REGIMES OF POOL BOILING

The phenomenon of boiling is characterized by different regimes. Nukiyama, in 1934, actually distinguished the existence of various regimes of boiling. Drew and Miller extended the theory further in 1937.

Let us, for instance, consider the case of *pool boiling* of water, over a heated flat plate or a hot wire submerged in a pool of water, at t_{sat}. Fig. 5.2 represents the type of heat transfer data obtained for such a boiling process. In this figure the *heat flux*, (q/A), is plotted against the *temperature excess* $\Delta t = (t_w - t_{sat})$.

Generally speaking, we may classify three types of boiling which occur in six regimes as shown in Fig. 5.2 and given in Table 5.1.

Table 5.1 Regimes of pool boiling

No.	*Type of boiling*	*Regime No.*	*Type of Regime*
(i)	Interface evaporation	1	Free convection boiling
(ii)	Nucleate boiling	2	Bubbles form but condense as they rise.
		3	Bubbles rise to interface and cause rapid evaporation.
(iii)	Film boiling	4	Partial nucleate boiling with unstable film.
		5	Stable film boiling.
		6	Radiation predominant.

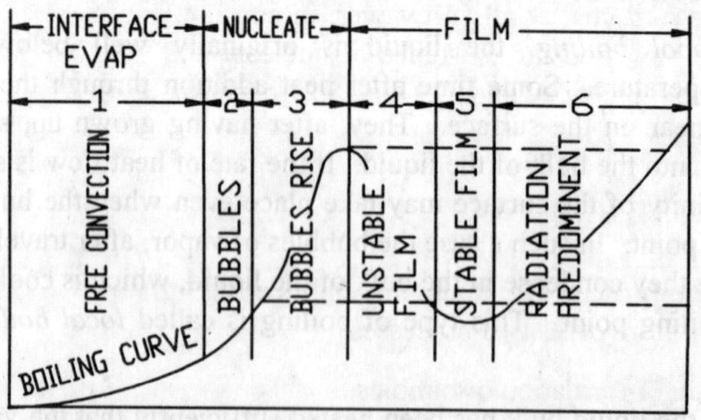

Fig. 5.2 Regimes of Pool Boiling

(i) As the wall surface temperature t_w is raised, Δt is increased, convection currents circulate in the liquid and the steam is produced by evaporation at the liquid surface. This is represented as *regime* 1 and called *interface evaporation*. Here only liquid is in contact with the heated surface and the heat transfer is only due to *free convection*.

(ii) With further increase in Δt the energy level of the liquid adjacent to the surface becomes high at a number of *favoured spots* where *vapor bubbles* are formed. They rise above the metal surface but condense before reaching the liquid surface. This is known as *regime* 2. As the temperature is increased further, up to point A (Fig. 5.2), the bubbles become more numerous. Now, the liquid is hot, so as not to allow any condensation of the bubbles which rise to the free liquid surface and thus help rapid evaporation. This is known as *regime* 3. Both the regimes 2 and 3 fall in the category of *nucleate boiling*.

(iii) Beyond point A representing the *critical heat flux*, further increase in *excess temperature* results in *film boiling*. Here the number of bubbles formed is so large that they almost cover the whole of the metal surface and provide an insulating effect. It counteracts the beneficial effect of agitation by bubbles and results in decrease in the heat flux. The vapor film is not stable here in *regime* 4, as under the action of circulating currents it collapses and reforms rapidly. In *regime* 5 the film is stable and the heat flow is lowest. For values of Δt beyond 550 C (*regime* 6) the temperature of metal surface is quite high and the heat transfer occurs predominantly by *radiation*, thereby increasing the heat flux. Hence the boiling is more stable in this *regime* where it is called stable *film boiling*.

Note the significance of the *critical heat flux point* A, the peak of the curve. The boiling beyond this point is unstable unless we proceed upto point B, where the metal surface temperature may be as high as above melting point. Within the points A and B of the curve the boiling process can not remove heat equal to the heat input corresponding to the increased t_w. This increases further the value of t_w, which in turn decreases q/A and thus, if heat supplied is not suddenly reduced, the metal is likely to be damaged or even melted. The point A is therefore called *burn-out point*. The exact knowledge about this point has assumed considerable importance recently, due to the present trend of using high heat fluxes.

It should further be noted that if the wall surface is heated by some isothermal device like condensing vapor on the other side (Δt does not exceed a certain value) the film boiling may not be expected at all. However, in constant heat flux heating like that by electrical input we may find the existence of nucleate and film boiling simultaneously, corresponding to say points C and D (Fig. 5.2) occuring at different locations of the surface.

5.4 THE NUCLEATE BOILING

The *nucleate boiling* is of maximum engineering importance and needs special attention for study. It involves two processes.

(a) The *nucleation* or formation of bubbles.

(b) The subsequent *growth* and *motion* of these bubbles.

(a) Nucleation

The bubbles, preferentially get formed at those active *nucleii* on the heating surface where the temperature and nature of the surface are favourable. Usually nucleation starts when t_w is only slightly higher than t_s. It is accepted that bubbles originate at small cone-like cavities in the heating surface.

The bubble formation is greatly influenced by the nature of heating surface which affects the *angle of contact* between the bubble and the surface. The angle of contact is in fact a measure of *wettability* of the surface with the particular fluid. Three typical shapes of vapor bubbles may be observed as shown in Fig. 5.3.

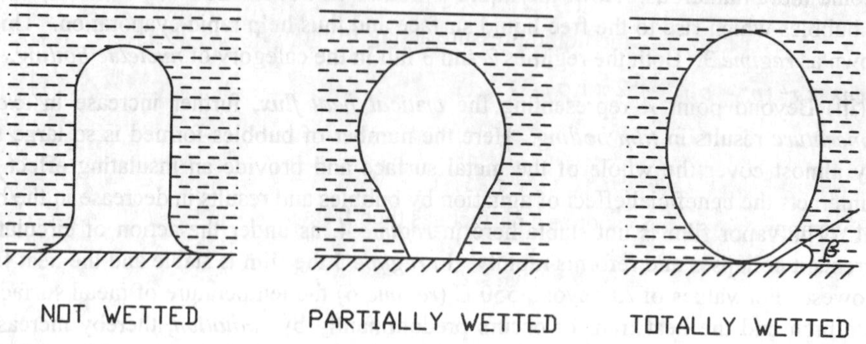

Fig. 5.3 Bubble in Contact with Surface

(i) Unwetted surface : The bubbles spread out forming a wedge between heating surface (smooth) and liquid.

(ii) *Partially wetted surface* : The contact angle is small than that for unwetted surface. The heat transfer area is large due to the fact that the area covered by vapor is reduced.

(iii) *Totally wetted surface* : The surface is rough and the roughness counteracts the surface tension so that the metal surface is fully wetted. Since the area covered by insulating vapor film is smallest it represents the most favorable circumstance for efficient heat transfer.

In order to study the dynamic equilibrium of the bubble, let us consider a bubble of radius r, shown in Fig. 5.4. Let the bubble be divided into two halves by plane 1-1.

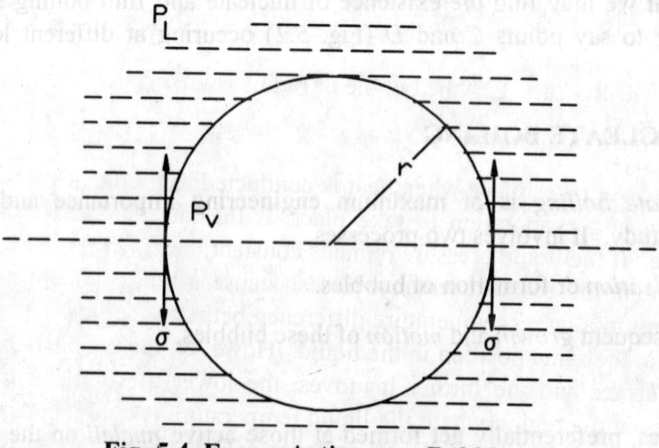

Fig. 5.4 Forces Acting on a Bubble

Now the force on the cross sectional area in plane 1...1 ,

$$F = 2\pi r \sigma$$

where σ = surface tension force per unit length.

This force should be equal to the force due to vapor and liquid pressures ;

$$= \pi r^2 (p_v - p_l)$$

where p_v = the vapor pressure acting from the inside of bubble.

p_l = the liquid pressure acting from outside of bubble.

$$\text{Hence; } \pi r^2 (p_v - p_l) = 2 \pi r \sigma$$

$$\text{or} \quad p_v - p_l = \frac{2\sigma}{r} \tag{5.1}$$

If we approximate the vapor as a perfect gas then from Clayperon Equation ;

$$\frac{dp}{p} = \frac{h''' - h''}{RT^2} dT$$

$$\text{or} \quad \frac{dp}{dT} = \frac{(h''' - h'')}{RT^2} p = \frac{pL}{RT^2}$$

{where latent heat of evaporation, L = (h''' – h'')}

$$\text{Now} \quad \frac{dp}{dT} = \frac{p_v - p_l}{t_v - t_{sat}}$$

$$\text{or} \quad p_v - p_l = (t_v - t_{sat}) \frac{dp}{dT}$$

$$\approx (t_v - t_{sat}) \frac{pL}{RT^2}$$

$$= \frac{2\sigma}{r}$$

$$\text{or} \quad (t_v - t_{sat}) = \frac{2\sigma RT_{sat}^2}{Lpr} \tag{5.2}$$

Thus if $(t_l - t_{sat}) > (t_v - t_{sat})$ the bubble of radius r will grow otherwise it will collapse.

(b) Bubble growth and Motion

Bubble growth takes place when heat is conducted to the liquid-vapor interface from the liquid, causing evaporation to take place at the interface and increasing the total vapor volume. If the liquid pressure remains constant, Eq. (5.1) requires that the pressure inside the bubble be reduced which will also cause a reduction in vapor temperature. This will result in a larger temperature difference between the liquid and the vapor if the bubble stays at the same position in the liquid. However, the bubble is likely to rise from the heated surface and the further it moves, the lower the liquid temperature. If the bubble moves into a region where the liquid temperature is below that of the vapor, heat will go out of the bubble, and it will collapse. Thus the bubble growth may reach a balance at some location in the liquid, or if the liquid is superheated enough, the bubble may rise to the surface before being dissipated.

Fritz proposed an empirical relation for bubble diameter, D_b at detachment from the heating surface as given below :

$$D_b = C_d \beta \sqrt{\frac{2 g_c \sigma}{g(\rho_l - \rho_v)}} \tag{5.3}$$

where C_d = 0.0148 for water

β = angle of contact as shown in Fig. 5.3.

Zuber also proposed the following relation :

$$f D_b = 1.18 \left(\frac{\tau_c}{\tau_c + \tau_d}\right) \left\{\frac{g_c g \sigma (\rho_l - \rho_v)}{\rho_l}\right\}^{1/4} \tag{5.4}$$

where τ_c = time for which bubble is attached to surface

τ_d = time that elapses before the next bubble is formed.

f = frequency of bubble formation

5.5 HEAT TRANSFER DURING BOILING

We shall now discuss the correlation of the data available on pool boiling heat transfer. The phenomena of boiling being more complex than that associated with single phase convection, cannot be represented by analytical-mathematical relations. The experimental results however have been correlated and are available for further applications.

(a) Heat Transfer in Free Convection Boiling

During the interface evaporation, the heat transfer is solely due to free convection ; hence it is possible to calculate the heat transfer coefficient by the relation :

$$Nu = \phi\,(Gr)\,\psi\,(Pr)$$

The function ϕ will depend upon the geometry of the heating surface.

(b) Heat Transfer in Nucleate Boiling

As discussed earlier, the heat is transferred in forced convection primarily by turbulent mixing of hot and cold fluid particles. The heat transfer data may be correlated by the following relation :

$$Nu = \phi\,(Re)\,\psi\,(Pr)$$

In this equation *Re* is a measure of the mixing and turbulent motion of the particles. Since the mixing of particles in nucleate boiling is predominantly due to *agitation* caused by *bubbles*, the turbulence due to ordinary convection currents is not important. Hence it will be appropriate to define a new *Reynolds Number*, Re_b which is a measure of agitation of liquid in nucleate boiling ;

$$Re_b = \frac{D_b G_b}{\mu_l} \tag{5.5}$$

where D_b = Bubble diameter

G_b = Mass velocity of bubbles per unit area

μ_l = Liquid viscosity.

Hence $(Nu)_{boiling} = \phi(Re_b)\,\psi(Pr)$ (5.6)

Using experimental data for pool boiling Rohsenhow modified Eq. (5.6) by means of simplifying assumptions and proposed the final correlating equation as given below :

$$\frac{c_l(t_w - t_{sat})}{(Pr_l)^s L} = C_{sf}\left\{\frac{(q/A)}{\mu_l L}\sqrt{\frac{g_c\,\sigma}{g(\rho_l - \rho_v)}}\right\}^{1/3} \qquad (5.7)$$

where c_l = Specific heat of liquid.

q/A = Heat flux.

g_c = Unit conversion factor.

g = Acceleration due to gravity.

ρ_v, ρ_l = Density of vapor and liquid respectively.

σ = Surface tension.

Pr_l = Prandtl number of saturated liquid.

μ_l = Viscosity of saturated liquid.

C_{sf} = An empirical constant (values given in Table 5.3),

s = Exponent. It is 1.0 for water and 1.7 for other liquids.

L = Latent heat of evaporation

Note that Eq. (5.7) holds good for clean surfaces. For a contaminated surface the exponent of Pr_l has been found to vary between 0.8 and 2.0.

Adams reported good agreement with the correlation Eq. (5.7) on the basis of his experiments on pool boiling of water on 0.025 in (0.635 mm) diameter platinum wire.

Table 5.2 : Values of Vapor-Liquid Surface Tension for Water

Saturation temperature, °C	$\sigma \times 10^3$, N/m
0	75.5
20	72.9
40	69.5
60	66.1
80	62.7
100	58.9
150	48.7
200	37.8
250	26.1
300	14.3
350	3.6

Table 5.3 : Values of Coefficient C_{sf} for Various Liquid-Surface Combinations

Fluid-Heating Surface Combination	C_{sf}
Water-copper	0.013
Carbon tetrachloride-copper	0.013
35% K_2CO_3-copper	0.0054
n-Butyl alcohol-copper	0.00305
50% K_2CO_3-copper	0.00275
Isopropyl alcohol-copper	0.00225
n-Pentane-chromium	0.015
Water-platinum	0.013
Benzene-chromium	0.010
Water-brass	0.0060
Water-emery polished copper	0.0128
Water-ground and polished stainless steel	0.0080
Water-chemically etched stainless steel	0.0133
Water-mechanically polished stainless steel	0.0132
Water-emery polished and paraffin treated copper	0.0147
Water-scored copper	0.0068
Water-Teflon pitted stainless steel	0.0058
n-Pentane-emery polished copper	0.0154
n-Pentane -lapped copper	0.0049
n-Pentane-emery rubbed copper	0.0074
n-Pentane-emery polished nickel	0.0127
Ethyl alcohol-chromium	0.027

(c) Heat Transfer in Forced Circulation Boiling

The above method of correlation of data for nucleate pool boiling may be extended to boiling of fluids flowing inside tubes or ducts by forced convection.

(i) Rohsenhow proposed superimposition of the boiling flux q_b/A over the convective flux q_c/A;

$$(q/A)_{total} = (q/A)_b + (q/A)_c$$

$$q_{total} = (q_b + q_c) \tag{5.8}$$

Here $$q_c = A\,h\,(t_w - t_{liquid})$$

The value of *h* is calculated by equations given in Chapter 4.

(ii) McAdams suggested the following relation for low pressure boiling water :

$$(q/A) = 2.253\,(\Delta T)^{3.96}\ \text{W/m}^2$$

$$(\text{for } 0.2 < p < 0.7\ \text{MN/m}^2) \tag{5.9}$$

For higher pressure we may use ;

$$(q/A) = 283.2\,(\Delta T)^3\,p^{4/3}\ \text{W/m}^2$$

$$(\text{for } 0.7 < p < 14\ \text{MN/m}^2) \tag{5.10}$$

In both relations ΔT is in °C and *p* in Meganewton per square meter.

(d) Maximum Heat Flux in Nucleate Boiling

The maximum heat flux represented by point A (Fig. 5.2) in nucleate boiling is of great interest to us as the surface temperatures in film boiling, near this (q/A) value, are higher than those in nucleate boiling. The condition of *maximum* or *critical heat flux* visualizes that bubbles form almost an insulating blanket over the surface.

(i) Rohsenhow and Griffith have correlated pool boiling data at peak heat flux by the following equation ;

$$\frac{(q/A)_{max}}{L\rho_v} = K g^{1/4} \left(\frac{\rho_l}{\rho_v} - 1\right)^{0.6} \tag{5.11}$$

(where K is a certain constant)

Later, Zuber developed an expression for the peak heat flux on the assumption that at this condition the vapor coming out of the vapor film is in the form of pulsating jet or vortex sheets. The relative velocity at the interface produces an instability causing the jet to break up into a sphere. The expression proposed by him is as given below ;

$$\frac{(q/A)_{max}}{L\rho_v} = \frac{\pi}{24} \left\{\frac{\sigma(\rho_l - \rho_v) g g_c}{\rho_v^2}\right\}^{1/4} \left(\frac{\rho_l}{\rho_l + \rho_v}\right)^{1/2} \tag{5.12}$$

(ii) Another expression recommended for the peak heat flux is ;

$$\left(\frac{q}{A}\right)_{max} = \frac{\pi}{24} \rho_v^{1/2} L \left[\sigma g (\rho_l - \rho_v) g_c\right]^{1/4} \tag{5.12a}$$

(e) Heat Transfer in Film Boiling

(i) *In film boiling* the vapor is generated due to heat conduction and radiation from the heating surface through the vapor film. Zuber developed the following equation representing minimum heat flux during film boiling ;

$$(q/A)_{min} = C \rho_v L \left\{\frac{g(\rho_l - \rho_v)}{\rho_l + \rho_v}\right\}^{1/2} \left\{\frac{g_c \sigma}{g(\rho_l - \rho_v)}\right\}^{1/4} \tag{5.13}$$

(where C is constant)

(ii) For the *stable film* regime, Bromley arrived at the following relationship,

$$h_f = C' \left\{\frac{k_v^3 \rho_v (\rho_l - \rho_v) g \lambda'}{D_o \mu_v \Delta t}\right\}^{1/4} \tag{5.14}$$

where C' = An empirical constant (0.62)

k_v = Thermal conductivity of saturated vapor

D_o = Diameter of tube (outside)

$$\lambda' = L\left(1 + \frac{0.68 \Delta t C_v}{L}\right)$$

$\Delta t = (t_w - t_{sat})$

C_v = Specific heat of saturated vapor.

L = Latent heat of vaporisation

The properties of the vapor are to be evaluated at $(t_w + t_{sat})/2$.

(iii) In addition radiation contributes to the heat transfer and increases the vapor film thickness, reducing the effective contribution of conduction. The total heat transfer coefficient is given by ;

$$h = h_f \left(\frac{h_f}{h}\right)^{1/3} + h_r$$

Where h_r, the radiative conductance may be evaluated by (Eq. 1.11). If we assume emissivity of the liquid to be unity and that of the surface as ε then,

$$h_r = \sigma\varepsilon\,(T_w^4 - T_{sat}^4)/(T_w - T_{sat})$$

where σ = Stefan-Boltzmann constant.

(iv) Some simple relations for estimating the boiling on the outside of submerged surfaces at atmospheric pressure are given below.

Surface	$\frac{q}{A}$, kW/m²	h, W/m²K	Eq. no.
Horizontal	$\frac{q}{A} < 16$	$1042(\Delta T)^{1/3}$	(5.15)
	$16 < \frac{q}{A} < 240$	$5.56(\Delta T)^3$	(5.16)
Vertical	$\frac{q}{A} < 3$	$537(\Delta T)^{1/7}$	(5.17)
	$3 < \frac{q}{A} < 63$ $(\Delta T = T_w - T_{sat}, °C)$	$7.96(\Delta T)^3$	(5.18)

The above values of *h* may be modified to take into account the effect of pressure by the relation ;

$$\left(\frac{h_p}{h_{p1}}\right) = \left(\frac{p}{p_1}\right)^{0.4} \tag{5.19}$$

where *p* is the system pressure and p_1 denotes standard atmospheric pressure.

(v) For forced convection local boiling inside vertical tubes we may use

$$h = 2.54\,(\Delta T)^3\, e^{p/1.551}\ \text{W/m}^2\text{K} \tag{5.20}$$

where ΔT = temperature difference between surface and saturated liquid, °C.

p = pressure, MN/m²

The above equation is valid for p = 5 to 170 atm.

Example 5.1

Calculate the heat transfer coefficient for water boiling at 1 atm pressure in a pan with a flat copper bottom which is maintained at 115°C.

Solution :

For water at 100°C,

$$\rho_l = 958.4 \text{ kg/m}^3, \ \rho_v = 0.598 \text{ kg/m}^3$$

$$C_l = 4211 \text{ J/kg K}, \mu_l = 2.78 \times 10^{-4} \text{ kg/m s}$$

$$P_{rl} = 1.75, L = 2257 \text{ J/g} = 2257 \times 10^3 \text{ J/kg}$$

and $\sigma = 58.9 \times 10^{-3}$ N/m, $C_{sf} = 0.0128$ (for polished copper-water)

From Eq. (5.7) with s = 1,

$$\frac{C_l(t_w - t_{sat})}{Pr_l L} = C_{sf}\left[\frac{q/A}{\mu_l L}\sqrt{\frac{g_c\sigma}{g(\rho_l-\rho_v)}}\right]^{\frac{1}{3}}$$

so

$$\frac{q}{A} = \left[\frac{C_l(t_w-t_{sat})}{Pr_l LC_{sf}}\right]^3 \frac{\mu_l L}{\sqrt{\frac{g_c\sigma}{g(\rho_l-\rho_v)}}}$$

$$= \left[\frac{(4211)(15)}{(1.75)(2257)(1000)(0.0128)}\right]^3 \times$$

$$(2.78 \times 10^{-4})(2257 \times 10^3) \div \sqrt{\frac{1\times 58.9\times 10^{-3}}{(9.8)(958.4-0.598)}}$$

$$= 4.92 \times 10^5 \text{ J/m}^2\text{s}$$

$$h = \frac{q/A}{t_w - t_{sat}} = \frac{4.92\times 10^5}{15}$$

$$= 3.28 \times 10^4 \text{ W/m}^2\text{K}$$ Ans.

Example 5.2

Estimate the maximum heat flux obtainable with nucleate pool boiling on a clean brass surface for water at 10 atm.

Solution :

(a) At 10 atm we have for water,

$$\sigma = 42.2 \times 10^{-3} \text{ N/m}, L = 2013 \text{ kJ/kg}$$

and $\rho_v = 5.22$ kg/m³, $\rho_l = 886.1$ kg/m³

Now,

$$(q/A)_{max} = \frac{\pi}{24}\rho_v^{1/2} L\,[\sigma g(\rho_l - \rho_v)\, g_c]^{1/4} \quad (5.12a)$$

or $\left(\dfrac{q}{A}\right)_{max} = \left(\dfrac{\pi}{24}\right)(5.22)^{1/2}(2013) \times [(42.2 \times 10^{-3})(9.8)(886.1 - 5.22)(1)]^{1/4}$

$= 2.63 \times 10^6 \text{ W/m}^2 = 2630 \text{ kW/m}^2$ **Ans.**

(b) If we use Eq. (5.12), the above result should be multiplied by the factor,

$$\left(\frac{\rho_l}{\rho_l + \rho_v}\right)^{1/2} = \left(\frac{886.1}{886.1 + 5.22}\right)^{1/2} = 0.997$$

Then,

$$\left(\frac{q}{A}\right)_{max} = 2.63 \times 10^6 \times 0.997 = 2620 \text{ k W/m}^2$$ **Ans.**

Example 5.3

What is the maximum heat flux in nucleate boiling with saturated water at 2 atm pressure if the gravitational field is equivalent to 1/10 that of the earth.

Solution :

At 2 atm pressure,

$$t_{sat} = 120.5\ °C\ ;\ \rho_l = 943.5 \text{ kg/m}^3,$$

$$\rho_v = 1.13 \text{ kg/m}^3\ ;\ L = 2201 \text{ kJ/kg}$$

$$\sigma = 54.7 \times 10^{-3} \text{ N/m.}$$

and $g = 9.8/10 = 0.98 \text{ m/s}^2.$

Using Eq. (5.12a),

$$\left(\frac{q}{A}\right)_{max} = \frac{\pi}{24}\rho_v^{1/2} L\,[\sigma g(\rho_l - \rho_v) g_c]^{1/4}$$

$$= \frac{\pi}{24}(1.13)^{1/2}(2201) \times [(54.7 \times 10^{-3})(0.98)(942.37)(1)]^{1/4}$$

$$= 816 \text{ kW/m}^2$$ **Ans.**

Example 5.4

Water at 5 atm flows inside a 2.5 cm diameter tube under forced convection local boiling conditions. The tube wall temperature is 20°C above the saturation temperature. What is the heat transfer in 1.0 m length of tube.

Solution :

$$\Delta T = 20°C$$

$$p = 5 \times 1.0132 \times 10^5 \text{ N/m}^2$$

$$= 0.507 \text{ MN/m}^2$$

From Eq. (5.20)

$$h = (2.54)\,(20)^3\, e^{0.507/1.551}$$
$$= 28170 \text{ W/m}^2\text{K}$$
$$q = hA\,(\Delta T)$$
$$= (28{,}170)\,(\pi \times 0.025 \times 1.0)\,(20)$$
$$= 44{,}267 \text{ W/m}$$

Ans.

Example 5.5

A horizontal copper tube of 0.5 cm OD is placed in a pool of water at 100°C. The outside surface temperature of the tube is 232°C. Calculate the heat transfer coefficient if film boiling occurs.

Solution :

From Eq. (5.14),

$$h_f = 0.62\left[\frac{k_v^{\,3}\,\rho_v\,(\rho_l-\rho_v)\,g\lambda'}{D_o\,\mu_v\,\Delta t}\right]^{1/4}$$

Now, $\rho_l = 958.4 \text{ kg/m}^3,\ \rho_v = 0.6 \text{ kg/m}^3,\ \mu_v = 12.1 \times 10^6 \text{ Ns/m}^2$

$k_v = 0.0249 \text{ W/mK},\ C_v = 2034 \text{ J/kg K}$

and $L = 2257 \text{ kJ/kg},\ \lambda' = 2.257 \times 10^6 + 0.68\,(2034)\,(132)$

$= 2.44 \times 10^6 \text{ J/kg}$

So,
$$h_f = 0.62\left[\frac{(0.0249)^3\,(0.6)(958.4-0.6)(9.8)(2.44\times 10^6)}{(0.005)(12.1\times 10^6)(132)}\right]^{1/4}$$

$$= 250 \text{ W/m}^2\text{ K}$$

Ans.

5.6 PHENOMENA OF CONDENSATION

The *condensation* of vapor is the reverse phenomenon of the *evaporation* of liquid. Heat energy is released during the former while it is absorbed during the later process. When a saturated vapor is brought in contact with a surface at a lower temperature, heat is received by the surface from the vapor and *condensation* occurs. The heat exchange is equivalent to the *latent heat*. There are two distinct ways by which condensation may appear to be occuring :

(a) Film wise condensation.

(b) Drop wise condensation.

Normally, the condensed liquid spreads out over the whole surface and a continuous *film* is formed. The vapor condenses at the liquid interface because of the heat transferred through the liquid film. The condensate liquid flows downwards under the influence of gravity. At low vapor velocity the liquid film may be very thick and the heat transfer through the film may be primarily due to conduction. The rate of heat flow to the wall thus depends upon the thickness of the film and therefore on the rate of condensation and that of removal of condensate from the surface. Since condensate film plays an important role during the whole phenomena, the same is called *film condensation.* This is the type which is usually found in engineering practice.

If, on the other hand, liquid does not wet the surface, but the *droplets* form, and run down the surface coalescing as they travel, the phenomena is called *drop wise condensation.* The mechanism of this type is not well understood nor there is enough knowledge about the prediction of associated heat transfer. However, it is noticed that the rate of heat transfer in dropwise condensation is about 15 to 20 times that for film condensation under same temperature difference. Hence dropwise condensation is very desirable but the conditions under which it occurs are difficult to achieve. It usually occurs on those surfaces which are contaminated by substances which prevent the condensate from wetting the whole surface.

We shall deal with the elementary analysis of film condensation only.

5.7 FILM CONDENSATION ON A VERTICAL SURFACE

Nusselt, in 1916, obtained theoretical relations for calculating *heat transfer coefficient* for *film condensation* for pure vapors over plane vertical surface. This was the first attempt to analyse the phenomena theoretically, and has led to further research work in the field. He made the following assumptions.

(i) The vapor is pure and at temperature t_{sat}.

(ii) The surface temperature t_w is uniform.

(iii) The temperature distribution in the liquid is linear.

(iv) The velocity profile through the film is parabolic and is given by ;

$$V_x = V_\infty \left(\frac{2y}{\delta} - \frac{y^2}{\delta^2} \right)$$

(v) No shear stress exists at the liquid vapor interface.

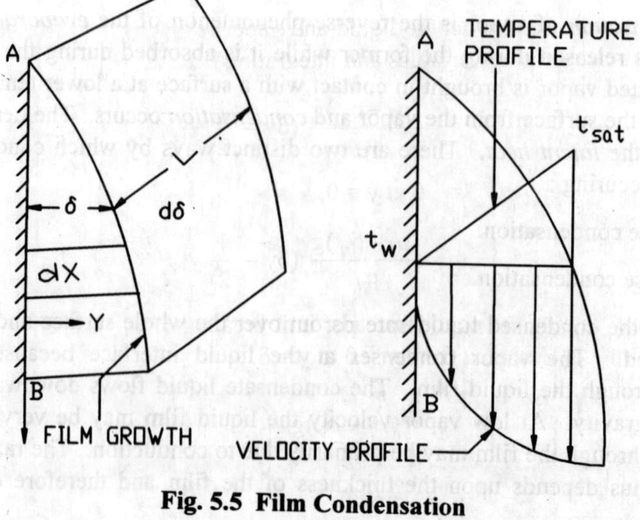

Fig. 5.5 Film Condensation

Let AB represent the cool vertical surface, in Fig. 5.5. The film thickness is zero at the upper edge A, and increases with distance down the surface. Let thickness be δ at a distance x from the top.

Consider an element of length dx and unit width in the horizontal direction. The heat released during condensation flows only by conduction through the film. The amount of heat flowing through the element is given by ;

$$\frac{q}{A} = \frac{k(T_{sat} - T_w)}{\delta} \tag{5.21}$$

Since $$A = 1 \times dx,$$

$$q = k\,dx\,(T_{sat} - T_w)/\delta$$

Also, $$q = h_x\,dx\,(T_{sat} - T_w)$$

hence the local heat transfer coefficient h_x is given by ;

$$h_x = \frac{k}{\delta} \tag{5.22}$$

(a) Determination of h_x During Film Condensation

Considering the forces acting on the element, it is assumed that pressure and acceleration forces are negligible, and the flow of liquid is laminar and is caused by gravity alone.

The downward force acting on the liquid at a distance greater than y from the surface is $(\delta - y)\,dx\,\rho_l\,g/g_c$. Assuming that the vapor outside the condensate layer is in hydrostatic balance ($dp/dx = \rho_v\,g/g_c$), an upward force equal to $(\delta - y)\,dx\,\rho_v\,g/g_c$ will be present due to pressure difference between the upper and lower faces of the element. The remaining force will be the viscous shear $(\mu_l\,dV/dy)\,dx$ at the vertical plane y. Hence, under steady state conditions,

$$(\delta - y)(\rho_l - \rho_v)g = \mu_l \frac{dV}{dy} \tag{5.23}$$

Here the subscripts l and v denote liquid and vapor respectively. The velocity V at y can be obtained by separating variables and integrating Eq. (5.23) :

$$V_y - V_{y=0} = \frac{(\rho_l - \rho_v)g}{\mu_l}\left(\delta y - \frac{1}{2}y^2\right)$$

Since $$V = 0 \text{ at } y = 0,$$

$$V = \frac{(\rho_l - \rho_v)g}{\mu_l}\left(\delta y - \frac{1}{2}y^2\right) \tag{5.24}$$

The mass flow rate of the condensate per unit breadth, Γ is found by integrating the local mass flow rate $\rho_l V dy$ at x, from $y = 0$ to $y = \delta$.

$$\Gamma = \int_0^{\delta} \frac{(\rho_l - \rho_v)g}{\mu_l}\left(\delta y - \frac{y^2}{2}\right) dy$$

or $$\Gamma = \frac{\rho_l(\rho_l - \rho_v)\delta^3}{3\mu_l} g \tag{5.25}$$

Since the film thickness is gradually increasing the condensation of the liquid simultaneously increases.

The change in *condensate flow* rate with the thickness of the condensate layer is ;

$$\frac{d\Gamma}{d\delta} = \frac{g\rho_l(\rho_l - \rho_v)}{\mu_l}\delta^2$$

or $$d\Gamma = \frac{g\rho_l(\rho_l - \rho_v)}{\mu_l}\delta^2 \, d\delta \tag{5.26}$$

Heat transfer through the condensate layer is solely by conduction. Assuming a linear temperature gradient, the average enthalpy change of the vapor in condensing and subcooling upto the average liquid temperature of the *condensate film* is ;

$$\Delta h = h_{fg} + \frac{1}{\Gamma}\int_0^\delta (\rho_l V)\, c_{pl}(T_{sat} - T)dy$$

Since, $$\frac{T_{sat} - T}{T_{sat} - T_w} = \frac{\delta - y}{\delta} = \left(1 - \frac{y}{\delta}\right)$$

the expression for Δh is given by ;

$$\Delta h = h_{fg} + \frac{3\mu_l}{\rho_l(\rho_l - \rho_v)\delta^3 g}\int_0^\delta \frac{\rho_l(\rho_l - \rho_v)g}{\mu_l}\left(\delta y - \frac{1}{2}y^2\right) \times c_{pl}(T_{sat} - T_w)\left(1 - \frac{y}{\delta}\right)dy$$

or $$\Delta h = h_{fg}\frac{3}{\delta^3}c_{pl}(T_{sat} - T_w)\int_0^\delta \left(\delta y - \frac{3}{2}y^2 + \frac{1}{2}\frac{y^3}{\delta}\right)dy$$

$$= h_{fg} + \frac{3}{\delta^3}c_{pl}(T_{sat} - T_w)\frac{\delta^3}{8}$$

$$= h_{fg} + \frac{3}{8}c_{pl}(T_{sat} - T_w)$$

Since the rate of heat transfer to the wall is ;

$$\frac{q}{A} = \frac{k(T_{sat} - T_w)}{\delta}$$

we get, for steady state conditions, $q/A = \Delta h \frac{d\Gamma}{dx}$

hence

$$\frac{q}{A} = \frac{k(T_{sat} - T_w)}{\delta} = \left\{h_{fg} + \frac{3}{8}c_{pl}(T_{sat} - T_w)\right\}\frac{d\Gamma}{dx} \tag{5.27}$$

Equating the expressions for $d\Gamma$ from Eqs. (5.26) and (5.27) we get ;

$$\delta^3\, d\delta = \frac{k\mu_l (T_{sat} - T_w)}{g\rho_l(\rho_l - \rho_v) h'_{fg}}\, dx \tag{5.28}$$

where $$h'_{fg} = h_{fg} + \frac{3}{8} c_{pl}(T_{sat} - T_w) \tag{5.29}$$

Integrating Eq. (5.28) between the limits $\delta = 0$ at $x = 0$ and $\delta = \delta$ at $x = x$, the film thickness at x is given by ;

$$\delta = \left(\frac{4\mu_l kx(T_{sat} - T_w)}{g\rho_l(\rho_l - \rho_v) h'_{fg}}\right)^{\frac{1}{4}} \tag{5.30}$$

Since $h_x = k/\delta$,

so $$h_x = \left(\frac{\rho_l(\rho_l - \rho_v) g h'_{fg} k^3}{4\mu_l x(T_{sat} - T_w)}\right)^{\frac{1}{4}} \tag{5.31}$$

The local Nusselt number at x is therefore given by ;

$$Nu_x = \frac{h_x x}{k} = \left(\frac{\rho_l(\rho_l - \rho_v) g h'_{fg} x^3}{4\mu_l k(T_{sat} - T_w)}\right)^{\frac{1}{4}} \tag{5.32}$$

(b) Average Value of h During Film Condensation

Note that h_x decreases and the film thickness increases with x. The average value of h for a plate of height L is obtained by integrating h_x over the plate and dividing by the area. If we take unit width,

$$\bar{h} = \frac{1}{L}\int_0^L h_x\, dx = \frac{4}{3} h_l \tag{5.33}$$

or $$\bar{h} = 0.943 \left(\frac{\rho_l(\rho_l - \rho_v) g h'_{fg} k^3}{\mu_l L(T_{sat} - T_w)}\right)^{\frac{1}{4}} \tag{5.34}$$

For a surface inclined at an angle ϕ with the horizontal, the average coefficient is ;

$$\bar{h} = 0.943 \left(\frac{\rho_l(\rho_l - \rho_v) g h'_{fg} k^3 \sin\phi}{\mu_l L(T_{sat} - T_w)}\right)^{\frac{1}{4}} \tag{5.35}$$

An analysis by Rohsenow which is in better agreement with experimental results for $Pr > 0.5$ and $c_{pl}(T_{sat} - T_w)/h'_{fg} < 1.0$ shows that h'_{fg} may be replaced by $[h_{fg} + 0.68\, c_{pl}(T_{sat} - T_w)]$.

(c) Effect of Shear Stress at Interface

Chen took into account the shear and momentum effects at the interface and recommends a correction factor given by ;

$$\frac{h'}{h} = \left(\frac{1+0.68\,A+0.02\,AB}{1+0.85\,B-0.15\,AB}\right)^{0.25} \quad (5.36)$$

(for $1 < Pr_l < 0.005$)

where, h' = corrected heat transfer coefficient

h = heat transfer coefficient from Eq. (5.34)

$$A = \frac{c_l\,(T_{sat} - T_w)}{h_{fg}} < 2$$

$$B = \frac{k_l\,(T_{sat} - T_w)}{\mu_l\, h_{fg}} < 20$$

Example 5.5

A vertical tube 40 mm dia and 1 m long is used for condensing dry steam at atmospheric pressure. The tube surface temperature t_w = 60 C.

(a) Determine the mass of condensate.

(b) If the tube is held in horizontal position, will there be any change in mass of condensate ? If yes, find the value.

Solution :

(a) t_{sat} = 100 C ; t_w = 60 C

hence the effective temperature of film t_f is given by :

$$t_f = t_w + 0.25\,(t_{sat} - t_w)$$

$$= 60 + 0.25\,(100 - 60) = 70°C$$

The property values of saturated water at 70 C are :

k = 0.675 W/mK

= 2.43 kJ/hr m K

ρ_l = 971.8 kg/m^3

h_{fg} = 2500 kJ/kg

μ_l = 36.2×10^{-6} kgf sec/m^2

= 355×10^{-6} kg/m sec

The tube may be treated as a plate of area π DL, and Eq. (5.34) may be applied with $h'_{fg} = h_{fg}$ and neglecting ρ_v in comparison to ρ_l

$$h = 0.943 \times 1.20 \left\{\frac{(971.8^2)(9.806 \times 3600^2)(2500)(2.43)^3}{(355 \times 10^{-6} \times 3600)(1)(100-60)}\right\}^{\frac{1}{4}}$$

{multiplied by 1.20 due to McAdams, sec Eq. (5.46)}

$$= 20{,}000 \text{ kJ/m}^2 \text{ hr K}$$

Hence the quantity of steam condensed ;

$$= \frac{h(\pi DL)(t_{sat} - t_w)}{h_{fg}}$$

$$= \frac{(20{,}000)(\pi \times 0.04 \times 1)(100-60)}{2500}$$

$$= 43.5 \text{ kg/hr}$$

(b) Since the path of condensate flow is shorter and the film is thinner, *h* for tubes placed in horizontal position will be much higher. Hence the mass of steam condensed will increase.

In this case Eq. (5.37) on page 347 will be applicable :

$$h\_ = 0.725 \left\{ \frac{971.8^2 (9.806 \times 3600^2)(2500)(2.43)^3}{(355 \times 10^{-6} \times 3600)(0.04)(100-60)} \right\}^{\frac{1}{4}}$$

$$= 27{,}000 \text{ kJ/m}^2 \text{ hr K}$$

Hence condensate mass ;

$$= \frac{(27000) \times (\pi \times 0.04 \times 1)(40)}{2500}$$

$$= 58.4 \text{ kg/hr}$$ Ans.

Example 5.6

Saturated steam at 34×10^3 N/m^2 (0.34 bar) condenses on a 1 m long vertical plate whose surface temperature is 60°C. Calculate the value of h_x at 1/3 m from the top. Also find the average value of the heat transfer coefficient and the height for which the condensate film is laminar.

Solution :

At a pressure of 0.34 bar the data is as below :

$$T_{sat} = 71.8°\text{C} \qquad c_{pl} = 4.188 \text{ kJ/kg K}$$

$$\rho_v = 0.21 \text{ kg/m}^3 \qquad \rho_l = 976.62 \text{ kg/m}^3$$

$$k = 0.668 \text{ W/mK} \qquad \mu_l = 3.998 \times 10^{-4} \text{ Ns/m}^2$$

$$h_{fg} = 2329 \text{ kJ/kg}$$

$$h'_{fg} = 2329 + \frac{3}{8}(4.188)(71.8 - 60) = 2347 \text{ kJ/kg}$$

Note that the physical properties have been evaluated at T_{sat} for convenience.
From Eq. (5.34)

$$h = 0.943 \left(\frac{\rho_l (\rho_l - \rho_v) g h'_{fg} k^3}{\mu_l L (T_{sat} - T_w)} \right)^{\frac{1}{4}}$$

$$= 0.943\left(\frac{(976.62)(976.41)(9.8)(2347\times10^3)(0.668)^3}{(3.998\times10^{-4})(11.8)}\right)^{1/4}$$

$$= 5760 \text{ W/m}^2\text{K}$$ Ans.

From Eq. (5.31) $$h_x = \left(\frac{\rho_l(\rho_l-\rho_v)gh'_{fg}k^3}{4\mu_l x(T_{sat}-T_w)}\right)^{\frac{1}{4}}$$

$$= \left(\frac{(976.62)(976.41)(9.8)(2347\times10^3)(0.668)^3}{(4)(3.998\times10^{-4})(1/3)(11.8)}\right)^{1/4}$$

$$= 5677.9 \text{ W/m}^2\text{ K}$$ Ans.

From Eqs. (5.42) and (5.25) the critical Re will be ;

$$Re_f = \frac{4\Gamma}{\mu_f} = \frac{4\rho_l(\rho_l-\rho_v)\delta^3 g}{3\mu_l^2} = 2000$$

so $$\delta^3 = \frac{6000\,\mu_l^2}{4\rho_l(\rho_l-\rho_v)g}$$

$$= \frac{(6000)(3.998\times10^{-4})^2}{(4)(976.62)(976.41)(9.8)} = 2.565\times10^{-11}$$

or $$\delta = 0.000295 \text{ m}$$

Now, $$\delta = \left(\frac{4\mu_l k x(T_{sat}-T_w)}{g\rho_l(\rho_l-\rho_v)h'_{fg}}\right)^{\frac{1}{4}} \quad (5.30)$$

hence the height x for which film is laminar is ;

$$x = \frac{g\rho_l(\rho_l-\rho_v)h'_{fg}\delta^4}{4\mu_l k(T_{sat}-T_w)}$$

$$= \frac{(9.8)(976.62)(976.41)(2347\times10^3)(2.95\times10^{-4})^4}{(4)(3.998\times10^{-4})(0.668)(11.8)}$$

$$= 13.2 \text{ m}$$ Ans.

Example 5.7

Steam at 1 atm pressure condenses on a 50 cm by 50 cm vertical plate. The plate temperature is 96°C. Find the heat transfer coefficient and the mass of steam condensed per hour.

Solution :

The effective temperature of the film is $t_w + 0.25\,(t_{sat} - t_w) = 96 + 0.25\,(100 - 96) = 97°C$.

At 97°C, ρ_l = 960 kg/m^3; k = 0.68 W/m K; μ_l = 2.82 × 10^{-4} kg/ms, and at 100°C, h_{fg} = 2255 kJ/kg. For simplicity let $h'_{fg} \approx h_{fg}$ and $\rho_l (\rho_l - \rho_v) \approx \rho_l^2$.

We have first to find Re_f which depends on the mass flow of condensate. This, in turn, depends on h. Let us assume laminar flow. From Eq. (5.34), using the simplifications given above we have,

$$h = 0.943 \left[\frac{\rho_l^2 \, h_{fg} \, g \, k^3}{L \, \mu_l \, (T_{sat} - T_w)} \right]^{\frac{1}{4}}$$

$$= 0.943 \left[\frac{(960)^2 \, (2255 \times 10^3)(9.8)(0.68)^3}{(0.5)(2.82 \times 10^{-4})(100 - 96)} \right]^{\frac{1}{4}}$$

$$= 9740 \text{ W/m}^2\text{K}$$

Hence, from Eq. (5.45), Reynolds number is,

$$Re_f = \frac{4 h L (T_{sat} - T_w)}{h_{fg} \, \mu_l}$$

$$= \frac{(4)(9740)(0.5)(100 - 96)}{(2255 \times 10^3)(2.82 \times 10^{-4})} = 122.5 < 2000$$

The flow is therefore liminar.

Now, $q = h A (T_{sat} - T_w)$

$$= (9740)(0.5)^2 (100 - 96)$$

$$= 9740 \text{ W.}$$

Mass flow of condensate is,

$$m = \frac{q}{h_{fg}} = \frac{9740}{2255 \times 10^3} = 4.32 \times 10^{-3} \text{ kg/s}$$

$$= 15.55 \text{ kg/hr}$$

Ans.

5.8 FILM CONDENSATION ON TUBES

(i) The above analysis may also be applied to the inside and outside surfaces of vertical tubes if the tube diameter is large compared to film thickness. However, the results are not applicable to inclined tubes.

The average heat transfer coefficient of a pure saturated vapor condensing on the outside of a horizontal tube is evaluated by ;

$$h = 0.725 \left[\frac{\rho_l (\rho_l - \rho_v) \, g \, h'_{fg} \, k^3}{D \, \mu_l \, (T_{sat} - T_w)} \right]^{\frac{1}{4}} \tag{5.37}$$

If condensation occurs on a horizontal tube bank with n tubes placed directly over one another in the vertical direction, the heat transfer coefficient may be calculated after replacing D by nD in Eq. (5.37).

(ii) Chen suggested that additional condensation occurs on the liquid layer between the tubes because the liquid film is sub - cooled. Chen's analysis gives ;

$$h = 0.728 \left\{1+\frac{0.2\,C_{pl}\,(T_{sat}-T_w)}{h_{fg}}(n-1)\right\} \times \left[\frac{g\,\rho_l\,(\rho_l-\rho_v)\,k^3\,h'_{fg}}{nD\,\mu_l\,(T_{sat}-T_w)}\right]^{\frac{1}{4}} \tag{5.38}$$

This equation is in good agreement with experimental results if

$$[(n-1)\,C_{pl}\,(T_{sat}-T_w)/h_{fg}] < 2$$

In Eq. (5.38) h will be in W/m^{2}K if the other quantities are in the following units :

T_{sat} = temperature of saturated vapor, °C

T_w = wall surface temperature, °C

L = height of plane surface, m

D = tube diameter, m

μ_l = viscosity of liquid, kg/m s (or Ns/m^2 since $g_c = 1$)

h_{fg} = latent heat of condensation, J/kg

h_{fg}' = $h_{fg} + (3/8)\,C_{pl}\,(T_{sat} - T_w)$, J/kg

g = acceleration due to gravity, m/s^2

ρ_v = density of vapor, kg/m^3

ρ_l = density of liquid, kg/m^3

k = thermal conductivity of liquid, W/mK

C_{pl} = specific heat of liquid, J/kg °C

The physical properties of the liquid film should be preferably computed at an effective film temperature T_f given by ;

$$T_f = T_w + 0.25\,(T_{sat} - T_w) \tag{5.39}$$

Example 5.9

Steam at 0.4 bar and 350 K is condensing on 0.013 m OD tube having an average wall temperature of 325 K. Calculate the unit surface conductance for this tube in (a) horizontal position (b) vertical position.

Solution :

The property values of steam at 337°K are ;

ρ_v = 0.25 m^3/kg μ_l = 4.48×10^{-4} Ns/m^2

k = 0.661 W/mK ρ_l = 980.9 kg/m^3

h_{fg} = 2.35×10^3 kJ/kg C_{pl} = 4.184 kJ/kg °C so $$h_{fg}' = 2350 + (3/8)(4.18)(25) = 2356 \text{ J/kg}$$

(a) For tube in horizontal position Eq. (5.37) is applicable, hence

$$h = 0.725 \left[\frac{\rho_l (\rho_l - \rho_v) g h'_{fg} k^3}{D \mu_l (T_{sat} - T_w)} \right]^{1/4}$$

$$= 0.725 \left[\frac{980.9 (980.9 - 0.25)(9.81)(2356 \times 1000)(0.661)^3}{(0.013)(4.48 \times 10^{-4})(350 - 325)} \right]^{1/4}$$

$$= 10{,}680 \text{ W/m}^2\text{K}$$ Ans.

(b) The tube may be treated as a vertical plate of area π DL. The average value of unit surface conductance will be given by Eq. (5.34) as h = 4239 W/m²K Ans.

5.9 CONDENSATE - FILM REYNOLDS NUMBER

We may define the Reynolds number for a condensate film as

$$R_{ef} = \frac{D_h \rho_l V}{\mu_f} = \frac{4 A \rho_f V}{P \mu_f} \tag{5.40}$$

where the subscript f denotes the liquid film and

D_h = hydraulic diameter, m

A = flow area, m²

P = shear or wetted perimeter, m

V = average flow velocity, m/s

Since the mass flow rate m at a particular section of the condensate film is ;

$$m = \rho_l AV$$

so $$R_{ef} = \frac{4m}{P \mu_f} \tag{5.41}$$

For a vertical plate of unit width P = 1 ; for a vertical tube P = πD. The Reynolds number is sometimes expressed in terms of the mass flow per unit width of the plate Γ. So that

$$R_{ef} = \frac{4\Gamma}{\mu_f} \tag{5.42}$$

In calculating the Reynolds number the mass flow may be related to the total heat transfer and the heat transfer coefficient by ;

$$q = hA (T_{sat} - T_w) = m h'_{fg} \tag{5.43}$$

where A is the total surface area for heat transfer.

Thus $$m = \frac{q}{h'_{fg}} = \frac{hA(T_{sat} - T_w)}{h'_{fg}}$$

$$R_{ef} = \frac{4hA(T_{sat} - T_w)}{h'_{fg} P \mu_f} \tag{5.44}$$

Since A = LW and P = W, where L and W are the height and width of the plate, respectively, we get,

$$R_{ef} = \frac{4hL(T_{sat} - T_w)}{h'_{fg} \mu_f} \tag{5.45}$$

The laminar condensation equations given above agree well with experimental data when the film remains smooth. In practice, ripples develop in the film for R_{ef} as low as 30 or 40. When this occurs, the actual values of *h* may be 20 percent higher than those predicted by Eq. (5.35). McAdams has suggested that the right hand side of Eq. (5.35) should be multiplied by 1.2 for design purposes. The resulting equation for vertical plates will then be ;

$$h = 1.13 \left[\frac{\rho_l (\rho_l - \rho_v) g h'_{fg} k^3}{L \mu_l (T_{sat} - T_w)} \right]^{1/4} \tag{5.46}$$

If the condensing vapor is superheated, the preceding equations may be used to calculate *h*, provided that the heat flow is calculated on the basis of the temperature difference between the surface and the saturation temperature corresponding to the system pressure.

5.10 CONDENSATION NUMBER

It is often convenient to express the heat transfer coefficient directly in terms of R_{ef}. Including the effect of inclination we may write the equation for the heat transfer coefficient as ;

$$h = C \left[\frac{\rho_l (\rho_l - \rho_v) k^3 g \sin\phi \, h'_{fg}}{\mu_l L (T_{sat} - T_w)} \right]^{1/4} \tag{5.47}$$

where C is a constant and is substituted for a plate or cylinder. From Eq. (5.43) we get ;

$$T_{sat} - T_w = \frac{m h'_{fg}}{hA} \tag{5.48}$$

Substituting Eq. (5.48) in Eq. (5.47) and solving for *h* we get;

$$h^{3/4} = C \left[\frac{\rho_l (\rho_l - \rho_v) g \sin\phi \, k^3 \, A / L}{\mu_l m} \right]^{1/4}$$

$$= C \left[\frac{\rho_l (\rho_l - \rho_v) g k^3}{\mu_l^2} \times \frac{\mu_l P}{4m} \times \frac{4 \sin\phi \, A / P}{L} \right]^{1/4}$$

Hence,

$$h = C^{4/3} \left[\frac{\rho_l (\rho_l - \rho_v) g k^3}{\mu_l^2} \times \frac{\mu_l P}{4m} \times \frac{4 \sin\phi \, A / P}{L} \right]^{1/3} \tag{5.49}$$

Let us now define a new dimensionless group, the *Condensation number*, C_o as ;

$$C_0 = h \left(\frac{\mu_l^2}{k^3 \rho_l (\rho_l - \rho_v) g} \right)^{1/3} \tag{5.50}$$

then we get, from Eqs. (5.49) and (5.41)

$$C_0 = C^{4/3} \left(\frac{4 \sin\phi\, A / P}{L} \right)^{1/3} R_{ef}^{-1/3} \tag{5.51}$$

(i) For a vertical plate A/PL = 1, by using C = 0.943 we get ;

$$C_0 = 1.47\, R_{ef}^{-1/3} \text{ (for } R_{ef} < 1800) \tag{5.52}$$

(ii) For a horizontal cylinder A/PL = π, and using C = 0.725

$$C_0 = 1.514\, R_{ef}^{-1/3} \text{ (for } R_{ef} < 1800) \tag{5.53}$$

(iii) For $R_{ef} > 1800$, an empirical correlation by Kirkbride may be used :

$$C_0 = 0.0077\, R_{ef}^{0.4} \tag{5.54}$$

5.11 EFFECT OF TURBULENCE AND HIGH VAPOR VELOCITY ON FILM-WISE CONDENSATION

The relationships discussed till now are based on the assumption that the condensate flow is laminar hence they do not apply to *turbulent flow* of condensate. Strictly speaking the turbulent flow of liquid is difficult to achieve on horizontal tubes, but may be easily established over the lower portion of the vertical surfaces. The average heat transfer coefficient becomes larger in such cases.

(a) Colburn suggested that the condensate flow is turbulent for $R_{ef} > 2000$. According to his analysis the *local heat transfer coefficient* for *turbulent flow* of condensate is given by ;

$$h_x = 0.056 \left(\frac{4\Gamma}{\mu_l} \right)^{0.2} \left(\frac{k^3 \rho_l^2 g}{\mu_l^2} \right)^{\frac{1}{3}} (Pr_l)^{1/2} \tag{5.55}$$

(b) Another approximation in the Nusselt's analysis is that the frictional drag at vapor liquid interface is negligible. If the uncondensed vapor flows at a high velocity this assumption may not be valid. Analysis of this problem was carried on by Carpenter and Colburn and they suggested that in the presence of *high vapor velocity* the transition from laminar to turbulent flow occurs at $R_{ef} \approx 300$. They proposed the following correlation for condensation in vertical tubes at high vapor velocities up to 170 m/s.

$$(h/c_{pl}\, G_m)\, Pr_l^{0.5} = 0.046 \sqrt{\rho_l / \rho_v}\; f \tag{5.56}$$

where G_m = mean value of mass velocity of vapor, kg/s

f = pipe friction coefficient evaluated at the average vapor velocity.

The value of G_m may be calculated by :

$$G_m = \sqrt{\frac{G_1^2 + G_1 G_2 + G_2^2}{3}} \tag{5.57}$$

where G_1 = mass velocity at top of tube, kg/s

G_2 = mass velocity at bottom of tube, kg/s

A modification of Eq. (5.56) has been proposed by Soliman :

$$(h/c_{pl}\, G_m)\, Pr_l^{0.35} = 0.036\, \sqrt{\rho_l / \rho_v}\; f \tag{5.58}$$

All physical properties of the liquid in Eqs. (5.55) and (5.58) should be evaluated at a reference temperature of $(0.25\, T_{sat} + 0.75\, T_w)$.

(c) For condensation on the outside of a horizontal tube, the effect of vapor velocity may be taken into account by using the following correlation :

$$h' = \left[\frac{1}{2} h_s^2 + \left(\frac{1}{4} h_s^4 + h^4\right)^{\frac{1}{2}}\right]^{\frac{1}{2}} \tag{5.59}$$

where h' is the corrected heat transfer coefficient, taking into account the effect of vapor shear, h is the uncorrected value given by Eq. (5.37), and h_s the contribution of the vapor shear to the heat transfer coefficient. The value of h_s is calculated by ;

$$\frac{h_s D}{k} = 0.9 \left(\frac{\rho_l V_\infty D}{\mu_l}\right)^{0.5} \quad \text{for } \frac{\rho_l V_\infty D}{\mu_l} < 10^6 \tag{5.60}$$

and

$$\frac{h_s D}{k} = 0.59 \left(\frac{\rho_l V_\infty D}{\mu_l}\right)^{0.5} \quad \text{for } \frac{\rho_l V_\infty D}{\mu_l} > 10^6 \tag{5.61}$$

(V_∞ is the vapor velocity approaching the tube.)

5.12 FILM CONDENSATION INSIDE HORIZONTAL TUBES

The above discussion has referred to condensation on exterior surfaces where the vapor and condensate flows are not restricted by the dimensions of the flow channel. Condensation inside tubes is encountered in condensers for refrigeration and air conditioning systems. The overall flow rate of the vapor strongly influences the heat transfer rate in forced convection condensation systems. Further, liquid accumulation on the walls affects the vapor flow rate.

For condensation of refrigerants at low vapor velocities inside horizontal tubes Chato recommends :

$$h = 0.555 \left[\frac{\rho_l (\rho_l - \rho_v)\, g\, k^3\, h_{fg}'}{\mu D (T_{sat} - T_w)}\right]^{1/4} \tag{5.62}$$

(for $Re_v = D\, G_v / \mu_v < 35000$)

The vapor Reynolds number, Re_v is evaluated at inlet conditions to the tube. For higher vapor flow rates we may use ;

$$\frac{hD}{k} = 0.026\, Pr_l^{1/3}\, Re_m^{0.8} \tag{5.63}$$

where Re_m is the mixture Reynolds number given by

$$Re_m = \frac{D}{\mu_l}\,[G_l + G_v\,(\rho_l/\rho_v)^{1/2}] \tag{5.64}$$

$$(\text{for } Re_v = DG_v/\mu_v > 20{,}000 \text{ and } Re_l = DG_l/\mu_l > 5000)$$

The mass velocities for the vapor and liquid, G_v and G_l, are calculated as if each occupied the entire flow area.

PROBLEMS

5.1 Calculate the average heat flux from a 12 mm diameter tube to water boiling outside at 100 C. The tube is immersed in water and is held at 500 C by electrical heating.

5.2 A copper container with a flat bottom is used to boil water at 100 C. The temperature of the bottom of the container is 115 C. Calculate *h*.
[Ans. 3720 W/m^2K]

5.3 The heat input to a vertical, thin-walled pipe 5 cm in diameter is 750 kcal/hr for 1.5 m of pipe. Water flows slowly through the tube and is evaporated under atmospheric conditions. Estimate *h* for the boiling water.

5.4 Calculate h between a horizontal heating surface and boiling water if the heat flux is 35,000 W/m^2 and the pressure is 15 atm.

5.5 A copper tube 1.8 cm OD is submerged in water at 15 C, and refrigerant at an average temperature of –12 C is pumped through it. The average temperature of the outside surface of tube is – 35°C. Neglecting the tube resistance and assuming the value of *h* at the outer surface to be 82 W/m^2K calculate the initial rate of ice formation on the outside surface of the tube. [Ans. 1.3 gm/hr]

5.6 The following data was obtained for boiling water :

Pressure : 760 mm Hg, ΔT : 9.45 C

q/A : 50,000 W/m^2, β : 47°

Calculate the bubble departure diameter using the following equations :

(a) Fritz equation ;

$$D = 0.0148\,\beta\left[\frac{2g_c\sigma}{g(\rho_L-\rho_v)}\right]^{1/2}$$

(b) Zuber equation ;

$$D\,(\text{ft}) = \left[\frac{6g_c\sigma}{g(\rho_L-\rho_v)}\times\frac{k_L\,\Delta T}{q_w}\right]^{1/3}$$

For water the following values may be used :

$$\rho_L = 960 \text{ kg/m}^3, k_L = 0.685 \text{ W/mK}$$

$$\sigma = 6.0 \times 10^{-3} \text{ kg/m}, \rho_v = \frac{1}{1.674} \text{ kg/m}^3$$

[Ans. 2.44 mm ; 163 mm]

5.7 Find the heat transfer coefficient for steam at 60°C condensing on a vertical 15 cm high plate at 58 C.

5.8 Calculate h for steam condensing at 95 C on a horizontal pipe 10 cm in dia. The pipe surface is at 85 C.

5.9 Freon-12 at 60 C condenses on the outer surface of a vertical tube 90 cm long and 2.5 cm in diameter held at 57 C. Calculate the rate of condensation per hour assuming the following properties for Freon-12

$$\rho = 1280 \text{ kg/m}^3, \mu = 0.9 \text{ kg/m-hr}$$

$$L = 130 \text{ kJ/kg}, k = 0.072 \text{ W/m K}$$

5.10 Calculate the average heat transfer coefficient for film condensation of water at 100 kPa for (a) vertical surface 1.5 m high ; (b) the outside surface of a 1.6 cm OD vertical tube 1.5 m long ; (c) the outside surface of a 1.6 cm OD horizontal tube 1.5 m long and (d) a 10 tube vertical bank of 1.6 cm OD horizontal tubes 1.5 m long. Assume that the surface temperatures in all cases are 10°C below saturation temperature.

5.11 Atmospheric steam condenses on a square array of 100 tubes, 1.27 cm OD. Calculate the mass of steam condensed per unit length of tubes if the tube wall temperature is 98°C.

[Ans. h = 12,540 W/m^{2}K ; 159.7 kg/m-hr]

5.12 A condenser is required to handle 10,000 kg/hr of steam at 0.1 kg/cm^2 pressure. The heat transfer surface is made of horizontal brass tubes 3 cm ID, 1 mm thick, and 3 m long. 400,000 kg/hr of cooling water at 25 C circulates through the tubes with a velocity of 21 m/sec. How many tubes are required in the condenser.

5.13 A horizontal tube 15 mm OD is exposed to saturated steam at 100 C. The outer tube wall is maintained at 99 C, by passing a liquid metal through the tube. Calculate (a) the condensing coefficient (b) mass of steam condensed per meter of tube length per min.

CHAPTER 6

Thermal Radiation

6.1 NATURE OF THERMAL RADIATION

In preceding chapters we have discussed the transport of energy by conduction and by convection. These two modes of energy transport required the existence of a *material medium*. For conduction of heat, there must be temperature inequalities at adjacent points in the material medium. For convection of heat there must be a fluid free to move and transport energy with it. In this chapter we will discuss the third mechanism of energy transport, *radiation*. For radiation no material medium is required and energy can be transported through regions of space devoid of any matter.

Another difference between radiation and conduction should be noted. In the process of conduction, heat flows only from a high temperature body to a low temperature body. If a third body colder than the other two, is interposed between them, then heat flows from both the first two bodies to the third body, which is the coldest of the three. But radiation may travel from a hot body through a cold non-absorbing medium leaving it unaffected or through a partially absorbing medium and then reach a warmer body again. For example, the radiation from the sun, which possesses a very high temperature, travels through the cold upper atmosphere to reach the warmer earth's surface.

The exact nature of radiation is still not fully understood. Two theories have been proposed to explain the phenomenon of radiation :

(a) According to the *wave theory*, all space is believed to be filled up by a hypothetical medium called *ether*. The phenomenon of radiation occurs because of the propagation of *electromagnetic waves* through the ether. Since every substance is always in contact with the ether, any substance can set up wave motion in the ether as a result of vibration of its molecules. These waves are called *electromagnetic waves*. All electromagnetic waves travel with the same speed which is equal to the speed of light. A hot body emits electromagnetic waves at the expense of its *stored energy*. When these waves fall on a second body, they are partly reflected and partly converted back into *stored energy* at the surface or inside the absorbing body.

(b) According to the *quantum theory* when the temperature of a solid body is raised, some of the constituent atoms and molecules are raised to *excited states* of higher energy. There is a tendency for these atoms or molecules to return spontaneously to lower energy states. When this occurs, energy is emitted in the form of electromagnetic radiation. Because the emitted radiation may result from change in the electronic, vibrational, and rotational states of the atoms and molecules, the radiation will be distributed over a wide range of energy values. However, the energy values of the radiation so emitted are not continuous but in the form of successive and discrete quantities called *quanta*. The quantities are of different sizes. The energy E of a quantum may be represented by :

$$E = h\nu \tag{6.1}$$

where h = Planck's constant.

ν = the frequency of the emitted quantum.

The higher the frequency, the larger the energy of quantum. Also, the higher the temperature of emitter, the larger the frequency of the quantum.

The term *photon* is often used for the above quantum of radiation. A photon is believed to be a particle having no mass, having energy $h\nu$, and momentum $h\nu/c$ where c is the velocity of light.

Although energy is transmitted by radiation at all frequencies (or wavelengths) manifestation in the form of *heating effects* occurs only over a very limited range of wavelengths or frequencies. Radiation belonging to this narrow wavelength range is termed *thermal radiation*. Thermal radiation extends over the range of wavelengths from 0.1 μ* to 100 μ or from 1000°A to 1000,000°A. Note that the range of *light radiation* (which can be sensed by the human eye) extends only from 0.35 μ to 0.8 μ or from 3500°A to 8000°A.

6.2 SPECTRUM OF ELECTROMAGNETIC RADIATION

A complete spectrum of electromagnetic radiation is given in Fig. 6.1. The various mechanisms responsible for the different types of radiation are indicated in Table 6.1.

Table 6.1 : Mechanisms for different types of radiation

Type of Radiation	*Wave length*	*Mechanism*
Y-rays	10^{-4} to 1.0	Displacement of nucleons in atomic nucleus.
X-rays	0.1 to 600	Displacement of outer electrons of an atom.
Ultraviolet	200 to 5000	Displacement of outer electrons of an atom.
Visible	3,500 to 8,000	Displacement of outer electrons of an atom.
Infrared	8×10^3 to 5×10^6	Molecular vibrations and rotations.
Radio waves (Radar, television, etc.)	10^6 to 10^{14}	Electrical conductor carrying alternating current.

As mentioned earlier, at any given temperature the quantity of radiation emitted per unit wavelength is different at different wavelengths. For some radiation at a certain temperature, the distribution of the energy radiated at various wavelengths is shown in Fig. 6.2. The ordinate in this figure represents the *monochromatic emissive power*, that is

* 1μ (micron) = 10^{-6} m; 1°A = 10^{-10} m = 10^{-4} μ

the energy radiated per unit time per unit area of the radiating surface per unit wavelength range and the abscissa represents the *wavelength*, λ. The curve shows that the major part of radiation is emitted within a narrow wavelength range on both sides of the wavelength at which the monochromatic emissive power is maximum. For example, a body at 1000 C emits most of its radiation between 1 and 20 μ. On the other hand, the sun whose surface temperature is nearly 5600 C emits 90 % of its radiation between 0.1 and 3 μ.

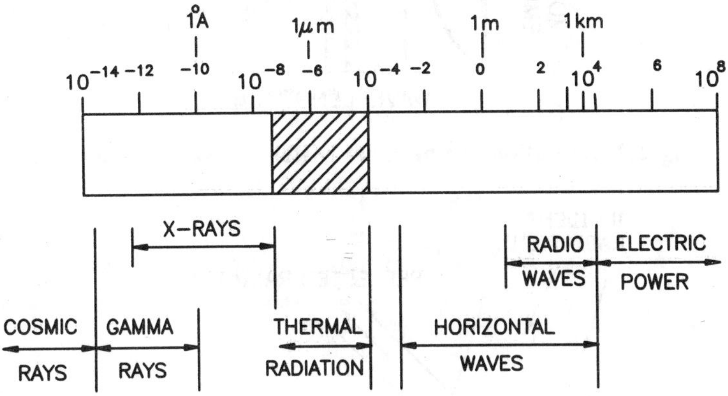

Fig. 6.1 (a) Electromagnetic Spectrum

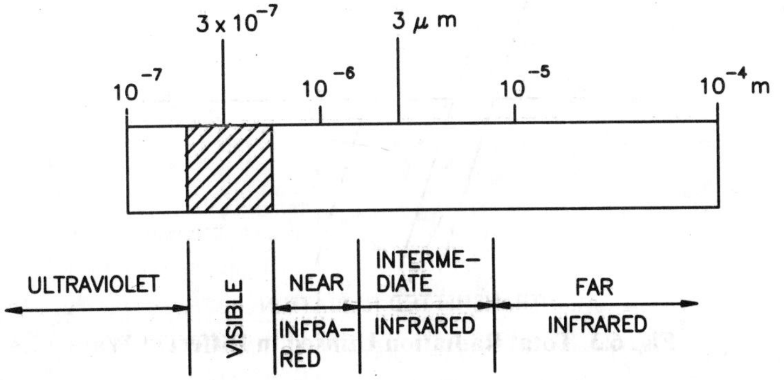

Fig. 6.1 (b) Range of Thermal Radiation

6.3 ABSORPTION, TRANSMISSION, REFLECTION AND EMISSION OF RADIATION

Radiation impinging on the surface of a body may be partly absorbed, partly transmitted and partly reflected, as shown schematically in Fig. 6.3.

The fraction of the incident radiation absorbed is called the *absorptivity*, α. Similarly the fraction of the incident radiation reflected is called the *reflectivity*, ρ and the fraction of radiation transmitted is called the *transmissivity*, τ.

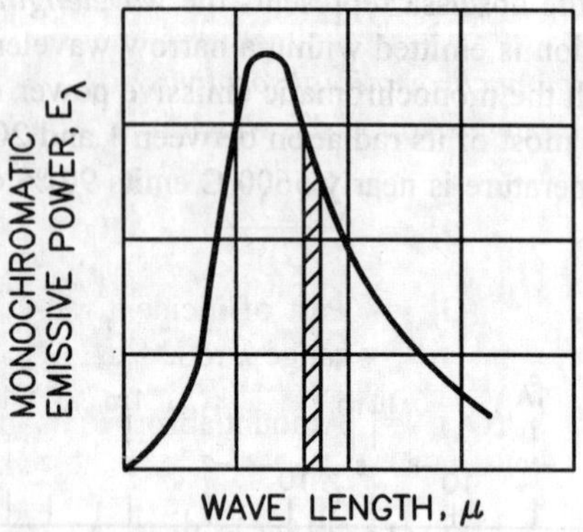

Fig. 6.2 Radiation of Energy at Different Wavelengths

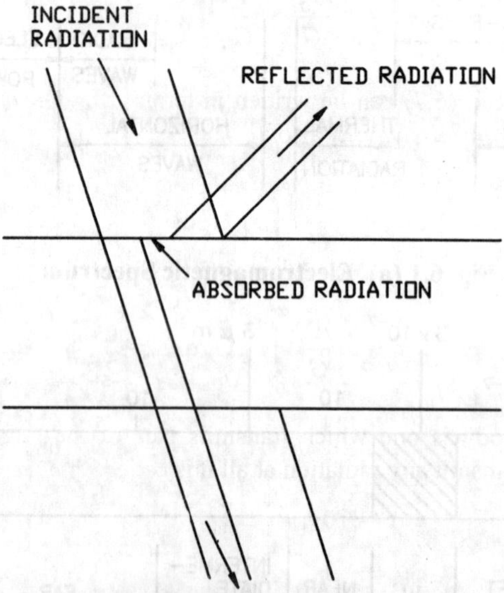

Fig. 6.3 Total Radiation Utilised in Different Ways

If G denotes the total incident radiation per unit time per unit area of a surface, and G_α, G_ρ and G_τ represent respectively the amount of radiation absorbed, reflected and transmitted :

$$\text{then} \quad \alpha = \frac{G_\alpha}{G} \tag{6.2}$$

$$\rho = \frac{G_\rho}{G} \tag{6.3}$$

$$\tau = \frac{G_\tau}{G} \tag{6.4}$$

also, $$G_\alpha + G_\rho + G_\tau = G$$

hence $$\alpha + \rho + \tau = 1 \tag{6.5}$$

We can also define *monochromatic absorptivity*, α_λ as the fraction of the incident energy in the wavelength range λ to $\lambda + d\lambda$, which is absorbed.

$$\alpha_\lambda = \frac{(G_\alpha)_\lambda}{G_\lambda} \tag{6.6}$$

where G_λ = Part of incident radiation G which is in spectral range λ to $\lambda + d\lambda$.

$(G_\alpha)_\lambda$ = Amount absorbed in spectral range λ to $\lambda + d\lambda$

Since $$G_\alpha = \int_0^\infty (G_\alpha)_\lambda \, d\lambda = \int_0^\infty \alpha_\lambda \, G_\lambda \, d\lambda$$

or $$\alpha = \frac{G_\alpha}{G} = \frac{1}{G} \int_0^\infty \alpha_\lambda \, G_\lambda \, d\lambda \tag{6.7}$$

Equations similar to Eq. (6.7) can be written in terms of *monochromatic reflectivity*, ρ_λ and *monochromatic transmissivity*, τ_λ also :

$$\rho = \frac{1}{G} \int_0^\infty \rho_\lambda \, G_\lambda \, d\lambda \tag{6.8}$$

and $$\tau = \frac{1}{G} \int_0^\infty \tau_\lambda \, G_\lambda \, d\lambda \tag{6.9}$$

(i) A *transparent* body is one which transmits part of the radiation falling on its surface. If it does not transmit any radiation at all it is called an *opaque* body.

For an opaque body, $\tau = 0$

Thus; $$\alpha + \rho = 1 \tag{6.10}$$

When radiation waves strike the surface of an opaque body, part of the incident waves are reflected and the remainder absorbed by a very thin layer of material (approximately 1 μ thick). Since almost all engineering materials (solids and liquids) are thick enough to cover this layer, they may be considered opaque for all practical purposes.

(ii) A body which neither reflects nor transmits any part of the incident radiation but absorbs all of it, is called a *black body.*

So for a black body ;

$$\rho = 0 \, ; \tau = 0$$

thus, $$\alpha = 1 \tag{6.11}$$

In reality a *perfect black body* ($\alpha = 1$) does not exist. However the *concept* of a black body is of great importance in radiation study, similar to the concept of ideal gas in thermodynamics.

(iii) Analogous to the definition of a black body, *white body* is one which reflects all the incident radiation and does not absorb or transmit any part of it.

For a white body ;

$$\alpha = 0; \tau = 0$$

$$\text{thus} \qquad \rho = 1 \qquad (6.12)$$

If the absorptivity of a surface does not vary with temperature and wavelength of the incident radiation, it is termed *gray body.*

For a gray body ;

$$\alpha = (\alpha)_\lambda = \text{constant} \qquad (6.13)$$

If the absorptivity of a surface varies with the wavelength of radiation waves it is termed *coloured body.*

6.4 MONOCHROMATIC EMISSIVE POWER OF BLACK BODY, PLANCK'S DISTRIBUTION LAW

A black body does not emit the same amount of radiant energy at different temperatures. At a *given wavelength* the radiant energy emitted by a black body increases as temperature is increased. Likewise, at a given *temperature*, the amount of radiation emitted by a black body varies with the wavelength.

The energy emitted by a surface in *all directions* at a given wavelength is called the *monochromatic emissive power* of the surface, E_λ. The monochromatic emissive power of a *black body* is denoted by $(E_\lambda)_b$. As pointed out above $(E_\lambda)_b$ or (E_λ) increases with increase in temperature of the surface.

Wien proposed the following formula for the monochromatic emissive power of a black body ;

$$(E_\lambda)_b = 2\pi c^2 h\lambda^{-5} \exp\left(-\frac{ch}{\lambda kT}\right) \qquad (6.14)$$

Another formula for $(E_\lambda)_b$ was proposed by Rayleigh-Jeans,

$$(E_\lambda)_b = 2\pi c k T \lambda^{-4} \qquad (6.15)$$

None of the above two formulae agrees with experiments over the entire wavelength range. At very short wavelengths Wien's formula gives results in reasonable agreement with experiments but at large wavelengths it is far in error. Similarly, at sufficiently long wavelengths Reyleigh-Jeans formula gives correct results but at shorter wavelengths it gives much higher values than those obtained by measurement.

In 1900, Max Planck using his quantum theory, derived the following expression for $(E_\lambda)_b$;

$$(E_\lambda)_b = 2\pi c^2 h \frac{\lambda^{-5}}{\left(\exp\dfrac{ch}{\lambda kT}\right) - 1} \qquad (6.16)$$

In Eqs. (6.14), (6.15) and (6.16) the symbols used have the following meaning :

c = Velocity of light, 2.998×10^{10} cm/sec.

h = Planck's quantum constant, 6.6237×10^{-34} joule-sec.

k = Boltzmann constant, 1.3803×10^{-23} joule/K.

λ = Wavelength, cm.

T = Absolute temperature, K.

The value of $(E_\lambda)_b$ is then obtained in terms of joules per cm^2 per sec per cm wavelength range ;

$$(E_\lambda)_b = \left[\frac{joule}{cm^2 - s - cm}\right] \quad (6.17)$$

If we like, we can also express $(E_\lambda)_b$ in calories per cm^2 per sec per micron wavelength range. Equation (6.16) is often expressed as :

$$(E_\lambda)_b = C_1 \frac{\lambda^{-5}}{\left(\exp\frac{c_2}{T\lambda}\right) - 1} \quad (6.18)$$

where $C_1 = 2\pi c^2 h = 37.404 \times 10^{-17}$ Jm^2/s

$= 37.404 \times 10^{-6}$ erg cm^2/sec.

$C_2 = ch/k = 1.4387 \times 10^{-2}$ m K

Figure 6.4 shows *Planck's distribution law* of for several temperatures. Here $(E_\lambda)_b$ is expressed in cal/sec-cm^2-cm and λ is in microns. Planck's law is found to be in close agreement with experiments. It correctly predicts the entire energy versus wavelength curve and the shift of the maximum towards shorter wavelengths at higher temperatures.

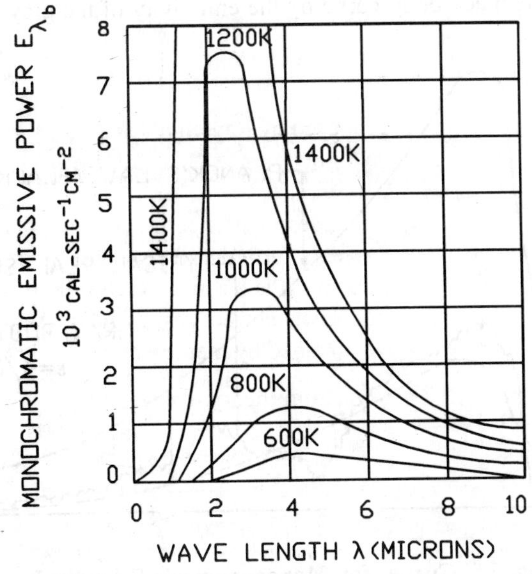

Fig. 6.4 Planck's Distribution Law

It is interesting to note that Planck's relation i.e. Eq. (6.16) reduces to Wien's formula Eq. (6.14) at small values of λ and to Rayleigh-Jeans formula Eq. (6.15) at large values of λ.

(i) If λ is small or λ << 1 ; $\left(\exp\frac{ch}{\lambda kT}\right) >> 1$.

Hence Eq. (6.15) reduces to ;

$$(E_\lambda)_b = \frac{2\pi c^2 h\lambda^{-5}}{\exp(ch/\lambda kT)}$$

$$= 2\pi c^2 h \lambda^{-5} \exp(-ch/\lambda kT) \qquad (6.14)$$

(ii) If λ is large or λ >> 1; $\frac{ch}{\lambda kT} << 1$, and exp (ch/λkT) can be expended in a series:

$$\exp(ch/\lambda kT) = 1 + \frac{ch}{\lambda kT} + \frac{c^2 h^2}{2\lambda^2 k^2 T^2} + \ldots\ldots\ldots + \ldots$$

$$\approx 1 + \frac{ch}{\lambda kT}$$

or

$$(E_\lambda)_b \approx \frac{2\pi c^2 h\lambda^{-5}}{1+\frac{ch}{\lambda kT}-1} = 2\pi c k T \lambda^{-4} \qquad (6.15)$$

It should be carefully noted that Planck's distribution law holds for a hypothetical black body. Real surfaces show marked deviation from Planck's law. This can be clearly seen from Fig. 6.5 which shows the monochromatic emissive power curves (i) for a black body (obtained from Planck's law), (ii) for a gray body (obtained by multiplying the ordinates of the black body curve by the emissivity of the gray body) and (iii) for a typical real surface.

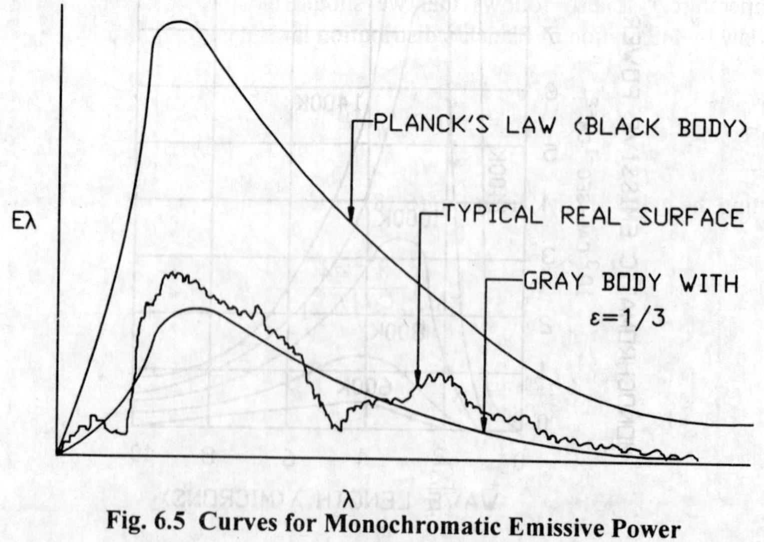

Fig. 6.5 Curves for Monochromatic Emissive Power

6.5 TOTAL EMISSIVE POWER; STEFAN-BOLTZMANN LAW

The total emissive power E of a surface is defined as the total radiant energy emitted by the surface *in all directions* over the entire wavelength range per unit surface area per unit time. The value of E is generally expressed in cal/sec-cm^2 or W/m^2.

Experiments show that the total emissive power of a given radiating surface depends on the *temperature* of the surface. Stefan, using the experimental data of Dulong and Petit and of Tyndall, discovered in 1879 that the total emissive power of a radiating surface is proportional to the fourth power of the absolute temperature of the surface.

In 1884, Boltzmann furnished a theoretical proof of the empirical equation of Stefan. This proof is based on principles of thermodynamics. Boltzmann's proof was confined to *black* radiating surfaces. The result of the above work is called the *Stefan and Boltzmann law* for the total emissive power of a black body. This law was mentioned in Section1.8 and states that :

$$E_b = \sigma T^4 \qquad (6.19)$$

{This equation is similar to Eq. (1.4)}

where $\sigma = 0.567 \times 10^{-4}$ erg/cm^2-sec-K^4

$= 56.7 \times 10^{-9}$ J/m^2-sec-K^4

$= 56.7 \times 10^{-9}$ W/m^2-K^4

$= 0.204 \times 10^{-3}$ J/m^2-hr-K^4

$= 1.355 \times 10^{-12}$ cal/cm^2-sec-K^4

$= 487.6 \times 10^{-10}$ kcal/m^2-hr-K^4

$= 17.14 \times 10^{-10}$ btu/ft^2-hr-R^4

It follows from the definition of the total emissive power that at any temperature the value of E_b is represented by the total area under the curve of $(E_\lambda)_b$ in Fig. 6.4 pertaining to that temperature. It also follows that we should be able to deduce the Stefan-Boltzmann law by integration of Planck's distribution law given by Eq. (6.18).

$$E_b = \int_0^\infty (E_\lambda)_b \, d\lambda \qquad (6.20)$$

Substituting the value of $(E_\lambda)_b$ from Eq. (6.18) ;

$$E_b = C_1 \int_0^\infty \frac{\lambda^{-5} \, d\lambda}{e^{C_2/\lambda T} - 1} \qquad (6.21)$$

Now, let $$\frac{C_2}{\lambda T} = x$$

or $$\lambda = \frac{C_2}{xT} \qquad (a)$$

or $$d\lambda = -\frac{C_2}{x^2 T}\,dx \qquad \text{(b)}$$

Substituting Eqs. (a) and (b) in Eq. (6.21) and noting that the limits of integration are now from ∞ to 0 since x = 1/λ ;

$$E_b = -C_1 \int_{\infty}^{0} \frac{(x^5 T^5) C_2}{C_2^5 (e^x - 1) x^2 T}\,dx$$

$$= \frac{C_1 T^4}{C_2^4} \int_0^{\infty} \frac{x^3}{e^x - 1}\,dx$$

Now, $$x^3 (e^x - 1)^{-1} = x^3 (e^{-x} + e^{-2x} + e^{-3x} + \ldots\ldots)$$

so $$E_b = \frac{C_1 T^4}{C_2^4} \int_0^{\infty} x^3 (e^{-x} + e^{-2x} + e^{-3x} + \ldots\ldots)\,dx$$

also, $$\int_0^{\infty} x^3 e^{-nx}\,dx = \frac{3!}{n^{(3+1)}} = \frac{3!}{n^4}$$

hence $$E_b = \frac{C_1 T^4}{C_2^4} \left\{ \frac{3!}{1^4} + \frac{3!}{2^4} + \frac{3!}{3^4} + \ldots\ldots \right\}$$

$$= \frac{C_1 T^4}{C_2^4} \left(\frac{6\pi^4}{90} \right) = \frac{\pi^4}{15} \frac{C_1}{C_2^4} T^4$$

$$= \frac{\pi^4 \times 37.404 \times 10^{-17}}{15 \times (1.4387)^4 \times 10^{-8}} T^4 \text{ (erg/cm}^2\text{-sec)}$$

$$= 0.567 \times 10^{-4}\, T^4 \text{ erg/cm}^2\text{-sec}$$

or $$E_b = 0.567 \times 10^{-7}\, T^4 \text{ J/m}^2 \text{ sec} \qquad (6.22)$$

The Eq. (6.22) may also be written as :

$$E_b = \sigma T^4 \text{ (The Stefan-Boltzmann law)} \qquad (6.19)$$

6.6 EMISSIVITY

The *emissivity*, ε of a surface is defined as the ratio of the emissive power of the surface to the emissive power of a hypothetical black body at the same temperature. The emissivity of a substance may depend both on temperature and wavelength. Sometimes emissivity refers to radiation in all directions and sometimes in a given direction only. For this reason, it is convenient to define the following emissivities ;

(a) Monochromatic emissivity

(b) Total emissivity

(c) Normal total emissivity

(a) *Monochromatic emissivity*, ε_λ, is the ratio of the monochromatic emissive power of a surface to the monochromatic emissive power of a black body at the same wavelength and temperature.

$$\varepsilon_\lambda = \frac{(E_\lambda)}{(E_\lambda)_b} \tag{6.23}$$

(b) *Total emissivity* ε, is the ratio of the total emissive power of a surface to the total emissive power of a black body at same temperature.

$$\varepsilon = \frac{E}{E_b} \tag{6.24}$$

(c) *Normal total emissivity* ε_n, is the ratio of the *normal component* of the total emissive power of a surface (E_n) to the normal component of the total emissive power of a black body $(E_b)_n$ at the same temperature.

$$\varepsilon_n = \frac{E_n}{(E_b)_n} \tag{6.25}$$

It would be evident from Eqs. (6.23), (6.24) and (6.25) that for a black body.

$$(\varepsilon_\lambda)_b = 1; \ \varepsilon_b = 1; \ (\varepsilon_n)_b = 1 \tag{6.26}$$

Note that the *emissivity* of a surface is a *property* of the surface. It depends only on the nature or characteristics of the surface and is independent of the nature or wavelength of the impinging radiation waves. The absorptivity of a surface, on the other hand, is not a property because it is dependent on the nature of the incident radiation.

For example suppose a substance has absorptivity of 0.5 for radiation waves in the wavelength range 5μ to 10μ. If this substance is now surrounded by another body which emits radiation in the range 50μ to 100μ only, none of the incident radiation falling on the above substance will be absorbed by it.

On the other hand if the surrounding body is a black body emitting radiation of all wavelengths, 50% of the incident radiation of wavelength range 5μ to 10μ will be absorbed. For this reason, the absorptivity of substance at a given temperature, is usually measured when the substance is in thermal equilibrium with a black body at the same temperature. This absorptivity is called the *equilibrium absorptivity* of the substance.

6.7 KIRCHOFF'S LAW

This law was established in 1860 and is one of the most important laws in radiation study. *Kirchoff's law* states that at any temperature the ratio of the total emissive power E to the absorptivity α is a constant for all substances which are in thermal equilibrium with their environment. Suppose we have three bodies in thermal equilibrium with each other, then :

$$\frac{E_1}{\alpha_1} = \frac{E_2}{\alpha_2} = \frac{E_3}{\alpha_3} = \text{constant} \tag{6.27}$$

If third body is a black body;

$$\frac{E_1}{\alpha_1} = \frac{E_2}{\alpha_2} = \frac{E_b}{\alpha_b} = \frac{E}{\alpha} = \text{constant} \tag{6.28}$$

or

$$\frac{E_1}{E_b} = \alpha_1; \ \frac{E_2}{E_b} = \alpha_2; \ (\alpha_b = 1) \tag{6.29}$$

since

$$\frac{E_1}{E_b} = \varepsilon_1; \ \frac{E_2}{E_b} = \varepsilon_2$$

$$\alpha_1 = \varepsilon_1 \ ; \ \alpha_2 = \varepsilon_2$$

Thus,

$$\alpha = \varepsilon$$

It follows that if a body is in thermal equilibrium with its environment, its absorptivity α is equal to its emissivity ε.

Kirchoff's law is also valid for monochromatic radiation. In other words, at a given temperature and wavelength, the ratio of the monochromatic emissive power E_λ to the monochromatic absorptivity α_λ is also a constant for all substances in thermal equilibrium with their environment.

$$\frac{(E_\lambda)_1}{(\alpha_\lambda)_1} = \frac{(E_\lambda)_2}{(\alpha_\lambda)_2} = \frac{(E_\lambda)_b}{(\alpha_\lambda)_b} = \text{constant} \tag{6.31}$$

since $(\alpha_\lambda)_b = 1$

$$\frac{(E_\lambda)_1}{(E_\lambda)_b} = (\alpha_\lambda)_1$$

$$\frac{(E_\lambda)_2}{(E_\lambda)_b} = (\alpha_\lambda)_2$$

but,

$$\frac{(E_\lambda)_1}{(E_\lambda)_b} = (\varepsilon_\lambda)_1 \text{ and } \frac{(E_\lambda)_2}{(E_\lambda)_b} = (\varepsilon_\lambda)_2$$

Hence,

$$(\alpha_\lambda)_1 = (\varepsilon_\lambda)_1; \ (\alpha_\lambda)_2 = (\varepsilon_\lambda)_2$$

or

$$\alpha_\lambda = \varepsilon_\lambda \tag{6.32}$$

Thus, if a body is in thermal equilibrium with its environment, its monochromatic absorptivity α_λ is equal to its monochromatic emissivity ε_λ.

We can prove Kirchoffs law as follows. Suppose that two small bodies B_1 and B_2 of surface areas A_1 and A_2 are placed in a large evacuated enclosure perfectly insulated from its surroundings. Radiation will be exchanged between the two bodies and the enclosure walls until equilibrium is attained and both bodies and the enclosure walls have reached the same temperature. Then the rate at which a body emits radiation must equal the rate at which it absorbs radiation. Let G be the rate at which radiant energy from the wall falls per unit area of each of the bodies, α_1 and α_2 the absorptivities and ε_1 and ε_2 the emissivities of the bodies B_1 and B_2 respectively.

An energy balance will give,

$$A_1 \, G \, \alpha_1 = A_1 \, E_1$$

$$A_2 \, G \, \alpha_2 = A_2 \, E_2$$

or

$$\frac{E_1}{\alpha_1} = \frac{E_2}{\alpha_2} = \frac{E}{\alpha} = \frac{E_b}{\alpha_b} = E_b = \text{constant} \tag{6.28}$$

To prove that E_λ/α_λ = constant, we can proceed as follows. Consider two large parallel plates 1 and 2 as shown in Fig. 6.6. Plate 1 emits and absorbs radiation waves of wavelength λ only. Plate 2 emits and absorbs rays of all wavelengths. The outer surfaces of both plates are assumed to be perfect reflectors. If the two plates are in thermal equilibrium, each must absorb as much as it emits.

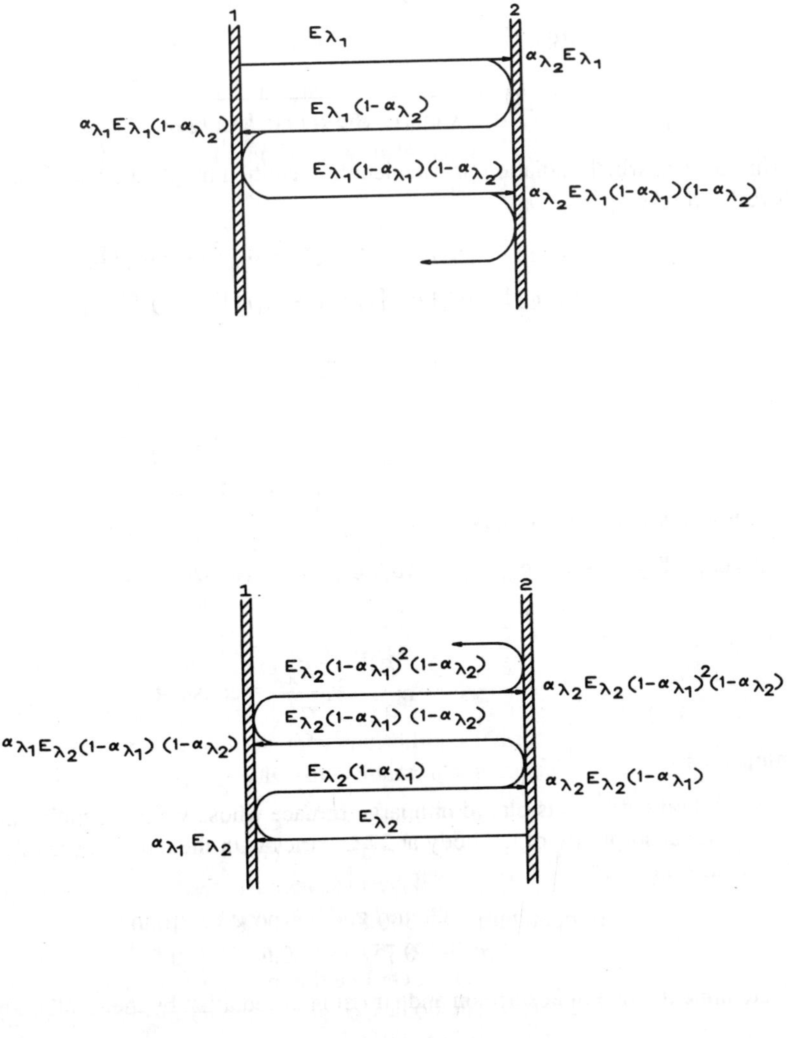

Fig. 6.6 Proof of Kirchoff's Law

Now, plate 1 absorbs radiation of wavelength λ only. Hence radiation of all wavelengths other than λ received by it from plate 2 must be completely reflected and reabsorbed by 2. Thus it is only radiation of wavelength λ emitted by 2 which is partly absorbed and partly reflected by 1. Since the temperature of 2 does not vary it follows that it absorbs as much radiation of wavelength λ as it emits. Thus radiation of wavelength λ emitted by 2 equals radiation of wavelength λ emitted by 1 and absorbed by 2 plus radiation of wavelength λ emitted by 2, reflected by 1 and reabsorbed by 2.

Now, considering unit area of each surface, (Fig. 6.6), we have ;

(i) Rate of emission of radiation of wavelength λ by plate 2 = $(E_\lambda)_2$.

(ii) Rate of absorption by plate 2 of radiation emitted by plate 1

$$= (\alpha_\lambda)_2 (E_\lambda)_1 + (\alpha_\lambda)_2 (E_\lambda)_1 \{1 - (\alpha_\lambda)_1\} \{1 - (\alpha_\lambda)_2\} + \ldots$$

$$= (\alpha_\lambda)_2 (E_\lambda)_1 + [1 + \{1 - (\alpha_\lambda)_1\} \{1 - (\alpha_\lambda)_2\} + \ldots]$$

$$= (\alpha_\lambda)_2 (E_\lambda)_1 \left(\frac{1}{1 - \{1 - (\alpha_\lambda)_1\} \{1 - (\alpha_\lambda)_2\}} \right)$$

(iii) Rate at which radiation of wavelength λ emitted by plate 2 and reflected by plate 1 is reabsorbed by plate 2

$$= (\alpha_\lambda)_2 (1 - \alpha_{\lambda 1}) E_{\lambda 2} + (\alpha_\lambda)_2 (1 - \alpha_{\lambda 1})^2 (1 - \alpha_{\lambda 2}) E_{\lambda 2} + \ldots$$

$$= (\alpha_\lambda)_2 (1 - \alpha_{\lambda 1}) E_{\lambda 2} [1 + (1 - \alpha_{\lambda 1}) (1 - \alpha_{\lambda 2}) + \ldots]$$

$$= (\alpha_\lambda)_2 (1 - \alpha_{\lambda 1}) E_{\lambda 2} \left(\frac{1}{1 - (1 - \alpha_{\lambda 1})(1 - \alpha_{\lambda 2})} \right)$$

hence,

$$E_{\lambda 2} = \frac{\alpha_{\lambda 2} E_{\lambda 1} + \alpha_{\lambda 2} (1 - \alpha_{\lambda 1}) E_{\lambda 2}}{1 - (1 - \alpha_{\lambda 1})(1 - \alpha_{\lambda 2})}$$

Transposing the terms we have ;

$$E_{\lambda 2}\, \alpha_{\lambda 1} + E_{\lambda 2}\, \alpha_{\lambda 2} - E_{\lambda 2}\, \alpha_{\lambda 1}\, \alpha_{\lambda 2} = \alpha_{\lambda 2}\, E_{\lambda 1} + \alpha_{\lambda 2}\, E_{\lambda 2} - \alpha_{\lambda 1}\, \alpha_{\lambda 2}\, E_{\lambda 2}$$

or

$$E_{\lambda 2}\, \alpha_{\lambda 1} = E_{\lambda 1}\, \alpha_{\lambda 2}$$

or

$$\frac{E_{\lambda 1}}{\alpha_{\lambda 1}} = \frac{E_{\lambda 2}}{\alpha_{\lambda 2}} = \frac{E_{\lambda b}}{\alpha_{\lambda b}} = \text{constant} \tag{6.31}$$

Example 6.1

A small body at 27 C is placed in a large furnace whose walls are maintained at 1000 K. The total absorptivity of the body at 27 C varies with the temperature of the incident radiation as follows :

Temperature	=	300 K	500 K	1000 K
α	=	0.75	0.6	0.5

Determine the rate of absorption and emission of radiation by the small body.

Solution :

Radiation falling on the small body is emitted by the furnace walls at 1000 K. The absorptivity of the body for this radiation is 0.5. Hence the rate of energy absorption is ;

$$= 0.5 \times (56.7 \times 10^{-9})(1000)^4$$

$$= 0.2835 \times 10^5 \text{ J/s-m}^2 \qquad \text{Ans.}$$

The emissivity of surface equals its absorptivity at 27 C. Thus $\varepsilon = 0.75$. Hence the rate of emission of radiant energy is ;

$$= 0.75 \times (56.7 \times 10^{-9})(300)^4$$

$$= 345 \text{ J/s-m}^2 \qquad \text{Ans.}$$

6.8 BLACK BODY

In practice, we can devise a black body by making a cavity with small opening. For example, consider a large hollow sphere or cylinder with a small hole in the wall, and maintained at a uniform temperature (Fig. 6.7). Any radiation entering through the small hole is partly absorbed and partly reflected diffusively at the inside surface of the sphere. The reflected radiation will repeatedly strike the interior surface of the sphere and each time a part of it will be absorbed. By the time the original beam reaches the hole once again and escapes, it would be so weak that energy leaving the sphere would be negligible. This would be true regardless of the value of the absorptivity of the wall. Thus the small hole above acts like a black body because practically all of the radiation incident upon it is absorbed.

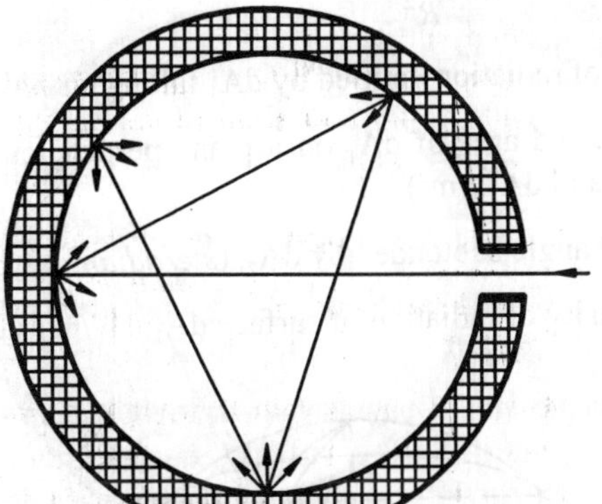

Fig. 6.7 Concept of a Black Body

Next we will show that the radiation streaming out of the hole is black body radiation. The radiation emitted by the interior surface of the sphere is absorbed and reflected many times so that it fills the sphere uniformly. Suppose we place a black body of area A_b inside the sphere, the temperature of this black body being the same as that of the interior surface of the sphere. The body receives radiation incident on the unit area of the sphere walls, i.e., the ratio of radiation incident on unit area of surface of the black body is the same. Let this rate be denoted by G. The black body absorbs all the incident radiation and in turn emits radiation at the rate E_b per unit area. Since the temperature of the black body and the interior surface of the sphere are equal, the black body must emit radiation at the same rate at which it absorbs incident radiation.

Hence, $$A_b G = A_b E_b$$

or $$G = E_b$$

This means that the radiation falling on the black body per unit area and time is equal to the total emissive power of a black body at the temperature of the enclosure. Thus the irradiation (rate at which energy falls on a surface per unit area) inside the sphere equals E_b. If we make a small hole in the wall of the sphere it will not appreciably disturb this condition and the radiation streaming out of this hole will be black body radiation.

Note carefully, however, that it is the area of the hole alone which can be considered a black body; the outside and inside *surfaces* and the sphere itself are *not* black bodies. A sphere like the one shown in Fig. 6.7 is also called an *isothermal enclosure.*

6.9 INTENSITY OF RADIATION

The *intensity of radiation* is defined as the rate of emission of radiation of all wavelengths in a *given direction* from a surface, per unit solid angle and per unit projected area of the surface on a plane perpendicular to the direction of radiation. It is expressed in W/m²-steradian.

Consider the radiation emitted by a small area dA_1 of a surface located at the centre of a hemisphere of radius R (Fig. 6.8). The amount of this radiation falling on a portion of the hemispherical surface dA_2 per unit time is then given by;

$$dq_{1-2} = I_1 (dA_1 \cos \theta) \frac{dA_2}{R^2} \tag{a}$$

where dq_{1-2} = rate of radiation emitted by dA_1 falling on dA_2, (W)

$dA_1 \cos \theta$ = projected area of dA_1 on a plane perpendicular to the line joining dA_1 and dA_2, (m^2)

dA_2/R^2 = solid angle subtended by dA_2, (*steradian*)

I = intensity of radiation of surface dA_1, (W/m²-steradian).

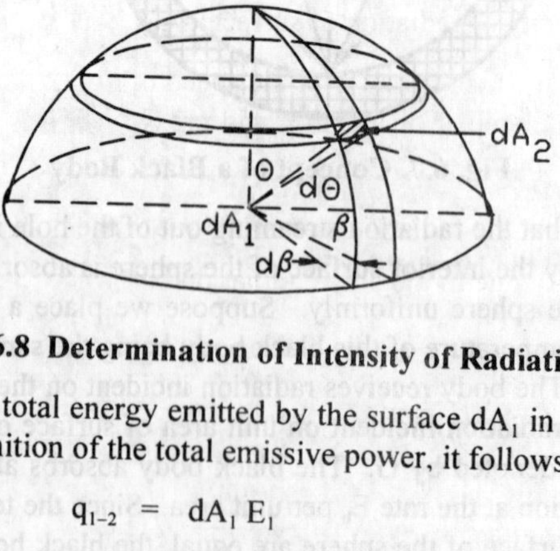

Fig. 6.8 Determination of Intensity of Radiation

Let q_{1-2} represent the total energy emitted by the surface dA_1 in all directions per unit time. Recalling the definition of the total emissive power, it follows that,

$$q_{1-2} = dA_1 E_1 \tag{b}$$

Since *all* the energy emitted by dA_1 must fall on the surface of the hemisphere, we also get;

$$q_{1-2} = \int I_1 (dA_1 \cos \theta) \frac{dA_2}{R^2}$$

$$= I_1\, dA_1 \int \cos\theta \frac{dA_2}{R^2}$$

Now, $$dA_2 = R^2 \sin\theta\, d\theta\, d\beta$$

or $$q_{1-2} = I_1\, dA_1 \int_0^{\pi/2} \int_{\beta=0}^{\beta=2\pi} (\cos\theta \sin\theta)\, d\theta\, d\beta$$

$$= I_1\, dA_1 (2\pi) \int_0^{\pi/2} (\cos\theta \sin\theta)\, d\theta$$

or $$q_{1-2} = \pi I_1\, dA_1 \qquad \text{(c)}$$

Equating Eqs. (b) and (c),

$$E_1\, dA_1 = \pi I_1\, dA_1 \text{ or } E_1 = \pi I_1$$

thus $$E = \pi I \qquad (6.33)$$

Thus, the total emissive power of a surface is π times the intensity of radiation.

6.10 SPACE DENSITY OF RADIATION

The radiant energy per unit volume in a stream of radiation is called the *radiation density*. Suppose that the radius R of the hemisphere in Fig. 6.8 is made very large. Now we may regard all the rays from dA_1 to dA_2 as being parallel. Let us assume that dA_1 is a *black* surface and let dq_{1-2} represent the energy emitted by dA_1 per second in the *direction* of dA_2. Then,

$$dq_{1-2} = I_b\, dA_1 \cos\theta \frac{dA_2}{R^2}$$

(where I_b is the intensity of radiation of surface dA_1)

Since the radiation waves travel with the speed of light, the above energy dq_{1-2} will be contained in a parallelepiped of length *c* and cross-sectional area $dA_1 \cos\theta$. Thus, the volume of parallelepiped is ;

$$dV = c\, dA_1 \cos\theta$$

The energy density due to the above radiation any where inside the volume dV is then given by ;

$$\frac{dq_{1-2}}{dV} = du = \frac{I_b\, dA_1 \cos\theta\, dA_2}{R^2} \times \frac{1}{c\, dA_1 \cos\theta}$$

$$= \frac{I_b}{c} \frac{dA_2}{R^2}$$

Note that this equation gives the energy density in the space above dA_1 due to the radiation emitted by dA_1 in the direction of dA_2. Since radiation is leaving dA_1 in all directions the total energy density near the surface will be,

$$u = \int du = \frac{I_b}{cR^2} \int dA_2$$

$$= \frac{I_b}{cR^2} 2 \pi R^2$$

so $$= \frac{2 \pi I_b}{c} = \frac{2 E_b}{c} \quad (6.34)$$

The energy density in an *isothermal enclosure* will be *twice that* in front of a black surface from radiation emitted by that surface. This is because energy streams are travelling in all directions in an enclosure whereas above a flat surface radiant energy travels in only a hemisphere of directions. The energy in an enclosure is thus given by ;

$$u = \frac{4 \pi I_b}{c} = \frac{4 E_b}{c} \quad (6.35)$$

6.11 RADIATION PRESSURE

Suppose a stream of radiation in a vacuum falls normally on the surface of the body. Let ω be the mean energy density in the oncoming waves. Then,

Energy reaching 1 cm^2 surface per second ;

$$= \omega \times 1 \times c = \omega c$$

Momentum reaching 1 cm^2 surface per second ;

$$= \frac{\omega c}{c} = \omega$$

Thus the waves bring to the surface ω units of momentum per unit area per unit time. This momentum is directed normally towards the surface. If the waves are absorbed by the surface, it receives this momentum and thus experiences a pressure *p* equal to the change of momentum per unit area per unit time. Hence, p = ω.

Next let us consider that the oncoming waves make an angle θ with the normal to the surface RP (Fig. 6.9). Now, the energy received by 1 cm^2 area of the surface in 1 second is that which fills a parallelepiped of length c and cross-sectional area PQ = 1 cos θ = cos θ cm^2.

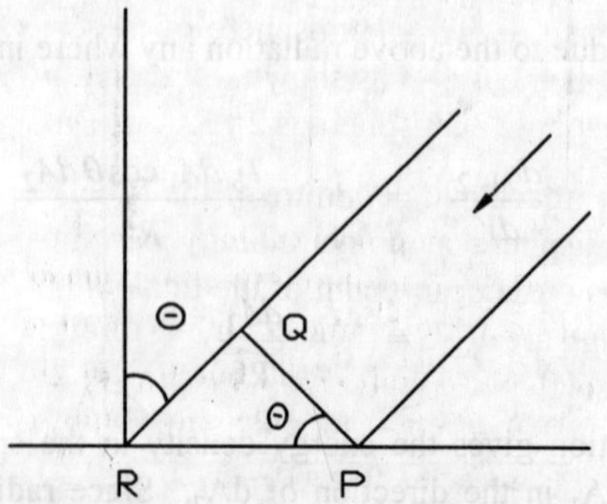

Fig. 6.9 Radiation Incident on a Surface

Hence, energy reaching 1 cm^2 surface in 1 second,

$$= (c)(\cos\theta)\,\omega = \omega\, c\cos\theta$$

Momentum reaching 1 cm^2 surface in 1 second,

$$= \frac{\omega\, c\cos\theta}{c}$$

$$= \omega\cos\theta$$

Component of above momentum normal to surface,

$$= (\omega\cos\theta)(\cos\theta) = \omega\cos^2\theta$$

If the radiation is entirely absorbed,

$$p = \omega\cos^2\theta \qquad \text{(a)}$$

The same expression holds for the pressure caused by the *emission* of a beam at an angle θ, or for the additional pressure caused by occurrence of reflected beam. If an incident beam is *specularly* reflected from a surface at an angle of incidence θ, the total pressure on the surface is $2\omega\cos^2\theta$.

Finally, let radiation be streaming towards a surface and also away from it in all directions, as in an isothermal enclosure.

Such distribution of radiation is equivalent to a large number of beams of plane waves, all of equal intensity, with their directions of propagation distributed equally in direction. Suppose there are N beams in all (including incoming and outgoing beams) and let the energy density due to any *one* beam be ω. Then,

Total energy density in front of the surface,

$$= \omega N = u$$

Radiation pressure due to one beam having incident angle θ

$$= \omega\cos^2\theta$$

Total radiation pressure due to N beams ;

$$= \Sigma\omega\cos^2\theta = p \qquad \text{(b)}$$

or

$$p = \Sigma\omega\cos^2\theta$$

In order to find $\Sigma\cos^2\theta$, let us refer to Fig. 6.10. Point O is located on the surface. With O as centre we have drawn a hemisphere of radius R. Of this hemisphere let us consider a ring-shaped element *Q-S* as shown. The area of this element is $2\pi R^2\sin\theta\, d\theta$ whereas the area of the whole hemisphere is $2\pi R^2$.

Now, the lines of approach and departure of the *N* beams, if drawn through O will cut the surface of the hemisphere in points *equally distributed* over its surface. Let *dN* represent the number of these beams (out of the total number *N*) whose lines of approach or departure pass through area *Q-S*. Then ;

$$\frac{dN}{N} = \frac{2\pi R^2\sin\theta\, d\theta}{2\pi R^2} = \sin\theta\, d\theta$$

or

$$dN = N\sin\theta\, d\theta$$

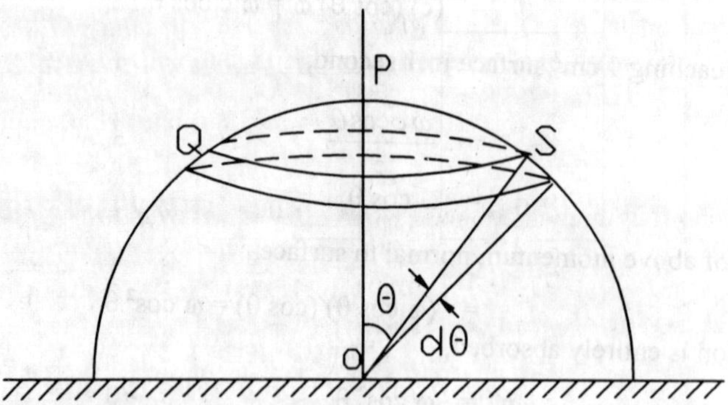

Fig. 6.10 Determination of Radiation Pressure

The value of $\cos^2\theta$ is the same for the above dN beams. Hence their contribution to $\Sigma\cos^2\theta$ is $\cos^2\theta\, dN$.

Thus, for dN beams;

$$\Sigma \cos^2\theta = \cos^2\theta\, dN = N \cos^2\theta \sin\theta\, d\theta$$

or for N beams;

$$\Sigma \cos^2\theta = \int \cos^2\theta\, dN$$

$$= N \int_0^{\pi/2} (\cos^2\theta \sin\theta\, d\theta) = \frac{N}{3}$$

Eq. (b) then gives,

$$p = \omega \Sigma \cos^2\theta = \frac{\omega N}{3}$$

Since $\omega N = u$, we have ;

$$p = \frac{u}{3} \tag{6.36}$$

Thus, the pressure on the walls of an isothermal enclosure equals one-third of the radiation density at any interior point.

6.12 DERIVATION OF STEFAN-BOLTZMANN LAW FROM THERMODYNAMICS

Boltzmann considered a Carnot engine using radiation (assumed to behave as a perfect gas) as the working medium. Figure 6.11 shows an engine consisting of a non-conducting cylinder and a frictionless piston along with the p-V diagram of the cycle. Radiation may enter or leave the cylinder through an opening in the cylinder head.

Suppose that the cylinder is evacuated and an isothermal enclosure at T_1 is placed opposite to the opening in the head. The cylinder will fill up with radiation from T_1 to a density u_1. The pressure of radiation will then $p_1 = u_1/3$ and the volume will be V_A and the piston will remain stationary at position A.

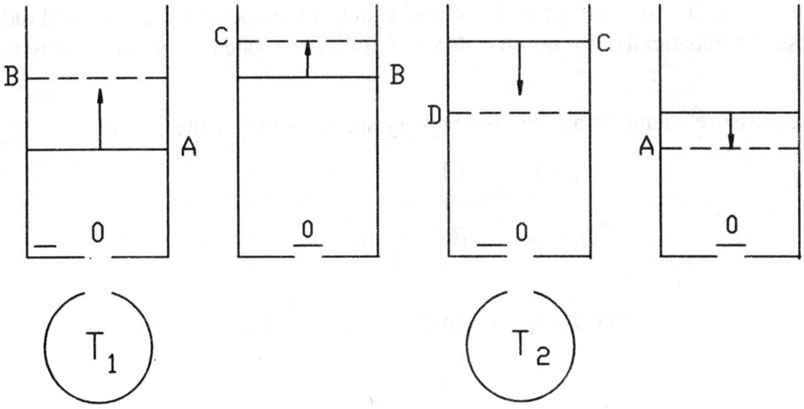

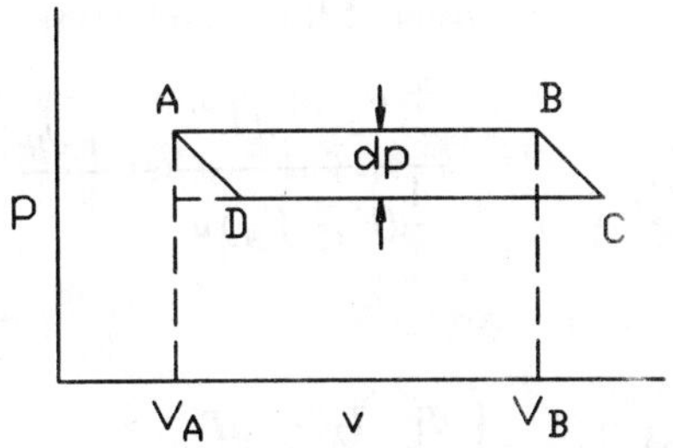

Fig. 6.11 A Carnot Engine Working due to Thermal Radiation

(a) Now, let the piston move slowly upward to position B, the pressure remaining constant by addition of radiation from T_1. For this process 1-2,

Increase in energy within cylinder $\Delta U = u_1 (V_B - V_A)$

Work done by system $\Delta W = p_1 (V_B - V_A)$

Radiation energy entering system ;

$$\Delta Q_1 = \Delta U + \Delta W = \left(u_1 + \frac{u_1}{3}\right) (V_B - V_A)$$

or

$$\Delta Q_1 = \frac{4}{3} u_1 (V_B - V_A)$$

(b) When the piston has reached position B, let the opening be covered, enclosure T_1 removed, and the piston allowed to move to position C. During this process of adiabatic expansion work is done by the system at the expense of internal energy. As a result the temperature of radiation inside the cylinder decreases to T_2, the energy density to u_2 radiation pressure to $p_2 = u_2/3$.

(c) The opening is once again uncovered and another isothermal enclosure at T_2 placed against it. The piston is allowed to travel backwards from C to D the pressure remaining constant at p_2. During this isothermal process radiant energy is transferred from the system to enclosure T_2. This amount is given by ;

$$\Delta Q_2 = \frac{4}{3} u_2 (V_D - V_C)$$

(d) Position D is selected in such a manner that, if the opening is closed and radiation compressed adiabatically, the initial volume V_A will be reached when the density reaches u_1.

If the pressure change is assumed to be very small we may write ;

$$T_1 - T_2 = dT$$

$$p_1 - p_2 = dp = \frac{1}{3} du$$

or

$$(\Delta W)_{cycle} = dp (V_B - V_A) = \frac{1}{3} (V_B - V_A) du$$

or

$$(\eta)_{cycle} = \frac{(\Delta W)_{cycle}}{\Delta Q_1}$$

$$= \frac{\frac{1}{3}(V_B - V_A) du}{\frac{4}{3}(V_B - V_A) u} = \frac{1}{4}\frac{du}{u}$$

But, for a Carnot cycle,

$$(\eta)_{cycle} = \frac{T_1 - T_2}{T_1} = \frac{dT}{T}$$

or

$$\frac{1}{4}\frac{du}{u} = \frac{dT}{T} \tag{6.37}$$

Integrating Eq. (6.37),

$$u = k_1 T^4 \tag{6.38}$$

where k_1 = constant

Since

$$E_b = \frac{c}{4} u$$

$$E_b = \frac{c k_1}{4} T^4$$

or

$$E_b = \sigma T^4 \tag{6.19}$$

Eq. (6.19) is the *Stefan-Boltzmann* law.

6.13 WIEN'S DISPLACEMENT LAW

In 1893, Wien discovered a relationship between the temperature of a black body and the wavelength at which the maximum value of the monochromatic emissive power occurs. It would be evident from Fig. 6.4 that if the temperature of a black body is increased, the wavelength λ_{max} corresponding to maximum monochromatic emissive power becomes smaller. Wien's displacement law states that ;

$$(\lambda_{max}) T = \text{constant} \tag{6.39}$$

It was originally deduced by using the Doppler principle and the first law of thermodynamics. Consider a sphere filled up with radiation (Fig. 6.12). Suppose that the wall of the sphere may move back and forth causing the sphere volume V to increase or decrease. Let the wall move with velocity υ. Then, according to the Doppler principle, the wavelength λ of a wave normal to the surface increases to $\lambda(1 + \upsilon/c)$ on reaching the wall and further increases to $\lambda(1 + \upsilon/c) \times (1 + \upsilon/c)$ after each reflection. Here c is the velocity of light.

Thus,

$$d\lambda = \lambda\left(1+\frac{\upsilon}{c}\right)\left(1+\frac{\upsilon}{c}\right) - \lambda$$

$$= \lambda\left[1+\frac{2\upsilon}{c}+\frac{\upsilon^2}{c^2}-1\right]$$

$$\approx \lambda\left(\frac{2\upsilon}{c}\right) \qquad \text{(a)}$$

or

$$\frac{d\lambda}{\lambda} = \frac{2\upsilon}{c} \qquad \text{(b)}$$

In most cases the wave may not be travelling normal to the surface but in a direction making an angle θ with normal, as in Fig. 6.12. In this case the component of υ in the direction of wave travel is $\upsilon \cos \theta$. Hence,

$$\frac{d\lambda}{\lambda} = \frac{2\upsilon\cos\theta}{c} \qquad \text{(c)}$$

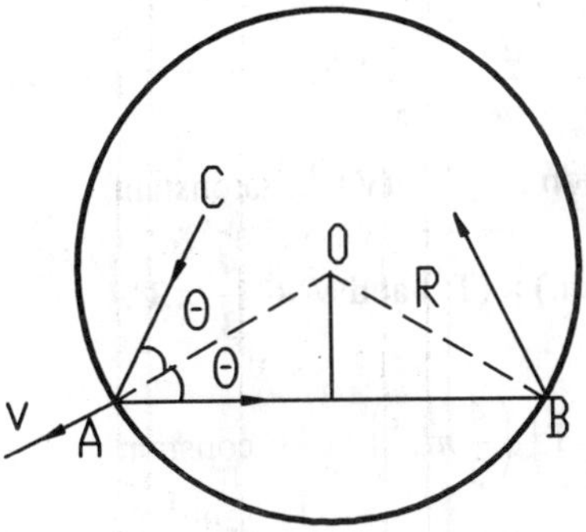

Fig. 6.12 Derivation of Wein's Displacement Law

In order to find the number of wave reflections per unit time, observe that the distance travelled by the wave between successive reflections is ;

$$AB = 2\,R\cos\theta$$

Therefore the time interval between successive reflections is $(2\,R\cos\theta)/c$. Hence the number of reflections per unit time is $c/(2\,R\cos\theta)$. Therefore the change in wavelength per unit time is given by;

$$\frac{d\lambda}{\lambda} = \frac{2\upsilon\cos\theta}{c} \times \frac{c}{2R\cos\theta}$$

or
$$\frac{d\lambda}{\lambda} = \frac{\upsilon}{R} \tag{d}$$

Since υ = change in R in unit time, we get ;

$$\frac{d\lambda}{\lambda} = \frac{dR}{R} \tag{e}$$

Now, applying the first law of thermodynamics to the radiation inside the sphere we obtain ;

$$dQ = dU + p\,dV$$

If the process is assumed to be adiabatic ; dQ = 0

and
$$dU + p\,dV = 0$$

Now, U = u V ; and $p = \frac{u}{3}$

or
$$d(u\,V) + \frac{u}{3}\,dV = 0$$

$$V\,du + u\,dV + \frac{u}{3}\,dV = 0$$

$$V\,du + \frac{4}{3}\,u\,dV = 0$$

$$du + \frac{4}{3}\,u\,\frac{dV}{V} = 0$$

or
$$\frac{du}{u} + \frac{4}{3}\,\frac{dV}{V} = 0 \tag{f}$$

Integrating this expression ; $uV^{4/3}$ = constant

Since u = (a constant) × (T^4) and $V = \frac{4}{3}\pi R^3$,

$$T^4\left(\frac{4}{3}\pi R^3\right)^{4/3} = \text{constant}$$

or
$$RT = \text{constant}$$

or
$$\frac{dR}{R} = -\frac{dT}{T} \tag{g}$$

From Eqs. (e) and (g)
$$\frac{d\lambda}{\lambda} = -\frac{dT}{T}$$

or
$$\lambda T = \text{constant} \tag{6.39a}$$

Equation (6.39 a) is the general expression for *Wien's displacement law*. Note that so far we have not shown that λ in Eq. (6.39a) corresponds to the maximum value of the monochromatic emissive power.

Aliter I

After Planck's Law for $(E_\lambda)_b$ had been formulated, it was also possible to deduce Wien's law from the same. If T is held constant, the condition for $(E_\lambda)_b$ to be maximum is ;

$$\frac{d(E_\lambda)_b}{d\lambda} = 0$$

Since

$$(E_\lambda)_b = \frac{C_1 \lambda^{-5}}{(e^{C_2/\lambda T} - 1)} \tag{6.18}$$

$$\frac{d(E_\lambda)_b}{d\lambda} = \frac{-5C_1 \lambda^{-6}}{(e^{C_2/\lambda T} - 1)} + \frac{C_1 \lambda^{-5} e^{C_2/\lambda T}}{(e^{C_2/\lambda T} - 1)^2} \times \frac{C_2}{\lambda^2 T} = 0$$

or

$$-5\lambda + \frac{e^{C_2/\lambda T}}{(e^{C_2/\lambda T} - 1)} \times \frac{C_2}{T} = 0$$

or

$$\frac{C_2}{5\lambda T} + \left\{e^{(-C_2/\lambda T)} - 1\right\} = 0$$

The root of the above equation is ;

$$\frac{C_2}{\lambda T} = 4.965$$

Hence,

$$(\lambda_{max})\, T = \frac{C_2}{4.965} = \text{constant} \tag{6.39b}$$

Substituting $C_2 = 1.4387$ cm K in Eq. (6.39b) we get ;

$$(\lambda_{max})\, T = 0.2884 \text{ cm K} \tag{6.39c}$$

or

$$(\lambda_{max})\, T = 2884\ \mu\text{ K} \tag{6.39d}$$

Aliter 2

Further, Wien's displacement law can be stated in a different way. Let us focus attention on a particular spectral range from λ_1 to $\lambda_1 + d\lambda_1$ containing energy $(u_\lambda)_1\, d\lambda_1$ per unit volume in an enclosure at T_1. Suppose an adiabatic expansion of the radiation inside the enclosure occurs which lowers the temperature to T_2. According to Eq. (6.39a), this expansion will change the spectral range λ_1 to $(\lambda_1 + d\lambda_1)$ to a new range λ_2 to $(\lambda_2 + d\lambda_2)$ in such a way that ;

$$\frac{\lambda_2}{\lambda_1} = \frac{\lambda_2 + d\lambda_2}{\lambda_1 + d\lambda_1} = \frac{T_1}{T_2}$$

$$\frac{d\lambda_2}{d\lambda_1} = \frac{T_1}{T_2} \tag{h}$$

At the same time the energy originally contained in the spectral range λ_1 to $(\lambda_1 + d\lambda_1)$ and now contained in the range λ_2 to $(\lambda_2 + d\lambda_2)$ decreases *in the same ratio* as the total energy in the enclosure. Hence if the new value u_λ is $(u_\lambda)_2$ then ;

$$\frac{u_{\lambda 2}\, d\lambda_2}{u_{\lambda 1}\, d\lambda_1} = \frac{u_2}{u_1} = \frac{T_2^4}{T_1^4} \qquad \text{(i)}$$

Combining Eqs. (h) and (i) we get ;

$$\frac{u_{\lambda 2}}{u_{\lambda 1}} = \frac{T_2^4\, d\lambda_1}{T_1^4\, d\lambda_2} = \frac{T_2^4\, T_2}{T_1^4\, T_1} = \frac{T_2^5}{T_1^5}$$

Since $(u_{\lambda 2}/u_{\lambda 1}) = (E_{\lambda 2}/E_{\lambda 1})$ we get,

$$\frac{E_{\lambda 2}}{E_{\lambda 1}} = \frac{T_2^5}{T_1^5} \qquad (6.40)$$

$$\frac{E_{\lambda 2}}{T_2^5} = \frac{E_{\lambda 1}}{T_1^5} = \text{constant} \qquad (6.41)$$

The Eq. (6.41) is another form of the Wien's displacement law which can therefore be expressed by any of the following statements :

(a) For a black body, if the values of E_λ at two different temperatures are compared, not at the same wavelength but at wavelengths inversely proportional to T, then their values are proportional to the *fifth* power of the absolute temperature.

(b) For a black body, if the wavelength is varied in such manner that λT is constant than E_λ/T^5 has the same value at all temperatures.

Since E_λ/T^5 is a function of λT, not depending otherwise on T, if we plot E_λ/T^5 versus λT, a single curve will be obtained which would be valid at all temperatures. Such a plot is shown in Fig. 6.13. From this curve the value of E_λ at any value of λ and T can be obtained

Example 6.2

Assuming the sun to be a black body, emitting radiation with maximum intensity at $\lambda = 0.5\ \mu$ estimate the surface temperature of the sun and the heat flux at its surface.

Solution :

From Wien's displacement law Equation (6.39c),

$$T = \frac{0.2884}{\lambda_{max}} = \frac{0.2884}{0.5 \times 10^{-4}} = 5{,}760 \text{ K} \qquad \text{Ans.}$$

also, $(E)_{sun} = \sigma T^4$

$$= 56.7 \times 10^{-9}\,(5760)^4 = 62.412 \times 10^6 \text{ J/s m}^2$$

$$= 22.7 \times 10^{10} \text{ J/hr m}^2 \qquad \text{Ans.}$$

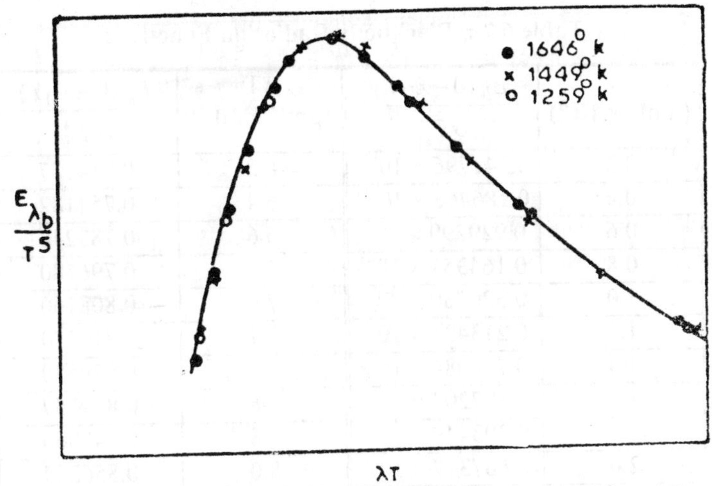

Fig. 6.13 Wien's Displacement Law shown as $(E_\lambda)_b/T^5$ Vs λT

Total Black Body Emission

In many cases we have to determine the total black body emission in a spectral band between wavelength λ_1 and λ_2. For this purpose we must first find the black body emission $E_b(0 - \lambda_1 T)$, in the interval $\lambda = 0$ to $\lambda = \lambda_1$ at temperature T.

$$\int_0^{\lambda_1} E_{b\lambda}\,(T)\,d\lambda = E_b\,(0 - \lambda_1 T)$$

Dividing both sides by σT^4,

$$\frac{E_b\,(0-\lambda_1 T)}{\sigma T^4} = \int_0^{\lambda_1 T} \frac{E_{b\lambda}}{\sigma T^5}\, d(\lambda T) \qquad (6.42)$$

Equation (6.42) gives the fraction of the total black body radiation which falls between $\lambda = 0$ to $\lambda = \lambda_1$ as a function of λT. The values are given in Table 6.2.

Example 6.3

A particular glass transmits 90% of incident radiation in the wavelength range 0.31 to 3.1 μm and it is opaque at longer and shorter wavelengths. What percentage of solar radiation will be transmitted by the glass. The sun may be assumed to radiate as a blackbody at 5800 K.

Solution :

$$\lambda_1 T = 0.31 \times 5800 = 1798 \text{ μm K}$$

$$\lambda_2 T = 3.1 \times 5800 = 17{,}980 \text{ μm K}$$

From Table 6.2 and using Eq. (6.42) we get ;

$$\left\{\int_0^{1798} E_{b\lambda}\, d\lambda\right\} \div \int_0^{\infty} E_{b\lambda}\, d\lambda = 0.039 = 3.9\ \%$$

Table 6.2 : Blackbody Radiation Functions

λT (μmK × 10^{-3})	$\frac{E_b(0-\lambda_1 T)}{\sigma T^4}$	λT (μmK × 10^{-3})	$\frac{E_b(0-\lambda_1 T)}{\sigma T^4}$
0.2	0.341796 × 10^{-26}	6.2	0.754187
0.4	0.186468 × 10^{-11}	6.4	0.754182
0.6	0.929299 × 10^{-7}	6.6	0.783248
0.8	0.164351 × 10^{-4}	6.8	0.796180
1.0	0.320780 × 10^{-3}	7.0	0.808160
1.2	0.213431 × 10^{-2}	7.2	0.819270
1.4	0.779084 × 10^{-2}	7.4	0.829580
1.6	0.197204 × 10^{-1}	7.6	0.839157
1.8	0.393449 × 10^{-1}	7.8	0.848060
2.0	0.667347 × 10^{-1}	8.0	0.856344
2.2	0.100897	8.5	0.874666
2.4	0.140268	9.0	0.890090
2.6	0.183135	9.5	0.903147
2.8	0.227908	10.0	0.914263
3.0	0.273252	10.5	0.923775
3.2	0.318124	11.0	0.931956
3.4	0.361760	11.5	0.939027
3.6	0.403633	12	0.945167
3.8	0.443411	13	0.955210
4.0	0.480907	14	0.962970
4.2	0.516046	15	0.969056
4.4	0.548830	16	0.973890
4.6	0.579316	18	0.980939
4.8	0.607597	20	0.985683
5.0	0.633786	25	0.992299
5.2	0.658011	30	0.995427
5.4	0.680402	40	0.998057
5.6	0.701090	50	0.999045
5.8	0.720203	75	0.999807
6.0	0.737864	100	1.000000

and $$\left\{\int_0^{17980} E_{b\lambda}\, d\lambda\right\} \div \int_0^{\infty} E_{b\lambda}\, d\lambda = 0.98 = 98\ \%$$

Thus (98 – 3.9) = 94.1 percent of the total incident solar radiation is in the wavelength range 0.31 to 3.1 μm and (0.9 × 94.1) = 84.7 percent of the solar radiation is transmitted.

6.14 LAMBERT'S COSINE LAW

This law, formulated in 1860, states that for a *diffuse* radiating surface the total emissive power E_θ in any direction is proportional to cosine of the angle of emission θ.

i.e. $E_\theta \propto \cos\theta$

$$E_\theta = (\text{a constant})(\cos\theta)$$

If E_n is the total emissive power in the normal direction we get ;

$$\frac{E_\theta}{E_n} = \frac{\cos\theta}{\cos 0}$$

or $$E_\theta = E_n \cos\theta \tag{6.43}$$

Equation (6.43) is true only for a diffuse radiating surface, that is a surface which does not reflect incident radiation as a mirror reflects light but from which the reflected radiation waves are dispersed equally in all directions. Diffuse radiation would seldom occur from perfectly smooth surfaces but it occurs from most practical surfaces which contain millions of small irregularities.

For surfaces which obey the cosine law it can be shown that the intensity of radiation I_θ in any direction is the same ;

$$I_\theta = I_n = \text{constant} \tag{6.44}$$

To demonstrate the above result, consider the rate of emission of radiation by a surface dA in the normal direction. This rate equals dA E_n. Suppose that this radiation is contained within a solid angle dω. Then the above rate is also given by I_n dA (cos 0) dω. Hence,

$$dA\, E_n = I_n\, dA\, d\omega$$

or $$E_n = I_n\, d\omega \tag{a}$$

Next, consider the rate of emission of radiation from dA in direction θ. This rate is given by dA E_θ and also by I_θ (dA cos θ) dω. Note that the radiation is even now contained within the solid angle dω. Thus,

$$dA\, E_\theta = I_\theta\, dA \cos\theta\, d\omega$$

or $$E_\theta = I_\theta \cos\theta\, d\omega \tag{b}$$

From Eqs. (a) and (b) ;

$$\frac{E_\theta}{E_n} = \frac{I_\theta \cos\theta}{I_n}$$

Using Lambert's cosine law Eq. (6.43) ;

$$\frac{E_n \cos\theta}{E_n} = \frac{I_\theta \cos\theta}{I_n}$$

or $$I_\theta = I_n \tag{6.44}$$

6.15 RADIATION CHARACTERISTICS OF REAL SURFACES

The radiation from real surfaces differs from black body radiation in several respects. These differences are clearly revealed if we plot the monochromatic absorptivity α_λ versus λ of various surfaces. Fig.6.14 shows α_λ versus λ for polished copper, commercial aluminum and anodized aluminum. The curves for magnesium oxide, Portland cement, and top soil are shown in Fig. 6.15.

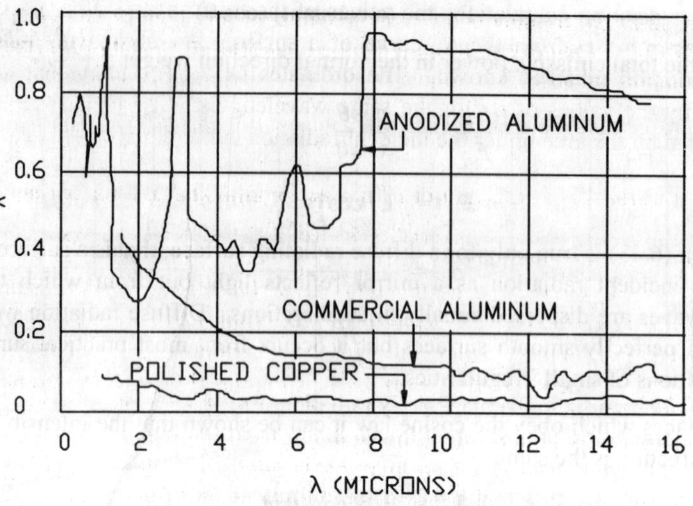

Fig. 6.14 α_λ Vs λ curves for Conductors

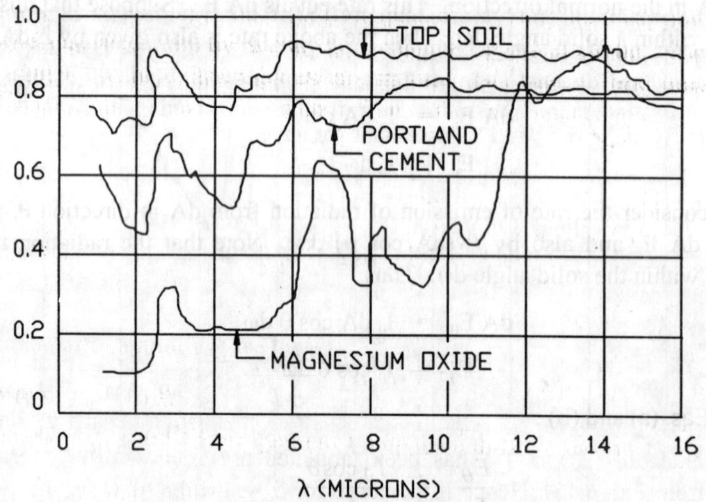

Fig. 6.15 α_λ Vs λ curves for Non-Conductors

The following points may be noted from these figures ;

(i) Solids which are good conductors of electricity (such as polished aluminum and copper) generally have low values of absorptivity (or emissivity) and are therefore good reflectors. They radiate very little energy. The value of α_λ or ε_λ generally decreases while the value of ρ_λ increases as the wavelength is increased.

(ii) Solids which are non-conductors of electricity (concrete, cement, etc.) generally have high values of absorptivity and consequently their emissivity is also high. Their reflectivity is relatively low. Moreover the value of α_λ or ε_λ generally increases with increase in λ. Note that when aluminum is anodized so that a non-conducting oxide coating is formed on the surface, its radiation characteristics are greatly changed.

For most radiation calculations the values of the total absorptivity α is required. In order to determine α from the α_λ curve of a surface, the spectral distribution of the incident radiation must be known. The ordinates of the incident radiation curve are multiplied by the values of α_λ for the same wavelengths. The area under the resulting curve divided by the area under the incident radiation curve gives the value of α.

$$\alpha = \frac{\int_0^\infty \alpha_\lambda E_\lambda \, d\lambda}{\int_0^\infty E_\lambda \, d\lambda} \tag{6.45}$$

If the incident radiation is black body radiation, Eq. (6.45) becomes ;

$$\alpha = \frac{\int_0^\infty \alpha_\lambda (E_\lambda)_b \, d\lambda}{\int_0^\infty (E_\lambda)_b \, d\lambda} \tag{6.46}$$

Since $\int_0^\infty (E_\lambda)_b \, d\lambda = \sigma T^4$ and T is constant we have ;

$$\alpha = \int_0^\infty \alpha_\lambda \frac{(E_\lambda)_b}{\sigma T^4} \frac{d(\lambda T)}{T}$$

or

$$\alpha = \int_0^\infty \alpha_\lambda \frac{(E_\lambda)_b}{\sigma T^5} \, d(\lambda T) \tag{6.47}$$

The value of $\{(E_\lambda)_b/\sigma T^5\}$ has been tabulated as function of λT. It can also be obtained from Fig. 6.13. Hence if the curve for α_λ is available, we can assume suitable increments of λT, tabulate the individual products $\{\alpha_\lambda (E_\lambda)_b/\sigma T^5\} \times \Delta(\lambda T)$ and then sum these products. This sum equals α.

The total absorptivity of various materials for incident black radiation at different temperatures is shown in Fig. 6.16. For non-conductors, the absorptivity *decreases* markedly with increase in temperature. Conductors (aluminum and graphite) on the other hand, show slow but definite *increase* in absorptivity with increasing temperature.

The emissivity of many surfaces varies also with direction. In other words, these surfaces do not obey the Lambert's cosine law as a black surface does. In Fig. 6.17 the directional emissivities ε_θ of several surfaces are shown plotted in polar diagrams. For a surface obeying the cosine law ε_θ would be constant and its emissivity curve would be a semicircle. It is evident from Fig. 6.17 that ;

(a) In the case of non-conductors such as wood, glass, paper, oxide films etc., ε_θ is nearly constant upto $\theta = 75°$, but beyond this value of θ; ε_θ drops off sharply.

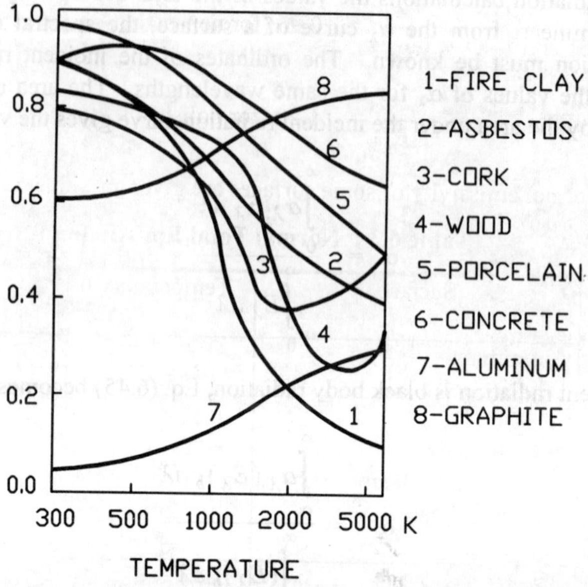

Fig. 6.16 Total Absorptivity for Different Materials

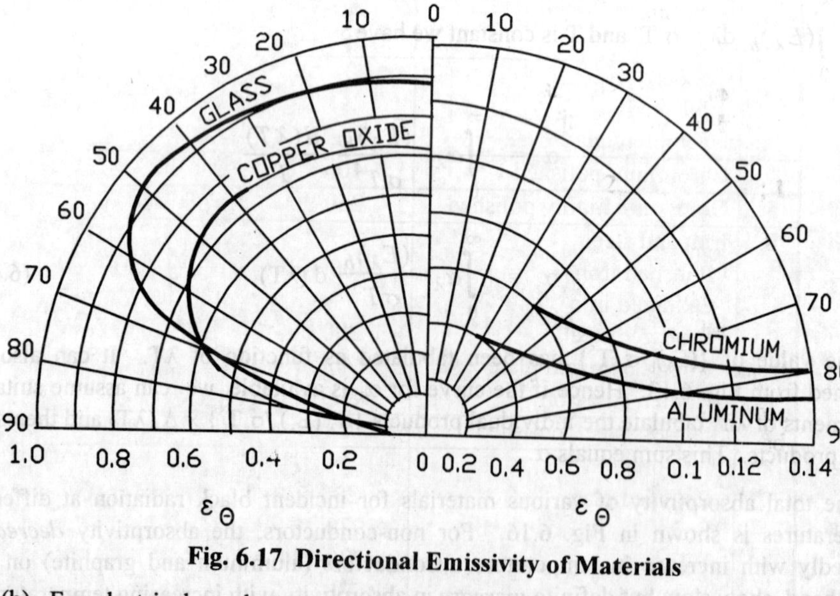

Fig. 6.17 Directional Emissivity of Materials

(b) For electrical conductors such as metals, ε_θ increases with angle θ and reaches maximum value very close to $\theta = 90°$.

For example the emissivity of polished chromium is only 0.06 in the normal direction but increases to 0.14 when viewed from an angle $\theta = 80°$. For this reason the total hemispherical emissivity of polished conductors may be 1.15 to 1.20 times the normal emissivity.

For engineering calculation it may be assumed that ;

$$\frac{\varepsilon}{\varepsilon_n} = 0.96 \text{ (for non-metals)} \tag{6.48}$$

$$\frac{\varepsilon}{\varepsilon_n} = 1.20 \text{ (for polished metals)} \tag{6.49}$$

The total normal emissivity of some surfaces are given in Table 6.3.

Table 6.3 : Normal Total Emissivities

No.	Surface	Temperature, °C	ε_n
1.	Aluminum		
	Highly polished	225-600	0.039-0.057
	Polished plate	100	0.095
	Rough plate	40	0.055-0.07
	Commercial sheet	100	0.09
	Oxidized plate	200	0.11
	Roofing	40	0.22
2.	Brass		
	Polished	100	0.06
	Rolled plate	20	0.06
	Dull plate	50-350	0.22
	Oxidized	200-600	0.10
3.	Copper		
	Polished	120	0.023
	Commercial	20	0.072
4.	Chromium, polished	40-1100	0.08-0.36
5.	Gold, pure highly polished	250-900	0.018-0.035
6.	Iron and steel		
	Polished iron	400-1000	0.144-0.377
	Oxidized iron	100	0.736
	Oxidized steel	200-600	0.72
	Rough oxidized plate	40-360	0.94-0.97
	Molten cast iron	1300-1400	0.29
7.	Stainless steel, polished	100	0.074
8.	Platinum		
	Polished plate	230-630	0.054-0.104
	Strip	930-1630	0.12-0.17
	Wire	230-1270	0.073-0.182
	Filament	30-1230	0.036-0.192
9.	Nickel		
	Nickel oxide	650-1250	0.59-0.86
	Nichrome wire, bright	50-1000	0.65-0.79
	Nichrome wire, oxidized	50-550	0.95-0.98
10.	Silver polished	230-630	0.02-0.032
11.	Tungston filament	3310	0.32

Table 6.3 (contd.) : **Normal Total Emissivities**

No.	Surface	Temperature, °C	ε_n
12.	Zinc		
	Commercial	230-530	0.045-0.053
	Galvanized sheet, bright	30	0.23
	Galvanized sheet, oxidized	25	0.28
13.	Carbon filament	1200	0.526
14.	Asbestos paper	40-350	0.93-0.945
15.	Concrete tiles	1000	0.63
16.	Red brick, rough	20	0.93
17.	Refractory brick	600-1000	0.80-0.90
18.	Paper	20	0.93
19.	Plaster (rough lime)	10-90	0.91
20.	Roofing paper	20	0.91
21.	Ashphalt	40	0.93
22.	Marble white	40	0.95
23.	Ice	0	0.27
24.	Water	0-100	0.95-0.963
25.	Glass	20	0.94
26.	Paints		
	Aluminized lacquer	100	0.52
	Black lacquer	20	0.82
	White enamel	20	0.90

6.16 RADIATION HEAT EXCHANGE BETWEEN BLACK SURFACES

Heat exchange by radiation between two surfaces depends not only on the emission, absorption, and reflection characteristics of the surfaces but also on their geometrical arrangement. For this reason an analysis of the radiant heat exchange becomes somewhat complicated. Further complication may arise if a partially absorbing and emitting medium is present in between the surfaces. The usual method of approach is to first consider *idealized* systems and then make suitable modifications so as to approximate the actual systems.

We shall therefore consider first the case of two *black* surfaces separated by a *non-absorbing* medium, such as air. Suppose we have to determine the radiant heat exchange between two surfaces A_1 and A_2 as shown in Fig.6.18. Let us first evaluate the heat exchange between an elemental area ΔA_1 of surface A_1 and an elemental area ΔA_2 of surface A_2.

The rate at which radiant energy leaving ΔA_1 impinges on ΔA_2 is equal to the product of the intensity of radiation I_{b1} of surface A_1, the projected area of ΔA_1 on a plane perpendicular to the line joining ΔA_1 and ΔA_2, and the solid angle which ΔA_2 subtends with centre on ΔA_1.

Now, projected area of ΔA_1 normal to line joining ΔA_1 and ΔA_2,

$$= \Delta A_1 \cos\theta_1$$

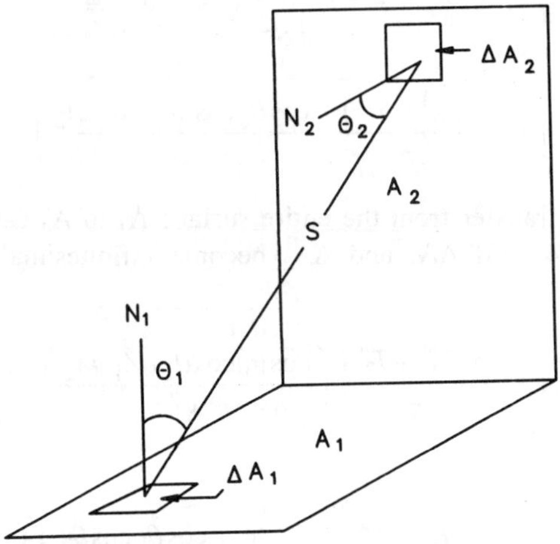

Fig. 6.18 Thermal Radiation between Two Black Bodies

Solid angle subtended by ΔA_2 with centre on ΔA_1

$$\Delta\omega_2 = \frac{\Delta A_2 \cos\theta_2}{S^2}$$

Hence rate at which energy emitted by ΔA_1 impinges on ΔA_2,

$$= I_{b1}\, \Delta A_1 \cos\theta_1 \frac{\Delta A_2 \cos\theta_2}{S^2}$$

Since A_2 is a black surface, all of the above energy is absorbed. Hence, rate at which energy emitted by ΔA_1 is absorbed by ΔA_2 is given by,

$$\Delta q_{12} = \frac{I_{b1} \cos\theta_1 \cos\theta_2\, \Delta A_1\, \Delta A_2}{S^2} \tag{6.50}$$

Similarly, the rate at which radiation energy emitted by ΔA_2 is absorbed by ΔA_1 is given by,

$$\Delta q_{21} = \frac{I_{b2} \cos\theta_1 \cos\theta_2\, \Delta A_1\, \Delta A_2}{S^2} \tag{6.51}$$

Therefore, the net transfer of radiant energy from ΔA_1 to ΔA_2 is,

$$(\Delta q_{12})_{net} = \Delta q_{12} - \Delta q_{21}$$

$$= (I_{b1} - I_{b2}) \frac{\cos\theta_1 \cos\theta_2\, \Delta A_1\, \Delta A_2}{S^2} \tag{6.52}$$

Now, $$I_{b1} = \frac{E_{b1}}{\pi} \text{ and } I_{b2} = \frac{E_{b2}}{\pi} \tag{6.33}$$

hence, substituting Eq. (6.33) in Eq. (6.52) we get ;

$$(\Delta q_{12})_{net} = \frac{(E_{b1} - E_{b2})\cos\theta_1 \cos\theta_2 \,\Delta A_1 \,\Delta A_2}{\pi S^2} \qquad (6.53)$$

or

$$(\Delta q_{12})_{net} = \frac{\sigma(T_1^4 - T_2^4)}{\pi}\left(\frac{\cos\theta_1 \cos\theta_2 \,\Delta A_1 \,\Delta A_2}{S^2}\right) \qquad (6.54)$$

The net rate of heat transfer from the entire surface A_1 to A_2 can be determined by integration of Eq. (6.54). If ΔA_1 and ΔA_2 become infinitesimally small, Eq.(6.54) reduces to ;

$$(dq_{12})_{net} = \frac{\sigma(T_1^4 - T_2^4)}{\pi}\left(\frac{\cos\theta_1 \cos\theta_2 \,dA_1 \,dA_2}{S^2}\right) \qquad (6.55)$$

Integrating Eq. (6.55),

$$(q_{12})_{net} = \frac{\sigma(T_1^4 - T_2^4)}{\pi}\int_{A2}\int_{A1}\left(\frac{\cos\theta_1 \cos\theta_2 \,dA_1 \,dA_2}{S^2}\right) \qquad (6.56)$$

Equation (6.56) gives the net rate of radiant heat transfer from surface A_1 to surface A_2.

6.17 THE GEOMETRICAL FACTOR

(i) If we integrate Eq. (6.50), we obtain,

$$q_{12} = \frac{\sigma T_1^4}{\pi}\int_{A2}\int_{A1}\left(\frac{\cos\theta_1 \cos\theta_2 \,dA_1 \,dA_2}{S^2}\right) \qquad (6.57)$$

Now the rate at which energy is emitted by A_1 in all directions is given by $\sigma A_1 T_1^4$. Since only a part of this energy falls on A_2 it is evident that ;

$$\frac{q_{12}}{A_1 \sigma T_1^4} < 1$$

or

$$\frac{1}{A_1 \pi}\int_{A1}\int_{A2}\frac{\cos\theta_1 \cos\theta_2 \,dA_1 \,dA_2}{S^2} < 1 \qquad (6.58)$$

Note that the quantity on the left hand side of Eq. (6.58) represents the ratio of the energy emitted by A_1 and intercepted by A_2 to the total energy emitted by A_1 in all directions. The ratio involves only the geometry of the system and is independent of T_1 and T_2. This ratio is given the name *geometrical factor* or *configuration factor* and is denoted by F_{12} given by ;

$$F_{12} = \frac{1}{A_1 \pi}\int_{A1}\int_{A2}\frac{\cos\theta_1 \cos\theta_2 \,dA_1 \,dA_2}{S^2} \qquad (6.59)$$

Note that F_{12} is the configuration factor of surface A_1 with respect to surface A_2 and it represents that fraction of the total energy emitted by A_1 which is intercepted by A_2. Substituting Eq. (6.59) in Eq. (6.57).

$$q_{12} = A_1 F_{12} \sigma T_1^4 \qquad (6.60)$$

Similarly we can obtain,

$$q_{21} = A_2 F_{21} \sigma T_2^4 \quad (6.61)$$

where,

$$F_{21} = \frac{1}{A_2 \pi} \int_{A1} \int_{A2} \frac{\cos\theta_1 \cos\theta_2 \, dA_1 \, dA_2}{S^2} \quad (6.62)$$

From Eqs. (6.60) and (6.61) we get ;

$$(q_{12})_{net} = (A_1 F_{12} \sigma T_1^4 - A_2 F_{21} \sigma T_2^4) \quad (6.63)$$

(ii) We shall now demonstrate the interesting result i.e.

$$A_1 F_{12} = A_2 F_{21}$$

Suppose that $T_1 = T_2$. In this case $(q_{12})_{net}$ must be zero. But Eq. (6.63) must be valid even when $T_1 = T_2$.

Hence, $$(q_{12})_{net} = 0 = \sigma T_1^4 (A_1 F_{12} - A_2 F_{21})$$

Since σ and T_1 are both non-zero quantities, it follows that,

$$A_1 F_{12} = A_2 F_{21} \quad (6.64)$$

The Eq. (6.64) is often called the *reciprocity theorem.*

Using Eq. (6.64) in Eq. (6.63) we obtain finally ;

$$(q_{12})_{net} = A_1 F_{12} \sigma (T_1^4 - T_2^4) = A_2 F_{21} \sigma (T_1^4 - T_2^4) \quad (6.65)$$

For the general case of heat exchange between two black surfaces we can write Eq. (6.65) in the form,

$$q_{net} = A F \sigma (T_1^4 - T_2^4) \quad (6.66)$$

where A = area of one of the surfaces.

F = the geometrical factor of this surface with respect to the other.

The actual determination of F by analytical means is rather difficult except in some simple cases. Graphical methods have been used to determine F for configurations generally encountered in practice. Figure 6.19 gives the configuration factors for black parallel planes (discs, rectangles and squares) placed directly opposed. Geometric factors for adjacent rectangles normal to each other are given in Fig. 6.20. Values of F for surface element dA_1 and parallel rectangle above it are given in Fig. 6.21.

Note that the geometric factor F_{12} defined above is for black surfaces. If the surface is not black it will not absorb all the radiation falling on it. However, if emissivities of the two surfaces are constant (gray surfaces) and do not depend on the angle of emission, F_{12} will still represent the ratio of the energy emitted by A_1 and intercepted by A_z to the total energy emitted by A_1 in all directions.

It also follows from the definition of F that if heat exchange occurs between several surfaces, then ;

$$F_{11} + F_{12} + F_{13} + \ldots = 1 \quad (6.67)$$

If from any part of a surface one *can not see* any other part of the same surface (flat or convex surface) then ;

$$F_{11} = 0$$

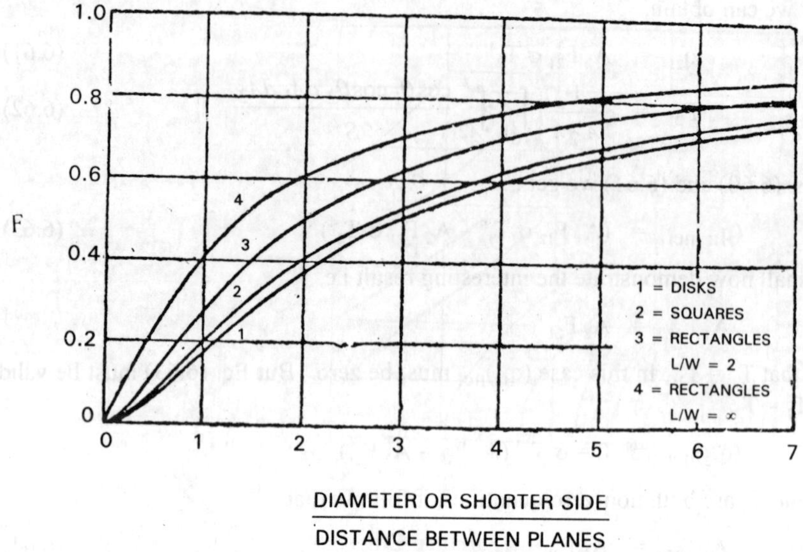

Fig. 6.19 Geometric Factors for Parallel Planes

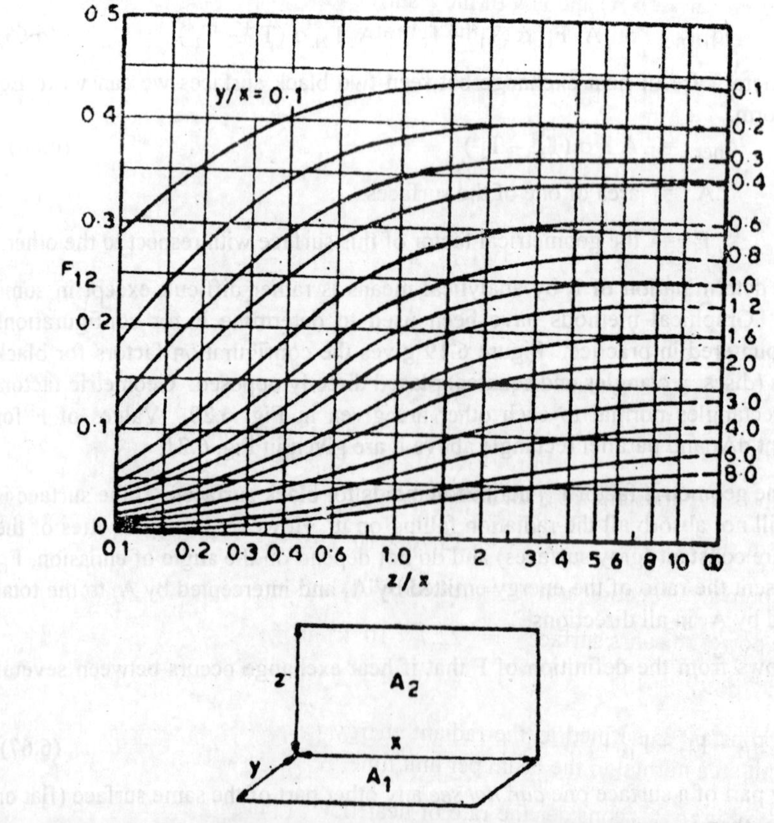

Fig. 6.20 Geometric Factors for Rectangles Normal to Each Other

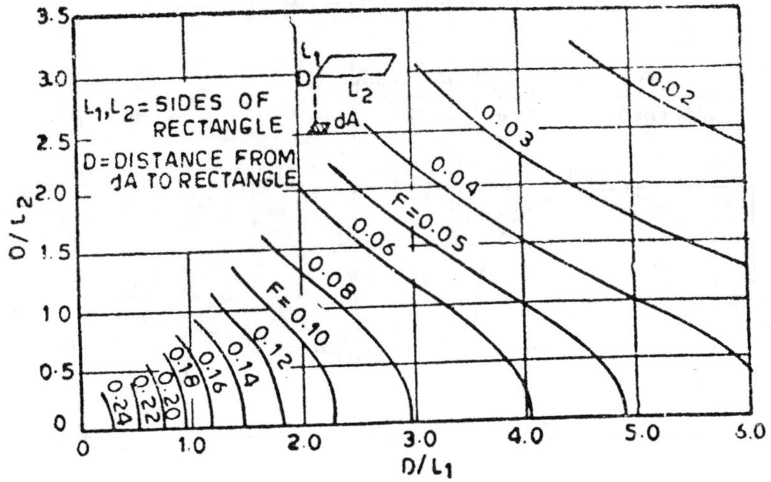

Fig. 6.21 Geometric Factors for a Surface Element and Parallel Rectangle

The net radiation from a black surface A_i in an enclosure consisting of several black surfaces can also be evaluated by using the geometrical factors F_{i1}, F_{i2} etc. Since the radiation exchange between A_i and any of the *n* surrounding surfaces (1, 2, 3 ... n) is not affected by the presence of other surfaces, the net rate of heat flow from A_i is ;

$$(q_i)_{net} = A_i F_{i1} (E_{bi} - E_{b1}) + A_i F_{i2} (E_{bi} - E_{b2}) + \ldots + A_i F_{in} (E_{bi} - E_{bn})$$

$$(q_i)_{net} = \sum_{k=1}^{k=n} A_i F_{ik} (E_{bi} - E_{bk}) \qquad (6.68)$$

If the surface A_i is convex ;

$$\Sigma F_{ik} = 1 \qquad (6.69)$$

otherwise,

$$F_{i1} + \Sigma F_{ik} = 1 \qquad (6.70)$$

Example 6.4

Calculate the solar constant from the following data ;

(i) Diameter of sun D_1 = 13.76×10^5 km

(ii) Distance between sun and earth = 14.86×10^7 km

(iii) Emissive power of sun's surface = 22.7×10^7 kJ/hr-m^2

Solution :

The solar constant is defined as the radiant energy from the sun striking the earth's surface per unit area normal to the beam per unit time.

Referring to Fig. 6.22, consider the rate of heat flow from an elemental area dA_1 on the surface of the sun to an area dA_2 on earth's surface. This rate is given by,

$$dq_{12} = \frac{I_1\, dA_1 \cos\theta_1\, dA_2 \cos\theta_2}{R^2}$$

or

$$\frac{dq_{12}}{dA_2 \cos\theta_2} = \left(\frac{E_I}{\pi}\right) \frac{dA_1 \cos\theta_1}{R^2} \qquad \text{(a)}$$

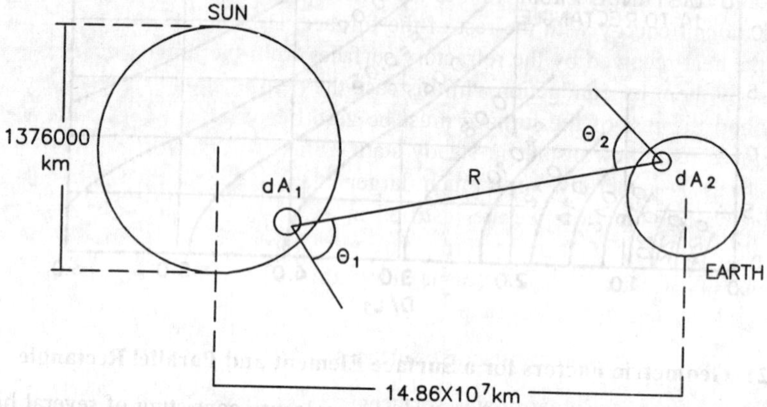

Fig. 6.22 Example 6.3

Since the projected area of A_2 normal to the line joining dA_1 and dA_2 is $dA_2 \cos\theta_2$, the RHS of Eq. (a) gives the rate at which energy radiated by dA_1 falls on unit area of earth's surface. But this unit area is also receiving radiation from other parts of sun's surface. Hence if we integrate the RHS of Eq. (a), we will get the solar constant ;

$$\text{Solar constant} = \int \frac{E_1}{\pi R^2}\, dA_1 \cos\theta_1$$

$$= \frac{E_1}{\pi R^2}\left(\frac{\pi}{4} D_1^2\right)$$

$$= \frac{22.7\times 10^7}{(14.86\times 10^7)^2} \times \frac{1}{4} \times (13.76 \times 10^5)^2$$

$$= 4960 \text{ kJ/hr-m}^2$$

Ans.

6.18 RADIANT HEAT EXCHANGE BETWEEN BLACK SURFACES CONNECTED BY REFRACTORY SURFACES

In many practical cases, radiant heat exchange between two surfaces will be affected by the nature of the *connecting* surfaces. If the connecting surfaces are *not black* surfaces, computation of heat exchange may generally be quite difficult. However, if the connecting surfaces are made of *refractory* material and are perfectly *reradiating* surfaces, the radiant heat exchange can be determined by the procedure explained below. The above condition is closely satisfied in an industrial furnace.

An industrial furnace is an enclosure which may be considered to be made up of the following surfaces :

(a) Surfaces constituting a heat source (fuel bed, electric resistors, etc.),

(b) Surfaces constituting a heat sink (boiler tubes, row of billets, etc.),

(c) Interconnecting refractory surfaces

Now, the interconnecting refractory surfaces, receive heat by convection from the furnace gases and lose heat by conduction to the exterior of the furnace. These surfaces also exchange radiant energy with the rest of the furnace, i.e., source and sink surfaces. Suppose that the heat received by the refractory surfaces from the furnace gases is equal to the heat lost by them by conduction. In this case the net radiation exchange between the refractory and the rest of the furnace must be zero because the temperature of the refractory surfaces remains constant in steady state. Since the radiation falling on the refractory surfaces is generally very much larger than the difference between gas convection and wall conduction, we are justified in assuming that the net radiant heat exchange at the refractory surfaces is zero. The refractory surfaces may therefore be regarded as non-flux surfaces as far as radiant energy transfer is concerned. Suppose that the source and sink surfaces are black, and

A_1, A_3, A_5 = areas of source surfaces

A_2, A_4, A_6 = areas of sink surfaces

A_P, A_Q, A_R, A_S = areas of refractory surfaces.

Now, the rate of radiation emitted by A_1 is $A_1 E_{b1}$. All of this energy must ultimately reach and be absorbed either by A_2, A_3, A_4, A_5, A_6 or returned to A_1 for absorption. None of it can be absorbed by the no-flux surfaces unless an equal quantity is emitted by them. We can therefore define a new geometric factor $\overline{F}$ in such a way that $A_1 \overline{F_{12}}$ would represent the fraction of $A_1 E_{b1}$ which reaches A_2 either directly or by the assistance of the non-flux surfaces. Then,

(i) Rate of direct radiant transfer from A_1 to A_2, = $F_{12} A_1 E_{b1}$

(ii) Rate of direct plus refractory-assisted radiant transfer from A_1 to A_2, = $\overline{F_{12}} A_2 E_{b1}$

(iii) Rate of direct radiant transfer from A_2 to A_1, = $F_{21} A_2 E_{b2}$

(iv) Rate of direct plus refractory-assisted radiant transfer from A_2 to A_1, = $\overline{F_{21}} A_2 E_{b2}$

(v) Net rate of direct plus refractory-assisted radiant transfer from A_1 to A_2,

$$(q_{12})_{net} = q_{1R2} = A_1 \overline{F_{12}} E_{b1} - A_2 \overline{F_{21}} E_{b2} \tag{6.71}$$

Since $q_{1R2} = 0$ when $T_1 = T_2$ or $E_{b1} = E_{b2}$, it follows that,

$$A_1 \overline{F_{12}} = A_2 \overline{F_{21}} \tag{6.72}$$

or

$$q_{1R2} = A_1 \overline{F_{12}} (E_{b1} - E_{b2})$$

$$= A_2 \overline{F_{21}} (E_{b1} - E_{b2}) \tag{6.73}$$

Note that in the given case the geometric factors obey the equation :

$$F_{11} + F_{12} + F_{13} + F_{14} + F_{15} + F_{16} + F_{1P} + F_{1Q} + F_{1R} + F_{1S} = 1 \tag{6.74}$$

Further, the factors $\overline{F}$ obey the equation ;

$$\overline{F_{11}} + \overline{F_{12}} + \overline{F_{13}} + \overline{F_{14}} + \overline{F_{15}} + \overline{F_{16}} = 1 \quad (6.75)$$

The factor $\overline{F}$ has been determined exactly for some simple cases and may be approximated for other cases. Figure 6.23 gives the values of $\overline{F}$ for directly opposed black parallel planes (discs, rectangles, squares) connected by refractory walls.

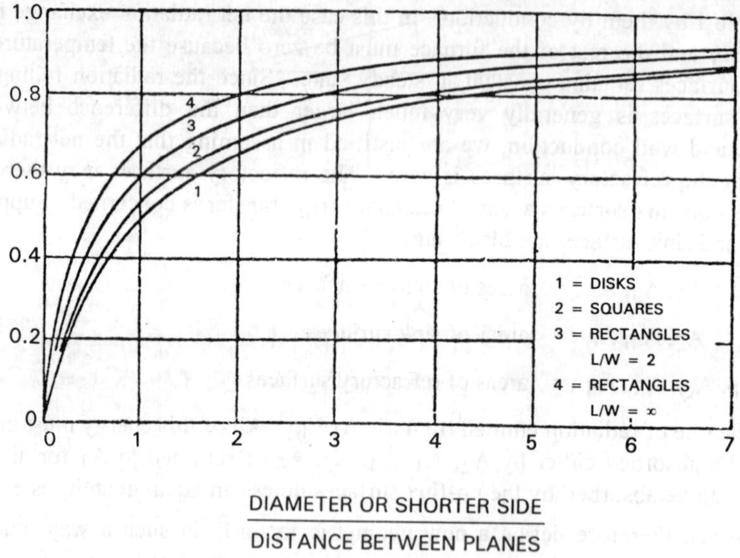

Fig. 6.23 Values of $\overline{F}$ for Opposed Planes

An expression for $\overline{F}_{12}$ can be easily obtained for the following case. Suppose the system consists of a single source surface A_1, single sink surface A_2 and the refractory surfaces A_P, A_Q, A_R etc., are so arranged in the enclosure that all elements of these surfaces are equally irradiated by A_1, that is, each element of the refractory surface intercepts per unit area the same fraction of energy emitted by A_1, similarly the refractory surfaces are equally irradiated by A_2. We can group all the refractory surfaces into a single surface of area A_R.

Let us now consider the radiation emitted by A_1 in unit time. This amount is $A_1 E_{b1}$. Of this quantity, the fractions reaching A_2 and A_R *directly* are as follows ;

Fraction reaching A_2 directly $= F_{12}$

Fraction reaching A_R directly $= F_{1R}$

A_2, being black, completely absorbs the fraction F_{12}. A_R being a no-flux surface must get rid of the fraction F_{1R} by reradiation or reflection, sending fractions F_{R1} to A_1, F_{R2} to A_2, and F_{RR} to itself for a second reflection or reradiation which again distributes between A_1, A_2 and A_R. It would be clear from Fig. 6.24 that ;

The fraction of the original energy emitted by A_1 which reaches A_2 *through the assistance* of A_R is,

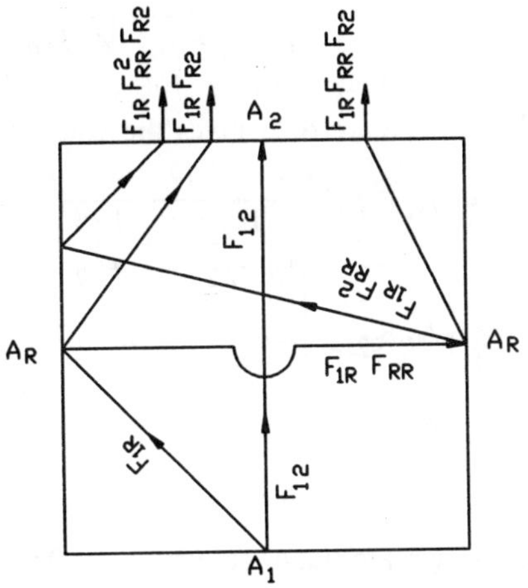

Fig. 6.24 Radiant Heat Exchange between A_1, A_2 and Refractory Surfaces grouped as A_R

$$= F_{1R}F_{R2} + F_{1R}F_{RR}F_{R2} + F_{1R}(F_{RR})^2F_{R2} + \ldots$$

$$= F_{1R}F_{R2}(1 + F_{RR} + F_{RR}^2 + F_{RR}^3 + \ldots)$$

$$= \frac{F_{1R}F_{R2}}{1-F_{RR}}$$

Hence, by definition,

$$\overline{F_{12}} = F_{12} + \frac{F_{1R}F_{R2}}{1-F_{RR}} \quad \text{(a)}$$

or $$A_1\overline{F_{12}} = A_1F_{12} + \frac{A_1F_{1R}F_{R2}}{1-F_{RR}}$$

or $$A_1\overline{F_{12}} = A_1F_{12} + \frac{F_{R2}}{\frac{1}{A_1F_{1R}} - \frac{F_{RR}}{A_1F_{1R}}} \quad \text{(b)}$$

Since,

$$F_{11} + F_{12} + F_{1R} = 1$$

$$F_{RR} + F_{R1} + F_{R2} = 1$$

or $$F_{RR} = 1 - F_{R1} - F_{R2}$$

Hence, Eq. (b) gives,

$$A_1\overline{F_{12}} = A_1F_{12} + \frac{F_{R2}}{\frac{1}{A_1F_{1R}} - \frac{1}{A_1F_{1R}} + \frac{F_{R1}}{A_1F_{1R}} + \frac{F_{R2}}{A_1F_{1R}}}$$

$$= A_1 F_{12} + \frac{1}{\dfrac{F_{R1}}{A_1 F_{1R} F_{R2}} + \dfrac{1}{A_1 F_{1R}}}$$

Since $A_1 F_{1R} = A_R F_{R1}$,

$$A_1 \overline{F_{12}} = A_1 F_{12} + \frac{1}{\dfrac{1}{A_R F_{R2}} + \dfrac{1}{A_1 F_{1R}}}$$

Also $A_R F_{R2} = A_2 F_{2R}$,

hence
$$A_1 \overline{F_{12}} = A_1 F_{12} + \frac{1}{\dfrac{1}{A_1 F_{1R}} + \dfrac{1}{A_2 F_{2R}}} \quad (6.76)$$

If the source and sink surfaces are flat or convex, $F_{11} = 0$ and $F_{22} = 0$. Hence,

$$F_{12} + F_{1R} = 1$$

and
$$F_{21} + F_{2R} = 1$$

or
$$A_1 \overline{F_{12}} = A_1 F_{12} + \frac{1}{\dfrac{1}{A_1 (1 - F_{12})} + \dfrac{1}{A_2 (1 - F_{21})}} \quad (6.77)$$

or
$$A_1 \overline{F_{12}} = A_1 F_{12} + \frac{A_1 A_2 (1 - F_{12})(1 - F_{21})}{A_2 - A_2 F_{21} + A_1 - A_1 F_{12}}$$

$$= \{A_1^2 F_{12} + A_1 A_2 F_{12} - 2 A_1^2 F_{12}^2 + A_1 A_2 - A_1 A_2 F_{12} - A_1 A_2 F_{21} + A_1 A_2 F_{12} F_{21}\} \div (A_1 + A_2 - 2 A_1 F_{12})$$

Simplifying and using $A_1 F_{12} = A_2 F_{21}$,

$$A_1 \overline{F_{12}} = \frac{A_1 (A_2 - A_1 F_{12}^2)}{A_1 + A_2 - 2 A_1 F_{12}} \quad (6.78)$$

or
$$\overline{F_{12}} = \frac{A_2 - A_1 F_{12}^2}{A_1 + A_2 - 2 A_1 F_{12}} \quad (6.79)$$

Thus, Eq. (6.79) enables $\overline{F_{12}}$ to be determined if F_{12} is known.

Example 6.5

Two black discs of diameter 0.5 m are placed directly opposite at a distance of 1 m. The discs are maintained at 1000 K and 500 K respectively. Calculate the heat flow between the discs : (a) when no other surfaces are present, (b) when the discs are connected by right-cylindrical black no-flux surface.

Solution :

(a)
$$(q_{12})_{net} = A_1 F_{12} \sigma (T_1^4 - T_2^4) \quad (6.65)$$

$$A_1 = \frac{\pi}{4}(0.5)^2 = \frac{0.25\pi}{4} \text{ m}^2$$

To find F_{12} let us refer to Fig. 6.19, curve 1 ;

For $\frac{\text{diameter}}{\text{distance}} = \frac{0.5}{1} = 0.5$

we get, $F_{12} = 0.06$

so $(q_{12})_{net} = 0.25\,(\pi/4)\,(0.06)\,(2.04 \times 10^{-4}) \times (1000^4 - 500^4)$

$= 2.25 \times 10^6$ J/hr = 2250 kJ/hr Ans.

(b) $(q_{12})_{net} = A_1\, \overline{F_{12}}\, \sigma\, (T_1^4 - T_2^4)$

From Fig. (6.23) curve 1, $\overline{F_{12}} = 0.34$

or $(q_{12})_{net} = 0.25\,(\pi/4)\,(0.34)\,(2.04 \times 10^{-4}) \times (1000^4 - 500^4)$

$= 12.75 \times 10^6$ J/hr Ans.

6.19 RADIANT HEAT EXCHANGE BETWEEN NON-BLACK (GRAY) SURFACES

In the preceding sections we considered radiation between *black* surfaces. The assumption that the surface is black simplifies the problem because all of the incident radiation is absorbed. There are, however, many problems involving surfaces which are not black. In this case the analysis becomes very difficult unless the surfaces are considered to be *gray*. The radiation exchange between gray surfaces would evidently depend upon the temperatures of the surfaces, the geometry of the enclosure, and the emissivities of the surfaces. For this case, Hottel has defined another configuration factor F_g such that the net exchange from A_1 to A_2 is given by ;

$$(q_{12})_{net} = A_1 (F_g)_{12} (E_{b1} - E_{b2}) = A_2 (F_g)_{21} (E_{b1} - E_{b2}) \quad (6.80)$$

or

$$(q_{12})_{net} = A_1 (F_g)_{12}\, \sigma\, (T_1^4 - T_2^4) = A_2 (F_g)_{21}\, \sigma\, (T_1^4 - T_2^4) \quad (6.81)$$

If only one source and one sink surface is present in an enclosure, it is found that F_g is given by ;

$$(F_g)_{12} = \frac{1}{\dfrac{1}{\overline{F_{12}}} + \left(\dfrac{1}{\varepsilon_1} - 1\right) + \dfrac{A_1}{A_2}\left(\dfrac{1}{\varepsilon_2} - 1\right)} \quad (6.82)$$

Here ε_1 and ε_2 are the emissivities of the surfaces A_1 and A_2 respectively.

Equation (6.82) is valid both when A_1 and A_2 are *connected by refractory surfaces* and when A_1 and A_2 are not *connected by* any intermediate surface. When there are no connecting surfaces $\overline{F_{12}} = F_{12}$ and Eq. (6.82) yields ;

$$(F_g)_{12} = \frac{1}{\dfrac{1}{F_{12}} + \left(\dfrac{1}{\varepsilon_1} - 1\right) + \dfrac{A_1}{A_2}\left(\dfrac{1}{\varepsilon_2} - 1\right)} \quad (6.82a)$$

Equation (6.82) can be further simplified for some simple geometrical arrangements mentioned below :

(a) Infinite Parallel Planes (Gray)

In this case the planes are not connected by any intermediate surface. Therefore $(F_g)_{12}$ is given by Eq. (6.82a). Here F_{12} represents the configuration factor of surface A_1 with respect to surface A_2 when both of these surfaces are black.

For infinite parallel black planes, it is evident that all the radiation emitted by one plane must reach and be absorbed by the other plane. Hence $F_{12} = F_{21} = 1$. Also $A_1 = A_2$.

$$(F_g)_{12} = \frac{1}{\frac{1}{1} + \left(\frac{1}{\varepsilon_1} - 1\right) + \left(\frac{1}{\varepsilon_2} - 1\right)}$$

or

$$(F_g)_{12} = \frac{1}{\frac{1}{\varepsilon_1} + \frac{1}{\varepsilon_2} - 1} \qquad (6.82b)$$

Aliter

Equation (6.82b) can also be obtained directly by analysis as follows.

Referring to Fig. 6.25, and considering unit area of each plane we find ;

$$q_{12} = \alpha_2 E_1 + \alpha_2 E_1 (1 - \alpha_1)(1 - \alpha_2) + \alpha_2 E_1 (1 - \alpha_1)^2 (1 - \alpha_2)^2 + \ldots\ldots$$

$$= \alpha_2 E_1 [1 + (1 - \alpha_1)(1 - \alpha_2) + (1 - \alpha_1)^2 (1 - \alpha_2)^2 + \ldots\ldots]$$

$$= \alpha_2 E_1 \left[\frac{1}{1 - (1 - \alpha_1)(1 - \alpha_2)}\right]$$

$$= \varepsilon_2 E_1 \left[\frac{1}{1 - (1 - \varepsilon_1)(1 - \varepsilon_2)}\right]$$

or

$$q_{12} = \varepsilon_2 \varepsilon_1 E_{b1} \left[\frac{1}{1 - (1 - \varepsilon_1)(1 - \varepsilon_2)}\right] \qquad (a)$$

so

$$q_{21} = \varepsilon_1 \varepsilon_2 E_{b2} \left[\frac{1}{1 - (1 - \varepsilon_1)(1 - \varepsilon_2)}\right] \qquad (b)$$

Therefore the net rate of radiation transfer from A_1 to A_2 per unit area of both is obtained from Eqs. (a) and (b),

$$(q_{12})_{net} = (E_{b1} - E_{b2}) \left[\frac{\varepsilon_1 \varepsilon_2}{1 - (1 - \varepsilon_1)(1 - \varepsilon_2)}\right]$$

$$= (E_{b1} - E_{b2}) \left[\frac{\varepsilon_1 \varepsilon_2}{1 - 1 + \varepsilon_1 + \varepsilon_2 - \varepsilon_1 \varepsilon_2)}\right]$$

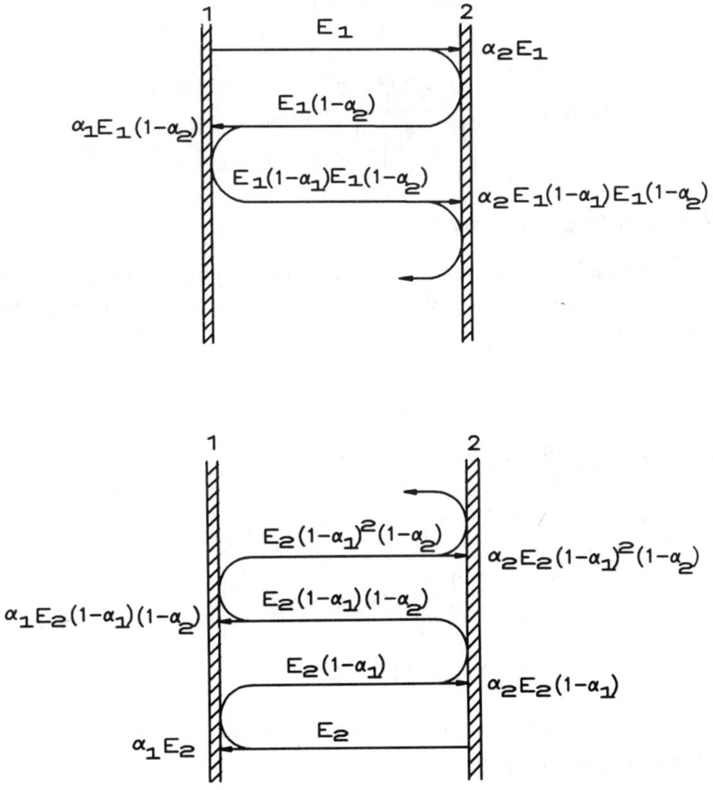

Fig. 6.25 Determination of the value of $(F_g)_{12}$

or
$$(q_{12})_{net} = (E_{b1} - E_{b2})\left[\frac{1}{\dfrac{1}{\varepsilon_1}+\dfrac{1}{\varepsilon_2}-1}\right] \tag{c}$$

Comparing Eqs. (c) and (6.80) with $A_1 = 1$ we have,

$$(F_g)_{12} = \frac{1}{\dfrac{1}{\varepsilon_1}+\dfrac{1}{\varepsilon_2}-1} \tag{6.82b}$$

(b) Concentric Cylinders or Spheres

Here again no connecting surfaces are present and $\overline{F_{12}} = F_{12}$. Since the inner cylinder or sphere of area A_1 sees only the outer surface and not itself, $F_{12} = 1$. Hence,

$$(F_g)_{12} = \frac{1}{\dfrac{1}{1}+\left(\dfrac{1}{\varepsilon_1}-1\right)+\dfrac{A_1}{A_2}\left(\dfrac{1}{\varepsilon_2}-1\right)}$$

or
$$(F_g)_{12} = \frac{1}{\frac{1}{\varepsilon_1} + \frac{A_1}{A_2}\left(\frac{1}{\varepsilon_2} - 1\right)} \quad (6.82c)$$

and
$$(q_{12})_{net} = \frac{A_1 \sigma (T_1^4 - T_2^4)}{\frac{1}{\varepsilon_1} + \frac{A_1}{A_2}\left(\frac{1}{\varepsilon_2} - 1\right)}$$

(c) Small Body in Large Enclosure

Practical examples of this case include a pipe carrying steam in a large room or a thermocouple bed located inside a duct for measurement of fluid temperature. It is clear that if we designate the small body by suffix 1 then ;

$$\overline{F_{12}} = F_{12} = 1$$

Also $A_1 << A_2$ and $\frac{A_1}{A_2} = 0$.

Hence,
$$(F_g)_{12} = \frac{1}{1 + \left(\frac{1}{\varepsilon_1} - 1\right)} = \varepsilon_1 \quad (6.82b)$$

and
$$(q_{12})_{net} = A_1 \varepsilon_1 \sigma (T_1^4 - T_2^4)$$

Table 6.4 gives the value of the factors F_{12}, $(F_g)_{12}$ and $\overline{F_{12}}$ to be used in cases commonly encountered in practice.

Example 6.6

Compute the direct radiation heat exchange between a concrete floor ($\varepsilon = 0.6$) 4.5 m × 6 m and the plastered wall ($\varepsilon = 0.9$) 3 m high and adjacent to the 6 m side. The floor is at 67 C and the wall is at 27 C.

Solution :

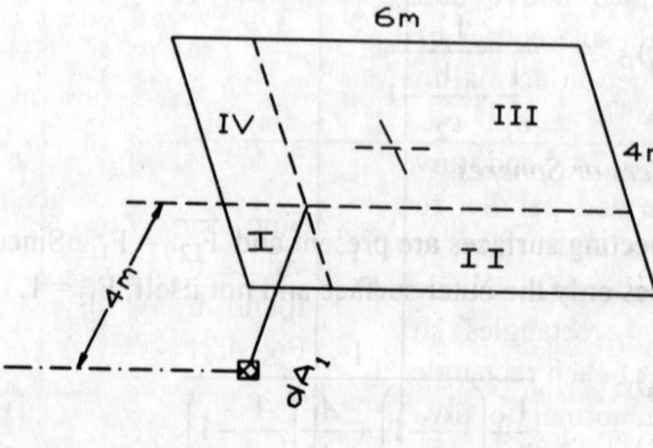

Fig. 6.26 Example 6.6

$$(q_{12})_{net} = A_1 (F_g)_{12} \sigma (T_1^4 - T_2^4)$$

$$A_1 = 4.5 \times 6 = 27 \text{ m}^2$$

From Table 6.4

$$(F_g)_{12} = F_{12} \varepsilon_1 \varepsilon_2$$

To find F_{12} refer to Fig. 6.20.

$$\frac{y}{x} = \frac{4.5}{6} = 0.75 \text{ and } \frac{z}{x} = \frac{3}{6} = 0.50$$

Table 6.4 Radiation Between Solid Surfaces $[q = A_1 (F_g)_{12} \sigma (T_1^4 - T_2^4)]$

No.	Description of surface	Area A_1	F_{12}	$\overline{F_{12}}$	$(F_g)_{12}$
1.	Infinite parallel planes.	A_1 or A_2	1	F_{12}	$\dfrac{1}{\frac{1}{\varepsilon_1} + \frac{1}{\varepsilon_2} - 1}$
2.	Small enclosed body 1 in large enclosure 2. Body 1 cannot see itself.	A_2	1	F_{12}	ε_1
3.	Concentric spheres or cylinders Relatively large convex body completely enclosed. Suffix 1 denotes smaller surface.	A_1	1	F_{12}	$\dfrac{1}{\frac{1}{\varepsilon_1} + \frac{A_1}{A_2}\left(\frac{1}{\varepsilon_2} - 1\right)}$
4.	Equal parallel squares, rectangles or discs, directly opposed.	A_1 or A_2	Fig. 4.19	$\frac{1+F_{12}}{2}$	$\dfrac{1}{\frac{1}{F_{12}} + \frac{1}{\varepsilon_1} + \frac{1}{\varepsilon_2} - 2}$
5.	Equal parallel squares, rectangles or discs directly opposed and connected by refractory surface.	A_1 or A_2	Fig. 4.19	Fig 4.23	$\dfrac{1}{\frac{1}{\overline{\overline{F_{12}}}} + \frac{1}{\varepsilon_1} + \frac{1}{\varepsilon_2} - 2}$
6.	Two rectangles with common side at right angles to each other.	A_1 or A_2	Fig. 4.20	-	$F_{12} (\varepsilon_1 \varepsilon_2)$ (appr.)
7.	Surface element dA_1 and rectangular surface above and parallel to it with one corner of rectangle contained in normal to it.	DA_1	Fig. 4.21	-	$F_{12} (\varepsilon_1 \varepsilon_2)$
8.	Surface element dA_1 and any rectangle above and parallel to it. (Draw normal to dA_1 to intersect the rectangle. Divide rectangle into 4 rectangles so that one corner of each rectangle is contained in normal to dA_1. Then treat as in item 7 above).	dA_1	Sum of four values of F found from Fig. 4.21	-	$F_{12} (\varepsilon_1 \varepsilon_2)$

Hence, $F_{12} = 0.185$

or $(q_{12})_{net} = (27)(0.185)(0.6)(0.9)(2.04 \times 10^{-7})(340^4 - 300^4)$

$= 2896$ kJ/hr Ans.

Example 6.7

A wall having an emissivity of 0.9 is 6 m wide and 4 m high. It is at a distance of 4 m from a 20 cm × 20 cm opening in a furnace wall. The centre line of the opening lies 1 m lower and 1 m left of centre of the wall. The furnace temperature is 1500 C and that of the wall is 27 C. Calculate the radiation heat transfer.

Solution :

$$(q_{12})_{net} = A_1 (F_g)_{12}\, \sigma\, (T_1^4 - T_2^4) \quad (6.81)$$

$$A_1 = dA_1 = \frac{20 \times 20}{100^2} = 0.04 \text{ m}^2$$

From Table 6.4, we get $(F_g)_{12} = F_{12}\, \varepsilon_1\, \varepsilon_2$

We can consider the furnace opening to act as a black body. Hence $\varepsilon_1 = 1$. To find F_{12}, divide the wall into 4 rectangles as shown in Fig. 6.27. The sides of these rectangles are :

I : 1m × 2m; II : 1m × 4m

III : 3m × 4m; IV : 2m × 3m

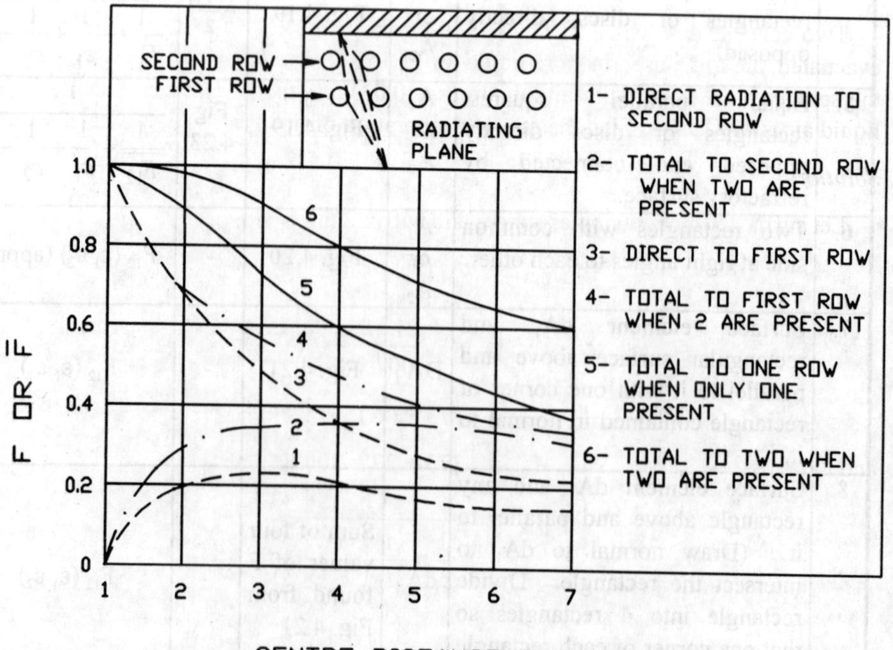

Fig. 6.27 Example 6.7

(i) Rectangle I :

$$\frac{D}{L_1} = \frac{4}{1} = 4 \text{ and } \frac{D}{L_2} = \frac{4}{2} = 4$$

$$F = 0.033$$

(ii) Rectangle II :

$$\frac{D}{L_1} = \frac{4}{1} = 4 \text{ and } \frac{D}{L_2} = \frac{4}{4} = 1$$

$$F = 0.05$$

(iii) Rectangle III :

$$\frac{D}{L_1} = \frac{4}{3} = 1.33 \text{ and } \frac{D}{L_2} = \frac{4}{4} = 1$$

$$F = 0.12$$

(iv) Rectangle IV :

$$\frac{D}{L_1} = \frac{4}{2} = 2 \text{ and } \frac{D}{L_2} = \frac{4}{3} = 1.333$$

$$F = 0.08$$

Thus,

$$F_{12} = 0.033 + 0.05 + 0.12 + 0.08 = 0.283$$

$$(F_g)_{12} = 0.283 \times 1 \times 0.9 = 0.2547$$

$$(q_{12})_{net} = (0.04)\,(0.2547)\,(2.04 \times 10^{-7})\,(1773^4 - 300^4)$$

$$= 20.55 \times 10^3 \text{ kJ/hr}$$ Ans.

Example 6.8

Two concentric spheres, 21 cm and 30 cm in diameter, with the space between them evacuated, are to be used to store liquid air (– 153 C) in a room at 27 C. The surfaces of the spheres are flushed with aluminum (ε = 0.03) and the latent heat of vaporization of liquid air is 209 kJ/kg. Find the rate of evaporation of liquid air.

Solution :

Let suffix 1 denote the inner sphere and suffix 2 the outer sphere.

$$(q_{12})_{net} = A_1 (F_g)_{12}\, \sigma\, (T_1^4 - T_2^4)$$

$$A_1 = \pi \left(\frac{21}{100}\right)^2 = 0.1386 \text{ m}^2$$

$$T_1 = 273 + (-153) = 120 \text{ K}$$

and

$$T_2 = 273 + 27 = 300 \text{ K}$$

$$(F_g)_{12} = \frac{1}{\frac{1}{\varepsilon_1} + \frac{A_1}{A_2}\left(\frac{1}{\varepsilon_2} - 1\right)}$$

$$= \frac{1}{\frac{1}{0.03} + \left(\frac{21}{30}\right)^2 \left(\frac{1}{0.03} - 1\right)} = 0.0204$$

or $$(q_{12})_{net} = (0.1386)(0.0204)(2.04 \times 10^{-7})[(120)^4 - (300)^4]$$

$$= -4.536 \text{ kJ/hr}$$

or $$(q_{21})_{net} = 4.536 \text{ kJ/hr}$$

Rate of evaporation

$$= \frac{4.536}{209} = 0.0216 \text{ kg/hr} \quad \text{Ans.}$$

Example 6.9

The floor of a furnace consists of carborundum muffle (ε = 0.65) 4m × 4m. The heat from the muffle is transferred to a row of 10 cm diameter tubes placed at 25 cm centre to centre distance, parallel to the floor. The tubes are backed by a well insulated refractory roof. The tubes fill the furnace top, the area of which equals that of the floor. The distance from muffle top to the row of tubes is 4 m. The muffle surface temperature is 1500 K and the tubes are at 600 K. The side walls of the chamber are perfectly reradiating. The emissivity of the tubes is 0.8. Find the radiation heat transfer from the muffle to the tubes.

Solution :

This problem is somewhat complicated because of the presence of the refractory furnace top. If this were not present, the system would consist of one *source surface* (floor) and one *sink surface* (tubes) connected by reradiating walls. However, the solution can be found by replacing the refractory-backed tubes by an equivalent surface located just below the tubes.

Suppose that an imaginary black plane 1 of area A_1, equal to the area of the floor A_f, is located just below the tubes. Let this plane be at the same temperature as the tubes. Now, the rate at which radiation emitted by the tubes is absorbed by this imaginary plane, or the rate at which radiation emitted by the plane is absorbed by the tubes, is given by ;

$$q_{t1} = q_{1t} = A_t (F_g)_{t1} \sigma T_t^4 = A_1 (F_g)_{1t} \sigma T_1^4 \quad \text{(a)}$$

The factor $(F_g)_{1t}$ represents the configuration factor from A_1 to A_t when both A_1 and A_t are gray and are connected by reradiating wall (the refractory top above the tubes).

Hence, from Eq. (6.82),

$$(F_g)_{1t} = \frac{1}{\frac{1}{\overline{F}_{1t}} + \left(\frac{1}{\varepsilon_1} - 1\right) + \frac{A_1}{A_t}\left(\frac{1}{\varepsilon_t} - 1\right)}$$

From Fig. 6.27, curve 5,

$$\overline{F}_{1t} = 0.8$$

$$\frac{A_1}{A_t} = \frac{\text{centre distance}}{\text{tube perimeter}} = \frac{25}{10\pi} = 0.795$$

and $$\varepsilon_1 = 1; \ \varepsilon_t = 0.8$$

so
$$(F_g)_{1t} = \frac{1}{\dfrac{1}{0.8}+\left(\dfrac{1}{1}-1\right)+0.795\left(\dfrac{1}{0.8}-1\right)} = 0.69$$

Since the rate of energy emission by the refractory-backed tubes to the black plane A_1 is $\{A_1\,(F_g)_{1t}\,\sigma\,T_1^4\}$ from Eq. (a), it follows that the tube system is equal in total emissive power to a plane of area A_1 and of emissivity $(F_g)_{1t} = 0.69$. We can therefore replace the tube system by a gray plane of area A_1 and emissivity $\varepsilon_1 = 0.69$. The new furnace system now consists of the muffle floor and an equal and parallel gray floor of emissivity 0.69, at temperature $T_1 = T_t$, at a distance of 4 m and connected by reradiating wall. For this system,

$$(F_g)_{f1} = \frac{1}{\dfrac{1}{\overline{F}_{f1}}+\left(\dfrac{1}{\varepsilon_f}-1\right)+\dfrac{A_1}{A_f}\left(\dfrac{1}{\varepsilon_1}-1\right)}$$

$$\overline{F}_{f1} = 0.53 \text{ (from Fig. 6.23)}$$

or
$$(F_g)_{f1} = \frac{1}{\dfrac{1}{0.53}+\left(\dfrac{1}{0.65}-1\right)+\left(\dfrac{1}{0.69}-1\right)} = 0.348$$

or
$$(q_{f1})_{net} = \sigma\,A_f\,(F_g)_{f1}\,(T_f^4 - T_1^4)$$
$$= (0.204 \times 10^{-3})\,(16)\,(0.348)\,[(1500)^4 - (600)^4]$$
$$= 556 \times 10^7 \text{ J/hr}$$ Ans.

6.20 ELECTRICAL NETWORK METHOD FOR SOLVING RADIATION PROBLEMS

Heat transfer by radiation can also be determined by first reducing the actual system to an equivalent *electrical network* and then solving this network. In order to make use of this technique let us first define two new terms.

(i) The *irradiation G*, is defined as the total energy incident on a surface per unit area per unit time (kJ/hr m^2).

(ii) The *radiosity J*, is defined as the total energy leaving a surface per unit area and per unit time (kJ/hr m^2).

Consider a gray opaque surface of emissivity ε and absorptivity α. The total energy leaving the surface is the sum of the energy emitted by the surface and the energy reflected by it out of the incident energy.

Hence,
$$J = E + \rho G = \varepsilon E_b + (1 - \alpha)\,G$$

hence
$$J = \varepsilon E_b + (1 - \varepsilon)\,G \tag{6.83}$$

and
$$G = \frac{J - \varepsilon E_b}{1 - \varepsilon} \tag{6.84}$$

Note that for a black surface $\rho = 0$ and $\varepsilon = 1$;

Hence $$J = E_b$$

It follows from the definition of G and J that the net rate at which radiation leaves a surface of area A is,

$$q_{net} = A\,(J - G) \tag{6.85}$$

Combining Eqs. (6.84) and (6.85) we get ;

$$q_{net} = A\left[J - \frac{J - \varepsilon E_b}{1-\varepsilon}\right]$$

or

$$q_{net} = A\,\frac{\varepsilon}{1-\varepsilon}\,(E_b - J) = \frac{E_b - J}{(1-\varepsilon)/A\varepsilon} \tag{6.86}$$

Equation (6.86) states that the net rate at which radiation leaves a gray surface can be regarded as the *current flow* when a potential difference ($E_b - J$) exists across a resistance of $(1-\varepsilon)/A\varepsilon$. Thus we see that the effect of the emissivity and reflectivity of the surface can be taken into account by connecting a potential $E_b = \sigma T^4$ through a resistance $(1-\varepsilon)/(A\varepsilon)$, or a conductance $A\varepsilon/(1-\varepsilon)$, to a potential J. The potential J is determined by the surface and the rest of the enclosure. Now the effect of system geometry on the net direct radiation between two surfaces of radiosities J_1 and J_2 is the same as for similar black surfaces; it can therefore be expressed in terms of the factor F. Hence, the *direct* radiation exchanged between any two opaque gray surfaces A_1 and A_2 is given by :

$$(q_{12})_{net} = A_1 F_{12}(J_1 - J_2) = A_2 F_{21}(J_1 - J_2) \tag{6.87}$$

Equation, (6.86) and (6.87) form the basis for evaluating the radiant heat exchange between gray surfaces in an enclosure by means of an *equivalent network.* We shall now show the network for some special cases.

(a) System Consisting of Two Gray Surfaces

This is represented by network of Fig. 6.28 where the values of resistances as well as conductances are shown. The total resistance R_t between E_{b1} and E_{b2} is,

$$R_t = \frac{1-\varepsilon_1}{A_1\varepsilon_1} + \frac{1}{A_1 F_{12}} + \frac{1-\varepsilon_2}{A_2\varepsilon_2}$$

or

$$(q_{12})_{net} = \left\{\frac{1}{\dfrac{1-\varepsilon_1}{A_1\varepsilon_1} + \dfrac{1}{A_1F_{12}} + \dfrac{1-\varepsilon_2}{A_2\varepsilon_2}}\right\}(E_{b1} - E_{b2})$$

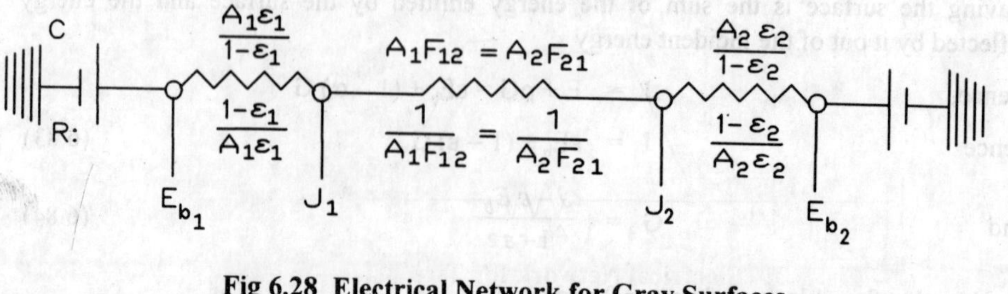

Fig 6.28 Electrical Network for Gray Surfaces

Rearranging the terms we get,

$$(q_{12})net = A_1 (E_{b1} - E_{b2}) \frac{1}{\frac{1}{F_{12}} + \left(\frac{1}{\varepsilon_1} - 1\right) + \frac{A_1}{A_2}\left(\frac{1}{\varepsilon_2} - 1\right)}$$

If we compare this equation with Eq. (6.80), we obtain Eq. (6.82a) again :

$$(F_g)_{12} = \frac{1}{\frac{1}{F_{12}} + \left(\frac{1}{\varepsilon_1} - 1\right) + \frac{A_1}{A_2}\left(\frac{1}{\varepsilon_2} - 1\right)} \qquad (6.82a)$$

(b) System Consisting of Two Black Surfaces

As mentioned earlier, for a black surface $J = E_b$. Hence, the equivalent network is as shown in Fig. 6.29. The net radiation transfer is given by ;

$$(q_{12})net = A_1 F_{12} (E_{b1} - E_{b2})$$

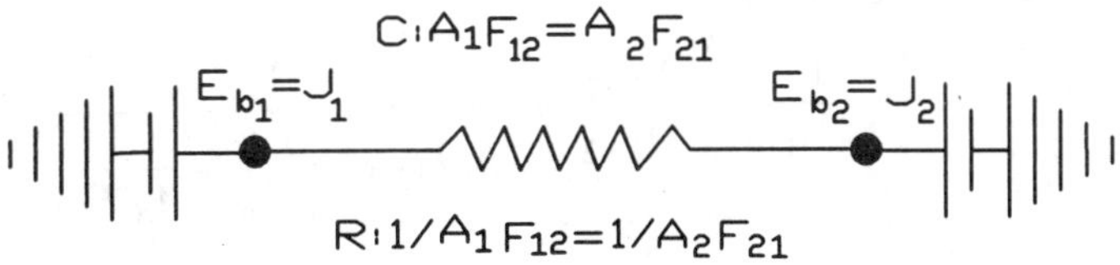

Fig. 6.29 Electrical Network for Two Black Surfaces

(c) System Consisting of Two Black Surfaces Connected by a Single Refractory Surface

Figure 6.30 represents the network for a system consisting of two black surfaces 1 and 2 which are connected by a single reradiating surface R. Note that the surface R is not connected to any potential because the *net* radiation transfer from this surface is zero. The total resistance between E_{b1} and E_{b2} is given by ;

$$\frac{1}{R_t} = A_1 F_{12} + \frac{1}{\frac{1}{A_1 F_{1R}} + \frac{1}{A_R F_{R2}}}$$

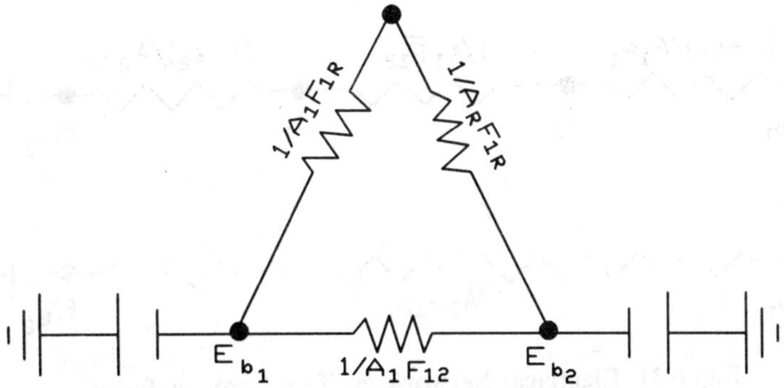

Fig. 6.30 Electrical Network for Two Black Surfaces Connected by a Refractory Surface

Since $F_{1R} = (1 - F_{12})$, $A_R F_{R2} = A_2 F_{2R}$ and $F_{2R} = (1 - F_{21})$,

$$\frac{1}{R_t} = A_1 F_{12} + \frac{1}{\dfrac{1}{A_1(1-F_{12})} + \dfrac{1}{A_2(1-F_{21})}}$$

$$\text{or } (q_{12})_{net} = (E_{b1} - E_{b2})\left\{A_1 F_{12} + \frac{1}{\dfrac{1}{A_1(1-F_{12})} + \dfrac{1}{A_2(1-F_{21})}}\right\}$$

Since $(q_{12})_{net}$ also equals $A_1 . \overline{F}_{12} (E_{b1} - E_{b2})$ we find,

$$A_1 \overline{F}_{12} = A_1 F_{12} + \frac{1}{\dfrac{1}{A_1(1-F_{12})} + \dfrac{1}{A_2(1-F_{21})}} \tag{6.77}$$

(d) System Consisting of Two Gray Surfaces Connected by a Single Refractory Surface

The Equivalent network for this system is shown in Fig. 6.31. The total resistance between E_{b1} and E_{b2} is now given by

$$R_t = \frac{1-\varepsilon_1}{A_1 \varepsilon_1} + \frac{1-\varepsilon_2}{A_2 \varepsilon_2} + \frac{1}{A_1 F_{12} + \dfrac{1}{\dfrac{1}{A_1 F_{1R}} + \dfrac{1}{A_R F_{R2}}}}$$

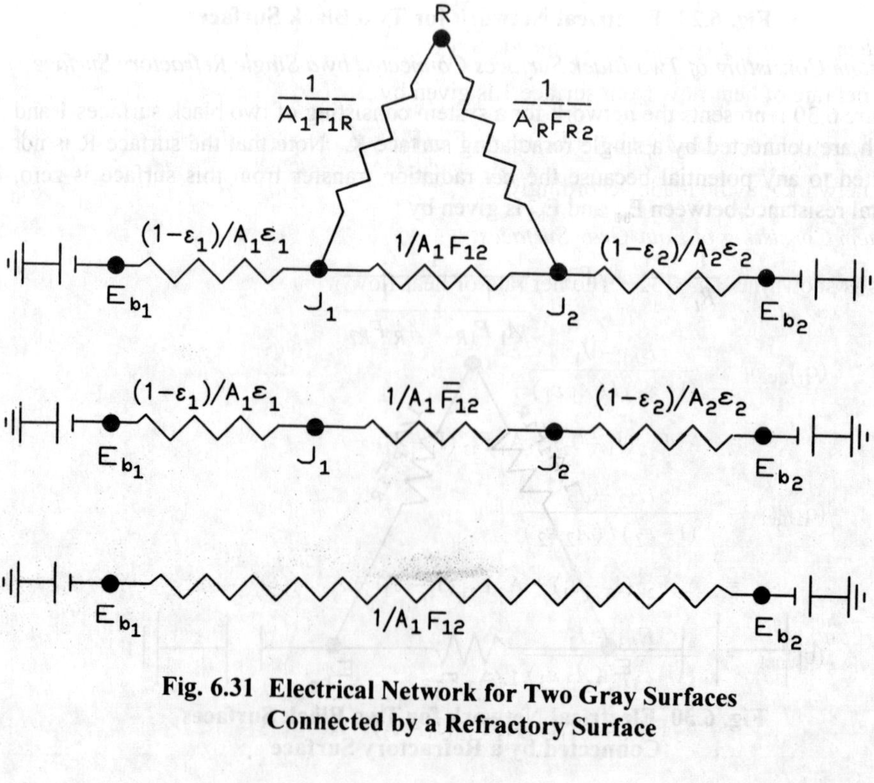

Fig. 6.31 Electrical Network for Two Gray Surfaces Connected by a Refractory Surface

$$= \frac{1-\varepsilon_1}{A_1\,\varepsilon_1} + \frac{1-\varepsilon_2}{A_2\,\varepsilon_2} + \frac{1}{A_1\,F_{12} + \dfrac{1}{\dfrac{1}{A_1\,(1-F_{12})} + \dfrac{1}{A_2\,(1-F_{21})}}}$$

From Eq. (6.77) we get,

$$A_1\,\overline{F}_{12} = A_1\,F_{12} + \frac{1}{\dfrac{1}{A_1\,(1-F_{12})} + \dfrac{1}{A_2\,(1-F_{21})}}$$

or

$$R_t = \frac{1-\varepsilon_1}{A_1\,\varepsilon_1} + \frac{1-\varepsilon_2}{A_2\,\varepsilon_2} + \frac{1}{A_1\,\overline{F}_{12}}$$

$$(q_{12})_{net} = (E_{b1} - E_{b2})\,\frac{1}{\dfrac{1-\varepsilon_1}{A_1\,\varepsilon_1} + \dfrac{1-\varepsilon_2}{A_2\,\varepsilon_2} + \dfrac{1}{A_1\,\overline{F}_{12}}}$$

$$(q_{12})_{net} = (E_{b1} - E_{b2})\,\frac{A_1}{\dfrac{1}{\overline{F}_{12}} + \left(\dfrac{1}{\varepsilon_1} - 1\right) + \dfrac{A_1}{A_2}\left(\dfrac{1}{\varepsilon_2} - 1\right)}$$

Since $(q_{12})_{net}$ is also equal to $A_1\,(F_g)_{12}\,(E_{b1} - E_{b2})$ we find that Eq. (6.82) is again obtained ;

$$(F_g)_{12} = \frac{1}{\dfrac{1}{\overline{F}_{12}} + \left(\dfrac{1}{\varepsilon_1} - 1\right) + \dfrac{A_1}{A_2}\left(\dfrac{1}{\varepsilon_2} - 1\right)} \qquad (6.82)$$

(e) System Consisting of Four Black Surfaces

The net rate of heat flow from surface 1 is given by :

$$(q_1)_{net} = A_1\,F_{12}\,(E_{b1} - E_{b2}) + A_1\,F_{13}\,(E_{b1} - E_{b3}) + A_1\,F_{14}\,(E_{b1} - E_{b4}) \qquad (6.88)$$

The network may be made accordingly.

(f) System Consisting of Four Gray Surfaces

This is shown in Fig. 6.32. The net rate of heat flow from the four surfaces is given by ;

$$(q_1)_{net} = \frac{E_{b1} - J_1}{(1-\varepsilon_1)/(A_1\,\varepsilon_1)}$$

$$= A_1\,F_{12}\,(J_1 - J_2) + A_1\,F_{13}\,(J_1 - J_3) + A_1\,F_{14}\,(J_1 - J_4) \qquad (6.89a)$$

$$(q_2)_{net} = \frac{E_{b2} - J_2}{(1-\varepsilon_2)/(A_2\,\varepsilon_2)}$$

$$= A_2\,F_{21}\,(J_2 - J_1) + A_2\,F_{23}\,(J_2 - J_3) + A_2\,F_{24}\,(J_2 - J_4) \qquad (6.89b)$$

$$(q_3)_{net} = \frac{E_{b3} - J_3}{(1-\varepsilon_3)/(A_3\,\varepsilon_3)}$$

$$= A_3\,F_{31}\,(J_3 - J_1) + A_3\,F_{32}\,(J_3 - J_2) + A_3\,F_{34}\,(J_3 - J_4) \qquad (6.89c)$$

$$(q_4)_{net} = \frac{E_{b4} - J_4}{(1-\varepsilon_4)/(A_4\varepsilon_4)}$$

$$= A_4 F_{41} (J_4 - J_1) + A_4 F_{42} (J_4 - J_2) + A_4 F_{43} (J_4 - J_3) \quad (6.89d)$$

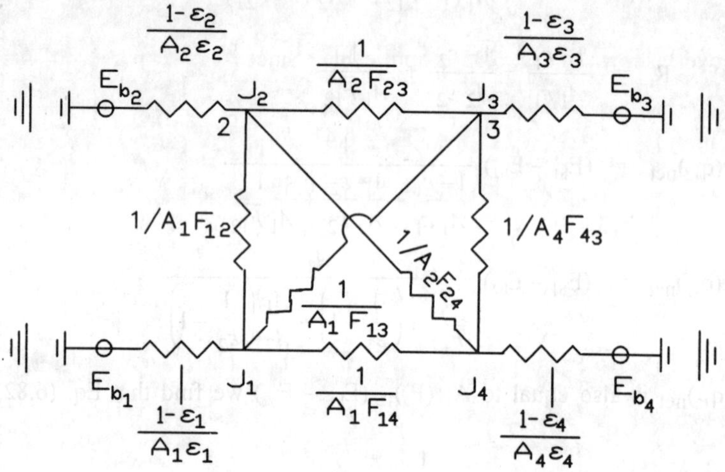

Fig. 6.32 Electrical Network for Four Gray Surfaces

(g) System Consisting of Three Gray Surfaces which see each other and nothing else.

The network for this case is shown in Fig. 6.33.

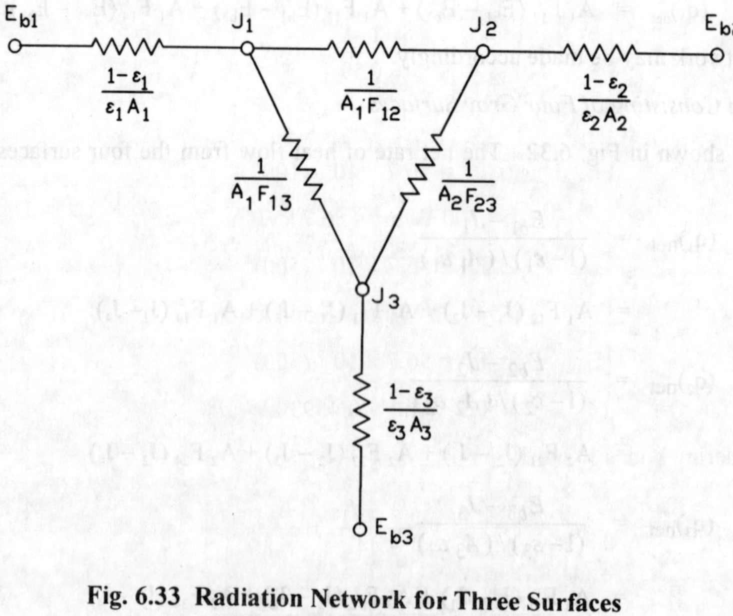

Fig. 6.33 Radiation Network for Three Surfaces (which see each other and nothing else)

Example 6.10

Two parallel square plates 1m × 1m are spaced 0.5 m apart. The plates are maintained at 1000 K and 500 K and their emissivities are 0.2 and 0.5 respectively. The plates are located in a very large room. The walls of the room are at 300 K. Find the net heat transfer to each plate and to the room considering only the plate surfaces facing each other.

Solution :

The network shown in Fig. 6.33 is applicable. Since A_3 is very large $(1 - \varepsilon_3)/\varepsilon_3 A_3 \approx 0$ and $E_{b3} = J_3$.

From Fig. 6.19, $F_{12} = F_{21} = 0.41$

So $F_{13} = 1 - F_{12} = 0.59$

$F_{23} = 1 - F_{21} = 0.59$

The resistances are then,

$$\frac{1-\varepsilon_1}{\varepsilon_1 A_1} = \frac{1-0.2}{0.2\times 1} = 4.0$$

$$\frac{1-\varepsilon_2}{\varepsilon_2 A_2} = \frac{1-0.5}{0.5\times 1} = 1.0$$

$$\frac{1}{A_1 F_{12}} = \frac{1}{1\times 0.41} = 2.44$$

$$\frac{1}{A_1 F_{13}} = \frac{1}{1\times 0.59} = 1.695$$

$$\frac{1}{A_2 F_{23}} = \frac{1}{1\times 0.59} = 1.695$$

Also,

$$E_{b1} = \sigma T_1^4 = 56.7 \times 10^{-9} (1000)^4$$
$$= 56{,}700 \text{ W/m}^2 = 56.7 \text{ kW/m}^2$$
$$E_{b2} = \sigma T_2^4 = 56.7 \times 10^{-9} (500)^4$$
$$= 3543.75 \text{ W/m}^2 = 3.544 \text{ kW/m}^2$$
$$E_{b3} = \sigma T_3^4 = 56.7 \times 10^{-9} (300)^4$$
$$= 459.3 \text{ W/m}^2 = 0.4593 \text{ kW/m}^2$$

Considering nodes J_1 and J_2.

$$\frac{E_{b1}-J_1}{4.0} + \frac{J_2-J_1}{2.44} + \frac{E_{b3}-J_1}{1.695} = 0$$

$$\frac{J_1-J_2}{2.44} + \frac{E_{b3}-J_2}{1.695} + \frac{E_{b2}-J_2}{1.0} = 0$$

Inserting the values of E_{b1}, E_{b2} and E_{b3} we get

$$J_1 = 13.06 \text{ kW/m}^2$$

$$J_2 = 4.58 \text{ kW/m}^2$$

So
$$q_1 = \frac{E_{b1} - J_1}{(1-\varepsilon_1)/\varepsilon_1 A_1} = \frac{56.7 - 13.06}{4.0}$$

$$= 10.91 \text{ kW} \qquad \textbf{Ans.}$$

$$q_2 = \frac{E_{b2} - J_2}{(1-\varepsilon_2)/\varepsilon_2 A_2} = \frac{3.544 - 4.58}{1}$$

$$= -1.04 \text{ kW} \qquad \textbf{Ans.}$$

The heat received by the room

$$q_3 = \frac{J_1 - J_3}{1/A_1 F_{13}} + \frac{J_2 - J_3}{1/A_2 F_{23}}$$

$$= \frac{13.06 - 0.4593}{1.695} + \frac{4.58 - 0.4593}{1.695}$$

$$= 7.434 + 2.431 = 9.165 \text{ kW} \qquad \textbf{Ans.}$$

Note total $q_3 = q_1 + q_2$

(h) Enclosures with Absorbing and Transmitting Media

We shall now consider an enclosure containing a medium which both absorbs and transmits radiation. Gases, plastics and various glasses are examples of such media. Consider Fig. 6.34 (a) where the space between the two plates is filled with a *gray gas* that does not reflect any of the incident radiation, but only absorbs or transmits the radiation. The properties of the medium (gas) are denoted by the subscript m. For this gray medium,

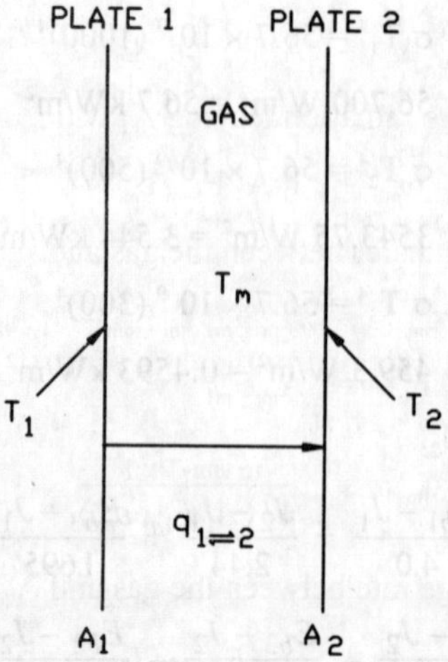

Fig. 6.34 (a) A Gas-Enclosure

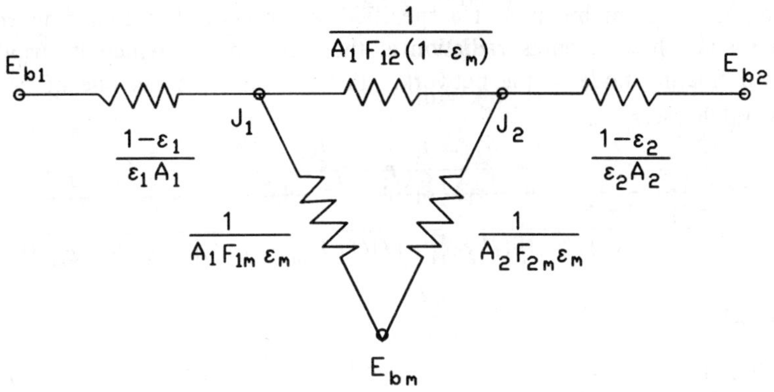

Fig. 6.34 (b) Electric Network for Radiation between Finite Planes Separated by a Gas

$$\alpha_m = \varepsilon_m$$

and $$\tau_m = 1 - \alpha_m = 1 - \varepsilon_m$$

Portion of the radiation leaving surface 1 that reaches surface 2 after passing through the gas is $J_1 A_1 F_{12} \tau_m$ and that from surface 2 which reaches surface 1 is $J_2 A_2 F_{21} \tau_m$. Therefore, the net rate of heat transfer between the two surfaces is

$$q_{1-2} = A_1 F_{12} \tau_m (J_1 - J_2)$$

$$= \frac{J_1 - J_2}{1/\{A_1 F_{12}(1-\varepsilon_m)\}} \quad (6.90)$$

Thus the equivalent resistance between nodal points J_1 and J_2 for this case will be $1/\{A_1 F_{12}(1-\varepsilon_m)\}$.

Radiant heat exchange occurs also between each of the surfaces and the gas. If the gas is at temperature T_m it emits radiation at the rate, $J_m = \varepsilon_m E_{bm}$. The portion of the energy emitted by the gas which reaches surface 1 is ;

$$A_m F_{m-1} J_m = A_m F_{m-1} \varepsilon_m E_{bm}$$

Similarly, the fraction of the radiation leaving A_1 which is absorbed by the gas is ;

$$J_1 A_1 F_{1-m} \alpha_m = J_1 A_1 F_{1-m} \varepsilon_m$$

Thus, the net rate of heat transfer between the gas and surface

$$q_{m-1} = A_m F_{m1} \varepsilon_m E_{bm} - J_1 A_1 F_{1m} \varepsilon_m$$

Since $$A_1 F_{1m} = A_m F_{m1}$$

$$q_{m-1} = \frac{E_{bm} - J_1}{1/(A_1 F_{1m} \varepsilon_m)} \quad (6.91)$$

Similarly, the heat exchange rate between the gas and A_2 is

$$q_{m-2} = \frac{E_{bm} - J_2}{1/(A_2 F_{2m} \varepsilon_m)} \quad (6.92)$$

The equivalent network is shown in Fig. 6.34 (b).

If the gas is not maintained at a specified temperature, but attains an equilibrium temperature at which it emits radiation at the same rate at which it absorbs it, E_{bm} becomes a floating node in the network. For this case the net rate of heat transfer between A_1 and A_2 is ;

$$q_{1-2} = \frac{\sigma(T_1^4 - T_2^4)}{\frac{1-\varepsilon_1}{\varepsilon_1 A_1} + \frac{1-\varepsilon_2}{\varepsilon_2 A_2} + \frac{1}{A_1\{F_{12}\,\tau_m + 1/[(1/F_{1m}\,\varepsilon_m) + (A_1/A_2\,F_{2m}\,\varepsilon_m)]\}}} \qquad (6.93)$$

Example 6.11

Two large parallel planes are at 1000 K and 500 K respectively, $\varepsilon_1 = 0.3$ and $\varepsilon_2 = 0.7$. The planes are separated by a gray gaseous medium having $\varepsilon_m = 0.2$. What is (a) the heat transfer rate between the two planes and (b) the temperature of the gas.

Solution :

(a) All shape factors are unity for large planes. Using unit area basis, the resistances are (Fig. 6.34b) :

$$\frac{1-\varepsilon_1}{\varepsilon_1} = \frac{0.7}{0.3} = 2.333$$

$$\frac{1-\varepsilon_2}{\varepsilon_2} = \frac{0.3}{0.7} = 0.429$$

$$\frac{1}{F_{12}(1-\varepsilon_m)} = \frac{1}{1(1-0.2)} = 1.25$$

and
$$\frac{1}{F_{1m}\,\varepsilon_m} = \frac{1}{F_{2m}\,\varepsilon_m} = \frac{1}{0.2} = 5.0$$

also, $E_{b1} = \sigma T_1^4 = 56{,}700$ W/m^2 and $E_b{}^2 = \sigma T_2^4 = 3544$ W/m^2

The equivalent resistance of the central triangle is,

$$R = \frac{1}{(1/1.25) + 1/(5.0+5.0)} = 1.111$$

so
$$\frac{q_{12}}{A} = \frac{E_{b1} - E_{b2}}{\sum R} = \frac{56700 - 3544}{2.333 + 1.111 + 0.429}$$

$$= \frac{53{,}156}{3.873} = 13{,}725 \text{ W/m}^2 \qquad \text{Ans.}$$

(b) The radiosities may be calculated as follows,

$$\frac{q}{A} = \frac{E_{b1} - J_1}{(1-\varepsilon_1)/\varepsilon_1} = \frac{J_2 - E_{b2}}{(1-\varepsilon_2)/\varepsilon_2} = 13{,}725 \qquad (6.90)$$

$$E_{b1} - J_1 = (13{,}725)(2.333) = 32{,}020$$

hence $J_1 = 56700 - 32020 = 24{,}680 \text{ W/m}^2$

$J_2 = 3544 + (0.429)(13725) = 9{,}432 \text{ W/m}^2$

For this particular network, considering node E_{bm},

$$[(J_1 - E_{bm})/5.0)] + [(J_2 - E_{bm})/5.0] = 0$$

$$E_{bm} = (J_1 + J_2)/2 = (24680 + 9432)/2 = 17{,}056 \text{ W/m}^2$$

so $\sigma T_m^4 = 17{,}056$

hence $T_m = 740.6 \text{ K}$ Ans.

6.21 RADIATION SHIELDS

Radiation shields are used to reduce radiant heat transfer between two surfaces. For example, the bulb of a thermometer or a thermocouple junction, used for measurement of fluid temperature, should be shielded in order to reduce radiation effects to a minimum. Otherwise the temperature indicated may involve some error.

When a metallic shield is inserted between two radiating surfaces the heat transfer by radiation is greatly reduced. To understand this result let us consider two parallel surfaces, each of area A, at temperature T_1 and T_2 respectively (Fig. 6.35). The net rate of heat transfer from surface 1 to 2 is ($F_{12} = 1$),

$$(q_{12})_{net} = A F_{12} \sigma (T_1^4 - T_2^4) = A\sigma (T_1^4 - T_2^4)$$

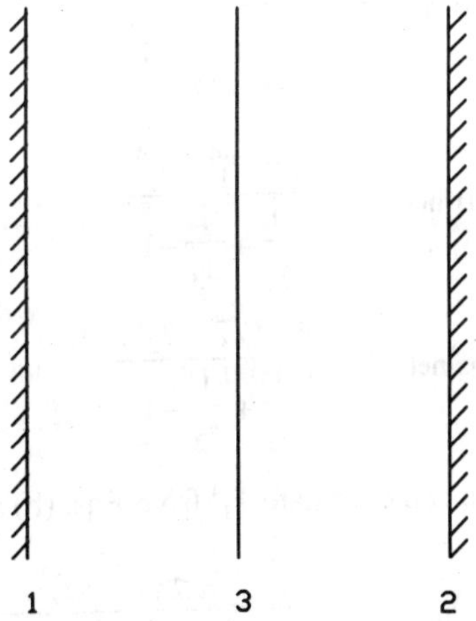

Fig. 6.35 Radiation Shield

Suppose that we interpose a thin shield of metal of area A between 1 and 2 parallel to them. Let the conductivity of the shield be very high so that both of its surfaces can be assumed to be at the same temperature. Further, let these surfaces be black. After steady state is reached let T_3 denote the temperature of the shield ; the temperature of surfaces 1 and 2 being maintained at T_1 and T_2 as before. Since the temperature T_3 remains constant it follow that :

$$(q_{13})_{net} = (q_{32})_{net}$$

Since $F_{13} = F_{32} = 1,$

$$A\sigma(T_1^4 - T_3^3) = A\sigma(T_3^4 - T_2^4)$$

or $$T_3^4 = \frac{1}{2}(T_1^4 + T_2^4)$$

thus, $$(q_{32})_{net} = A\sigma\left(\frac{T_1^4 + T_2^4}{2} - T_2^4\right)$$

$$= \frac{1}{2} A\sigma(T_1^4 - T_2^4)$$

$$(q_{13})_{net} = (q_{32})_{net} = \frac{1}{2}(q_{12})_{net}$$

Thus, by inserting one shield, the heat transfer rate is reduced to one-half of the original value. If two shields are placed, it can be shown that the heat transfer rate is reduced to one-third of original value, and with n shields :

$$(q)_{n\ shield} = \left(\frac{1}{n+1}\right)(q)_{without\ shield} \qquad (6.94)$$

If the surface 1 and 2 are not black but have emissivitities ε_1 and ε_2 and if a shield of emissivity ε_3 is interposed between them we can again follow the above procedure to determine the reduction in heat transfer rate.

Now, $$(q_{12})_{net} = \frac{A\sigma(T_1^4 - T_2^4)}{\frac{1}{\varepsilon_1} + \frac{1}{\varepsilon_2} - 1} \qquad (a)$$

$$(q_{13})_{net} = \frac{A\sigma(T_1^4 - T_3^4)}{\frac{1}{\varepsilon_1} + \frac{1}{\varepsilon_3} - 1} \qquad (b)$$

$$(q_{32})_{net} = \frac{A\sigma(T_3^4 - T_2^4)}{\frac{1}{\varepsilon_3} + \frac{1}{\varepsilon_2} - 1} \qquad (c)$$

Since $(q_{13})_{net} = (q_{32})_{net}$, we can eliminate T_3^4 from Eqs. (b) and (c) and obtain :

$$(q_{13})_{net} = \frac{A\sigma(T_1^4 - T_2^4)}{\left(\frac{1}{\varepsilon_1} + \frac{1}{\varepsilon_3} - 1\right) + \left(\frac{1}{\varepsilon_3} + \frac{1}{\varepsilon_2} - 1\right)} \qquad (d)$$

The ratio of radiant energy transfer with one shield to that without shield is obtained from Eqs. (a) and (d) ;

$$\frac{[(q_{12})_{net}]_{shield}}{[(q_{12})_{net}]_{no\ shield}} = \frac{\left(\frac{1}{\varepsilon_1} + \frac{1}{\varepsilon_2} - 1\right)}{\left(\frac{1}{\varepsilon_1} + \frac{1}{\varepsilon_3} - 1\right) + \left(\frac{1}{\varepsilon_3} + \frac{1}{\varepsilon_2} - 1\right)} \qquad (6.94a)$$

If $\varepsilon_1 = \varepsilon_2 = \varepsilon_3$ the above fraction is $\frac{1}{2}$, as in the case of black surfaces.

Example 6.12

(a) The temperature of a non-absorbing gas flowing in a large duct is measured by means of a thermocouple ($\varepsilon = 0.9$). The duct wall is at 200 C. The temperature indicated by the thermocouple is 500 C. If the value of *h* for convection heat transfer between thermocouple and gas is 140 W/m²K determine the true temperature of the gas.

(b) Suppose a thin cylindrical shield ($\varepsilon = 0.3$) having inside diameter 5 times the outer diameter of the couple is placed around the thermocouple. If thermocouple indicates 500 C what is the true temperature ? Assume h = 140 W/m²K for heat transfer between shield and gas.

Solution :

Let the gas temperature be denoted by T_g. and let suffix c and w denote the thermocouple and duct wall respectively. Now the rate of convective heat transfer from gas to thermocouple,

$$q_{gc} = h\, A_c\, (T_g - T_c)$$

and radiant heat transfer rate from couple to duct wall,

$$q_{cw} = A_c\, (F_g)_{cw}\, \sigma\, (T_c^4 - T_w^4)$$

Under steady state the above two rates must be equal otherwise the thermocouple will not denote a constant temperature. Hence,

$$h\, A_c\, (T_g - T_c) = A_c\, (F_g)_{cw}\, \sigma\, (T_c^4 - T_w^4)$$

$$(F_g)_{cw} = \varepsilon_c = 0.9$$

so $$(140)\,(T_g - 773) = (0.9)\,(56.7 \times 10^{-9})\,(773^4 - 473^4)$$

or $$T_g = 886 \text{ K}$$

or $$t_g = 613 \text{ C} \qquad \text{Ans.}$$

When the thermocouple is shielded then radiant heat exchange occurs between the thermocouple and the shield and between the shield and duct wall. In steady state, heat transfer rate by convection from gas to thermocouple equals the rate at which the couple transfers heat to the shield by radiation. Similarly, the rate at which the shield receives heat by convection from the gas plus the rate at which it receives radiation from the thermocouple equals the rate of radiant heat transfer from the shield to the duct walls. Thus, if the shield is denoted by *s*,

$$h\, A_c\, (T_g - T_c) = A_c\, (F_g)_{cs}\, \sigma\, (T_c^4 - T_s^4) \qquad \text{(a)}$$

and, $$h\,(2\,A_s)\,(T_g - T_s) + A_c\,(F_g)_{cs}\,\sigma\,(T_c^4 - T_s^4) = A_s\,(F_g)_{sw}\,\sigma\,(T_s^4 - T_w^4)$$

Dividing this equation by A_c, and putting $A_s = 5\, A_c$,

$$\{(h)\,(10)\,(T_g - T_s) + (F_g)_{cs}\,\sigma\,(T_c^4 - T_s^4)\} = 5\,(F_g)_{sw}\,\sigma\,(T_s^4 - T_w^4)$$

$$(10\,h)\,(T_g - T_s) = 5\,(F_g)_{sw}\,\sigma\,(T_s^4 - T_w^4) - (F_g)_{cs}\,\sigma\,(T_c^4 - T_s^4)$$

Now
$$(F_g)_{cs} = \frac{1}{\dfrac{1}{\varepsilon_c} + \dfrac{A_c}{A_s}\left(\dfrac{1}{\varepsilon_s} - 1\right)}$$

$$= \frac{1}{\dfrac{1}{0.9} + \dfrac{1}{5}\left(\dfrac{1}{0.3} - 1\right)} = 0.635$$

$$(F_g)_{sw} = \varepsilon_s = 0.3$$

Therefore,

$$(T_g - T_s) = \frac{56.7 \times 10^{-9}}{1400}\{5 \times 0.3\,(T_s^4 - 473^4) - 0.635\,(773^4 - T_s^4)\}$$

$$= \frac{5.67}{1400}\left[1.5\left\{\left(\frac{T_s}{100}\right)^4 - \left(\frac{473}{100}\right)^4\right\} - 0.635\left\{\left(\frac{7.73}{100}\right)^4 - \left(\frac{T_s}{100}\right)^4\right\}\right] \quad \text{(b)}$$

Equation (a) gives upon simplification,

$$T_g - T_c = \frac{0.635 \times 5.67}{140}\left[\left(\frac{T_c}{100}\right)^4 - \left(\frac{T_s}{100}\right)^4\right]$$

or
$$T_g - 773 = 0.0258\left[\left(\frac{773}{100}\right)^4 - \left(\frac{T_s}{100}\right)^4\right] \quad \text{(c)}$$

Equation (b) and (c) have to be solved for T_g and T_c. This is more conveniently done by trial. Assume $T_s = 480$ C. Then Eq. (c) gives,

$$T_g - 773 = 0.3$$

or
$$T_g = 782.3$$

Substituting this value in Eq. (b),

$$\text{LHS} = 29.3$$

and
$$\text{RHS} = 15.58$$

Since LHS ≠ RHS, we have to make a second trial. Let $T_s = 490$ C. From Eq. (c);

$$T_g - 773 = 5.16 \text{ or } T_g = 778.16$$

Substituting these values of T_g and T_s in Eq. (b)

$$\text{LHS} = 15.16 \text{ and RHS} = 17.1$$

Since LHS ≈ RHS, we need not make a third trial.

hence
$$T_g = 778 \text{ K} = 505 \text{ C}$$ **Ans.**

Note that the error in thermocouple indication is reduced from 113 C to only 5 C as a result of shielding.

6.22 RADIATION FROM GASES AND VAPOURS

Many common gases such as O_2, N_2, H_2 *dry air*, etc., and the mixtures of these gases, do not emit or absorb appreciable amounts of radiation at temperatures generally encountered in practice. They can therefore be considered transparent to thermal radiation.

On the other hand, gases and vapours such as CO_2, CO, H_2O, SO_2, NH_3, HCl, the hydrocarbons, and the alcohols emit and absorb sufficient amounts of radiant energy to merit consideration. A detailed discussion of the radiation characteristics of these gases and vapours requires too much space. We shall therefore confine ourselves to the radiation from CO_2 and H_2O which are always present in most industrial furnaces.

There are several important differences between radiation from solids and radiation from gases. Firstly, gases emit and absorb radiation only between *narrow ranges* or bands of wavelength whereas radiation from solids cover all wave lengths. The emission bands of CO_2 and H_2O at 800 C are shown in Fig. 6.36. The wavelength ranges of these bands are given in Table 6.5.

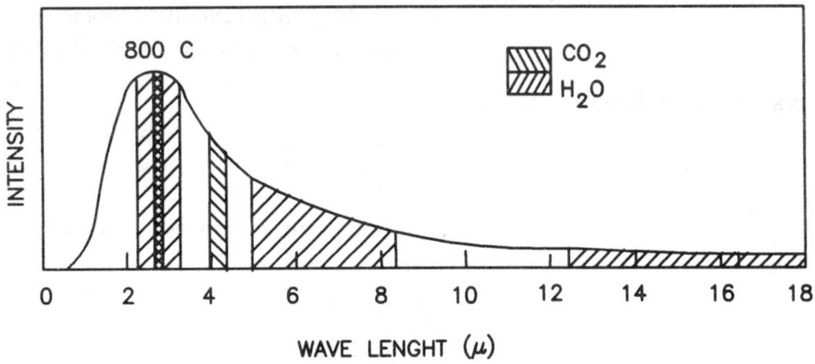

Fig. 6.36 Radiation Bands for CO_2, and H_2O

Table 6.5 Emission Bands of CO_2 and H_2O

Band Number	CO_2		H_2O	
	Wavelength range, μ	Band Width μ	Wavelength range, μ	Band Width μ
First	2.64 - 2.84	0.20	2.24 - 3.37	1.13
Second	4.01 - 4.80	0.79	4.80 - 8.50	3.7
Third	12.50-16.50	4.00	12.0 - 25.0	13.0

If we assume that within their selective emission bands these gases emit black-body radiation, the top of the bands will be bounded by the black-body radiation curve. At 1000 C, the total radiation emitted by CO_2 in its three bands in nearly 10 % of the total black-body radiation at the same temperature ; for H_2O it is 46 %. With increasing temperature the intensity of gas radiation does not increase as rapidly as that of black-body radiation. The reason is that, as the temperature is increased, the maximum intensity of black-body radiation occurs at shorter wavelength, and thus the percentage of black-body radiation falling within the emission bands of the gases is reduced.

The second important point regarding gas radiation is that radiation entering a gas volume is not absorbed within a small distance from the surface as is the case in most solids. Instead the intensity decreases slowly as the radiation passes through the gas. Hence the absorptivity of a gas volume depends not only upon its pressure and temperature but also upon its shape and size. This applies also to the emissivity of a gas volume.

6.23 ABSORPTIVITY OF GASES

Suppose monochromatic radiation of wavelength λ enters the surface of a gas volume. Let $I_{\lambda o}$ and $I_{\lambda x}$ denote the surface intensity of this radiation at the surface (x = 0) and at distance x from the surface, respectively. Now, it turns out that for a homogeneous gas volume, the decrease in monochromatic intensity with distance at a given point is directly proportional to the monochromatic intensity at the point :

$$\frac{dI_{\lambda x}}{dx} = -k_\lambda (I_\lambda)_x$$

$$d I_{\lambda x} = -k_\lambda (I_\lambda)_x \, dx \qquad (6.95)$$

where the proportionality constant k_λ is called the *monochromatic absorption coefficient* of the gas.

Integrating Eq. (6.95) from x = 0 to x = x,

$$(I_\lambda)_x = (I_\lambda)_0 \left[e^{-(k_\lambda x)}\right] \qquad (6.96)$$

If the thickness of gas layer is L, the monochromatic intensity of radiation leaving the gas will be ;

$$(I_\lambda)_L = (I_\lambda)_0 \left[e^{-(k_\lambda L)}\right] \qquad (6.97)$$

The intensity absorbed by the gas is given by the difference between the intensities of radiation entering and leaving the gas ;

$$\text{Intensity absorbed} = I_{\lambda 0} - I_{\lambda L} = I_{\lambda 0}\left(1-e^{-k_\lambda L}\right) \qquad (6.98)$$

or

$$\frac{\text{Intensity absorbed}}{\text{Original Intensity}} = \frac{I_{\lambda 0}\,(I-e^{-k_\lambda L})}{(I_\lambda)_0} = \left(1-e^{-k_\lambda L}\right) \qquad (6.99)$$

The quantity $\left(1-e^{-k_\lambda L}\right)$ therefore represents the monochromatic absorptivity of the gas $(\alpha_g)_\lambda$ at wavelength λ. According to Kirchoff's law, this also represents the monochromatic emissivity $(\varepsilon_g)_\lambda$ of the gas. Note that if L is very large $(\alpha_g)_\lambda = (\varepsilon_g)_\lambda = 1$, i.e. for very thick layers gas radiation approaches black body radiation within the wavelengths of its bands.

In order to evaluate $(\alpha_g)_\lambda$ for finite gas volume we should know k_λ and L. The absorption coefficient k_λ depends on the wavelength and temperature and pressure of the gas. The length L would depend upon the shape and size of the gas volume. Furthermore, even after $(\alpha_g)_\lambda$ is determined, a summation over all of the radiation bands is necessary in order to find out the total or effective absorptivity and emissivity.

To find out the effective absorptivity of a gas volume we may assume that the total intensity of radiation at a distance x from the surface is given by an expression similar to Eq. (6.96) ;

$$I_x = I_0 e^{-kx} \tag{6.100}$$

$$I_L = I_0 e^{-kL} \tag{6.101}$$

$$I_0 - I_L = I_0 (1 - e^{-kL}) \tag{6.102}$$

where I_0 = sum of all monochromatic intensities at x = 0.

I_x = sum of all monochromatic intensities at x = x.

k = total absorption coefficient.

The *effective absorptivity* of the gas is then given by :

$$\alpha_g = (1 - e^{-kL}) \tag{6.103}$$

Note that *k* is generally not equal to k_λ. Equation (6.103) has been verified by experiments of various investigators. However, precise determination of k is quite difficult. For this reason, the effective absorptivity of gases is generally calculated by using an approximate method developed by Hottel. This method is explained below.

Hottel evaluated the emissivities of various gases at different temperatures and pressures and presented his results in a set of graphs which are shown in Figs. 6.38 to 6.42. Hottel's graphs are strictly valid only for hemispherical gas volume of radius L radiating to an elemental surface at the centre of the base of the hemisphere, as shown in Fig. 6.37. However, for other gas shapes an effective beam length can be determined from Table 6.6. For approximate calculations ;

$$L = 3.4 \times \frac{\text{volume}}{\text{surface area}} \tag{6.104}$$

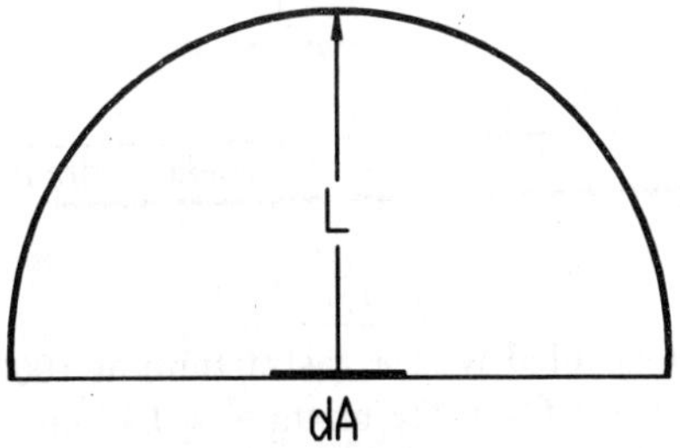

Fig. 6.37 Hottel's Graphs Valid for Hemispherical Volume Only

The curves in Figs. 6.38 and 6.39 give the *effective emissivity* of CO_2 and H_2O respectively at different temperatures for various values of the product pL where :

p = partial pressure, atm

L = mean beam length, meter

The curves are for a total pressure p_t = 1 atm. For CO_2 if the total pressure is different than 1 atm, the value of $(\varepsilon)_{CO_2}$ obtained from Fig. 6.38 should be multiplied by a correction factor $(C)_{CO_2}$ given in Fig. 6.40.

Table 6.6 Effective Beam Length of Radiation in Various Gas shapes

No.	Gas Shape	Characteristic Dimension, D	Beam Length, L
1.	Sphere, radiation to entire surface	Diameter	0.65 D
2.	Infinite cylinder radiating to walls	Diameter	0.95 D
3.	Cube, radiation to any face	Side	0.60 D
4.	Space between infinite parallel planes	Separating distance	1.8 D
5.	Right circular infinite cylinder radiating to elemental surface on centre of base	Diameter	D
6.	Right circular cylinder of height equal to diameter, radiating to whole surface	Diameter	0.60 D
7.	Right circular cylinder of height equal to diameter, radiating to elemental surface on centre of base	Diameter	0.71 D
8.	Circular cylinder of semi-infinite height radiating to entire base	Diameter	0.65 D
9.	Hemisphere radiating to element in centre of base	Radius	D
10.	Volume surrounding infinite tube bundle radiating to a single tube. (i) Equilateral triangle arrangement: S = 2 D S = 3 D (ii)Square arrangement	Tube dia D, distance between tube centres S	 3 (S - D) 3.8 (S - D) 3.5 (S - D)
11.	1 × 2 × 6 rectangular parallelepiped radiating to 2 × 6 faces 1 × 6 faces 1 × 2 faces all faces	Shortest edge	 1.18 D 1.24 D 1.18 D 1.20 D
12.	Any other shape	Mean hydraulic radius	4D

Example 6.13

A 1.8 m cubic chamber is filled with a gas mixture at 1000 C to total pressure of 2 atm. The partial pressure of CO_2 in the mixture is 0.3 atm. The other gases are all transparent to radiation. Find the emissivity of the gas volume.

Solution :

$$L = \frac{2}{3} \times 1.8 = 1.2 \text{ m}$$

$$pL = 0.3 \times 1.2 = 0.36 \text{ matm}$$

From Fig. 6.38, at 1000 C and pL = 0.36

$$(\varepsilon)_{CO_2} = 0.16 \text{ at } P_t = 1 \text{ atm}$$

From Fig. 6.40, at P_t = 2 atm, pL = 0.36 matm,

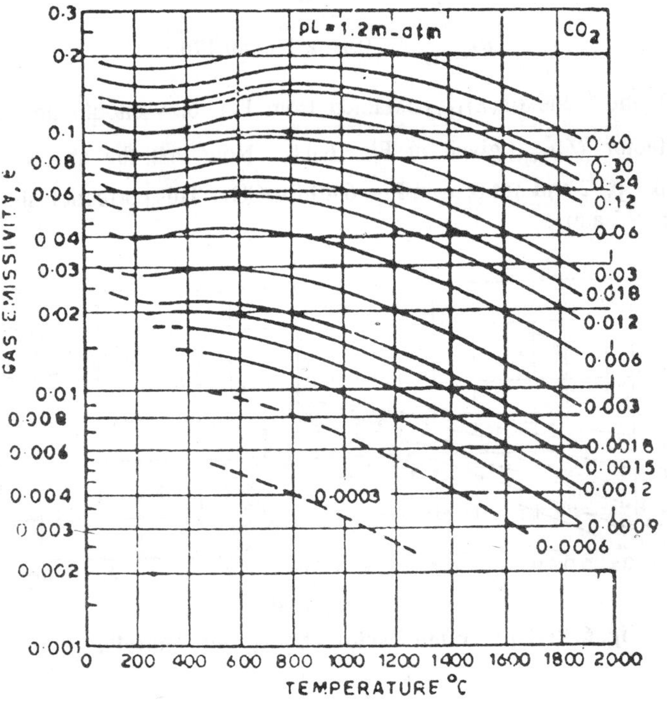

Fig. 6.38 Emissivity of CO_2, Hottel's Graph

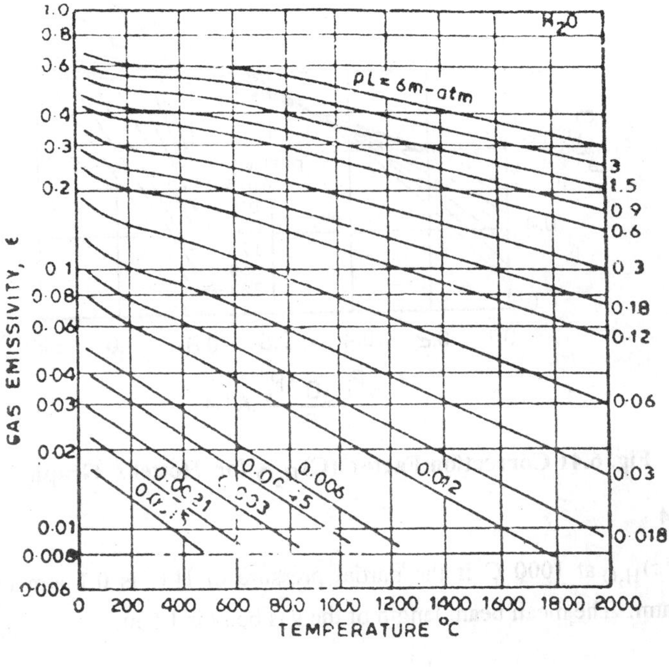

Fig. 6.39 Emissivity of H_2O, Hottel's Graph

$$(C)_{CO_2} = 1.10$$

or
$$(\varepsilon)_{CO_2} = 0.16 \times 1.1 = 0.176 \quad \text{Ans.}$$

For H_2O, the value of $(\varepsilon)_{H_2O}$ obtained from Fig. 6.39 should be multiplied by a correction factor $(C)_{H_2O}$ given in Fig. 6.41. Note that the abscissa of Fig. 6.41 represents $(p + p_t)/2$ and not p_t as in Fig. 6.40 for CO_2. This correction should be applied even when $p_t = 1$ atm.

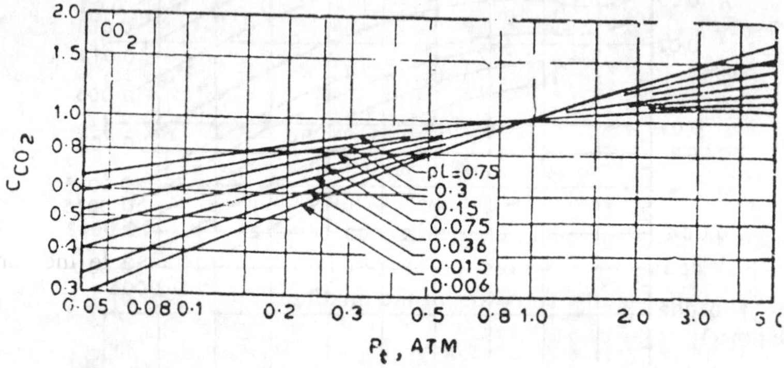

Fig. 6.40 Correction Factor $(C)_{CO_2}$ for Hottel's Graph

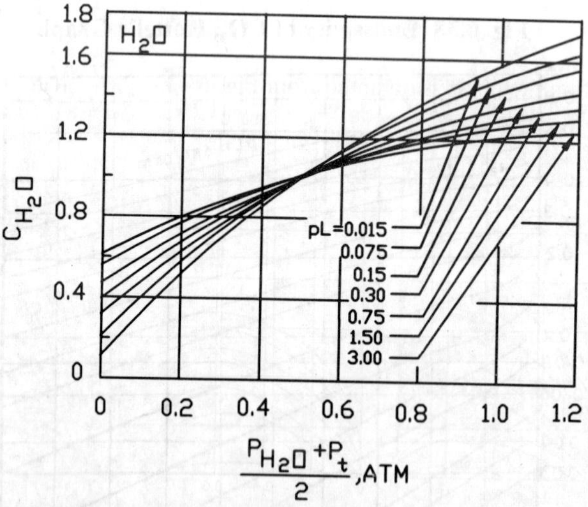

Fig. 6.41 Correction Factor $(C)_{H_2O}$ for Hottel's Graph

Example 6.14

Calculate $(\varepsilon)_{H_2O}$ at 1000 C if the partial pressure of H_2O is 0.1 atm and the total pressure is 2 atm. The mean beam length of the gas body is 1.5 m.

Solution :

$$pL = 0.1 \times 1.5 = 0.15 \text{ matm.}$$

From Fig. 6.39 at 1000 C and pL $= 0.15$

$$(\varepsilon)_{H_2O} = 0.133$$

Now,
$$\frac{p + p_t}{2} = \frac{0.1 + 2}{2} = 1.05$$

From Fig. 6.41, at pL = 0.15 and $\left(\frac{p + p_t}{2}\right) = 1.05$

$$(C)_{H_2O} = 1.45$$

or
$$(\varepsilon)_{H_2O} = 0.133 \times 1.45 = 0.192$$
Ans.

If both CO_2 and H_2O are present in a gas volume, a rough estimate of the effective emissivity of the mixture can be made by adding the emissivities of the two constituents. However, the value so obtained is somewhat on the higher side. This is because some of the emission bands of H_2O and CO_2 overlap ; hence a part of the radiation of H_2O will be absorbed by CO_2 before it leaves the gas volume and is emitted to the surrounding surfaces. Thus the emmissive power of the mixture is always less than the sum of the emissive powers of the constituents.

To determine accurately the emissivity of the mixture, the individual emissivities of the two gases should be found as explained above, added together and then a correction factor $\Delta\varepsilon$ should be subtracted from this sum. If suffix *m* denotes the mixture,

$$\varepsilon_m = [(\varepsilon)_{CO_2} (C)_{CO_2}] + [(\varepsilon)_{H_2O} (C)_{H_2O}] - \Delta\varepsilon \tag{6.105}$$

The correction factor $\Delta\varepsilon$ is obtained from Fig. 6.42.

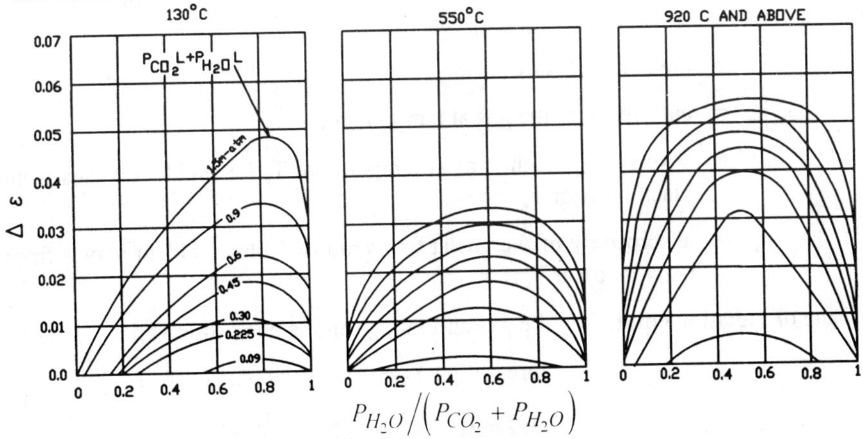

Fig. 6.42 Correction Factor for Mixture (Hottel's Graph)

Example 6.15

A gas mixture contains 11.2 % CO_2 and 10.4 % H_2O by volume. The total pressure is 1 atm and the gas temperature is 650 C. Calculate ε_m. The mean beam length may be assumed to be 0.14 m.

Solution :

$$P_{CO_2} = 0.112 \text{ atm} \; ; \; P_{H_2O} = 0.104 \text{ atm}$$

$$(P_{CO_2})L = 0.112 \times 0.14 = 0.0157 \text{ matm}$$

From Fig. 6.38 for CO_2 ;

$$\text{At } T_g = 923 \text{ K}, (P_{CO_2})L = 0.0157, \varepsilon_{CO_2} = 0.063$$

$$(\text{since } p_t = 1 \text{ atm}, C_{CO_2} = 1)$$

or

$$\varepsilon_{CO_2} C_{CO_2} = 0.063$$

From Fig. 6.39 and 6.41 for H_2O :

$$\text{At } T_g = 923 \text{ K}, (P_{H_2O})L = 0.0146, \varepsilon_{H_2O} = 0.039$$

$$\text{At } (P_{H_2O})L = 0.0146, \frac{P_{H_2O} + P_t}{2} = 0.552, C_{H_2O} = 1.08$$

so

$$\varepsilon_{H_2O} C_{H_2O} = 0.039 \times 1.08 = 0.042$$

$$\varepsilon_m = 0.063 + 0.042$$

$$= 0.105$$

Ans.

6.24 RADIANT HEAT EXCHANGE BETWEEN A GAS AND ITS ENCLOSURE

Let us now determine the radiant heat exchange between a gas at temperature T_g and the surface of a *black* enclosure at T_w. Suppose :

A = area of enclosure surface.

ε_g = emissivity of the gas at temperature T_g

α_g' = absorptivity of the gas at temperature T_g for incident waves from a black body at T_w.

α_g = absorptivity of the gas at temperature T_w for incident waves from a black body at T_w.

(i) Rate of radiation emitted by the gas and falling on the enclosure wall :

$$= (\text{area of gas}) (\text{emissivity of gas}) (\sigma T_g^4)$$

$$= A\varepsilon_g \sigma T_g^4$$

(ii) All of the above energy is absorbed by the enclosure walls which are black.

or

$$(q)_{gw} = \sigma A \varepsilon_g T_g^4 \quad (6.106a)$$

(iii) Similarly, the rate of energy emission by the walls ;

$$= \sigma A t_w^4$$

(iv) But all of this energy is not absorbed by the gas. The rate at which this energy is absorbed by the gas is ;

$$(q)_{wg} = (\sigma A T_w^4) \times \text{(absorptivity of gas at } T_g \text{ for incident radiation from black body at } T_w)$$

or
$$q_{wg} = \sigma A \alpha_g' T_w^4 \quad (6.106b)$$

(v) Hence, the net rate of heat transfer from the gas to the walls ;

$$(q_{gw})net = \sigma A (\varepsilon_g T_g^4 - \alpha_g' T_w^4) \quad (6.107)$$

Equation (6.107) is not convenient to use because the value of α_g' has to be evaluated on the basis of two temperatures, T_w and T_g. For this reason, an approximation can be made by using α_g (absorptivity of gas at T_w for incident radiation from black body at T_w) in place of α_g'. Thus Eq. (6.107) becomes ;

$$(q_{gw})net = \sigma A (\varepsilon_g T_g^4 - \alpha_g T_w^4) \quad (6.108)$$

If greater accuracy than that given by Eq. (6.108) is required, then Eq. (6.107) may be used for determining α_g' as follows. Evaluate the absorptivity at T_w and (pL) (T_w/T_g) instead of pL from Fig. 6.38 or 6.39. Apply correction factor from Figs. 6.40 or 6.41. Then multiply the result by $(T_g/T_w)^{0.65}$ for CO_2 and by $(T_g/T_w)^{0.45}$ for H_2O.

Note that the absorptivity of the gas is determined from the same curves which give the emissivity by following the same procedure. The correction factors C and $\Delta\varepsilon$ are valid for absorptivity also.

Example 6.16

If the gas mixture of Example 6.15 (L = 0.14 m) is confined in a black enclosure whose walls are at 210 C, find the net radiant heat transfer per m^2 of wall area.

Solution :

(a)
$$q = \sigma A (\varepsilon_g T_g^4 - \alpha_g T_w^4) \quad (6.108)$$

$$A = 1 \text{ m}^2$$

$$\varepsilon_g = 0.105 \text{ at } T_g = 923 \text{ K}$$

Now, α_g has to be found at T_w = 210 C.

From Fig. 6.38 for CO_2 :

At $T_w = 483$ K, $(P_{CO_2})\,L = 0.0157, \alpha_{CO_2} = \varepsilon_{CO_2} = 0.055, C_{CO_2} = 1$

or
$$\alpha_{CO_2} = \varepsilon_{CO_2} = 0.055$$

From Fig. 6.39 and 6.41 for H_2O.

$$\text{At } T_w = 483 \text{ K}, (P_{H_2O})\,L = 0.0146,$$

$$\alpha_{H_2O} = \varepsilon_{H_2O} = 0.065$$

$$C_{H_2O} = 1.08$$

or
$$(\alpha)_{H_2O}\,(C)_{H_2O} = 0.065 \times 1.08 = 0.07$$

From Fig. 6.42
$$\Delta\varepsilon = \Delta\alpha \approx 0$$

or $$\alpha_g = 0.055 + 0.07 = 0.125$$

Hence, $$q = (0.204 \times 10^{-3})\,(1)\,\{0.105\,(923)^4 - 0.125\,(483)^4\}$$
$$= 14.15 \times 10^6 \text{ J/hr m}^2$$ Ans.

(b) If use of Eq. (6.107) was desired we should calculate α_g as follows :

(i) $$(P_{CO_2})\,L \times \frac{T_w}{T_g} = 0.0157\left(\frac{483}{923}\right) = 0.0082$$

$$(P_{H_2O})\,L \times \frac{T_w}{T_g} = 0.0146\left(\frac{483}{923}\right) = 0.0076$$

(ii) From Fig. 6.38 for CO_2 :

At $$T_w = 483 \text{ K},$$

$$(P_{CO_2})\,L \times \frac{T_w}{T_g} = 0.0082,$$

and $$\alpha_{CO_2} = 0.044;\ C_{CO_2} = 1$$

or $$C_{CO_2}\,\alpha_{CO_2}\left(\frac{T_g}{T_w}\right)^{0.65} = 0.044\left(\frac{923}{483}\right)^{0.65} = 0.067$$

or $$(\alpha')_{CO_2} = 0.067$$

(iii) From Figs. 6.39 and 6.41 for H_2O :

At $T_w = 483$ K, $(P_{H_2O})\,L\,(T_w/T_g) = 0.0076$, $\alpha_{H_2O} = 0.042$, $C_{H_2O} = 1.08$

or $$C_{H_2O}\,\alpha_{H_2O}\left(\frac{T_g}{T_w}\right)^{0.45} = (1.08)\,(0.042)\left(\frac{923}{483}\right)^{0.45} = 0.061$$

(iv) Since $$\Delta\alpha = \Delta\varepsilon \approx 0,$$

$$\alpha_g' = 0.067 + 0.061 = 0.128$$

Now $$q = \sigma A\,(\varepsilon_g T_g^4 - \alpha_g' T_w^4) \qquad (6.107)$$
$$= (0.204 \times 10^{-3})\,(1)\,\{0.105\,(923)^4 - 0.128\,(483)^4\}$$
$$= 14.1 \times 10^6 \text{ J/hr m}^2$$ Ans.

If the inside surface of the enclosure is not black, the net rate of heat transfer from the gas to the walls may be computed from ;

$$q = \sigma A\,\{\varepsilon_w'\,[\varepsilon_g T_g^4 - \alpha_g T_w^4]\} \qquad (6.109)$$

where ε_w is the *effective surface emissivity*. Its value is usually taken to be the mean of ε_w and 1 in order to account for beams of radiation from gas to the walls which are reflected and then absorbed by the gas.

$$\varepsilon_w' = \frac{\varepsilon_w + 1}{2} \qquad (6.110)$$

6.15 MORE ILLUSTRATIVE EXAMPLES

Example 6.17

A light bulb having a tungston filament operates at 120 V. The electrical resistivity of tungston is 5.65×10^{-6} ohm-m. For a wire of diameter 0.1 mm and length 200 mm, determine the maximum temperature attained assuming that all the heat generated is transferred by radiation to the surroundings at 27°C. For tungston k = 180 W/m K and emissivity is 0.39.

Solution :

Rate of heat generation Q is given by,

$$Q = V^2/R$$

$$R = \rho L/A = (5.65 \times 10^{-6} \times 0.2)/(\pi/4)(0.0001)^2$$

$$= 144 \text{ ohm}$$

$$Q = (120)^2/144 = 100\text{W}$$

Since all the heat generated is transferred by radiation to the surroundings,

$$Q = q_R = A_1 (F_g)_{12} \sigma (T_1^4 - T_2^4)$$

where A_1 is the surface area of the wire and T_1 (°K) its surface temperature ; $(F_g)_{12} = \varepsilon_1 = 0.39$ (Table 6.4) and T_2 = temperature of the surroundings = 300 K.

Hence,
$$100 = (\pi \times 0.0001 \times 0.2)(0.39)(5.67 \times 10^{-8}) \times (T_1^4 - 300^4)$$

or
$$T_1 = 2913.1 \text{ K}$$

Now,
$$T_{max} - T_1 = \frac{Q' R^2}{4k} \qquad (2.82)$$

where Q' = heat generated per unit volume

$$= Q/\pi R^2 L$$

Hence,
$$T_{max} - T_1 = \frac{Q}{\pi R^2 L} \frac{R^2}{4k} = \frac{Q}{4\pi Lk}$$

$$= \frac{100}{(4\pi)(0.2)(180)} = 0.22$$

so
$$T_{max} = T_1 + 0.22 = 2913.1 + 0.22$$

$$= 2913.3 \text{ K} = 2640.3 \text{ °C}$$ Ans.

Example 6.18

The radiation from a furnace (black body radiation at 2500 K) is viewed from a glass window 10 cm × 20 cm. The emissivity of the glass is 0.35 up to 4 μm and 0.9 above 4 μm ; the transmissivity is 0.55 from 0.2 to 4μm. Determine the amounts of radiation absorbed and transmitted by the glass window.

Solution :

$$\sigma T^4 = 5.669 \times 10^{-3} \times (2500)^4 = 22{,}14{,}000 \text{ W}$$

$$\lambda_1 T = 0.2 \times 2500 = 0.5 \times 10^3 \ \mu\text{m K}$$

$$\lambda_2 T = 4 \times 2500 = 10 \times 10^3 \ \mu\text{m K}$$

From Table 6.2 we get ;

$$\frac{E_b\,(0-\lambda_1 T)}{\sigma T^4} = 0 \text{ (say)} ; \quad \frac{E_b\,(0-\lambda_2 T)}{\sigma T^4} = 0.914$$

Hence, total incident radiation on window (area = 0.1 × 0.2 = 0.02 m²) ;

$$0 \text{ to } \lambda_1 T = 0$$

$$0 \text{ to } \lambda_2 T = 0.914 \times 2214000 \times 0.02 = 40491.7 \text{ W}$$

and above

$$\lambda_2 T = (1 - 0.914) \times 2214000 \times 0.02 = 3797.3 \text{ W}$$

Radiation absorbed,

$$= (0.35 \times 40491.7) + (0.9 \times 3797.3)$$

$$= 14172.1 + 3417.6 = 17589.7 \text{ W}$$ **Ans.**

Radiation transmitted,

$$= 0.55\,(40491.7 - 0)$$

$$= 22270.4 \text{ W}$$ **Ans.**

Example 6.19

Derive an expression for the geometric factor F_{12} for an elementary area dA_1 and a circular disk of diameter D, parallel to dA_1 and at a distance L from it, the centres of the disk and dA_1 being on the same line.

Solution :

Consider an elementary area dA_2 of the disk consisting of an annular ring of radius r and thickness dr, shown in Fig. 6.43, so that $dA_2 = 2\pi r\, dr$.

$$F_{12} = \frac{1}{\pi} \int_{A_2} \frac{\cos\theta_1 \cos\theta_2 \, dA_2}{S^2} \qquad (6.59)$$

(Since $A_1 = dA_1$)

For $\theta_1 = \theta_2$;

$$\cos\theta_1 = \frac{L}{S} = \frac{L}{(L^2 + r^2)^{1/2}}$$

$$F_{12} = \frac{1}{\pi} \int_{r=0}^{D/2} \frac{L^2\, 2\pi r\, dr}{(L^2 + r^2)(L^2 + r^2)}$$

or

$$F_{12} = -\left[\frac{L^2}{(L^2 + r^2)}\right]_0^{D/2} = \frac{D^2}{4L^2 + D^2}$$ **Ans.**

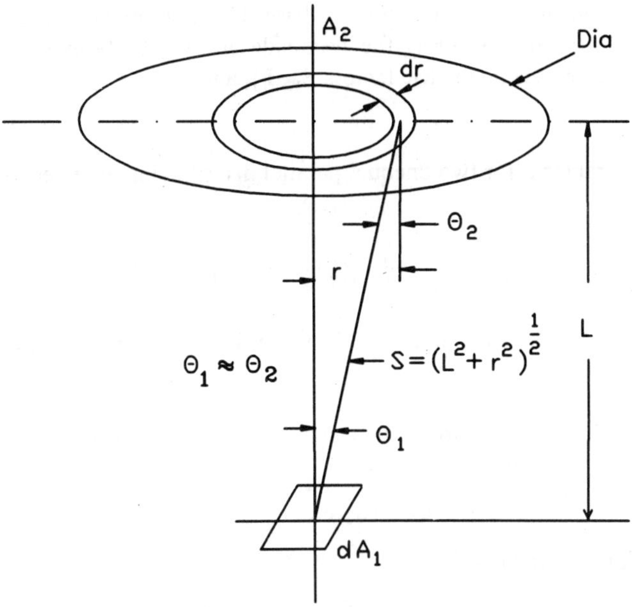

Fig. 6.43 Example 6.19

Example 6.20

A bowl has the shape of a hemisphere of radius R. Derive an expression for the geometric factor F_{11} of the inside surface of the bowl with respect to itself.

Solution :

Let us assume that the bowl is covered by a lid of radius R (surface area $A_2 = \pi R^2$), so that we have an enclosure. Let A_1 denote the inside surface area of the bowl. $A_1 = 2\pi R^2$.

$$F_{11} + F_{12} = 1$$

$$F_{21} + F_{22} = 1$$

Since area A_2 is flat, $F_{22} = 0$ and $F_{21} = 1$

we have $$F_{11} = 1 - F_{12}$$

$$A_1 F_{12} = A_2 F_{21}$$

and $$F_{12} = \frac{A_2}{A_1} F_{21}$$

so $$F_{11} = 1 - \frac{A_2}{A_1} \quad (1)$$

$$= 1 - \frac{\pi R^2}{2\pi R^2} = 0.5$$ Ans.

Example 6.21

A hollow frustum of a cone has top and base diameters of 60 cm and 120 cm and vertical height of 40 cm. Denoting the base side and top surfaces by A_1, A_2 and A_3 respectively, determine the geometric factors F_{12}, F_{13} and F_{22}.

Solution :

The geometric factor F for two unequal parallel disks having their centers on the same line is given by

$$F = \frac{1}{2a^2}[L^2 + a^2 + b^2 - \{(L^2 + a^2 + b^2)^2 - 4a^2 b^2\}^{1/2}]$$

where a is the radius of the smaller disk, b the radius of the larger disk, and L the distance between the disks.

Hence, $$F_{31} = \frac{1}{2(30)^2}[(40^2 + 30^2 + 60^2) - \{(40^2 + 30^2 + 60^2)^2 - 4(30)^2(60)^2\}^{1/2}]$$

$$= 0.653$$

From the reciprocity theorem,

$$A_1 F_{13} = A_3 F_{31}$$

or $$F_{13} = \frac{A_3}{A_1} F_{31} = \frac{1}{4} \times 0.653 = 0.163$$ Ans.

Now, $$F_{31} + F_{32} = 1$$

so $$F_{32} = 1 - F_{31} = 1 - 0.653 = 0.347$$

and $$A_2 F_{23} = A_3 F_{32}$$

so $$F_{23} = \frac{A_3}{A_2} F_{32}$$

$$= \frac{\pi(30)^2}{\pi(90)(50)}(0.347) = 0.069$$

Further, $$F_{11} + F_{12} + F_{13} = 1$$

Since surface 1 is flat, $F_{11} = 0$;

so $$F_{12} + F_{13} = 1$$

or $$F_{12} = 1 - F_{13} = 1 - 0.163 = 0.837$$ Ans.

Again, $$A_2 F_{21} = A_1 F_{12}$$

or $$F_{21} = \frac{A_1}{A_2} F_{12} = \frac{\pi(60)^2}{\pi(90)(50)}(0.837) = 0.67$$

since, $$F_{22} + F_{23} + F_{21} = 1$$

so $$F_{22} = 1 - F_{23} - F_{21} = 1 - 0.069 - 0.67 = 0.261$$ Ans.

Example 6.22

A bowl 20 cm in diameter and 12 cm in height contains cooked vegetables. The temperature of the bowl is 150 °C, and the surroundings are at 27 °C. Determine the net radiant heat transfer rate between the bowl and the surroundings. (a) if the bowl and surroundings behave like black bodies and (b) if the bowl has an emissivity of 0.8 and the surroundings have emissivity of 0.9.

Solution :

The net radiant heat exchange rate between the bowl and the surroundings is given by,

$$q = A_1 (F_g)_{12} \sigma (T_1^4 - T_2^4) \tag{6.65}$$

where suffix 1 denotes the bowl and 2 the surroundings.

Here,

$$A_1 = 2 (\pi/4) D_1^2 + \pi D_1 H_1$$
$$= 2 (\pi/4) (0.2)^2 + \pi (0.2) (0.12)$$
$$= 0.138 \text{ m}^2$$

(a) For this case,

$$(F_g)_{12} = F_{12} = 1 \text{ (Table 6.4)};$$

so

$$q = (0.138) (1) (5.6697 \times 10^{-8}) (423^4 - 300^4)$$
$$= 187.3 \text{ W} \qquad \text{Ans.}$$

(b) For this case,

$$(F_g)_{12} = \varepsilon_1 = 0.8$$

so

$$q = (0.138) (0.8) (5.6697 \times 10^{-8}) (423^4 - 300^4)$$
$$= 149.8 \text{ W} \qquad \text{Ans.}$$

Example 6.23

Three aluminum spheres having 20 cm, 40 cm and 60 cm internal diameters, are placed concentrically. The surface of the innermost sphere is maintained at 1000 K while that of the outermost sphere at 300 K. Assuming that the walls are extremely thin and that there is perfect vacuum in the annular space between the spheres, determine the steady state temperature of the intermediate sphere. The emissivity of each surface is 0.1.

Solution :

Let us denote the innermost sphere by suffix 1, the intermediate sphere by suffix 2 and the outer sphere by suffix 3. Under steady state,

$$q_{12} = q_{23}$$

so

$$A_1 (F_g)_{12} \sigma (T_1^4 - T_2^4) = A_2 (F_g)_{23} \sigma (T_2^4 - T_3^4)$$

or

$$T_1^4 - T_2^4 = \frac{A_2 (F_g)_{23}}{A_1 (F_g)_{12}} (T_2^4 - T_3^4) \tag{a}$$

$$A_2/A_1 = R_2^2/R_1^2 = (40/20)^2 = 4$$

For concentric spheres the shape factor is given by ;

$$(F_g)_{12} = \frac{1}{\dfrac{1}{\varepsilon_1} + \dfrac{A_1}{A_2}\left(\dfrac{1}{\varepsilon_2} - 1\right)} \qquad (6.82c)$$

so for our case,
$$(F_g)_{12} = \frac{1}{\dfrac{1}{0.1} + \dfrac{1}{4}\left(\dfrac{1}{0.1} - 1\right)} = 0.0816$$

and
$$(F_g)_{2\text{-}3} = \frac{1}{\dfrac{1}{\varepsilon_2} + \dfrac{A_2}{A_3}\left(\dfrac{1}{\varepsilon_3} - 1\right)} = \frac{1}{\dfrac{1}{0.1} + \left(\dfrac{40}{60}\right)^2\left(\dfrac{1}{0.1} - 1\right)}$$

$$= 0.0714$$

Thus, Eq. (a) gives ;

$$(1000^4 - T_2^4) = (4)\left(\frac{0.0714}{0.0816}\right)(T_2^4 - 300^4) = 3.5\,(T_2^4 - 300^4)$$

or
$$T_2 = 691.4 \text{ K}$$ Ans.

PROBLEMS

Note : (1 Nm = 1 W/s = 1 J)

6.1 What is the difference between total and equilibrium emissivity ?

6.2 When a body is said to be black, does it absorb radiation energy of all wave lengths. If no, then what is the range of wave lengths which it will absorb ?

6.3 A black body may be imitated by means of a hollow container with a small opening. Exactly which surface of the device can be considered as a black body ? Would such a black body have an emissivity of 1 ? Why ?

6.4 Is the absorptivity of a substance always equal to its emissivity ? If so, does this mean that energy absorbed is always equal to that emitted ? Does it also mean that the net gain of radiation energy is always equal to zero ? If not, is this a violation of Kirchoff's law ?

6.5 If a substance is opaque to light, is it also opaque to radiation of heat ?

6.6 Is the total emissive power of any substance always proportional to the fourth power of the absolute temperature ?

6.7 Plain glass transmits perfectly the solar energy in the range 0.2 μ to 3 μ and does not transmit any other wavelength. Tinted glass transmits perfectly in the range 0.5 μ to 1 μ and does not transmit at other wavelengths. If solar energy falls on the glass, compare the energy rates transmitted by the plain and tined glasses. The temperature of the sun may be taken as 5600 C.

6.8 What fraction of the solar energy is in the visible and infrared region if sun radiates as a black body at 5600 C.

6.9 A thin sphere of copper, with its internal surface highly oxidized, has a diameter of 15 cm. How small a hole must be made in the sphere to make an opening that will have an absorptivity of 0.9 ?

6.10 A device utilizing solar energy consists of a parabolic mirror that focuses the impinging sunlight on to a pyrex tube containing a high-boiling nearly black liquid. The liquid is circulated to a heat exchanger in which the heat energy is transferred to super-heated water at 25 atm pressure. Steam may be withdrawn and used to run an engine. The most efficient design requires a mirror 3 m in diameter to generate 2 kW when the axis of the mirror is pointed directly at the sun. What is the overall efficiency of the device ?

6.11 Compute the radiant energy loss from a 1 cm diameter opening in a thin-walled furnace located in a large enclosure, if the temperature within the furnace is 900 C and the surroundings are at 20 C.

[Ans. : 27.2 kJ/hr]

6.12 What is the net radiant exchange between the floor of a furnace 5 m by 4 m and a side wall 5 m by 2.5 m, if the temperatures of the floor and wall are 600 C and 300 C, respectively ? Assume both surfaces are black.

6.13 Two equal parallel black discs, 0.5 m diameter, are located 0.25 m apart and directly opposite each other. If the temperatures of the discs are 200 C and 50 C, calculate the net radiant heat transfer per square meter of surface area.

6.14 A shed is rectangular in shape with the floor 5 m by 10 m and the roof 2.5 m above the floor. The floor is heated by hot water running through coils. On cold winter days the wall and roof are at about –20 C. At what rate must heat be supplied through the floor in order to maintain the floor temperature at 25 C. (Assume that all the surfaces of the system are black).

6.15 Two parallel directly opposed black rectangular surfaces 1 m × 4 m are 1 m apart and connected by a refractory surface. If the two surface temperatures are 1000 C and 1250 C; what is (a) the direct radiation exchange between the two surfaces per square meter of area, and (b) the total radiation between the surfaces based on 1 m^2 area ?

6.16 Calculate the radiation exchange per m^2 of area between two infinite parallel black plates at 2000 C and 500 C.

6.17 Solve problem 6.16 if a third parallel black plate is located between the two plates. What will be the temperature of the centre plate ?

6.18 A brick wall with an emissivity of 0.8 is 6 m wide and 5 m high. It stands at a distance of 4 m from a 40 cm × 30 cm opening to a furnace in a boiler room. The centre line of the furnace opening lies in the same vertical plane as the centre line of the wall, but 1 m lower. The furnace operates at an average temperature of 1500 C while the wall temperature is constant at 27 C. Consider the opening as a black body. Calculate the net transfer by radiation only between the opening and the wall.

6.19 Two parallel surfaces of emissivities 0.9 and 0.8 respectively, 1 m in diameter, are separated by a distance of 0.5 m. If these two surfaces are connected by non-conducting but reradiating surfaces, what is net radiation exchange between them if the temperatures are 900 and 700 C ? What would be the value of the direct radiation exchange between the two surfaces ?

6.20 Draw an equivalent electrical network of the system in problem 6.19 and specify the values of the conductances.

6.21 Draw the analogous electrical circuit for radiation exchange between two parallel directly opposed gray surfaces of size 6 m by 9 m connected by refractory surfaces. If the emissivity of the lower surface is 0.5 and that of the upper is 0.6 and the distance between them is 4 m, specify the numerical value of each circuit constant. If the lower surface is at 60 C and the upper at 15 C, what is the net radiant exchange between them ?

6.22 A steam pipe of 12 cm OD is exposed without any insulation to the air in a large room. The saturated steam flowing through the pipe is at pressure of 10 atm. The temperature at the surface of the plastered wall of the room is 17 C. Calculate the net heat transfer rate by radiation between pipe and the wall in kJ/hr per unit length of the pipe.

6.23 A radiant heating system is installed in the plastered ceiling of a room 5 m long by 5 m wide by 3 m high. The temperature on the concrete floor is maintained at 27 C. Assume that no heat flows through the walls, which are coated with reradiating material. If the required heat supply of the room is 4.2×10^6 J/hr, determine the necessary temperature at the ceiling surface.

2.24 A person standing 10 m from a point heat source is subjected to a radiation intensity of 200×10^6 J/hr m^2. How far should he stand from the heat source to feel an intensity of half the previous value ?

2.25 A hot plate 20 cm in diameter is used to heat water in a flat-bottom cooking pan 10 cm in diameter. The surface of the hot plate has an emissivity of 0.4 and is at a temperature of 600 C. At the bottom of the cooking pan, the temperature is 100 C and the emissivity is 0.4. Compute the rate of radiant heat flow between the two surfaces if the vertical distance between them is (a) 2.5 cm (b) 1.5 cm.

2.26 20×10^6 J/hr is radiated from the ceiling of a room to the floor and the side walls. The dimensions of the ceiling are 4 m by 4 m ; of the side wall 4 m by 3 m. The emissivity of the ceiling is 0.7,of the floor 0.6, and of the side wall 0.64. Determine the mean temperature at the surface of the side wall and the floor if the surface temperature of the ceiling is maintained at 45 C with a heating system inside the ceiling. (Assume the temperature of the floor is the same as that of the side wall)

2.27 A thin, polished copper plate ($\varepsilon = 0.04$) is introduced as a radiation shield between two dull steel plates ($\varepsilon = 0.8$)

(a) Determine the fractional reduction in radiant energy transfer due to the presence of the shield.
(b) What is the fractional reduction if the copper shield becomes oxidized ($\varepsilon = 0.6$).

6.28 A mercury thermometer is situated in a room in which the air temperature is 15 C while the temperature of the walls of the room is 10 C. Assume that heat transfer coefficient for convection at the bulb of the thermometer to be 3.0 W/m^{2}K and the emissivity of the glass to be unity.

(a) What is the reading of the thermometer ?

(b) A long, thin, metallic shield ($\varepsilon = 0.2$) is placed around the bulb. What will the thermometer read with shield in place. Assume bulb and shield temperatures to be the same.

6.29 The total heat transfer coefficient including convection and radiation transfer is 30 W/m^{2}K for the outer surface of a pipe in a large enclosure. Assuming that the pipe surface is at 200 C and can be considered black, calculate the convective coefficient h if the walls of the enclosure are at 100 C.

6.30 The temperature of an air stream in a duct is being measured by means of a thermocouple. The thermocouple wires and junction are cylindrical, 0.125 cm in diameter, and extend across the duct perpendicular to the flow with the junction in the centre of the duct. Estimate the temperature of the gas stream from the following data obtained under steady conditions :

(i) Thermocouple junction temperature is 260 C.

(ii) Duct wall temperature is 150 C.

(iii) Emissivity of the thermocouple wires is 0.8.

(iv) Convection heat transfer coefficient from wire to air is 300 W/m^{2}K.

(v) The wall temperature is constant at the value given for 20 duct diameters upstream and downstream of the thermocouple installation. The thermocouple leads are positioned so that the effect of conduction losses on the junction temperature may be neglected.

CHAPTER
7

Heat Exchangers

7.1 INTRODUCTION

A heat exchanger is an apparatus or an equipment in which the process of heating or cooling occurs. The heat is transferred from one fluid being cooled to another fluid being heated. For instance, boilers, condensers, furnaces, evaporators, are all heat exchangers of some type designed for some specific purpose but primarily to transfer heat from one fluid to another.

The transfer of heat in an exchanger between two fluids could be carried out either by direct contact of the fluids or by transmission through a wall separating the fluids. The former types are called *direct contact* heat exchangers while the later as *regenerators, recuperators* or *surface exchangers*.

In *direct contact* heat exchangers the process of heat transfer occurs through direct contact and *mixing* of the hot and cold fluids. The heat transfer is usually accompanied by mass transfer in such cases. These heat exchangers are used in cases where mixing of two fluids is either harmless or desirable. Water cooling-towers and jet condensers are the two examples of such heat exchangers.

The *regenerators* are those devices in which hot and cold fluids alternately flow over the surface. The heat carried by the hot fluid is accumulated in the walls of the equipment and is then transferred to the cold fluid when it passes over the surface next.

More common exchangers are the *recuperators* in which both hot and cold fluids separated from each other by a wall, flow through the exchanger at the same time. The heat transfer process consists of convection between the fluid and wall, conduction through the wall and convection between the wall and the other fluid. In case the temperature difference between a wall and fluid is large, radiation heat exchange may also occur. Such exchangers are used where the cooling and heating fluids can not be allowed to mix. The economizer and air-preheater of boiler plant, radiator of a motor car, evaporator of an ice plant and milk chiller of a pasteurising plant all belong to this category.

7.2 HEAT EXCHANGER TYPES

Heat exchangers are generally classified according to the directions of flow of the hot and cold fluids with respect to each other, or according to the temperature distribution of the two fluids along the exchanger length. Thus, we may have the following types of heat exchanger :

(a) Parallel - flow exchanger

(b) Counter- flow exchanger

(c) Cross- flow exchanger

(d) Condenser or evaporators

(a) Parallel - flow Exchanger

In this type of heat exchanger the hot and cold fluids flow in the *same* direction, hence the name *parallel-flow*. Such an exchanger is shown schematically in Fig. 7.1 (a) and (b). In a parallel- flow exchanger, the temperature difference between the hot and cold fluids keeps on decreasing from inlet to exit as shown in Fig. 7.1 (c). Many devices, such as water heaters, oil heaters and oil coolers, etc., belong to this class.

(b) Counter- flow Exchangers

In this case, as schematically shown Fig. 7.2 (a) and (b) the two fluids flow through exchanger in opposite directions, hence the name *counter flow*. A common type of counter-flow, *shell and tube*, exchanger is shown in Fig. 7.3. The temperature distribution in a counter-flow exchanger is shown in Fig. 7.2 (c). It can be seen that the

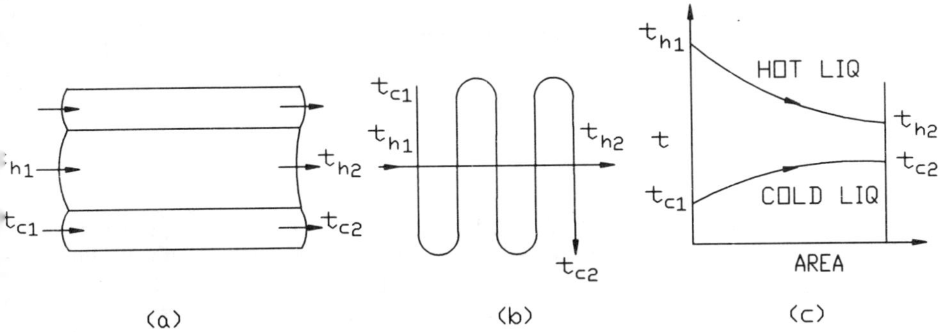

Fig. 7.1 Parallel Flow Heat Exchanger

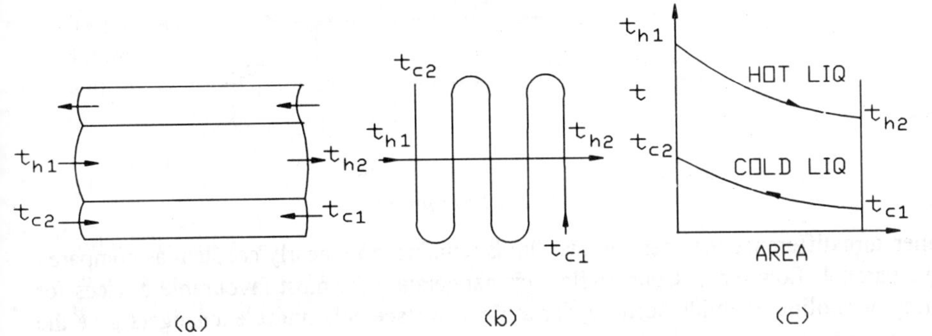

Fig. 7.2 Counter Flow Heat Exchanger

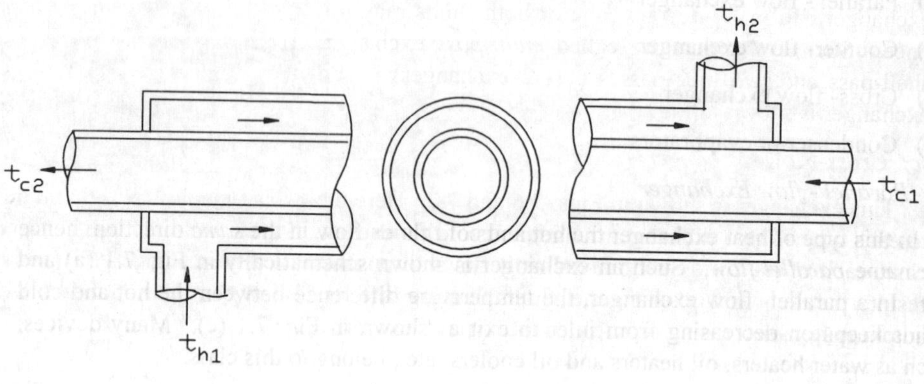

Fig. 7.3 Shell and Tube Type Exchanger

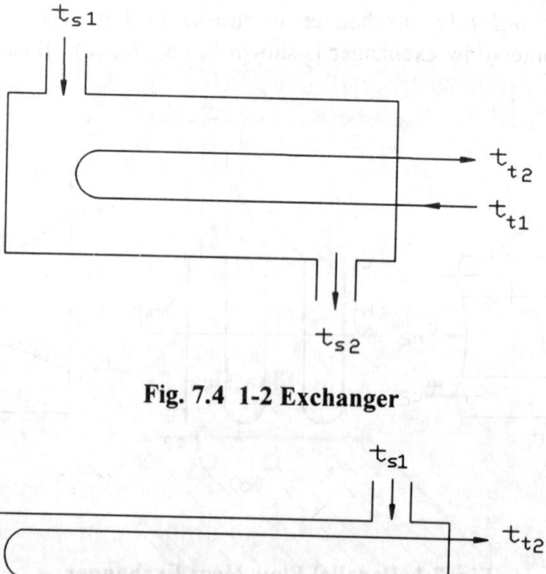

Fig. 7.4 1-2 Exchanger

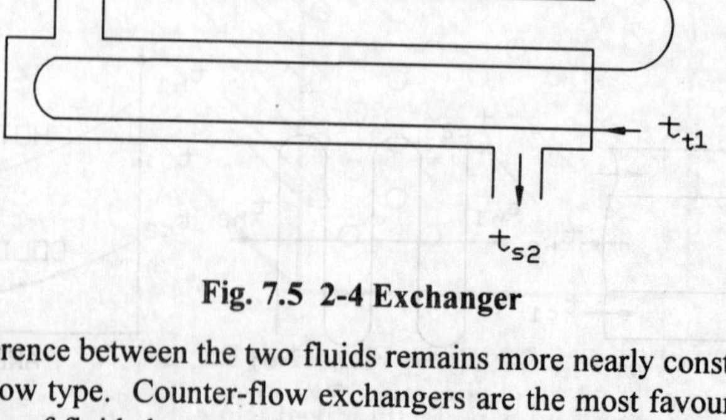

Fig. 7.5 2-4 Exchanger

temperature difference between the two fluids remains more nearly constant as compared to the parallel- flow type. Counter-flow exchangers are the most favourable devices for heating or cooling of fluids because, for a given surface area, these exchangers give the maximum heat transfer rate. This point will also be discussed later.

If the fluids flow through the exchanger only *once*, it is called a *single pass* heat exchanger. In many designs, one or both fluids may traverse the exchanger more than once. Such exchangers are called *multi-pass* exchangers. An exchanger having one shell-pass and two tube-passes (1 - 2 exchanger) is shown in Fig. 7.4 while a 2 - 4 exchanger is shown in Fig. 7.5.

(c) Cross-Flow Exchanger

This exchanger is shown in Figs. 7.6 and 7.7. Here the two fluids flow at right angles to each other. Two different arrangements of this exchanger are commonly used. In one case, each of the fluids is unmixed as it flows through the exchanger (Fig. 7.6). As a result, the temperatures of the fluids leaving the exchanger are not uniform. An automobile radiator is an example of this type of exchanger.

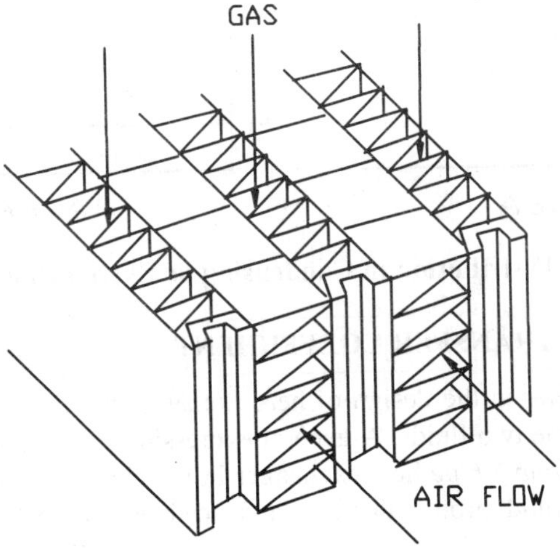

Fig. 7.6 Cross Flow Heat Exchanger

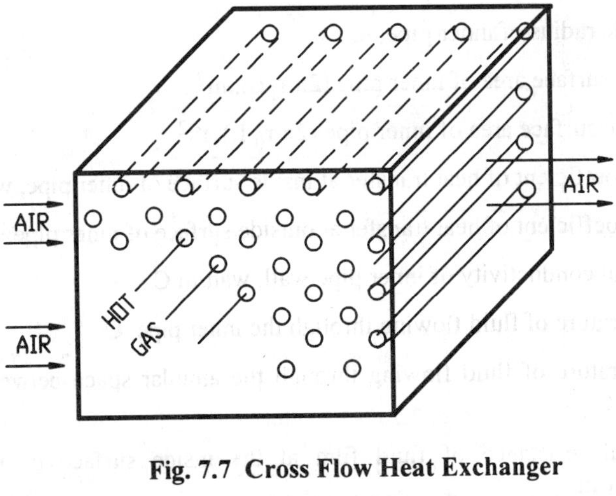

Fig. 7.7 Cross Flow Heat Exchanger

In the other case, one fluid is perfectly mixed while the other is unmixed as it flows through the exchanger. This type is shown in Fig. 7.7. The air flowing over the tube bank is mixed while the hot gases flowing inside the separate tubes are not mixed.

(d) *Condenser or Evaporator*

In a *condenser* the condensing fluid (hot fluid) remains at constant temperature throughout the exchanger while the temperature of the colder fluid gradually increases from inlet to exit. Similarly, in an *evaporator* the boiling fluid (cold fluid) remains at constant temperature while the hot fluid temperature gradually decreases. The temperature distribution in these exchangers are shown in Fig. 7.8 (a) and (b). Since the temperature of one of the fluids remains constant, it is immaterial whether the two fluids flow in the same direction or in opposite directions.

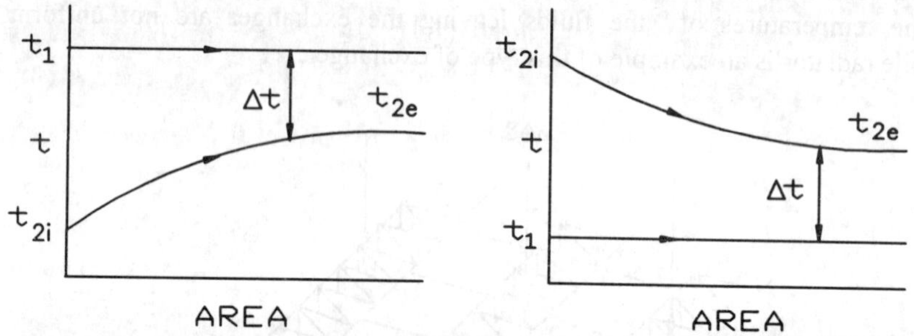

Fig. 7.8 Temperature Distribution in Condensing or Evaporating Heat Exchanger

7.3 OVERALL HEAT TRANSFER COEFFICIENT

An important parameter in the design of heat exchangers is the *overall heat transfer coefficient, U,* between the two fluids. A general expression for *U* can be easily obtained as follows. Consider a *double pipe* heat exchanger in which one fluid flows through the inner pipe and the other fluid through the annular space between the two pipes.

Let L = length of heat exchanger, m

r_i = inside radius of inner pipe, m

r_o = outside radius of inner pipe, m

A_i = inside surface area of inner pipe ($2\pi\, r_i\, L$), m^2

A_o = outside surface area of inner pipe ($2\pi\, r_o\, L$), m^2

h_i = film coefficient of heat transfer at inside surface of inner pipe, watt/m^2 C

h_o = film coefficient of heat transfer at outside surface of inner pipe, watt/m^2 C

k_w = thermal conductivity of inner pipe wall, watt/m C

t_i = temperature of fluid flowing through the inner pipe, C

t_o = temperature of fluid flowing through the annular space between the two pipes, C

R_i = thermal resistance of fluid film at the inside surface of inner pipe, m^2 C/watt

R_o = thermal resistance of fluid film at the outside surface of inner pipe, m^2 C/watt

R_w = thermal resistance of inner pipe wall, m^2 C/watt

(i) The rate of heat transfer between the two fluids is given by;

$$q = \frac{t_i - t_0}{\Sigma R}$$

where $\Sigma R = R_i + R_0 + R_w$

Since $R_i = \frac{1}{A_i h_i}$

$$R_w = \frac{ln(r_o / r_i)}{2\pi L k_w}$$

$$R_o = \frac{1}{A_o h_o}$$

hence
$$q = \frac{(t_i - t_o)}{\frac{1}{A_i h_i} + \frac{ln(r_o / r_i)}{2\pi L k_w} + \frac{1}{A_o h_o}} \tag{7.1}$$

(ii) If U_i and U_o denote respectively the overall heat transfer coefficient based on unit area of the inside and outside surfaces of the inner pipe, then

$$q = A_i U_i (t_i - t_o) = A_o U_o (t_i - t_o) \tag{7.2}$$

From Eqs. (7.1) and (7.2)
$$U_i = \frac{1}{\frac{1}{h_i} + \frac{A_i \, ln(r_o / r_i)}{2\pi L k_w} + \left(\frac{A_i}{A_o}\right)\frac{1}{h_o}} \tag{7.3}$$

and
$$U_o = \frac{1}{\left(\frac{A_o}{A_i}\right)\frac{1}{h_i} + \left(\frac{A_o}{2\pi L}\right)\frac{ln(r_o / r_i)}{k_w} + \frac{1}{h_o}} \tag{7.4}$$

(iii) Since $A_i = 2\pi r_i L$ and $A_o = 2\pi r_o L$, Eqs. (7.3) and (7.4) can also be written as ;

$$U_i = \frac{1}{\frac{1}{h_i} + \left(\frac{r_i}{k_w}\right) ln(r_o / r_i) + \left(\frac{r_i}{r_o}\right)\frac{1}{h_o}} \tag{7.5}$$

and
$$U_o = \frac{1}{\left(\frac{r_o}{r_i}\right)\frac{1}{h_i} + \left(\frac{r_o}{k_w}\right) ln(r_o / r_i) + \frac{1}{h_o}} \tag{7.6}$$

Typical values of over all heat transfer coefficients for various applications are given in Table 7.1.

7.4 EFFECT OF SCALE FORMATION

In most heat exchangers some scale formation will take place on one or both sides of the heat transfer surface after the heat exchanger has been in use for some time. This introduces two additional resistances in the heat flow path. Thus the total thermal resistance becomes;

Table 7.1 Over all heat transfer coefficients, U for various applications

	Heat Flow				U
	from	h_c (W/m^2)	to	h_o (W/m^{2}K)	(W/m^{2}K)
1.	Gas	5 to 15	(i) Stagnant gas	5 to 10	1 to 2
	(free convection)		(Room/outside air through glass)		
			(ii) Flowing gas, (super heaters)	10 to 100	3 to 10
			(iii) Flowing liquids		10 to 40
			(a) water	3,000 to 10,000	plus radiation
			(b) others	500 to 2,000	
			(combustion Chamber)		
			(iv) Boiling liquid		10 to 40
			(a) water	3,500 to 60,000	plus radiation
			(b) others	1,000 to 20,000	
			(Steam Boiler)		
2.	Gas	10 to 100	(i) Flowing gas, (heat exchangers)	10 to 100	10 to 30
	(flowing)				
			(ii) Stagnant liquid, (Gas Boilers)	50 1000	10 to 50
3.	Liquid	50 to 10,000	(i) Stagnant liquid	50 to 1,000	25 to 500
	(free convection)		(Oil bath for heating)		
			(ii) Flowing liquid		
			(a) water	3,000 to 10,000	500 to 1500
					(with Stirring)
			(b) others	500 to 2,000	500 to 1500
			(cooling coil)		
4.	Liquid (flowing)		(i) Stagnant gas	5 to 15	5 to 15
			(Radiator central heating)		
	(a) Water	3,000 to 10,000	(ii) flowing gas, (Gas cooling)	10 to 100	10 to 50
	(b) others	500 to 3000	(iii) Stagnant liquid	50 to1,000	
			(Heating coil in water/water)		
			(a) without Stirring		50 to 250
			(b) without Stirring		500 to 2800
			(iv) Flowing liquid		
			(a) water	3,000 to 10,000	
			(b) others	500 to 2,000	
			(Heat exchanger, (a) water/ water		900 to 2,500
			(b) water/other liquid)		200 to 1,000
			(v) Boiling liquid		
			(a) water	3,500 to 60,000	300 to 1,000
			(b) others	1,000 to 20,000	
			(evaporators for refrigeration)		
5.	Vapor				
	(condensing)				
	(a) water	5,000 to 30,000	(i) Stagement gas, (steam radiator)	5 to 15	5 to 20
	(b) others	1,000 to 4,000	(ii) Flowing gas, (air heators)	10 to 100	10 to 50
			(iii) Stagement liquid	50 to 1000	
			(steam jackets with stirrers		
			(a) water		300 to 1000
			(b) others)		150 to 500
			(iv) Flowing liquid	3,000 to 10,000	
			(a) water	500 to 2000	
			(b) others		
			(condensers, (a) steam/water		1000 to 4000
			(b) other vapor/water)		300 to 1000
			(v) Boiling liquid	35,00 to 60,000	
			(a) water	10,000 to 20,000	
			(b) others		
			(evaporators, (a) stream/water		1,500 to 6,000
			(b)stream/other liquid)		300 to 2,000

$$\Sigma R = R_i + R_{si} + R_w + R_{so} + R_o$$

where R_{si} = thermal resistance due to scale formation on inside surface of inner pipe, m^2 C/ watt.

R_{so} = thermal resistance due to scale formation on outside surface of inner pipe, m^2 C/watt.

Since it is difficult to ascertain accurately the thickness and thermal conductivity of the *scale* formed, the effect of the scale deposit on heat flow is generally taken into account by specifying an equivalent *scale heat transfer coefficient, h_s.*

The reciprocal of the scale heat transfer coefficient is called the *fouling factor*. If h_{si} and h_{so} denote the heat transfer coefficient for the scale formed on the inside and outside surface, respectively, of the inner pipe, then;

$$R_{si} = \frac{1}{A_i h_{si}} \tag{7.7}$$

$$R_{so} = \frac{1}{A_o h_{so}} \tag{7.8}$$

and

$$q = \frac{(t_i - t_o)}{\dfrac{1}{A_i h_i} + \dfrac{1}{A_i h_{si}} + \dfrac{ln(r_o / r_i)}{2\pi L k_w} + \dfrac{1}{A_o h_{so}} + \dfrac{1}{A_o h_o}} \tag{7.9}$$

or

$$U_i = \frac{1}{\dfrac{1}{h_i} + \dfrac{1}{h_{si}} + \dfrac{A_i\, ln(r_o / r_i)}{2\pi L k_w} + \left(\dfrac{A_i}{A_o}\right)\dfrac{1}{h_{so}} + \left(\dfrac{A_i}{A_o}\right)\dfrac{1}{h_o}} \tag{7.10}$$

or

$$U_i = \frac{1}{\dfrac{1}{h_i} + \dfrac{1}{h_{si}} + \left(\dfrac{r_i}{k_w}\right) ln(r_o / r_i) + \left(\dfrac{r_i}{r_o}\right)\dfrac{1}{h_{so}} + \left(\dfrac{r_i}{r_o}\right)\dfrac{1}{h_o}} \tag{7.11}$$

$$U_o = \frac{1}{\left(\dfrac{A_o}{A_i}\right)\dfrac{1}{h_i} + \left(\dfrac{A_o}{A_i}\right)\dfrac{1}{h_{si}} + \left(\dfrac{A_o}{2\pi L}\right)\dfrac{ln(r_o / r_i)}{k_w} + \dfrac{1}{h_{so}} + \dfrac{1}{h_o}} \tag{7.12}$$

or

$$U_o = \frac{1}{\left(\dfrac{r_o}{r_i}\right)\dfrac{1}{h_i} + \left(\dfrac{r_o}{r_i}\right)\dfrac{1}{h_{si}} + \left(\dfrac{r_o}{k_w}\right) ln(r_o / r_i) + \dfrac{1}{h_{so}} + \dfrac{1}{h_o}} \tag{7.13}$$

The fouling factors ($1/h_s$) for some representative applications are listed in Table 7.2

Example 7.1

Find the overall coefficient of heat transfer between water and oil if the water flows through a copper pipe 1.8 cm ID and 2.1 cm OD while the oil flows through the annulus between this pipe and a steel pipe. The water and oil side film coefficients are 4650 and 1280 W/mK. The fouling factors on the water and oil sides may be taken to be 0.000344 and 0.00086 respectively. The thermal conductivity of the tube wall is 349 W/mK.

Solution :

Let us find U based on outer area of inner pipe From Eq..(7.13);

Table 7.2 : Fouling Factors

Fluid	Fouling Factor $\left(\frac{1}{hs}\right)$ (m^2 K/W)
Distilled water	0.000086
Sea water	0.000172
Well water	0.000344
Treated boiler feed water	0.000172
Liquid gasoline	0.000086
Refrigerant liquid, brine or oil-bearing steam	0.000172
Refrigerant vapors	0.000344
Fuel oil and crude oil	0.00086
Diesel exhaust gas	0.00172
Quenching oil	0.0007
Alcohol vapours	0.00009
Steam, non-oil bearing	0.00009
Industrial air	0.0004

$$U = \frac{1}{\left(\frac{r_o}{r_i}\right)\frac{1}{h_i}+\left(\frac{r_o}{r_i}\right)\frac{1}{h_{si}}+\left(\frac{r_o}{k_w}\right)\ln(r_o/r_i)+\frac{1}{h_{so}}+\frac{1}{h_o}}$$

Now, $\frac{r_o}{r_i} = \frac{2.1}{1.8} = 1.167,\ r_o = 1.05 \times 10^{-2}$

$h_i = 4650$

$h_{si} = (1/0.000344)$

$h_{so} = (1/0.00086)$ and $h_o = 1280$

So $$U = \frac{1}{\left[\frac{1.167}{4650}+(1.167)(0.000344)+\frac{1.05\times10^{-2}}{349}\ln(1.167)+0.00086+\frac{1}{1280}\right]}$$

$= 435\ W/m^2\ K$ Ans.

7.5 LOGARITHMIC MEAN TEMPERATURE DIFFERENCE

In a heat exchanger the temperatures of the hot and cold fluids keep on changing from point to point along the length of the exchanger. The question therefore arises what value of the temperature difference should be used to compute the rate of heat flow. In other words, we have to find a *mean temperature difference* which, when multiplied by the overall coefficient of heat transfer and the appropriate area, will give the correct heat flow rate. An expression for this mean temperature difference for parallel-flow and counter-flow heat exchangers can be obtained as follows.

Let t_{h1}, t_{h2} = temperature of hot fluid at exchanger entrance and exit respectively, C

t_{c1}, t_{c2} = temperature of cold fluid at exchanger entrance and exit respectively, C

W_h, c_h = mass flow rate and specific heat respectively of hot fluid, kg/hr, kJ/kg C

W_c, c_c = mass flow rate and specific heat respectively of cold fluid, kg/hr, kJ/kg C

Referring to Fig 7.9 consider an infinitesimal length dx of the exchanger located at distance x from the entrance. Let t_h and t_c denote the temperature of hot and cold fluids at location x, also $(t_h + dt_h)$ and $(t_c + dt_c)$denote these temperatures at location $x + dx$. Then the rate of heat transfer dq through the infinitesimal area dA, corresponding to length dx, is given by

$$dq = U\,\mathrm{dA}\,(t_h - t_c) \qquad \text{(a)}$$

where U and dA both correspond to the outside surface of inner pipe (usual practice).

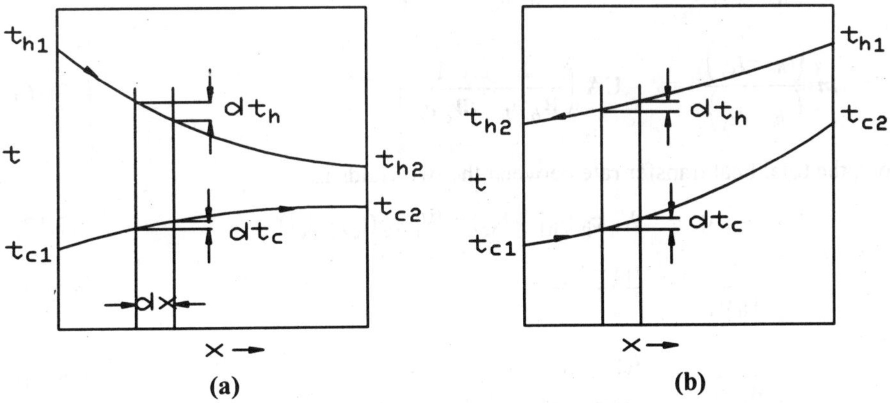

Fig. 7.9 Calculation of LMTD

If we take flow direction of cold fluid to be positive, the rate of heat flow for parallel-flow exchanger (Fig7.9 a) is given by ;

$$dq = W_c c_c\, dt_c = -W_h c_h\, dt_h \qquad \text{(b)}$$

and for counter flow arrangement (Fig.7.9 b) the heat flow is given by:

$$dq = W_c c_c\, dt_c = W_h c_h\, dt_h \qquad \text{(c)}$$

Note that in the positive direction dt_c and dt_h are both positive in the case of counter-flow whereas for parallel flow dt_c is positive and dt_h is negative.

Case I

Let us take the parallel-flow arrangement first [Fig. 7.9 (a)]. From Eq. (b);

$$dt_h = -\frac{dq}{W_h c_h}$$

and

$$dt_c = \frac{dq}{W_c c_c} \qquad (7.14)$$

so $$d(t_h - t_c) = -dq\left(\frac{1}{W_h c_h} + \frac{1}{W_c c_c}\right)$$

Substituting *dq* from Eq.(a) into Eq. (7.14),

$$d(t_h - t_c) = -UdA\,(t_h - t_c)\left(\frac{1}{W_h c_h} + \frac{1}{W_c c_c}\right)$$

or $$\frac{d(t_h - t_c)}{(t_h - t_c)} = -U\left(\frac{1}{W_h c_h} + \frac{1}{W_c c_c}\right)dA \qquad (7.15)$$

Integrating Eq.(7.15) from $x = 0$ to $x = L$ or $A = 0$ to $A = A$,

$$ln\,\frac{(t_h - t_c)_2}{(t_h - t_c)_1} = -UA\left(\frac{1}{W_h c_h} + \frac{1}{W_c c_c}\right)$$

or $$ln\,\frac{(t_{h_2} - t_{c_2})}{(t_{h_1} - t_{c_1})} = -UA\left(\frac{1}{W_h c_h} + \frac{1}{W_c c_c}\right) \qquad (7.16)$$

Now, the total heat transfer rate between the two fluids is

$$q = W_h c_h (t_{h1} - t_{h2}) = W_c c_c (t_{c2} - t_{c1}) \qquad (7.17)$$

or $$\frac{1}{W_h c_h} = \frac{t_{h1} - t_{h2}}{q} \qquad (7.18)$$

$$\frac{1}{W_c c_c} = \frac{t_{c2} - t_{c1}}{q} \qquad (7.19)$$

From Eqs.(7.16), (7.18),and (7.19),

$$ln\,\frac{(t_{h2} - t_{c2})}{(t_{h1} - t_{c1})} = -\frac{UA}{q}(t_{h1} - t_{h2} + t_{c2} - t_{c1})$$

$$= \frac{UA}{q}[(t_{h2} - t_{c2}) - (t_{h1} - t_{c1})]$$

or $$q = \frac{UA[(t_{h2} - t_{c2}) - (t_{h1} - t_{c1})]}{ln\left\{\dfrac{t_{h2} - t_{c2}}{t_{h1} - t_{c1}}\right\}} \qquad (7.20)$$

Since we wish to define the mean temperature difference Δt_m in such a way that :

$$q = U A \Delta t_m \qquad (7.21)$$

we get by comparing Eq. (7.20) and (7.21) ;

$$\Delta t_m = \frac{(t_{h2} - t_{c2}) - (t_{h1} - t_{c1})}{ln\left(\dfrac{t_{h2} - t_{c2}}{t_{h1} - t_{c1}}\right)} \qquad (7.22)$$

or $$\Delta t_m = \frac{(\Delta t)_2 - (\Delta t)_1}{ln\frac{(\Delta t)_2}{(\Delta t)_1}} \tag{7.23}$$

where $(\Delta t)_2$ and $(\Delta t)_1$ denote the difference between the temperatures of the hot and cold fluids at $A = A$ and $A = 0$ respectively.

Case : 2

Next let us consider the counter- flow case [Fig. 7.9 (b)].

From Eqs. (c) $$dt_h = \frac{dq}{W_h c_h} \text{ and } dt_c = \frac{dq}{W_c c_c}$$

or $$d(t_h - t_c) = dq\left(\frac{1}{W_h c_h} - \frac{1}{W_c c_c}\right). \tag{7.24}$$

From Eq. (a) and Eq. (7.24),

$$d(t_h - t_c) = UdA\,(t_h - t_c)\left(\frac{1}{W_h c_h} - \frac{1}{W_c c_c}\right)$$

or $$\frac{d(t_h - t_c)}{t_h - t_c} = U\left(\frac{1}{W_h c_h} - \frac{1}{W_c c_c}\right) dA \tag{7.25}$$

Integrating Eq. (7.25) from $A = 0$ to $A = A$,

$$ln\frac{(t_{h1} - t_{c2})}{(t_{h2} - t_{c1})} = UA\left[\frac{1}{W_h c_h} - \frac{1}{W_c c_c}\right] \tag{7.26}$$

Note that at $A = A$, the hot fluid at t_{h1} is entering the exchanger and cold fluid at t_{c2} is leaving the exchanger. Similarly at $A = 0$, the hot fluid at t_{h2} is leaving and the cold fluid at t_{c1} is entering the exchanger.

Since Eqs. (7.17), (7.18), (7.19) are applicable to this case also, we get ;

$$ln\frac{(t_{h1} - t_{c2})}{(t_{h2} - t_{c1})} = \frac{UA}{q}\{(t_{h1} - t_{h2}) - (t_{c2} - t_{c1})\}$$

$$= \frac{UA}{q}\{(t_{h1} - t_{c2}) - (t_{h2} - t_{c1})\}$$

or $$q = \frac{UA\{(t_{h1} - t_{c2}) - (t_{h2} - t_{c1})\}}{ln\left\{\frac{t_{h1} - t_{c2}}{t_{h2} - t_{c1}}\right\}} \tag{7.27}$$

and $$\Delta t_m = \frac{(t_{h1} - t_{c2}) - (t_{h2} - t_{c1})}{ln\left(\frac{t_{h1} - t_{c2}}{t_{h2} - t_{c1}}\right)} \tag{7.28}$$

or $$\Delta t_m = \frac{(\Delta t)_2 - (\Delta t)_1}{ln[(\Delta t)_2 / (\Delta t)_1]} \qquad (7.23)$$

where, as before, $(\Delta t)_2$ and $(\Delta t)_1$ represent the temperature difference between the hot and cold fluids at $A = A$ and $A = 0$ or at exit and entrance of the *cold fluid* respectively.

Note that the *same* expression Eq. (7.23 for) Δt_m is applicable to both parallel-flow and counter-flow arrangements provided that the correct values of $(\Delta t)_2$ and $(\Delta t)_1$ are taken.

Note also that if $W_h c_h = W_c c_c$, then $(t_{h1} - t_{h2}) = (t_{c2} - t_{c1})$ or $(t_{h1} - t_{c2}) = (t_{h2} - t_{c1})$. In this case the RHS of Eq. (7.27) become indeterminate. However, it would be evident that if $W_h c_h = W_c c_c$, the temperature difference between the two fluids remains constant throughout the exchanger. Hence the mean temperature difference is simply equal to $(t_{h1} - t_{c2})$ or $(t_{h2} - t_{c1})$

The mean temperature difference $(\Delta t)_m$ as given by Eq. (7.23) is called the *logarithmic mean temperature difference,* L.M.T.D. Note the following assumptions are implicit in the derivation of the expression for Δt_m :

(a) The overall coefficient of heat transfer U is constant throughout the exchanger length.

(b) The flow rates and specific heats of the two fluids are constant.

(c) The flow is steady.

Example 7.2

A single- pass steam condenser has to be designed to condense 35,000 kg/hr of dry saturated steam at 2.77 cm Hg pressure (saturation temperature = 27.6 C). The cooling water enters the tube at 16 C and leaves at 21 C. The tubes are 2.25 cm ID and 2.5 cm OD. The thermal conductivity of the tube material is 105 W/mK. The water and steam side film coefficients are 6960 and 5900 W/m²K respectively. . Determine the surface area of tubes required. Neglect scale resistance.

Solution :

In this case Eq. (7.13) simplifies to :

$$U = \frac{1}{\left(\frac{r_o}{r_i}\right)\frac{1}{h_i} + \left(\frac{r_o}{k}\right)\ln\left(\frac{r_o}{r_i}\right) + \frac{1}{h_o}}$$

$$= \frac{1}{\frac{2.5}{2.25}\left(\frac{1}{6960}\right) + \left(\frac{1.25}{105 \times 100}\right)\ln\left(\frac{2.5}{2.25}\right) + \frac{1}{5800}}$$

or $$U = 2900 \text{ W/m}^2\text{ K}$$

and $$\Delta t_m = \frac{(27.6-21)-(27.6-16)}{\ln(6.6/11.6)}$$

$$= \frac{5}{\ln(1.67)} = 8.85 \text{ C}$$

Now latent heat = 2450 kJ/kg

so $\quad q = 35,000 \times 2,450 \times 1,000$ J/hr

From Eq (7.21)

$$A = \frac{q}{U\Delta t_m}$$

$$= \frac{35,000 \times 2450 \times 1000}{2,900 \times 8.85 \times 3600} = 816 \text{ m}^2$$

Ans.

7.6 L.M.T.D. CORRECTION FACTORS

Expression given by Eq. (7.23) for the mean temperature difference is strictly applicable only to parallel- flow and counter-flow exchangers in which the two fluids *flow only once* across the exchanger length, that is, *single-pass* heat exchangers. If we are dealing with multi-pass parallel and counter flow exchangers or with cross flow exchangers, we cannot apply Eq (7.23).

The expressions giving the mean temperature difference for these exchangers are quite complicated. For this reason, the analysis of such exchangers is generally carried out as explained below.

The LMTD is first calculated for single-pass counter-flow exchanger by using Eq. (7.23). The $(\Delta t)_m$ so obtained is then multiplied by a *correction factor F* which takes into account the actual flow arrangement of the exchanger. Thus ;

$$q = UAF\,[(\Delta t)_m]_{\text{counter-flow}} \qquad (7.29)$$

The correction factors for different flow arrangements can be obtained from charts, such as that shown in Fig. 7.10 to 7.14. In these charts, the factor F is given as a function of two dimensionless variables called the *capacity ratio R* and the *temperature ratio*, θ.

The *capacity ratio*, R is defined as the ratio of the heat capacity of the fluid stream flowing through *tube* to the heat-capacity of the fluid stream flowing through the *shell*.

Thus, $$R = \frac{W_t c_t}{W_s c_s.} \qquad (7.30)$$

where the suffixes t and s denote the fluid flowing through the tube or shell respectively.

Since

$$W_t c_t (t_{t2} - t_{t1}) = W_s c_s (t_{s1} - t_{s2})$$

$$R = \frac{t_{s1} - t_{s2}}{t_{t2} - t_{t1}} \qquad (7.31)$$

The *temperature ratio*, θ is defined as:

$$\theta = \frac{t_{t2} - t_{t1}}{t_{s1} - t_{t1}} \qquad (7.32)$$

The charts giving values of F for some common arrangements of multipass and cross-flow exchangers are given in Figs.7.10 to 7.14. The use of these charts can be illustrated by means of examples.

Example 7.3

An exchanger is required to cool 5,000 kg/hr of oil at 70 C to 40 C with water entering at 15 C. The exit temperature difference between the two fluids should be 10°C. Find the ratio of area required for a 1-4 exchanger (one shell pass and four tube-passes) and a 2-4 exchanger. The cold fluid flows through the tube and the hot fluid through the shell.

Solution :

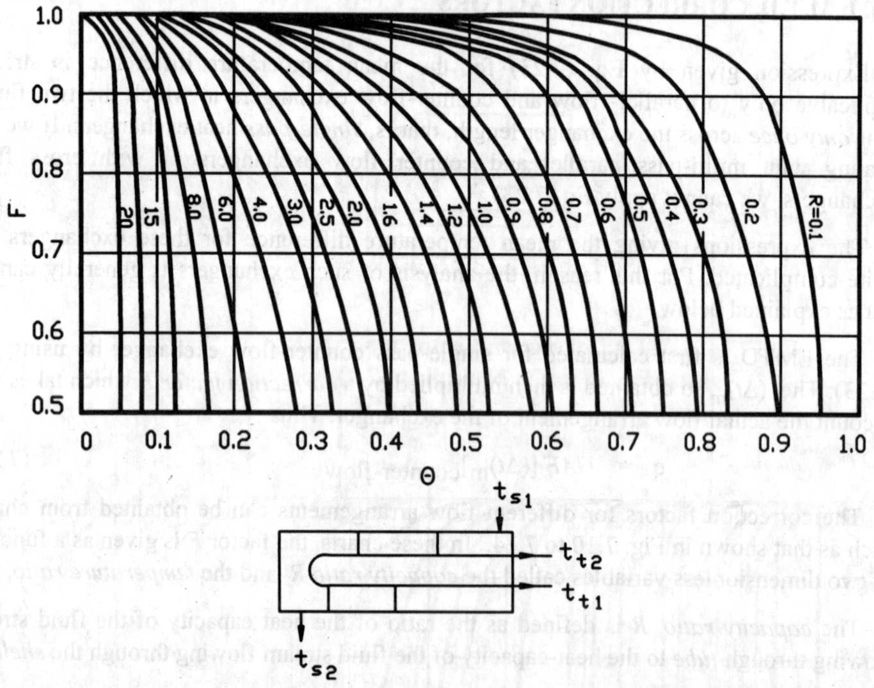

Fig. 7.10 The LMTD Correction Factor for Exchanger with 1 Shell-Pass and 2 (or multiples of 2) Tube - Passes

Here $\quad t_{h1} = 70\ C = t_{s1}$

$t_{h2} = 40\ C = t_{s2}$

and $\quad t_{c2} = t_{h2} - 10 = 40 - 10 = 30\ C = t_{t2}$

so $$R = \frac{t_{h1} - t_{h2}}{t_{c2} - t_{c1}} = \frac{70-40}{30-15} = 2$$

$$\theta = \frac{t_{c2} - t_{c1}}{t_{h1} - t_{c1}} = \frac{30-15}{70-15} = 0.273$$

For these values from Fig. 7.10,

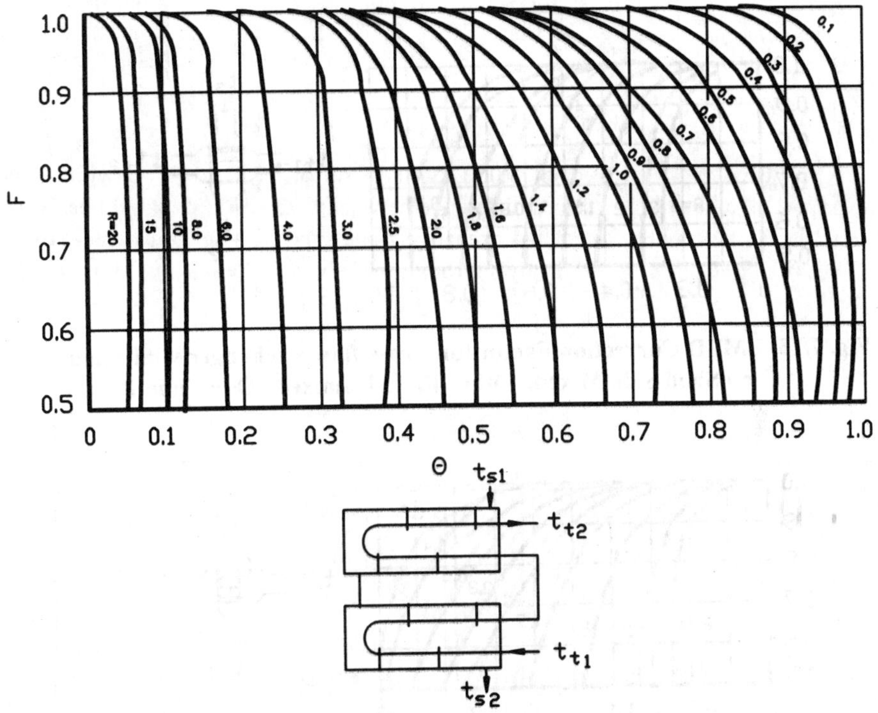

Fig. 7.11 LMTD Correction Factor for Exchanger having 2 Shell-Passes and a Multiple of 2 Tube-Passes

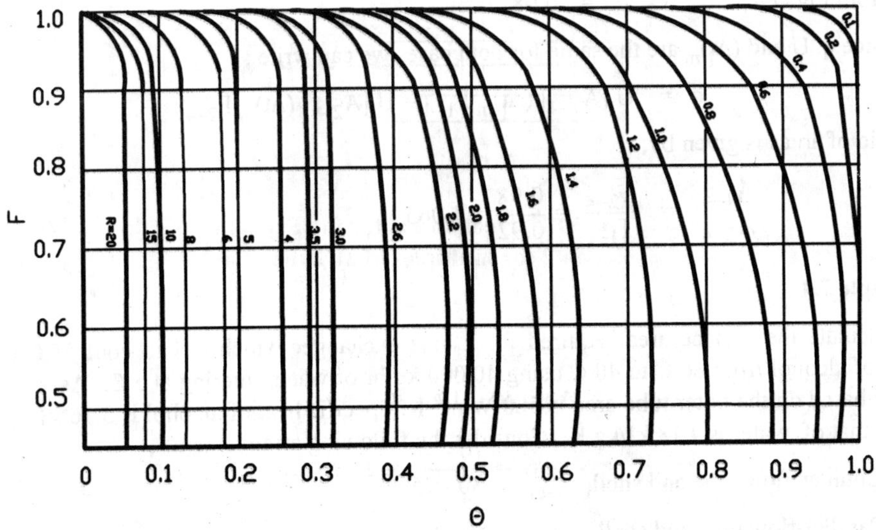

Fig. 7.12 LMTD Correction Factor for Exchanger having 3 Shell Passes and a Multiple of 6 Tube-Passes

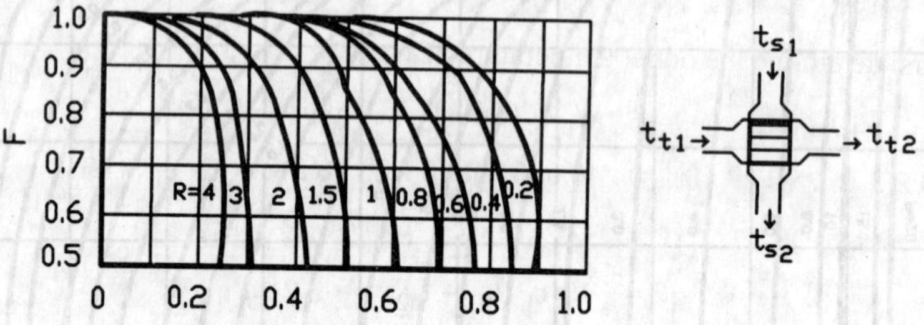

Fig. 7.13 LMTD Correction Factor for Cross-flow Exchanger with Fluid on Shell Side Mixed. Other Fluid Unmixed. One Tube-Pass

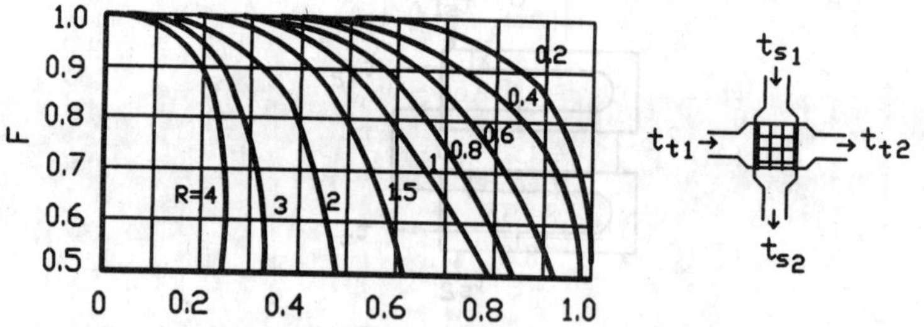

Fig. 7.14 LMTD Correction Factor for Cross-flow Exchanger with Both Fluids Unmixed. One Tube-Pass

$$F_{1-4} = 0.92 ;$$

and from Fig. 7.11, $F_{2-4} = 0.98.$

Since q, U and $(\Delta t)_m$ are the same for both cases we can write ;

$$q = U [A_{1-4}] (\Delta t)_m F_{1-4} = U [A_{2-4}] (\Delta t)_m F_{2-4}$$

so ratio of areas is given by,

$$\frac{A_{1-4}}{A_{2-4}} = \frac{F_{2-4}}{F_{1-4}} = \frac{0.98}{0.92} = 1.065 \qquad \text{Ans.}$$

Example 7.4

Calculate the surface area required for a heat exchanger which has to cool 55,000 kg/hr of alcohol from 66 C to 40 C using 40,000 kg/hr of water entering at 5 C. Assume that U based on the outer tube area is 580 W/m^2 K. Specific heat of alcohol is 3.76 kJ/kg K and that of water is 4.18 kJ/kg K. Consider the following arrangements ;

(a) Counter-flow tube and shell,

(b) Parallel-flow tube and shell,

(c) Reversed-current exchanger with 2 shell passes and 72 tube passes. Alcohol flows through the shell.

(d) Cross-flow, with one tube pass and one shell pass, shell side fluid mixed.

Solution :

Let us first determine the outlet temperature of water, t_{c2}.

$$W_h c_h (t_{h1} - t_{h2}) = W_c c_c (t_{c2} - t_{c1})$$

or $$55{,}000 \times 3.76 \times (66 - 40) = 40{,}000 \times 4.18 \times (t_{c2} - 5)$$

or $$t_{c2} = 37.2 \text{ C}$$

$$q = W_h c_h (t_{h1} - t_{h2})$$
$$= 55{,}000 \times 3.76 \times (66 - 40)$$
$$= 5{,}400{,}000 \text{ kJ/hr}$$

(a) $$(\Delta t_m)_{\text{counter flow}} = \frac{(\Delta t)_2 - (\Delta t)_1}{\ln\{(\Delta t)_2 / (\Delta t)_1\}} \quad (7.23)$$

$$(\Delta t)_2 = t_{h1} - t_{c2} = 66 - 37.2 = 28.8 \text{ C}$$
$$(\Delta t)_1 = t_{h2} - t_{c1} = 40 - 5 = 35 \text{ C}$$

so $$(\Delta t_m)_{\text{counter flow}} = \frac{28.8 - 35}{\ln \frac{28.8}{35}}$$

$$= \frac{6.2}{\ln 1.22} = 31.1\text{C}$$

or $$(A)_{\text{counter flow}} = \frac{q}{U(\Delta t_m)_{\text{counter flow}}}$$

$$= \frac{5{,}400{,}000 \times 10^3}{3600 \times 580 \times 31.1}$$

$$= 82.5 \text{ m}^2$$ Ans.

(b) $$(\Delta t)_2 = t_{h2} - t_{c2} = 40 - 37.2 = 2.8 \text{ C}$$
$$(\Delta t)_1 = t_{h1} - t_{c1} = 66 - 5 = 61 \text{ C}$$

or $$(\Delta t)_m = \frac{2.8 - 61}{\ln \frac{2.8}{61}}$$

$$= \frac{58.2}{3.08} = 18.9 \text{ C}$$

so $$(A)_{\text{parallel flow}} = \frac{5400{,}000 \times 10^3}{3600 \times 18.9 \times 580}$$

$$= 136 \text{ m}^2$$ Ans.

(c) $$(\Delta t_m)_{2-72} = (F_{2-72})\,(\Delta t_m)_{\text{counter flow}}$$

Now, $R = \frac{t_{s1} - t_{s2}}{t_{t2} - t_{t1}} = \frac{t_{h1} - t_{h2}}{t_{c2} - t_{c1}} = \frac{66-40}{37.2-5} = 0.81$

$$\theta = \frac{t_{t2} - t_{t1}}{t_{s1} - t_{t1}} = \frac{t_{c2} - t_{c1}}{t_{h1} - t_{c1}} = \frac{37.2-5}{66-5} = 0.528$$

From Fig. (7.10),

$$F_{2-72} = 0.965$$

So $A_{2-72} = \frac{q}{U(F_2 - {}_{72})(\Delta t_m)_{\text{counter flow}}}$

$$= \frac{5,400,000 \times 10^3}{3600 \times 580 \times 0.965 \times 31.1}$$

$$= 85.7 \text{ m}^2$$ Ans.

(d) From Fig. 7.13, for R = 0.81 and θ = 0.528 we get ;

$$(F)_{\text{cross flow}} = 0.87$$

or $(A)_{\text{cross flow}} = \frac{A_{\text{counter flow}}}{0.87}$

$$= \frac{82.5}{0.87}$$

$$= 95 \text{ m}^2$$ Ans.

7.7 MORE ILLUSTRATIVE EXAMPLES

It is now proposed to solve a few more examples illustrating the general calculations involved in the design of heat exchangers.

Example 7.5

Find the length of counter-flow heat exchanger to heat 4,000 kg/hr of oil from 10 to 20 C using hot water at 70 C. The hot water flows with a velocity of 75 cm/sec through a copper pipe of ID = 1.8 cm and OD = 2.1 cm. The oil flows through the annulus between this pipe and a 3 cm ID steel pipe. Neglect the tube wall resistance and scale resistance. The following properties of oil and water may be used.

	Water	*oil*
C_p *(kJ/kg-C)*	4.18	1.885
k *(W/mK)*	0.66	0.14
ρ *(kg/m³)*	1000	850
υ *(m²/sec)*	0.405×10^{-6}	0.72×10^{-5}

Solution :

(i) W_h = (area) (velocity) (density)

$$= \left(\frac{\pi}{4}\right)\left(\frac{1.8}{100}\right)^2 \left(\frac{75}{100} \times 3600\right) (1000)$$

$$= 690 \text{ kg/hr}$$

Now,

$$W_h c_h (t_{h1} - t_{h2}) = W_c c_c (t_{c2} - t_{c1})$$

or

$$690 \times 4.18 (70 - t_{h2}) = 4000 \times 1.885 (20 - 10)$$

$$(70 - t_{h2}) = \frac{4000 \times 1.885 \times 10}{690 \times 4.18} = 26.1$$

so

$$t_{h2} = 70 - 26.1 = 43.9 \text{ C}$$

From Eq. (7.28) ;

$$\Delta t_m = \frac{(t_{h1} - t_{c2}) - (t_{h2} - t_{c1})}{\ln\left(\frac{t_{h1} - t_{c2}}{t_{h2} - t_{c1}}\right)}$$

$$= \frac{(70 - 20) - (43.9 - 10)}{\ln(50 / 33.9)} = 40.25 \text{ C}$$

The next step is to calculate U. For this we need the values of h_i and h_o which can be obtained by first determining the Reynolds number for water-flow and oil-flow and then using appropriate forced convection equation.

(ii) Reynolds number for water flow is;

$$(Re)_{water} = \left(\frac{VD}{\nu}\right)$$

$$= \frac{(0.75 \times 3600)(1.8 \times 10^{-2})}{0.405 \times 10^{-6} \times 3600} = 33{,}400$$

(iii) Reynolds number for oil flow is ;

$$(Re)_{oil} = \left(\frac{VD}{\nu}\right)_{oil}$$

The velocity of oil is given by ;

$$V_{oil} = \frac{\text{volume flow per hr}}{\text{area of flow}}$$

or

$$V_{oil} = \frac{(4000 / 850)}{\frac{\pi}{4}(3^2 - 2.1^2) \times 10^{-4}}$$

$$= \frac{4.72}{3.61 \times 10^{-4}}$$

$$= 1.31 \times 10^4 \text{ m/hr}$$

(iv) The *equivalent* diameter of the annulus is (3 – 2.1) = 0.9,

or $$(Re)_{oil} = \frac{1.31\times10^4\times0.9\times10^{-2}}{0.72\times10^{-5}\times3600} = 4550$$

Since the flow of both water and oil is turbulent we can find h by using the Colburn equation.

$$\frac{h}{\rho V c\rho}\left(\frac{\mu c_p}{k}\right)^{2/3} = \frac{0.023}{(Re)^{0.2}}$$

(v) $$h_{oil} = h_o = \frac{0.023\,\rho V C_p}{(Re)^{0.2}\left(\frac{\mu c_p}{k}\right)^{2/3}}$$

$$= \frac{0.023\times850\times1.31\times10^4\times1.885}{(4550)^{0.2}\left(0.72\times10^{-5}\times850\times\frac{1.885\times1000}{0.14}\right)^{2/3}}$$

$$= 1010 \text{ J/sec m}^2\text{ K} = 1010 \text{ W/m}^2\text{ K}$$

(vi) Similarly,

$$h_{water} = h_i = \frac{0.023\times1000\times0.75\times3600\times4.18}{(33400)^{0.2}\left(0.405\times10^{-6}\times1000\times4.18\times1000/0.66\right)^{2/3}}$$

$$= 4830 \text{ W/m}^2\text{ K}$$

(vii) Thus $$U = \frac{1}{\left(\frac{r_o}{r_i}\frac{1}{h_i}\right)+\frac{1}{h_o}}$$

$$= \frac{1}{\frac{2.1}{1.8}\left(\frac{1}{4830}\right)+\frac{1}{1010}}$$

$$= \frac{1}{10^{-4}(2.42+9.9)} = \frac{10^4}{12.32}$$

$$= 812 \text{ W/m}^2\text{ K}$$

(viii) The rate of heat transfer is ;

$$q = W_c c_c (t_{c2} - t_{c1})$$

$$= 4000 \times 1.885 (20 - 10)$$

$$= 75200 \text{ kJ/hr}$$

(ix) so $$A = \frac{q}{U\Delta t_m}$$

$$= \frac{75200 \times 1000}{40.25 \times 812 \times 3600}$$

$$= 0.638 \text{ m}^2$$

or $$L = \frac{0.638}{\pi(2.1) \times 10^{-2}} = 9.67 \text{ m}$$ Ans.

Example 7.6

8,000 kg/hr of air at 105 C is cooled by passing it through a counter flow exchanger. Find the exit temperature of air if water enters at 15 C and flows at a rate of 7,500 kg/hr. $U = 145 \text{ W/m}^2 \text{ K}$. $A = 20 \text{ m}^2$.

Solution :

Since the exit temperature of neither fluid is known, we have to solve the problem by trial.

(i) Let us assume that the exit temperature of the cold fluid (water) is 35 C. Then,

$$q = 7500\,(35 - 15) \times 4.18$$

$$= 627{,}000 \text{ kJ/hr}$$

so $$(t_{h1} - t_{h2}) = \frac{6.27 \times 10^5}{8000 \times 1} = 78.3 \text{ C}$$

or $$t_{h2} = t_{h1} - 78.3 = 105 - 78.3 = 26.7 \text{ C}$$

since $$\Delta t_m = \frac{(t_{h1} - t_{c2}) - (t_{h2} - t_{c1})}{\ln\left(\dfrac{t_{h1} - t_{c2}}{t_{h2} - t_{c1}}\right)}$$

or $$\Delta t_m = \frac{(105 - 35) - (26.7 - 15)}{\ln\left(\dfrac{70}{11.7}\right)}$$

$$= \frac{70 - 11.7}{\ln(5.98)}$$

$$= \frac{58.3}{1.79} = 32.7 \text{ C}$$

or $$q = UA\,\Delta t_m$$

$$= (145 \times 3600 \times 20 \times 32.7)/1000$$

$$= 341{,}500 \text{ kJ/hr}$$

This value of q does not check with the value of 627,000 on the basis of assumed t_{c2}. Hence t_{c2} must be lower than 35 C.

(ii) Let us make a second trial with $t_{c2} = 30$ C.

$$q = 7500\,(30 - 15) \times 4.18$$
$$= 470{,}000 \text{ kJ/hr}$$
$$t_{h1} - t_{h2} = \frac{470{,}000}{8000 \times 1} = 58.8$$

or
$$t_{h2} = 100 - 58.1 = 41.2 \text{ C}$$
$$\Delta t_m = \frac{(105-30)-(41.2-15)}{\ln(75/26.2)}$$
$$= \frac{70-26.2}{\ln(2.86)} = 46.4 \text{ C.}$$

or
$$q = UA\,\Delta t_m$$
$$= (145 \times 3600 \times 20 \times 46.4)/1000$$
$$= 485{,}000 \text{ kJ/hr}$$

(iii) The value of q is slightly higher than the assumed value of 470,000 kJ. This means that the exit temperature of water will be a little higher than 30 C. A third trial will give t_{c2} = 30.8 C. Hence t_{h2} will be ;

$$t_{h2} = 43.3 \text{ C}$$ Ans.

Example 7.7

A Fr–12 condenser receives the hot gas at 80 C in a superheated state. The saturation temperature of Fr–12 is 40 C. The entering and leaving water temperatures are 20 C and 32 C respectively. The refrigerant liquid is subcooled to 32 C and the total heat rejection to water is 1150 kJ/min. If overall heat transfer coefficient U is 1160 W/m^2 K, calculate the area of the condenser coil required.

Solution:

It should be noted that in this type of heat exchanger, the temperature of both the fluids is varying and therefore logarithmic mean temperature difference (LMTD) will have to be calculated.

The temperature changes of water and Fr-12 may be represented graphically by Fig. 7.15. During the process of heat transfer the Fr-12 gas will reject heat in the following manner:

(a) To cool down to saturation temperature from 80 C to 40 C, (process *DC*)

(b) To liquify at 40 C (process *CB*)

(c) To subcool the liquid from 40 C to 32 C (process BA)

At the same time the water will receive heat and will get heated from 20°C to 32° C.

The condenser may be assumed to consist of three sections i.e. (i) desuperheater (ii) liquifier (iii) subcooler. Let the individual sections be so rejecting the heat that temperature of water at *B* and *C* is calculated as 20.25 C and 30 C.

Thus heat rejection in section DC ;

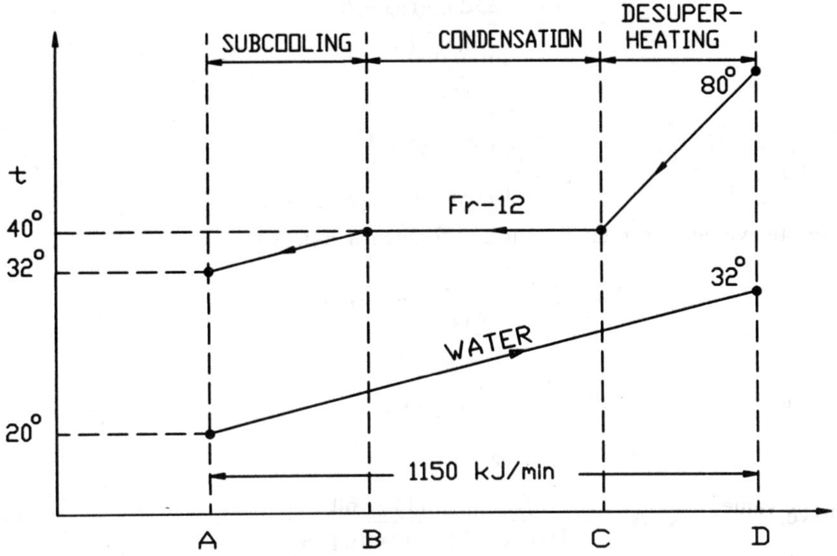

Fig. 7.15 Heat Transfer in a Condenser, Example 7.7

$$= \quad 1150 \times \frac{32-30}{32-20} = 192 \text{ kJ/min}$$

heat rejection in section DB ;

$$= \quad 1150 \times \frac{32-20.25}{32-20} = 1126 \text{ kJ/min}$$

Hence heat rejection during BC;

$$= \quad 1126 - 192 = 934 \text{ kJ/min}$$

and that during section AB = 24 kJ/min.

Therefore, the LMTD in section DC :

$$= \quad \frac{(80-32)-(40-30)}{\ln \frac{80-32}{40-30}}$$

$$= \quad \frac{48-10}{\ln 4.8} = \frac{38}{1.57} = 24.2 \text{ C}$$

The, LMTD in section CB ;

$$= \quad \frac{(40-30)-(40-20.25)}{\ln \frac{40-30}{40-20.25}}$$

$$= \quad \frac{10-19.75}{\ln \frac{10}{19.75}} = 14.4 \text{ C}$$

The LMTD in section BA ;

$$= \frac{(40-20.25)-(32-20)}{\ln\dfrac{40-20.25}{32-20}}$$

$$= \frac{19.75-12}{\ln\dfrac{19.75}{12}} = 16.3 \text{ C}$$

Therefore average mean temperature difference is approximately given by ;

$$\Delta t_m = \frac{1150}{\dfrac{192}{24.2}+\dfrac{934}{14.4}+\dfrac{24}{16.3}} = 15.5 \text{ C}$$

Since the area of the coil required is given by ;

$$Q = UA\,\Delta t_m$$

we have;

$$A = \frac{Q}{U\Delta t_m} = \frac{1150\times 60}{3.6\times 1160\times 15.5}$$

$$= 1.06 \text{ m}^2$$ *Ans.*

Example 7.8

A horizontal water heater is required to heat 15,000 kg/hr of water from 15 C to 85 C by means of steam condensing at 110 C. The heater is to consist of tubes having 1.9 cm OD and 1.65 cm ID.

Determine the number of tubes required and the length of each tube.

Solution:

(i) The rate of heat transfer is given by ;

$$q = 15{,}000 \times 4.18 \times (85-15) = 4400{,}000 \text{ kJ/hr}$$

Since the latent heat of steam at 110 C is 2240 kJ/kg, the steam flow rate is :

$$W = \frac{4400{,}000}{2240} = 1970 \text{ kg/hr}$$

The LMTD is given by ;

$$\Delta t_m = \frac{(110-85)-(110-15)}{\ln\dfrac{(110-85)}{(110-15)}}$$

or $\Delta t_m = 52.5$ C.

Hence we may take the mean or bulk temperature of water to be (110 – 52.5) = 57.5 C. Note that the arithmetic mean temperature of water will be (15 + 85)/2 = 50 C.

For water at 57.5 C ; $\mu = 1.88$ kg/ m-hr; $\rho = 985$ kg/m^3;

$C_p = 4.18$ kJ/kg-C ; $k = 0.66$W/mK = 2.39 kJ/m hr K

(ii) Let us assume that the velocity of water inside the tubes should not exceed 0.75 m/sec in order to keep the pressure drop reasonably low. This would require that the cross sectional area of the tubes should be (taking $\rho = 1000 kg/m^3$ for water),

$$A_t = \frac{15000}{1000} \times \frac{1}{0.75 \times 3600}$$

$$= \frac{1}{180} m^2$$

Cross-sectional area of 1 tube

$$= \frac{\pi}{4}(1.65)^2 \times 10^{-4}$$

$$= 2.14 \times 10^{-4} m^2$$

Hence, number of tubes required ;

$$= \left(\frac{1}{180}\right) \div (2.14 \times 10^{-4}) = 26$$

The value of Reynolds number for water flow is ;

$$Re = \frac{VD\rho}{\mu}$$

$$= \frac{(0.75 \times 3600)(0.0165) \times 985}{1.88}$$

$$= 23,400$$

Note that we have used the properties of water at the bulk temperature 57.5 C.

(iii) Since the water flow is turbulent, we may determine the water-side heat transfer coefficient h_1 by using McAdam Equation (4.42);

$$\frac{hD}{k_b} = 0.023\left(\frac{VD\rho}{\mu_b}\right)^{0.8}\left(\frac{\mu Cp}{k}\right)_b^{0.4}$$

Now $$\frac{VD\rho}{\mu_b} = 23,400$$

$$\left(\frac{\mu C_p}{k}\right)_b = \frac{1.88 \times 4.18}{2.39} = 3.3$$

So $$\frac{hD}{k_b} = 0.023 \times (23,400)^{0.8} (3.3)^{0.4}$$

$$= 0.023 \times 3125 \times 1.613$$

$$= 116$$

or $$h = \frac{116 \times 0.66}{0.0165} = 4640 W/m^2 K$$

(iv) The steam-side heat transfer coefficient h_2 may be determined by using Eq. (5.37) neglecting ρ_v and taking $h'_{fg} = h_{fg}$, and substituting ND for D because there are N tubes placed one above the other.

$$h = 0.725 \left\{ \frac{\rho^2 g h_{fg} k^3}{N \mu D (t_{sat} - t_w)} \right\}^{1/4}$$

In Eq. (5.37), the value of h_{fg} is taken at t_{sat} while the other properties are evaluated at $t_f = (t_{sat} + t_w)/2$. Since t_w is not known we have to use trial and error.

Let us assume that $t_w = 90$ C. Then ;

$$t_f = \frac{110+90}{2}\ 100 \text{ C.}$$

For liquid water at 100 C,

$$\rho = 956 \text{ kg/m}^3$$

$$C_p = 4.2 \text{ kJ/kg K}$$

$$k = 0.882 \text{ W/m K} = 2.57 \text{ kJ/hr m K}$$

$$\mu = 1.03 \text{ kg/hr-m}$$

Hence, if we assume N = 6, we get ;

$$h_2 = 0.725 \left\{ \frac{956^2 \times 9.81 \times 3600^2 \times 2250 \times (2.57)^3}{6 \times 1.03 \times 0.019 \times (110-90)} \right\}^{1/4}$$

$$= 26{,}800 \text{ kJ/hr m}^2 \text{ C} = 7{,}420 \text{ W/m}^2 \text{ K}$$

The value of t_w can now be calculated by making a heat balance and t_w checked with the assumed value:

$$h_1 A_1 (t_w - 57.5) = h_2 A_2 (110 - t_w)$$

$$(t_w - 57.5) = \frac{h_2 A_2}{h_1 A_1} (110 - t_w)$$

$$= \frac{7420}{4640} \left(\frac{1.9}{1.65} \right) (110 - t_w)$$

$$= 1.84 (110 - t_w)$$

so $$t_w = 92 \text{ C}$$

Since the value of t_w is slightly different from the assumed value of 90 C, a second trial can be made by assuming $t_w = 92$ C. Note that a 2°difference in t_w will change $(t_{sat} - t_w)$ from 20 C to 18 C and will thus change the value of h_2.

$$h_2 = 0.725 \left\{ \frac{956^2 \times 9.81 \times 3600^2 \times 2250 \times (2.57)^3}{6 \times 1.03 \times 0.019 \times (110-92)} \right\}^{1/4}$$

$$= 28{,}000 \text{ kJ/hr m}^2 \text{ K} = 7{,}800 \text{ W/m}^2 \text{ K}$$

The corresponding value of t_w is then ;

$$(t_w - 57.5) = \frac{7800}{4640}\left(\frac{1.9}{1.65}\right)(110 - t_w)$$

or $$t_w = 92.1 \text{ C}$$

Since this is very nearly equal to the assumed value of 92 C no further trial is needed. Note that we have neglected the resistance of the tube wall in above calculations.

(v) The outside surface area required is:

$$A_2 = \frac{q}{h_2(t_{sat} - t_w)}$$

$$= \frac{4400,000}{7800 \times 3.6 \times (110 - 92)} = 8.7 \text{ m}^2$$

Therefore, the length of 1 tube will be ;

$$\frac{8.7}{26 \times \pi \times 0.019} = 5.6 \text{ m}$$

Since this length is too much, we can use 2 passes so that the tube length is limited to 2.8 m. Thus there will be 2 passes, each pass having 26 tubes, 2.8 m long. Ans.

Example 7.9

3000 kg/hr of furnace oil is to be heated from 10 C to 90 C in a shell and tube type heat exchanger. The oil is to flow inside the tubes while steam at 120 C is to flow through the shell. The tubes are to be 1.9 cm OD and 1.65 cm ID. Design the heat exchanger.

Solution :

The bulk temperature of the oil may be taken as ;

$$t_b = \frac{10 + 90}{2} = 50 \text{ C}$$

At 50 C, the properties of furnace oil are ;

$$\rho = 900 \text{ kg/m}^3$$

$$\mu = 54 \text{ kg/hr-m}$$

$$C_P = 1.97 \text{ kJ/kg K}$$

$$k = 0.13 \text{ W/m K} = 0.47 \text{ kJ/hr m K}$$

The required rate of heat transfer is given by ;

$$q = 3000 \times 1.97 \times (90 - 10)$$

$$= 472,800 \text{ kJ/hr}$$

In order to keep the pressure drop low let us limit the velocity of oil to 5 cm/sec. Then, we may use Eq. (4.27) to determine the *oil-side heat transfer coefficient*, h_i.

$$\frac{hD}{k_b} = 1.86\left[\left(\frac{VD\rho}{\mu_b}\right)\left(\frac{\mu C_p}{k}\right)_b\left(\frac{D}{L}\right)\right]^{1/3}\left(\frac{\mu_b}{\mu_w}\right)^{0.14}$$

Let us assume that the tube surface temperature will be nearly 100 C and that D/L ≈ 150. Then (μ_w = 12 kg/hr-m);

$$\frac{hD}{k_b} = 1.86\left[\left(\frac{0.05\times3600\times0.0165\times900}{54}\right)\left(\frac{54\times1.97}{0.47}\right)\left(\frac{1}{150}\right)\right]^{1/3}\times\left(\frac{54}{12}\right)^{0.14}$$

$$= 1.86\left[(49.5)(228)\left(\frac{1}{150}\right)\right]^{1/3}\times(1.234) = 5.21$$

or

$$h_i = \frac{5.21\times0.13}{0.0165} = 41\ \text{W/m}^2\ \text{K}$$

(ii) The value of the steam-side heat transfer coefficient h_2 may be obtained by using Eq. (5.37). Since h_2 will be much larger than h_1 and U will depend mainly on h_1, it is not necessary to know h_2 very accurately. Hence we may use the assumed value of t_w = 100 C and evaluate the properties of water at t_f = (100 + 120)/2 = 110 C,

At 110 C, ρ = 950 kg/m^3

k = 0.73 W/m K = 2.64 kJ/hr m K

C_p = 4.2 kJ/kg K

μ = 0.97 kg/m-hr

At 120 C, h_{fg} = 2200 kJ/kg. Hence h_2 is from Eq. (5.37) (taking N = 10)

$$h_2 = 0.725\left\{\frac{\rho^2 g h_{fg} k^3}{\mu N D(t_{sat}-t_w)}\right\}^{1/4}$$

$$= 0.725\left\{\frac{950^2\times9.81\times3600^2\times2200\times2.64^3}{0.97\times10\times0.019\times(120-100)}\right\}^{1/4}$$

$$= 26{,}750\ \text{kJ/hr m}^2\ \text{K} = 7{,}420\ \text{W/m}^2\ \text{K}$$

(iii) The overall heat transfer coefficient, U based on the outside surface area is then (neglecting the tube wall resistance);

$$\frac{1}{U} = \frac{1}{h_2} + \frac{A_2}{A_1 h_1}$$

$$= \frac{1}{7420} + \frac{1.9}{1.65}\,\frac{1}{41}$$

or

$$U = 35.4\ \text{W/m}^2\ \text{K}$$

$$\Delta t_m = \frac{(120-90)-(120-10)}{\ln\left(\frac{30}{110}\right)} = \frac{80}{1.3} = 61.6\ \text{C}.$$

(iv) Hence, the outside surface area required is ;

$$A_2 = \frac{q}{U\Delta t_m}$$

or $$A_2 = \frac{472800}{35.4\times3.6\times61.6} = 60.23\ m^2$$

Total length of tubing required ;

$$= \frac{60.23}{\pi\times0.019} = 1009.5\ m$$

(v) Also, the number of tubes per pass;

$$= \left(\frac{3000}{900}\right) \div \left\{\frac{\pi(0.0165)^2}{4}\times0.05\times3600\right\} = 87$$

(vi) Length of each tube ;

$$= \frac{1009.5}{87} = 11.6$$

Since this length is too large, we may use 2 passes.

Hence, Length of each tube = 5.8 m

Number of tubes per pass = 87

Number of passes = 2 *Ans.*

Example 7.10

Design a vertical evaporator capable of evaporating 2000 kg/hr of water from a salt solution by using saturated steam at 115 C. The evaporator is to operate at 60 cm mercury pressure. The following data may be used :

Latent heat of steam = 2250 kJ/kg

Solution temperature = 94 C.

Heat supplied to solution per kg water evaporated = 2570 kJ

Use may be made of 5 cm OD, 4.5 cm ID brass tubes 1.5 m long. A scale coefficient $h_s = 29{,}000\ W/m^2\ K$ may be assumed for outside surface.

Solution :

(i) The inside film coefficient h_1 (and thence q/A_1) may be found by using Eq. (5.7) and the outside coefficient h_2 by using Eq. (5.34). Since both h_1 and h_2 will depend on the tube wall temperature t_w, the problem will have to be solved by trial and error. As a first trial, let us assume that t_w = 105 C. Then, from Eq. (5.7) ;

$$\frac{C_1(t_w - t_{sat})}{pr_l^{1.7}\ L} = C_{sf}\left\{\frac{q/A}{\mu_1\ L}\sqrt{\frac{g_c\,\sigma}{g(\rho_1-\rho_v)}}\right\}^{1/3}$$

$$t_w = 105 \text{ C}$$

$$t_{sat} = 94 \text{ C}$$

At 94 C, $\mu_l = 1.1$ kg/m-hr

$$L = 2280 \text{ kJ/kg}$$

$$\sigma = 6 \times 10^{-3} \text{ kg}_f\text{/m}$$

$$\rho_l = 960 \text{ kg/m}^3$$

$$C_p = 4.2 \text{kJ/kg K}$$

$$P_{rl} = 1.83$$

From Table 5.3, $C_{sf} = 0.006$. Hence,

$$\frac{4.2(105-94)}{(1.83)^{1.7} \times 2280} = 0.006 \left\{ \frac{\left(\frac{q}{A}\right)_1}{1.1 \times 2280} \sqrt{\frac{6 \times 10^{-3}}{960}} \right\}^{1/3}$$

where ρ_v has been neglected in comparison with ρ_l.

or
$$\left(\frac{q}{A_1}\right) = 182 \times 10^4 \text{ kJ/hr m}^2$$

Since,
$$\left(\frac{q}{A_1}\right) A_1 = \left(\frac{q}{A_2}\right) A_2$$

or
$$\left(\frac{q}{A_2}\right) = \left(\frac{q}{A_1}\right)\left(\frac{A_1}{A_2}\right) = 182 \times 10^4 \times \left(\frac{4.5}{5}\right)$$

$$= 16.38 \times 10^5 \text{ kJ/hr m}^2 \qquad \text{(a)}$$

(ii) Now, from Eq. (5.34),neglecting ρ_v and putting $h'_{fg} = h_{fg}$

$$h_2 = 0.943 \left\{ \frac{\rho^2 g h_{fg} k^3}{\mu l (t_{sat} - t_w)} \right\}^{1/4}$$

Substituting the property values (except h_{fg}) at $t_f = 110$ C, and $l = 1.5$ m,

$$h_2 = 0.943 \left\{ \frac{952^2 \times 9.81 \times 3600^2 \times 2230 \times 2.63^3}{0.94 \times 1.5 \times (115-105)} \right\}^{1/4}$$

$$= 2250 \text{ kJ/hr m}^2 \text{ K}$$

$$= 6200 \text{ W/m}^2 \text{ K},$$

Since $h_s = 29{,}000$,effective value of h_2 denoted by h_2' will be ;

$$\frac{1}{h'_2} = \frac{1}{h_2} + \frac{1}{h_s} = \frac{1}{6200} + \frac{1}{29{,}000}$$

or $$h'_2 = 5150 \text{ W/m}^2 \text{ K}$$

$$\frac{q}{A_2} = h'_2 (t_{sat} - t_w)$$

or $$\frac{q}{A_2} = 5150 (115 - 105) = 5.15 \times 10^5 \text{ W/m}^2$$

$$= 18.6 \times 10^6 \text{ kJ/hr m}^2$$

Since this value of (q/A_2) does not agree with the value of Eq. (a) as 16.38×10^5 found above, a second trial is required.

(iii) Let us now assume $t_w = 100$ C, then from Eq. (5.7),

$$\frac{4.2 \times 6}{1.83^{1.7} \times 2280} = 0.006 \left\{ \frac{(q/A_1)}{1.1 \times 2280} \sqrt{\frac{6 \times 10^{-3}}{960}} \right\}^{1/3}$$

or $$\frac{q}{A_1} = 29.4 \times 10^4 \text{ kJ/hr m}^2$$

or $$\frac{q}{A_2} = 29.4 \times 10^4 \times \frac{4.5}{5}$$

$$= 26.5 \times 10^4 \text{ kJ/hr m}^2 \qquad \text{(b)}$$

Also from Eq. (5.34), noting that a 5°C difference in t_w, which changes t_f by only 2.5 C, will have almost nil effect on the properties ;

$$h_2 = 0.94 \left\{ \frac{952^2 \times 9.81 \times 3600^2 \times 2230 \times 2.63^3}{0.94 \times 1.5 \times (115 - 100)} \right\}^{1/4}$$

or $$h_2 = 20400 \text{ kJ/hr m}^2 \text{ K}$$

$$h_2' = 17{,}000 \text{ kJ/hr m}^2 \text{ K}$$

or $$\frac{q}{A_2} = 17{,}000 \times 15 = 25.5 \times 10^4$$

Since this value is nearly equal to the value of Eq. (b) i.e. 26.5×10^4, no further trial is required and t_w may be taken as 100 C and h'_2 as 17,000 kJ/hr m^2 K

(iv) Hence,

$$A_2 = \frac{q}{h_2'(115 - t_w)}$$

$$= \frac{2000 \times 2250}{17000 \times 15}$$

$$= 17.65 \text{ m}^2$$

Number of tubes,

$$= \frac{17.65}{\pi \times 0.05 \times 1.5} = 75 \qquad \text{Ans.}$$

Example 7.11

A superheater of a large steam generator has the following data :

Steam output :	225,000 kg/hr
Steam entry to superheater is at :	60 bar dry and saturated
Steam leaves superheated at :	58 bars, 450 C
Hot gases (containing 4% moisture) entry :	1065 C
Hot gases leaving :	730 C

Determine (a) the gas side heat transfer area for parallel flow (b) the metal temperature at steam inlet. Given that 50 mm nominal size tubes are used and $(h)_{steam} = 348\ W/m^2\ K$ and $(h)_{gases} = 58\ W/m^2\ K$.

Solution :

$$\text{From steam tables } h_1 = 2790 \text{ kJ/kg}$$

$$h_2 = 3310 \text{ kJ/kg}$$

$$t_{sat} = 274.25 \text{ C at 60 bar}$$

$$q = 225{,}000\,(3310 - 2790) = 117 \times 10^6 \text{ kJ/hr}$$

Now

$$U_o = \frac{1}{\dfrac{1}{h_1} + \dfrac{D_o}{D_1}\left(\dfrac{1}{h_o}\right)} = \frac{1}{\dfrac{1}{58} + \dfrac{60.3}{56.65}\left(\dfrac{1}{348}\right)}$$

$$= 49.4 \text{ W/m}^2 \text{ K}$$

and

$$\Delta t_m = \frac{(1065 - 274.25) - (730 - 450)}{\ln(790.75/280)} = \frac{530.75}{1.03} = 515 \text{ C}$$

(a)

$$A_0 = \frac{q}{U\,\Delta t_m} = \frac{117 \times 10^6}{(49.4 \times 3.6) \times 515}$$

$$= 126{,}200 \text{ m}^2$$

Ans.

(b) Considering that the average values of U_o and h_1 apply , we may represent the metal temperature t_w as given by ;

$$58 \times (1065 - t_w) = 49.4\,(1065 - 274.25)$$

or

$$t_w = 395 \text{ C}$$

Ans.

Example 7.12

An electrical machine uses hollow square conductors from where heat generated due to I^2R loss (300 watt/meter length) is to be removed by blowing forced air as coolant both internally and externally. The conductors are 15.25 m long, and the internal dimensions are 5 cm × 5 cm . One half of the I^2R loss may be assumed removable from the core and the air flow in the core is 47,800 kg/m² hr. The entering air temperature is 25 C. Determine (a) the air discharge temperature and (b) the copper temperature at discharge point.

Solution:

(a) The heat flowing through the core;

$$q = \frac{300 \times 15.25}{2} \times 3.6 = 7700 \text{ kJ/hr}$$

Now, air flow area $= 5 \times 5 = 25 \text{ cm}^2$

so air flow $= 47{,}800 \times \frac{25}{100^2} = 119.5 \text{ kg/hr}$

$$\Delta t = \frac{7700}{119.5 \times 1.0} = 62.5 \text{ C}$$

Thus discharge temp. of air $= 62.5 + 25$

$= 87.5 \text{ C}$ *Ans.*

(b) $$Nu = 0.023 (Re)^{0.8} (Pr)^{0.4}$$

$$\frac{hD}{k} = 0.023 \left(\frac{GD}{\mu}\right)^{0.8} (Pr)^{0.4}$$

$$= 0.023 \left(\frac{47800 \times 5/100}{0.0706}\right)^{0.8} (0.698)^{0.4}$$

$$= 0.023 \times 4320 \times 0.867 = 84.5$$

so $$h = \frac{84.5k}{D} = \frac{84.5 \times 2.83 \times 10^{-2}}{5/100}$$

$$= 48 \text{ W/m}^2 \text{ K}$$

so $$\Delta t_m = \frac{q}{hA}$$

where $A = 15.25 \times \frac{(5+5+5+5)}{100} = 3.05 \text{ m}^2$

or $$\Delta t_m = \frac{7700}{48 \times 3.6 \times 3.05} = 14.5 \text{ C}$$

Thus the metal temperature at the exit

$= 87.5 + 14.5$

$= 102 \text{ C}$ *Ans.*

Example 7.13

A heat exchanger is to be designed . It is assumed that condensing vapor on shell side will be held at constant pressure. Gas will flow through the tubes in a single pass. The total gas flow rate m and the mass velocity G will be kept fixed. After selecting a tube diameter, D, the number of tubes N in parallel, the gas film coefficient h, the total heat transfer area A and the tube length L have been fixed. The effect of larger tube size on the design is to be explored.

Consider that G , m , q , and other physical properties are constant and determine the effect of increase in tube diameter on each of the following.

(a) number of tubes needed

(b) heat transfer coefficient obtained

(c) total heat transfer surface needed

(d) length of tubes needed

Solution :

(a) Both G and m for gas are constant

so $$\frac{m}{G} = \text{Area of flow} = \text{constant}$$

or $$\frac{\pi D^2}{4} \times N = \text{constant}$$

Hence if D increases, N decreases *Ans.*

(b) $$h = \frac{k}{D} \times 0.023 \left(\frac{GD}{\mu}\right)^{0.8} \left(\frac{C_P \mu}{k}\right)^{0.4} \quad (4.42)$$

Since k, G, μ are constants,

hence h varies as $(I/D)^{0.2}$

h decreases if D increases *Ans.*

(c) $$q = UA\,\Delta t_m$$

Since q and Δt_m are constant, UA is constant

Also U ≈ h and therefore it decreases as D increases.

Hence A should increase if D increases . *Ans.*

(d) $A = N(\pi DL)$

Let suffix prime indicate the values corresponding to increased D.

$$\frac{A'}{A} = \frac{N' \pi D' L'}{N \pi DL} = \frac{q/U' \Delta t_m}{q/U \Delta t_m} = \frac{U}{U'} = \frac{h}{h'} = \left(\frac{D'}{D}\right)^{0.2}$$

or $$\frac{L'}{L} = \frac{D}{D'} \times \frac{N}{N'} \times \left(\frac{D'}{D}\right)^{0.2}$$

Now $$N = \frac{4 \times A_{flow}}{\pi D^2}; \quad N' = \frac{4 \times A_{flow}}{\pi D'^2}$$

or $$\frac{N}{N'} = \left(\frac{D'}{D}\right)^2$$

so $$\frac{L'}{L} = \frac{D}{D'} \left(\frac{D'}{D}\right)^2 \left(\frac{D'}{D}\right)^{0.2} = \left(\frac{D'}{D}\right)^{1.2}$$

Therefore the length L increases if D increases. *Ans.*

7.8 HEAT EXCHANGER EFFECTIVENESS

The effectiveness of a heat exchanger, or temperature effectiveness ε, is defined as the ratio of the actual heat transfer rate to the maximum possible heat transfer rate. Theoretically, the maximum heat transfer rate will occur with a counterflow heat exchanger of very large area when either one or the other of the following conditions is attained :

(a) The exit temperature of the cold fluid equals the entry temperature of the hot fluid, i.e $t_{c2} = t_{h1}$:

(b) The exit temperature of the hot fluid equals the entry temperature of the cold fluid, i.e. $t_{h2} = t_{c1}$.

Whether condition (a) or (b) is attained depends on the heat capacities of the two streams. Since the following equality must be satisfied,

$$W_h c_h (t_{h1} - t_{h2}) = W_c c_c (t_{c2} - t_{c1})$$

It follows that if $W_h c_h > W_c c_c$ then ;

$$(t_{c2} - t_{c1}) > (t_{h1} - t_{h2})$$

Hence t_{c2} will reach t_{h1} earlier and condition (a) will prevail as can be seen in Fig. 7.16.

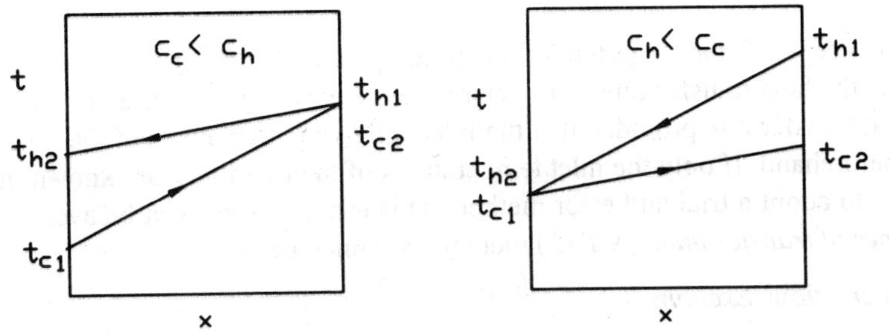

Fig. 7.16 Conditions for Evaluation of Heat Exchanger Effectiveness

Further , If $W_h c_h < W_c c_c$ then ;

$$(t_{h1} - t_{h2}) > (t_{c2} - t_{c1})$$

Hence t_{h2} will reach t_{c1} earlier as shown in Fig. 7.16.

If $W_c c_c < W_h c_h$

$$q_{max} = W_c c_c(t_{h1} - t_{c1}) = C_c (t_{h1} - t_{c1}) \quad (7.33)$$

If $W_h c_h < W_c c_c$

$$q_{max} = W_h c_h (t_{h1} - t_{c1}) = C_h (t_{h1} - t_{c1}) \quad (7.34)$$

$$[W_c c_c = C_c \text{ and } W_h c_h = C_h]$$

But the actual heat transfer for both cases is given by ;

$$q = C_c(t_{c2} - t_{c1}) = C_h(t_{h1} - t_{h2}) \quad (7.35)$$

Hence, (i)

$$\varepsilon = \frac{q}{q_{max}} = \frac{C_c(t_{c2} - t_{c1})}{C_c(t_{h1} - t_{c1})} = \frac{t_{c2} - t_{c1}}{t_{h1} - t_{c1}} \quad (7.36)$$

(if $C_c < C_h$)

and (ii)

$$\varepsilon = \frac{q}{q_{max}} = \frac{C_h(t_{h1} - t_{h2})}{C_h(t_{h1} - t_{c1})} = \frac{t_{h1} - t_{h2}}{t_{h1} - t_{c1}} \quad (7.37)$$

(if $C_h < C_c$)

The above expression for the effectiveness can also be put in a slightly different form. Let C_{min} represent the lower of the two heat capacities C_c and C_h. Then it follows from Eqs. (7.33) and (7.34) that ;

$$q_{max} = C_{min}(t_{h1} - t_{c1}) \quad (7.38)$$

Combining Eq. (7.35) with Eq. (7.38) we get ;

$$\varepsilon = \frac{C_c(t_{c2} - t_{c1})}{C_{min}(t_{h1} - t_{c1})} \quad (7.39)$$

and

$$\varepsilon = \frac{C_h(t_{h1} - t_{h2})}{C_{min}(t_{h1} - t_{c1})} \quad (7.40)$$

7.9 EFFECTIVENESS AND NUMBER OF TRANSFER UNITS (NTU)

The concept of the logarthmic mean temperature difference can be applied to determine the heat transfer rate or to determine the exchanger surface area required for a given heat transfer rate provided that the inlet and exit temperatures of the two fluids are known beforehand. If only the inlet temperatures of the two fluids are known, it becomes necessary to adopt a trial and error method. This trial and error can be avoided by using the *number of transfer units* (*N.T.U.*) method explained below.

(a) Counter - flow Exchanger

For this exchanger Eq. (7.25) gives ;

$$\frac{d(t_h - t_c)}{t_h - t_c} = U\left[\frac{1}{W_h C_h} - \frac{1}{W_c c_c}\right] dA \quad (7.25)$$

or

$$\frac{d(t_h - t_c)}{t_h - t_c} = \frac{U}{W_h c_h}\left[1 - \frac{W_h c_h}{W_c c_c}\right] dA \quad (7.41)$$

For brevity, let $W_h c_h = C_h$ and $W_c c_c = C_c$. Then Eq. (7.41) becomes ;

$$\frac{d(t_h - t_c)}{(t_h - t_c)} = \frac{U}{C_h}\left(1 - \frac{C_h}{C_c}\right) dA \quad (7.42)$$

Integrating Eq.(7.42) from $A = 0$ to $A = A$,

$$ln\frac{(t_{h1}-t_{c2})}{(t_{h2}-t_{c1})} = \frac{UA}{C_h}\left(1-\frac{C_h}{C_c}\right)$$

or

$$\frac{t_{h1}-t_{c2}}{t_{h2}-t_{c1}} = e^{(UA/C_h)(1-C_h/C_c)} \quad (7.43)$$

Inverting Eq. (7.43) ,

$$\frac{t_{h2}-t_{c1}}{t_{h1}-t_{c2}} = e^{-(UA/C_h)(1-C_h/C_c)} \quad (7.44)$$

Recall that the effectiveness of the heat exchanger is given by ;

(i) $$\varepsilon = \frac{t_{c2}-t_{c1}}{t_{h1}-t_{c1}} \text{ for } C_c < C_h$$

and (ii) $$\varepsilon = \frac{t_{h1}-t_{h2}}{t_{h1}-t_{c1}} \text{ for } C_h < C_c$$

Let us consider the two cases separately .

Case I :

$$C_h < C_c \text{ or } C_h = C_{min}$$

From Eq. (7.40) we get ;

$$t_{h1} - t_{h2} = \varepsilon\,(t_{h1} - t_{c1}) \quad (7.45)$$

or

$$t_{h2} = t_{h1} + \varepsilon\,(t_{c1} - t_{h1}) \quad (7.46)$$

Since

$$q = C_c\,(t_{c2} - t_{c1}) = C_h\,(t_{h1} - t_{h2})$$

$$t_{c2} = \frac{q}{C_c} + t_{c1}$$

$$= \frac{C_h}{C_c}(t_{h1} - t_{h2}) + t_{c1} \quad (7.47)$$

Substituting for $(t_{h1} - t_{h2})$ From Eq. (7.45)into Eq. (7.47)

$$t_{c2} = \frac{C_h}{C_c}\varepsilon\,(t_{h1} - t_{c1}) + t_{c1} \quad (7.48)$$

Equations (7.46) and (7.48) permit us to eliminate the exit temperatures of the two fluids $t_{h2}$and t_{c2} from Eq. (7.44). Substituting for t_{h2} and t_{c2} from Eq.(7.46) and (7.48) into Eq. (7.44) we obtain,

$$\frac{t_{h1}+\varepsilon(t_{c1}-t_{h1})-t_{c1}}{t_{h1}-\frac{C_h}{C_c}\varepsilon(t_{h1}-t_{c1})-t_{c1}} = e^{-(UA/C_h)(1-C_h/C_c)}$$

or

$$\frac{(t_{h1}-t_{c1})(1-\varepsilon)}{(t_{h1}-t_{c1})\left(1-\frac{C_h}{C_c}\varepsilon\right)} = e^{-(UA/C_h)(C_h/C_c)}$$

or
$$\frac{(1-\varepsilon)}{\left(1-\dfrac{C_h}{C_c}\varepsilon\right)} = e^{-(UA/C_h)(1-C_h/C_c)} \tag{7.49}$$

Solving Eq. (7.49),
$$\varepsilon = \frac{1-e^{-(UA/C_h)(1-C_h/C_c)}}{1-(C_h/C_c)e^{-(UA/C_h)(1-C_h/C_c)}} \tag{7.50}$$

Case 2 :

$$C_c < C_h$$

For this case by following similar steps as given above we get,

$$\varepsilon = \frac{1-e^{-(UA/C_c)(1-C_c/C_h)}}{1-\left(\dfrac{C_c}{C_h}\right)e^{-(UA/C_c)(1-C_c/C_h)}} \tag{7.51}$$

Inspection of Eqs (7.50) and (7.51) shows that both results can be expressed by a single expression if we write ;

$$\varepsilon = \frac{1-e^{-(UA/C_{min})(1-C_{min}/C_{max})}}{1-(C_{min}/C_{max})e^{-(UA/C_{min})(1-C_{min}/C_{max})}} \tag{7.52}$$

where C_{min} represents the smaller and C_{max} the larger of the two heat capacities C_c and C_h.

The dimensionless ratio AU/C_{min} is called the *number of transfer units, NTU.* Thus, for a counter-flow exchanger.

$$\varepsilon = \frac{1-e^{-NTU(1-C_{min}/C_{max})}}{1-(C_{min}/C_{max})e^{-NTU(1-C_{min}/C_{max})}} \tag{7.53}$$

(b) *Parallel flow Exchanger*

For this case Eq. (7.15) gives ;

$$\frac{d(t_h-t_c)}{(t_h-t_c)} = -U\left(\frac{1}{C_h}+\frac{1}{C_c}\right)dA$$

$$= -\frac{U}{C_h}\left(1+\frac{C_h}{C_c}\right)dA \tag{7.54}$$

Integrating Eq. (7.54) from $A = 0$ to $A = A$

$$ln\left(\frac{t_{h2}-t_{c2}}{t_{h1}-t_{c1}}\right) = -\frac{UA}{C_h}\left(1+\frac{C_h}{C_c}\right)$$

or
$$\frac{t_{h2}-t_{c2}}{t_{h1}-t_{c1}} = e^{-(UA/C_h)(1+C_h/C_c)} \tag{7.55}$$

From Eqs.(7.40) and (7.39), $$t_{h2} = t_{h1} - \frac{\varepsilon C_{min}(t_{h1} - t_{c1})}{C_h} \quad (7.56)$$

and $$t_{c2} = t_{c1} + \frac{\varepsilon C_{min}(t_{h1} - t_{c1})}{C_c} \quad (7.57)$$

Eliminating t_{h2} and t_{c2} from Eq. (7.55) with the aid of Eqs. (7.56) and (7.57),

$$\frac{1}{(t_{h1} - t_{c1})}\left[(t_{h1} - t_{c1}) - \varepsilon C_{min}(t_{h1} - t_{c1})\left(\frac{1}{C_h} + \frac{1}{C_c}\right)\right] = e^{-(UA/C_h)(1+C_h/C_c)}$$

$$1 - \varepsilon (C_{min})\left(\frac{1}{C_h} + \frac{1}{C_c}\right) = e^{-(UA/C_h)(1+C_h/C_c)}$$

Hence $$\varepsilon = \frac{1 - e^{(UA/C_h)(1+C_h/C_c)}}{C_{min}\left(\frac{1}{C_h} + \frac{1}{C_c}\right)} \quad (7.58)$$

(i) If $C_c > C_h$ then $C_{min} = C_h$, $C_{max} = C_c$ and Eq. (7.58) gives ;

$$\varepsilon = \frac{1 - e^{-(UA/C_{min})(1+C_{min}/C_{max})}}{1 + (C_{min} / C_{max})} \quad (7.59)$$

(ii) If $C_c < C_h$, then $C_{min} = C_c$, $C_{max} = C_h$ and Eq.(7.58) gives ;

$$\varepsilon = \frac{1 - e^{-(UA/C_{max})(1+C_{max}/C_{min})}}{1 + (C_{min} / C_{max})} \quad (7.60)$$

It may be rearranged to give;

$$\varepsilon = \frac{1 - e^{-(UA/C_{max})(1+C_{min}/C_{max})}}{1 + (C_{min} / C_{max})} \quad (7.59)$$

Since $UA/C_{min} = NTU$, the effectiveness for *both* cases may be expressed by :

$$\varepsilon = \frac{1 - e^{-NTU(1+C_{min}/C_{max})}}{1 + (C_{min} / C_{max})} \quad (7.60)$$

Note that the effectiveness for counter-flow and Parallel-flow exchangers is given by *different* expressions :

$$(\varepsilon)_{\text{counter-flow}} = \frac{1 - e^{-NTU(1-C_{min}/C_{max})}}{1 - (C_{min} / C_{max})e^{-NTU(1-C_{min}/C_{max})}} \quad (7.53)$$

$$(\varepsilon)_{\text{parallel-flow}} = \frac{1 - e^{-NTU(1+C_{min}/C_{max})}}{1 + (C_{min} / C_{max})} \quad (7.60)$$

Two limiting cases of Eqs. (7.53) and (7.60) are of interest.

Case 1 :

$$(C_{min}/C_{max}) \approx 0$$

In this case, for both counter-flow and parallel-flow exchangers we get ;

$$\varepsilon = 1 - e^{-NTU} \tag{7.61}$$

Note that in evaporators or condensers, one fluid remains at constant temperature throughout the exchanger. Hence its effective specific heat, and thus its heat capacity, is equal to infinity. In such cases $C_{max} = \infty$ and $C_{min}/C_{max} = 0$.

Case 2 :

$$(C_{min}/C_{max}) = 1$$

For this case, Eq. (7.53) becomes indeterminate.

(a) However, we can find the value of ε for *counter flow* heat-exchanger either by applying the calculus of limits to *RHS* of Eq. (7.53) or by the following method.

Since ; $$1 - (C_{min}/C_{max}) = 0, \text{NTU}\left(1 - \frac{C_{min}}{C_{max}}\right) << 1.$$

$$e^{-NTU(1 - C_{min}/C_{max})} = 1 - \text{NTU}\left(1 - \frac{C_{min}}{C_{max}}\right)$$

or $$\varepsilon = \frac{1 - \left[1 - NTU\left(1 - \frac{C_{min}}{C_{max}}\right)\right]}{1 - \frac{C_{min}}{C_{max}}\left[1 - NTU\left(1 - \frac{C_{min}}{C_{max}}\right)\right]}$$

$$= \frac{NTU\left(1 - \frac{C_{min}}{C_{max}}\right)}{\left(1 - \frac{C_{min}}{C_{max}}\right)\left(1 + \frac{C_{min}}{C_{max}} NTU\right)} = \frac{NTU}{1 + \left(\frac{C_{min}}{C_{max}}\right) NTU}$$

$$\varepsilon = \frac{\text{NTU}}{1 + \text{NTU}} \tag{7.62}$$

(b) For the case of *Parallel-flow* exchanger, when $(C_{min}/C_{max}) \approx 1$ we get

$$\varepsilon = \frac{1 - e^{-2(NTU)}}{2} \tag{7.63}$$

Kays and London have evaluated the effectiveness for many different types of heat exchangers of practical interest. Their results are available in the form of convenient graphs from which the effectiveness can be determined for given values of NTU and C_{min}/C_{max}. These graphs for several common arrangements are given in Figs. 7.17 to 7.21. Tables 7.3 and 7.4 give a summary of the relations for effectiveness and N. T. U.

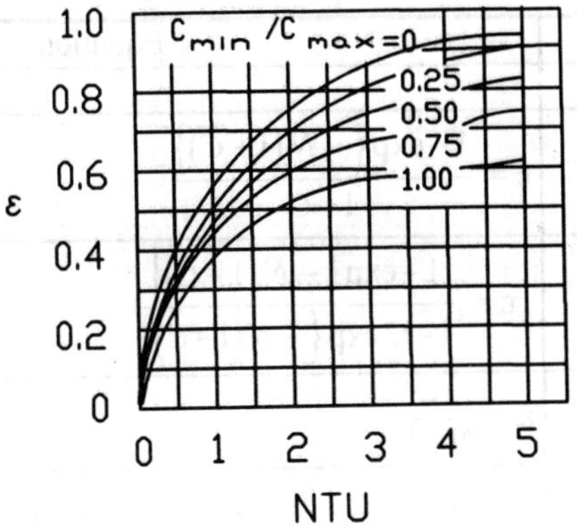

Fig. 7.17 Counter-Flow Heat Exchanger Effectiveness

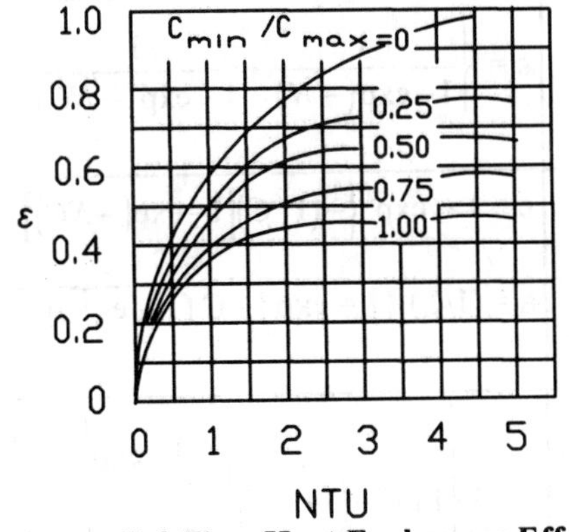

Fig. 7.18 Parallel-Flow Heat Exchanger Effectiveness

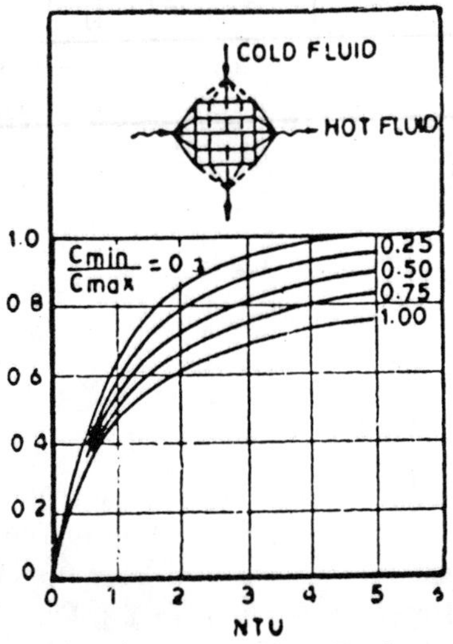

Fig. 7.19 Cross-Flow Heat Exchanger Effectiveness

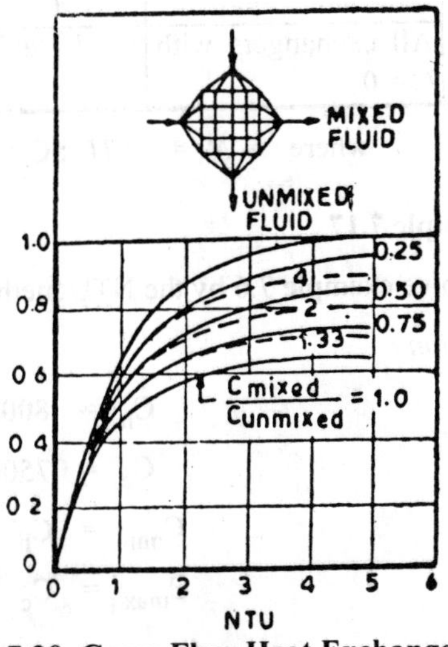

Fig. 7.20 Cross-Flow Heat Exchanger Effectiveness

Table 7.3 : Equations for Heat Exchanger Effectiveness

	Exchanger Type	Equation	
(i)	Double pipe		
	Parallel flow	$\varepsilon = \dfrac{1-\exp\{-N(1+C)\}}{1+C}$	(7.60)
	Counter flow	$\varepsilon = \dfrac{1-\exp\{-N(1-C)\}}{1-C\exp\{-N(1-C)\}}$	(7.53)
	Counter flow, C = 1	$\varepsilon = \dfrac{N}{N+1}$	(7.62)
(ii)	*Cross flow*		
	Both fluids unmixed	$\varepsilon = 1-\exp\left\{\dfrac{\exp(-CN^{0.78})-1}{CN^{-0.22}}\right\}$	(7.64)
	Both fluids mixed	$\varepsilon = \left\{\dfrac{1}{1-\exp(-N)}+\dfrac{1}{1-\exp(-NC)}-\dfrac{1}{N}\right\}^{-1}$	(7.65)
	C_{min} mixed, C_{max} unmixed	$\varepsilon = 1-\exp\left\{-(1/C)[1-\exp(-NC)]\right\}$	(7.66)
	C_{min} unmixed C_{max} mixed	$\varepsilon = (1/C)\{1-\exp[-C(1-e^{-N})]\}$	(7.67)
	Shell and tube; one shell pass; 2, 4, 6 tube passes	$\varepsilon = 2\left\{1+C+(1+C^2)^{1/4}\dfrac{1+\exp\left[-N(1+C^2)^{1/2}\right]}{1-\exp\left[-N(1+C^2)^{1/2}\right]}\right\}$	(7.68)
	All exchangers with C = 0	$\varepsilon = 1-e^{-N}$	(7.61)

where $N = NTU$; $C = C_{min}/C_{max}$

Example 7.17

Solve example 7.6 by the NTU method.

Solution :

$$C_h = 8000 \times 1.0$$

$$C_c = 7500 \times 4.18$$

or

$$C_{min} = C_h$$

$$C_{max} = C_c$$

$$\frac{C_{min}}{C_{max}} = \frac{8000}{7500 \times 4.18} = 0.256$$

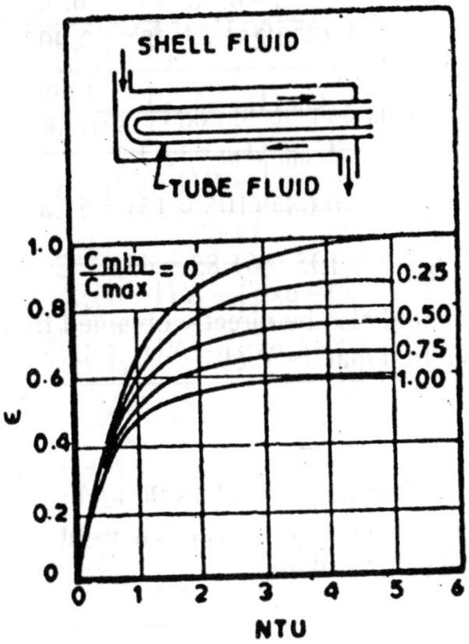

Fig. 7.21 Shell & Tube Heat Exchanger Effectiveness

Table 7.4 : Equations for NTU

	Exchanger Type	Equation	
(i)	Double pipe		
	Parallel flow	$N = -\ln[1-(1+C)\varepsilon]/(1+C)$	(7.69)
	Counter flow	$N = \frac{1}{C-1}\ln\left(\frac{\varepsilon-1}{C\varepsilon-1}\right)$	(7.70)
	Counter flow	$N = \varepsilon/(1-\varepsilon)$	(7.71)
(ii)	*Cross flow* :		
	C_{min} mixed, C_{max} unmixed	$N = -(1/C)\{1 + C\,ln(1-\varepsilon)\}$	(7.72)
	C_{min} unmixed, C_{max} mixed	$N = -ln\{1 + (1/C)\,ln(1 - C\varepsilon)\}$	(7.73)
	Shell and tube; one shell pass; 2, 4, 6. tube passes	$N = -(1+C^2)^{-1/2}\,ln[\{2/\varepsilon - 1 - C - (1+C^2)^{1/2}\} - \{2/\varepsilon - 1 - C + (1+C^2)^{1/2}\}]$	(7.74)
	All exchanges, C = 0	$N = -\ln(1-\varepsilon)$	(7.75)

Note.: $N = NTU$; $C = C_{min}/C_{max}$

$$NTU = \frac{UA}{C_{min}} = \frac{145 \times 3.6 \times 20}{8000 \times 1.0} = 1.3$$

From Eq.(7.53)
$$\varepsilon = \frac{1 - e^{-NTU(1-C_{min}/C_{max})}}{1 - \frac{C_{min}}{C_{max}} e^{-NTU(1-C_{min}/C_{max})}}$$

Now, $e^{-NTU(1-C_{min}/C_{max})} = e^{-1.3(1-0.256)} = 0.38$

so $$\varepsilon = \frac{1-0.38}{1-0.256\times0.38} = \frac{0.62}{0.903} = 0.687$$

From Eq. (7.40); $$\varepsilon = \frac{C_h(t_{h1}-t_{h2})}{C_{min}(t_{h1}-t_{c1})} = 0.687$$

so $$(t_{h1}-t_{h2}) = 0.687\,(105-15) = 61.83$$

or $$t_{h2} = 105-61.83 = 43.17\text{ C}$$ *Ans.*

Note that the value of ε could also be directly obtained from Fig. 7.17 for $(C_{min}/C_{max}) = 0.256$ and NTU = 1.3. We find that ε = 0.69

Example 7.15

A parallel flow heat exchanger has to cool 1,000 kg/hr of oil from 60 C to 35 C. The cooling water enters the exchanger at 15 C and leaves at 25 C. Determine the value of UA. The specific heat of oil is 2.1 kJ/kg K.

Solution :

The water flow (cold fluid) rate is given by ;

$$W_c\, c_c\,(t_{c2}-t_{c1}) = W_h\, c_h\,(t_{h1}-t_{h2})$$

$$W_c = \frac{1000\times2.1(60-35)}{4.2\times(25-15)} = 1250\text{ kg/hr}$$

Now, $$C_h = 1000\times2.1 = 2100$$

$$C_c = 1250\times4.2 = 5250$$

thus $$C_{min} = C_h = 2100$$

and $$\frac{C_{min}}{C_{max}} = \frac{2100}{5250} = 0.4$$

$$\varepsilon = \frac{C_h}{C_{min}}\left(\frac{t_{h1}-t_{h2}}{t_{h1}-t_{c1}}\right) = \frac{(60-35)}{(60-15)} = \frac{25}{45} = 0.55$$

From Fig. 7.18 ;

for $$\frac{C_{min}}{C_{max}} = 0.4 \text{ and } \varepsilon = 0.55$$

$$\frac{UA}{C_{min}} = 1.1$$

or $$UA = 1.1\times2100 = 2310\text{ kJ/hr K}$$

Example 7.16

A chemical (C_p = 3.3 kJ/kg K) flowing at the rate of 20,000 kg/hr enters a parallel flow heat exchanger at 120 C. The flow rate of cooling water (C_p = 4.186 kJ/kg K) is 50,000 kg/hr with an inlet temperature of 20 C. The heat transfer area is 10 m² and the overall heat transfer coefficient is 1050 W/m² K. Find the outlet temperatures of water and the chemical . (UPSC 92)

Solution :

The heat-capacities of hot and cold fluids are given by;

$$C_h = W_h C_h = (20{,}000/3600)\,(3.3 \times 1000) = 18{,}333.3 \text{ W/K}$$

$$C_c = W_c C_c = (50{,}000/3600)\,(4.186 \times 1000) = 58{,}138.9 \text{ W/K}$$

Now $C_{min} = C_h = 18333.3$

$C_{max} = C_c = 58138.9$

so $C_{min}/C_{max} = 0.3153$

and $NTU = UA/C_{min} = (1050)\,(10)/18333.3 = 0.5727$

The effectiveness of a parallel flow exchanger is given by ;

$$\varepsilon = \frac{1 - e^{-\left[1+\left(C_{min}/C_{max}\right)\right]NTU}}{1+\left(C_{min}/C_{max}\right)} \tag{7.60}$$

$$= \frac{1 - e^{-(1.3153)(0.5727)}}{1+0.3153} = 0.4025$$

Since $C_h < C_c$

$$\varepsilon = \frac{t_{h1} - t_{h2}}{t_{h1} - t_{c1}} = 0.4025 \tag{7.37}$$

or $t_{h1} - t_{h2} = \varepsilon\,(t_{h1} - t_{c1})$

$= (0.4025)\,(120 - 20) = 40.25$

hence, $t_{h2} = t_{h1} - 40.25 = 120 - 40.25 = 79.75°C$ *Ans.*

Now for the energy balance in heat exchanger,

$$C_h\,(t_{h1} - t_{h2}) = C_c\,(t_{c2} - t_{c1})$$

or $(t_{c2} - t_{c1}) = (t_{h1} - t_{h2})\,\dfrac{C_h}{C_c}$

$= (40.25)\,(0.3153) = 12.69$

so $t_{c2} = t_{c1} + 12.69 = 20 + 12.69 = 32.69°C$ *Ans.*

Example 7.17

Light lubricating oil (C_p = 2090 J/kg K) is cooled by allowing it to exchange energy with water in a small heat exchanger. The oil enters and leaves the heat exchanger at 375 K and 350 K, respectively, and flows at a rate of 0.5 kg/s. Water (C_p = 4177 J/kg K) at 280 K is available for cooling and its flow rate is 0.201 kg/s. Determine the required heat transfer area for counter flow operation. The overall heat transfer coefficient may be taken as 250 W/m^2 K. (UPSC 91).

Solution :

Energy balance yields.

$$q = W_h C_h (t_{h1} - t_{h2}) = W_c C_c (t_{c2} - t_{c1})$$

or $$(0.5)(2090)(375 - 350) = (0.201)(4177)(t_{c2} - 280)$$

or $$t_{c2} = 311.12 \text{ K}$$

Now, $$\Delta t_m = \frac{(t_{h1} - t_{c2}) - (t_{h2} - t_{c1})}{\ln\left(\frac{t_{h1} - t_{c2}}{t_{h2} - t_{c1}}\right)}$$

$$= \frac{(375 - 311.12) - (350 - 280)}{\ln(63.88/70)} = 66.9° \text{ K}$$

Thus $$A = \frac{q}{U\Delta t_m} = \frac{26125}{250 \times 66.9} = 1.562 \text{ m}^2$$ *Ans.*

Example 7.18

The following performance data refers to a cross-flow heat exchanger (both fluids unmixed).

	Air	Oil
Inlet temperature, (°C)	52	88
Outlet temperature (°C)	75	70
Mass flow rate (kg/hr)	1791.4	1080

The mean specific heat of air is 1.005 kJ/kg K and that of oil 2.13 kJ/kg K.

(a) Determine the UA product for the exchanger.

(b) Also calculate the exit temperatures of the two fluids if the inlet temperature of oil is increased to 95°C, the air inlet temperature and flow rates of the two fluids remaining unchanged.

Solution :

(a) With usual nomenclature, the heat transfer rate q is given by,

$$q = UA\, F\, \Delta t_m$$

(where F is the LMTD correction factor)

The LMTD, Δt_m for a counter flow heat exchanger, is

$$\Delta t_m = \frac{(t_{h1} - t_{c2}) - (t_{h2} - t_{c1})}{\ln\left(\frac{t_{h1} - t_{c2}}{t_{h2} - t_{c1}}\right)}$$

$$= \frac{(88 - 75) - (70 - 52)}{\ln(13/18)} = \frac{-5}{\ln(13/18)}$$

$$= 15.365°\text{C}$$

Now $$R = (t_{h1} - t_{h2})/(t_{c2} - t_{c1}) = 18/23 = 0.783$$

$$\theta = (t_{c2} - t_{c1})/(t_{h1} - t_{c1}) = 23/36 = 0.639$$

The correction factor F as found from Fig. 7.14, for these values of R and θ is 0.81.

Also, $$q = W_h C_h (t_{h1} - t_{h2}) = \frac{1080 \times 2130 \times 18}{3600} = 11502 \text{ W}$$

Hence, $$UA = \frac{q}{F \Delta t_m} = \frac{11502}{0.81 \times 15.365} = 924.2 \text{ m}^2$$ *Ans.*

(b) Let us first determine the NTU and effectiveness of the exchanger,

$$C_h = W_h c_h = (1080/3600)(2130) = 639.0 \text{ W/K}$$

$$C_c = W_c c_c = (1791.4/3600)(1005) = 500.1 \text{ W/K}$$

so $$C_{min} = C_c = 500.1 \; ; C_{max} = C_h = 639.0$$

$$NTU = UA/C_{min} = 924.2/500.1 = 1.848$$

$$C_{min}/C_{max} = 500.1/639.0 = 0.783$$

From Fig. 7.19, effectiveness is found as $\varepsilon = 0.64$

Now, $$\varepsilon = \frac{t_{c2} - t_{c1}}{t_{h1} - t_{c1}} = 0.64$$

So $$t_{c2} = t_{c1} + 0.64 (t_{h1} - t_{c1})$$

$$= 52 + 0.64 (95 - 52)$$

$$= 79.52°C$$ *Ans.*

Further, $$t_{h1} - t_{h2} = (C_c/C_h)(t_{c2} - t_{c1})$$

$$= (0.783)(79.52 - 52) = 21.5$$

or $$t_{h2} = t_{h1} - 21.5 = 95 - 21.5 = 73.5°C$$ *Ans.*

Example 7.19

In a parallel-flow, double-pipe, heat exchanger water flows through the inner pipe and is heated from 20°C to 70°C. Oil, flowing through the annulus is cooled from 200 to 100 °C.

(a) It is desired to cool the oil to a lower exit temperature by increasing the length of the heat exchanger. What is the minimum temperature to which oil may be cooled.

(b) What would be the exit temperature of oil if the existing exchanger is made to operate in a counterflow manner ?

(c) What is the ratio of the required length for counterflow to that for parallel flow if the oil exit temperature is t_{h2}.

Solution ;

(a) Energy balance equation is,

$$C_h (t_{h1} - t_{h2}) = C_c (t_{c2} - t_{c1})$$

or $$\frac{C_h}{C_c} = \frac{70-20}{200-100} = 0.5 = \frac{C_{min}}{C_{max}}$$

For a parallel flow heat exchanger the value of effectiveness is given by :

$$\varepsilon = \frac{1-e^{-NTU(1+C_{min}/C_{max})}}{1+C_{min}/C_{max}} \tag{7.60}$$

It will be maximum when exponential term is zero.

Hence $$(\varepsilon)_{max} = \frac{1}{1+(C_{min}/C_{max})} = \frac{1}{1+0.5} = \frac{1}{1.5}$$

Since, by definition $$\varepsilon = \frac{C_h}{C_{min}}\left(\frac{t_{h1}-t_{h2}}{t_{h1}-t_{c1}}\right) \tag{7.40}$$

$$= \frac{t_{h1}-t_{h2}}{t_{h1}-t_{c1}} = \frac{200-t_{h2}}{200-20}$$

or $$\frac{1}{1.5} = \frac{200-t_{h2}}{180} \text{ or } t_{h2} = 80°C$$ Ans.

(b) The heat transfer rate q may be expressed as,

$$q = UA\,(LMTD) = C_h\,(t_{h1} - t_{h2})$$

$$\frac{UA}{C_h} = \frac{t_{h1}-t_{h2}}{LMTD}$$

i. e. $$\frac{UA}{C_{min}} = \frac{t_{h1}-t_{h2}}{LMTD} \quad [C_h = C_{min}]$$

Now $$LMTD = \frac{(t_{h1}-t_{c1})-(t_{h2}-t_{c2})}{\ln\left(\dfrac{t_{h1}-t_{c1}}{t_{h2}-t_{c2}}\right)} = \frac{(200-20)-(100-70)}{\ln\left(\dfrac{180}{30}\right)}$$

$$= 83.72°C$$

Thus $$NTU = \frac{UA}{C_{min}} = \frac{t_{h1}-t_{h2}}{83.72} = \frac{200-10}{83.72} = 1.195$$

For NTU = 1.195 and C_{min}/C_{max} = 0.5, Fig. 7.17 gives the effectiveness of a counterflow exchanger to be,

$$\varepsilon = 0.625$$

$$t_{h1} - t_{h2} = \varepsilon\,(t_{h1} - t_{c1}) = 0.625 \times 180 = 112.5$$

or $$t_{h2} = t_{h1} - 112.5 = 200 - 112.5 = 87.5°C$$ Ans.

(c) If the oil exit temperature is t_{h2}, the cold fluid (water) exit temperature will be given by,

$$t_{c2} - t_{c1} = \frac{C_h}{C_c}(t_{h1} - t_{h2}) = 0.5\,(200 - t_{h2})$$

or $$t_{c2} = 20 + 100 - 0.5\,t_{h2} = 120 - 0.5\,t_{h2}$$

If the heat transfer rate is q, and suffixes pf and cf denote parallelflow and counterflow, respectively, then,

$$q = UA_{pf}(LMTD)_{pf} = UA_{cf}(LMTD)_{cf}$$

The ratio of lengths of the two types of heat exchangers is ;

$$\frac{L_{cf}}{L_{pf}} = \frac{A_{cf}}{A_{pf}} = \frac{(LMTD)_{pf}}{(LMTD)_{cf}}$$

$$(LMTD)_{pf} = \frac{(t_{h1}-t_{c1})-(t_{h2}-t_{c2})}{\ln\left(\frac{t_{h1}-t_{c1}}{t_{h2}-t_{c2}}\right)}$$

$$= \frac{180-(t_{h2}-120+0.5t_{h2})}{\ln\frac{180}{1.5t_{h2}-120}} = \frac{300-1.5t_{h2}}{\ln\left(\frac{180}{1.5t_{h2}-120}\right)} \qquad \text{(a)}$$

also

$$(LMTD)_{cf} = \frac{(t_{h1}-t_{c2})-(t_{h2}-t_{c1})}{\ln\left(\frac{t_{h1}-t_{c2}}{t_{h2}-t_{c1}}\right)} = \frac{(200-120+0.5t_{h2})-(t_{h2}-20)}{\ln\left(\frac{80+0.5t_{h2}}{t_{h2}-20}\right)}$$

or

$$(LMTD)_{cf} = \frac{100-0.5t_{h2}}{\ln\left(\frac{80+0.5t_{h2}}{t_{h2}-20}\right)} \qquad \text{(b)}$$

Dividing Eq. (a) by Eq. (b) we get,

$$\frac{(LMTD)_{pf}}{(LMTD)_{cf}} = \left(\frac{300-1.5t_{h2}}{100-0.5t_{h2}}\right) \times \frac{\ln[(80+0.5t_{h2})/(t_{h2}-20)]}{\ln[(180)/(1.5t_{h2}-120)]}$$

Thus the ratio of lengths of the two types of heat exchangers in terms of t_{h2} is given by ;

$$\frac{L_{cf}}{L_{pf}} = 3\,\frac{\ln[(80+0.5t_{h2})/(t_{h2}-20)]}{\ln[(180)/(1.5t_{h2}-120)]} \qquad \text{Ans.}$$

Example 7.20

The following data refer to a shell and tube heat exchanger : Shell ID = 90 cm; tube OD = 1.9 cm, ID = 1.65 cm ; tube length 5 m ; tube laid on 2.38 cm triangular pitch ; number of tubes = 1068 ; baffle spacing = 30 cm. The cold fluid enters the tubes at 20°C with a flow rate of 40 kg/s and makes 2 passes through the tubes. The hot fluid enters the shell at 70°C with a flow rate of 50 kg/s. The average thermal properties of the two fluids are as given below :

Cold fluid : $\rho = 997$ kg/m^3; $C_p = 4180$ J/kg K; $\nu = 0.832 \times 10^{-6}$ m^2/s; k = 0.613 W/mK; Pr = 5.66.

Hot fluid : $\rho = 985$ kg/m^3; $C_p = 4184$ J/kg K; $\nu = 0.478 \times 10^{-6}$ m^2/s; k = 0.651 W/mK; Pr = 3.02.

Determine the outlet temperatures of the two fluids and the pressure drops for both streams.

Solution :

(i) *Correlations to be used in solution*

For a shell and tube heat exchanger the following correlations may be used to determine the shell side heat transfer coefficient and pressure drop. Suffix *s* denotes shell and *e* denotes equivalent diameter of the shell.

$$Nu_s = \frac{h_s D_e}{k} = 0.36\, Re_s^{0.55}\, Pr^{1/3} \quad (7.76)$$

The above correlation may be used for Re_s from 2000 to 10^6. The fluid properties be evaluated at the mean bulk temperature.

For Re_s in the range 400 to 10^6, the pressure drop may be determined from the equation,

$$\Delta p_s = f(N_b + 1)\frac{D_s}{D_e}\frac{\rho V_s^2}{2} \quad (7.77)$$

where N_b is the number of baffles and $(N_b + 1)$ is the number of times the shell fluid crosses the tube bundle.

The friction factor *f* may be calculated from :

$$f = \exp\,[0.576 - 0.12 \ln Re_s] \quad (7.78)$$

(This equation takes into account the entrance and exit losses also).

The tube side pressure drop Δp_t may be determined from,

$$\Delta p_t = \frac{\rho V^2}{2}\left(\frac{f L N_p}{D_i} + 4 N_p\right) \quad (7.79)$$

where V is the fluid velocity, D_i and L the inside diameter and length of the tube, respectively, and N_p is the number of tube passes. The friction factor f may be obtained from the Moody diagram, given in Fig. 4.3.

(ii) Let us first determine the shell side heat transfer coefficient and pressure drop.

Shell side flow area is given by,

$$A_s = \frac{D_s (S_T - D_o) B}{S_T}$$

where B is the spacing between the baffles

so
$$A_s = \frac{(0.9)(2.38 - 1.9)10^{-2} \times 0.3}{2.38 \times 10^{-2}} = 0.05445 \text{ m}^2$$

$$\text{Shell fluid velocity } V_s = \frac{m_s}{\rho A_s} = \frac{50}{985 \times 0.05445} = 0.9322 \text{ m/s}$$

Shell equivalent diameter [see Fig. 4.4 (c)]

$$D_e = \frac{4[(s_T/2)(0.86S_T)-(\pi D_o^2/8)]}{(\pi D_o/2)}$$

$$= \frac{3.44 S_T^2 - \pi D_o^2}{\pi D_o} = \frac{3.44(2.38)^2 - \pi(1.9)^2}{\pi(1.9)} = 1.3362 \text{ cm}$$

Shell fluid Reynolds number,

$$Re_s = \frac{V_s D_e}{\nu} = \frac{0.9322 \times 1.3362 \times 10^{-2}}{0.478 \times 10^{-6}} = 26059$$

Now, $$Nu_s = 0.36\,(Re_s)^{0.55}\,(Pr_s)^{1/3} \qquad (7.7b)$$

$$= 0.36\,(26059)^{0.55}\,(3.02)^{1/3} = 139.7$$

so $$h_s = \frac{Nu_s k_s}{D_e} = \frac{139.7 \times 0.651}{1.3362 \times 10^{-2}} = 6804 \text{ W/m}^2\text{K}$$

also, shell side friction factor,

$$f_s = \exp[0.576 - 0.19 \ln Re_s]$$

$$= \exp[0.576 - 0.19 \ln 26059] = 0.2577$$

Shell side pressure drop,

$$\Delta p_s = f(N_b + 1)\frac{D_s}{D_e}\frac{\rho V_s^2}{2}$$

or $$\Delta p_s = (0.2577)(16+1)\left(\frac{0.9}{1.3362 \times 10^{-2}}\right)\frac{(985)(0.9322)^2}{2}$$

$[N_b = 5/0.3 = 16]$

$$= 1.26287 \times 10^5 \text{ N/m}^2$$

$$= 1.263 \text{ bar.}$$ Ans.

(iii) Next, we determine the heat transfer coefficient and pressure drop for the tube. Tube side flow area,

$$A_t = \left(\frac{1068}{2}\right)\left(\frac{\pi}{4}\right) D_i^2$$

$$= \left(\frac{1068}{2}\right)\left(\frac{\pi}{4}\right) \times (1.65)^2 \times 10^{-4} = 0.1141 \text{ m}^2$$

$$V_t = \frac{m_t}{\rho A_t} = \frac{40}{997 \times 0.1141} = 0.3516 \text{ m/s}$$

$$Re_t = \frac{V_t D_i}{\nu_t} = \frac{(0.3516)(1.65 \times 10^{-2})}{0.832 \times 10^{-6}} = 6973$$

Now, $$Nu_t = 0.023\,(Re_t)^{0.8}\,(Pr_t)^{0.4} \qquad (4.45)$$

$$= 0.023\,(6973)^{0.8}\,(5.66)^{0.4} = 54.64$$

so, $$h_t = \frac{(Nu_t)(k_t)}{D_i} = \frac{54.64 \times 0.613}{1.65 \times 10^{-2}} = 2030 \text{ W/m}^2\text{K}$$

Also, tube side heat transfer coefficient, based on tube outer surface area.

$$h_t' = h_t \frac{D_i}{D_o} = (2030)\left(\frac{1.65}{1.9}\right) = 1763 \text{ W/m}^2\text{K}$$

Tube side pressure drop,

$$\Delta p_t = \frac{\rho V_t^2}{2}\left(\frac{fL\,N_p}{Di} + 4\,N_p\right)$$

From Fig. (4.3) for $Re_t = 6973$,

$$f = 0.035$$

so, $$\Delta p_t = \frac{(997)(0.3576)^2}{2}\left(\frac{0.035 \times 5 \times 2}{0.0165} + 4 \times 2\right)$$

$$= 1862.2 \text{ N/m}^2 = 0.01862 \text{ bar}$$ Ans.

(iv) Neglecting tube wall thermal resistance, the overall heat transfer coefficient, U, based on outer tube area, is,

$$\frac{1}{U} = \frac{1}{h_s} + \frac{1}{h_t'} = \frac{1}{6804} + \frac{1}{1763}$$

or $$U = 1400.2 \text{ W/m}^2\text{K}$$

Heat exchanger surface area A is,

$$A = N_t \pi D_o L = 1068 \times \pi \times 0.019 \times 5$$

$$= 318.584 \text{ m}^2$$

Now, $$C_c = 40 \times 4180 = 167{,}200 \text{ W/K}$$

and $$C_h = 50 \times 4184 = 209{,}200 \text{ W/K}$$

$$C_{min} = C_c\,;\ C_{max} = C_h;\ C_{min}/C_{max} = 0.7992$$

so, $$NTU = UA/C_{min} = (1400.2)(318.584)/167{,}200$$

$$= 2.668$$

For $$NTU = 2.6688,\ C_{min}/C_{max} = 0.7992$$

$\varepsilon = 0.645$ is obtained from Fig. 7.21

so $$\varepsilon = \frac{C_c\,(t_{c2} - t_{c1})}{C_{min}\,(t_{h1} - t_{c1})} = 0.645$$

or $$t_{c2} = t_{c1} + 0.645\,(t_{h1} - t_{c1})$$

$$= 20 + 0.645\,(70 - 20) = 52.25°C$$ Ans.

and $$t_{h1} - t_{h2} = \frac{C_c}{C_h}(t_{c2} - t_{c1}) = 0.7992\,(52.25 - 20) = 25.8$$

or $t_{h2} = t_{h1} - 25.8 = 70 - 25.8 = 44.2°C$ Ans.

Note : In this example the shell side pressure drop is somewhat higher than permissible as the pressure drop should not exceed 0.7 bar. This can be remedied by increasing the baffle spacing which would reduce V_s as well as N_b causing a significant reduction in Δp_s.

PROBLEMS

(Units of k and h are W/mK and W/m^{2}K, respectively)

7.1 What is the overall heat transfer coefficient for a condenser tube (k = 90) having ID = 1.25 cm and OD = 2.5 cm. The steam and water side coefficients are 5,000 and 3,750 respectively.

7.2 What will be the value of U in problem 7.1 if scale resistances are taken into account. Well water is used for cooling and steam is oil-bearing.

7.3 In a multipass heat exchanger with 2 shell passes and 8 tube passes, 45,000 kg/hr of water are heated from 80 C to 150 C by means of hot exhaust gas (C_p = 1.05 kJ/kg K) entering at 340 C and leaving at 170 C. The total surface based on outer tube surface, is 900 m^2. Determine :

(a) LMTD considering the heat exchanger to be a simple counter flow type.

(b) Correction factor for actual arrangement.

(c) Actual LMTD

(d) Actual value of U.

7.4 1,000 kg/hr of oil (C_p = 2.1 kJ/kg C) is to be cooled from 65 C to 35 C. Find the minimum amount of water at 15 C required if we use ;

(a) Counter-flow exchanger,

(b) Parallel-flow exchanger.

7.5 100,000 kg/hr of water is heated from 40 C to 150 C by CO_2 entering at 430 C and leaving at 200 C. If U = 50, find the required area for ;

(a) Counter-flow exchanger,

(b) Cross-flow exchanger with gas side mixed.

7.6 A steam condenser handles 340,000 kg/hr of steam (x = 1.0). The inlet steam pressure is 3 cm Hg and the hot well temperature is 28 C. The cooling water enters the tubes at 15 C and leaves at 20 C. The tube wall thermal conductivity is 90. The tubes are 1.3 cm ID and 1.88 cm OD. The average water velocity inside the tubes is 2 m/sec. The steam side coefficient is 5,000. Find the outside tube area required.

7.7 A parallel-flow exchanger is required to heat 5,000 kg/hr of water from 20 C to 40 C with steam condensing at 120 C on the outside of the tubes. Tubes of 2.5 cm OD and 2.25 cm ID, 2.5 m long, are available. The steam side coefficient is 5,000 and k_w = 100. The velocity of water in each tube is 0.45 m/sec. Calculate the required number of tubes per pass and the number of passes.

7.8 A two-pass condenser has 32 copper tubes 3 cm ID and 4.5 cm OD, 5 m long. Steam at 1 atm condenses on the outside of the tubes. If 10,000 kg/hr of water are heated from 7 C to 80 C by condensation of 1800 kg/hr of steam, find :

(a) LMTD, (b) U, (c) Water side heat transfer coefficient.

7.9 A vapor condenser is required to remove 250,000 kcal/hr from a vapor at 150 C. The cooling fluid is water, which flows through 1.25 cm ID, 1.6 cm OD, 3 m long tubes at a flow Reynolds number of 10,000. The vapor side coefficient it 7500. The water enters at 15 C and leaves at 26 C. Determine the mass of water required per hour, the number of tubes per pass and the number of passes. Neglect tube wall resistance.

7.10 A heat exchanger heats 25,000 kg/hr of water entering at 30 C while cooling 20,000 kg/hr of water from 100 C to 80 C. U = 1400. Determine the area necessary for ;

(a) Parallel-flow arrangement,

(b) Counter-flow arrangement,

(c) 1-2 reversed current exchanger.

If, in case (b) above, the cold water enters at 20 C, what will be its outlet temperature when the inlet temperature and flow rate of the hot water remain unchanged.

7.11 In a counter-flow exchanger oil flowing through the tubes is cooled from 50 C to 34 C by water on the shell side. The inlet and outlet temperatures of water are 15 C and 29 C. The exchanger consists of an inner copper tube (3.25 cm OD, 0.32 cm thick) and an outer steel pipe (6 cm OD and 0.38 cm thick). The oil flow rate is 70 kg/min. Determine the length of the exchanger. For oil :

ρ = 910 kg/m^3 $\quad$ C_p = 2.0 kJ/kg K

k = 0.2 W/m K $\quad$ μ = 0.042 kg/m-sec

7.12 Calculate the length of 5 cm ID and 6 cm OD steel pipe (k = 55 W/mK) required to heat 7854 kg/hr of oil from 30 C to 90 C, if pipe is heated externally by steam condensing at 100 C. Steam side heat transfer coefficient is 9000 W/m^2 K. For oil:

ρ = 850 kg/m^3 $\quad$ C_p = 2.0 kJ/kg K

k = 0.19 W/m K $\quad$ μ = 22 kg/m-hr

7.13 A heat exchanger is required to heat 7000 kg/hr of water from 20 C to 93 C. Water flows through 90 tubes (k = 100 W/mK) 1 cm OD, 0.12 cm thick, and 4.5 m long. Steam at 100 C condenses on the outside of the tubes. If the value of h for the inner surface of the tubes is 3500 W/m^2 K, calculate the heat transfer coefficient for the outer surface.

7.14 A water heater used for heating water from 20 C to 84 C consists of 10 horizontal tubes 3.75 cm OD and 3.15 cm ID. Water flows through the tubes at an average velocity if 1.2 m/sec. Steam at 150 C condenses on the outside of the tubes. If h_1 = 7500 and h_o = 5800 W/m^{2}K, calculate the outer surface area required.

7.15 A counter-flow exchanger was constructed on the basis of U= 150. The area of the exchanger is 1 m^2. The water rate is 1,300 kg/hr and the inlet water temperature is 15 C. If this exchanger is now used to cool 550 kg/hr of oil (C_p = 2.0 kJ/kg-C) at an inlet temperature of 64 C, what will be the outlet temperatures of oil and water ?

7.16 9,000 kg/hr of oil (C_p = 2.1 kJ/kg K) are cooled from 80 C to 35 C by means of water entering at 25 C in a counter-flow exchanger. The exchanger area is 29 m^2 and U = 500. Determine the rate of flow of water.

7.17 Determine UA for a counter-flow exchanger to cool 2,000 kg/hr of oil (C_p = 2.1 kJ/kg K) from 70 C to 43 C, if the cooling water enters at 15 C and leaves at 28 C.

If the oil rate is reduced to 1000 kg/hr and the entering temperature is 75 C, what will be the oil temperature at exit ? The water rate and entering temperature remain the same.

7.18 The following data refer to a double-pipe parallel-flow heat exchanger having 2.25 cm ID and 2.6 cm OD inner pipe and 3.2 cm ID outer pipe of steel ;

High pressure steam entering inner tube : 400 C @ 1000 kg/hr

High pressure steam leaving inner tube : 385 C

Low pressure steam entering-outer pipe : 220 C @ 430 kg/hr

Low pressure steam leaving outer pipe : 397 C.

What will be the heat transfer rate if the steam flow in the inner tube is reversed; the entering temperatures remaining the same.

7.19 A counter flow heat exchanger is to be designed to cool 1000 kg/hr of oil (C_p = 2.1 kJ/kg K) from 100 C to 50 C by using water at 35 C. The maximum allowable exit temperature for water is 45 C. Determine the ratio $m_h C_h/m_c C_c$. Find the value of UA for given NTU = 1.5.

CHAPTER
8

Solar Thermal Collectors

8.1 THE SUN

The sun is the continuing source of energy of our planet. It is a sphere of intensely hot gaseous matter with a diameter of about 1.39×10^6 km. The earth revolves around the sun at an average distance of 1.5×10^8 km. Energy is believed to be generated within the sun in a thermonuclear fusion process in which hydrogen gets transformed into helium in the core (0.4×10^6 km dia) at temperatures of about 8 to 40×10^6 K. It is estimated that the generation of energy in the sun is at the rate of 3.7×10^{23} kW, out of which only a tiny fraction i.e. about 1.7×10^{14} kW reaches the earth.

The *solar energy* is emitted in the form of radiation which is the composite result of several layers which emit and absorb radiation of various wave lengths. It is however, adequate to consider the sun as a black body radiator at about 5762°K (the temperature of *photosphere*, a layer about a few hundred km thick outside the sun's surface).

8.2 SOLAR RADIATION

The extra-terrestrial radiation towards the earth varies throughout the year primarily because of the change in sun-earth distance. However the characteristics of the sun and its spatial relationship to earth result in a nearly fixed intensity of solar radiation outside the earth's atmosphere and is known as *solar constant.* The currently accepted value of the solar constant is 1383 W/m².

Figure 8.1 shows spectral distribution of solar radiation intensity upon a surface normal to sun's rays at the outer limit of atmosphere. The visible range is contained between wavelengths of 0.4 to 0.7 μ. Maximum intensity occurs around 0.5 μ. Accordingly the curve of Fig. 8.1 approximates emissions from a black body at 5762° K on the basis of Eq. (6.39 c) and Example 6.2. The area under the entire curve of Fig. 8.1 is the *solar constant.*

For terrestrial solar energy applications it is essential to account for the interaction of the solar energy during its passage through the atmosphere. When the solar radiation passes through the atmosphere a part of it gets *scattered* due to dry air molecules, water molecules and dust particles. Further, a part of the radiation gets absorbed by the ozone layer or by water vapour. The effect of atmospheric interactions, therefore, is to reduce the solar radiation, I reaching the earth.

Hence,

$$I = I_D + I_d + I_R \tag{8.1}$$

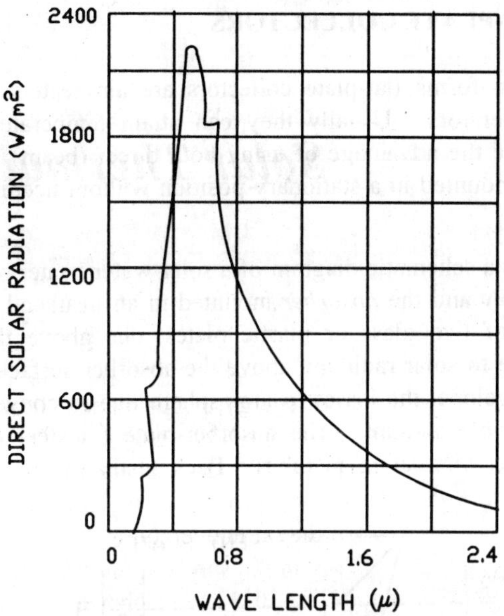

Fig. 8.1 Spectral Distribution of Solar Radiation Normal to Sun's Rays at Outer Limit of Atmosphere

where $I_D = I_N \cos\theta$

$\quad = $ incidence of direct radiation

$I_N = $ intensity of normal radiation

$I_d = $ incidence of diffused radiation

$I_R = $ incidence of reflected radiation

$\theta = $ angle of incidence of sun's rays.

The value of the terms on right hand side of Eq. (8.1) can be obtained by experiments and suitable derivations.

8.3 SOLAR THERMAL COLLECTORS

The *solar collector* is a device or equipment which receives and collects the solar radiant energy which may be transferred or converted into some other useful form depending upon the actual requirements. It is basically a form of heat exchanger though quite different from a conventional one, which accomplishes a fluid to fluid heat exchange at high rate without any predominance of radiation effects. In a solar collector, however, energy gets transferred from a distant source of radiation energy to a fluid, with a maximum expected heat flux of about 1100 W/m^2.

There may be two types of collectors viz. (a) *flat plate* collectors (b) *concentrating* collectors. In the former type the area absorbing solar radiation is the same as the area intercepting radiation. The concentrating collectors, on the other hand, have concave reflectors which concentrate the radiation falling on the total area of the reflector and direct it to the heat exchanger with a smaller surface area. The heat flux is accordingly increased. We shall discuss here the heat transfer aspects of *flat-plate solar collectors* only.

8.4 SOLAR FLAT-PLATE COLLECTORS

In most common forms flat-plate collectors are air heaters, water heaters or low pressure steam generators. Usually they can attain temperature upto 100°C above ambient. They have the advantage of using both direct (beam) and diffuse radiations; hence they can be mounted in a stationary position without needing orientation towards the sun.

Figure 8.2 shows a schematic diagram of a solar water heater of this type. It consists of the *cover assembly* and the *absorber* mounted in an insulated enclosure. The cover assembly is a set of two glass or plastic plates, one above the other, providing a transparent envelope to solar radiation above the absorber surface. Due to *glass house effect* within the envelope, the losses to atmosphere due to convection or radiation are reduced to a negligible amount. The absorber-plate transfers the solar radiation so received to the water circulating in tubes. Back insulation prevents heat loss to the surroundings.

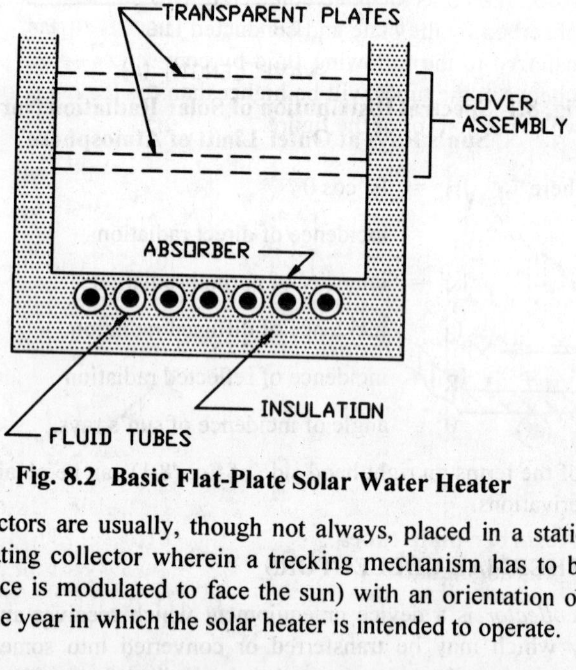

Fig. 8.2 Basic Flat-Plate Solar Water Heater

Flat -plate collectors are usually, though not always, placed in a stationary position (unlike a concentrating collector wherein a trecking mechanism has to be provided so that receiving surface is modulated to face the sun) with an orientation optimised for a particular time of the year in which the solar heater is intended to operate.

8.5 ENERGY BALANCE IN A FLAT-PLATE COLLECTOR

The incoming radiation on the collector can be balanced by the useful energy delivered to working fluid and various losses. Accordingly the energy equation for a flat-plate collector will be ;

$$A_c I_c (\tau\alpha)_c = q_u + q_l + q_s \tag{8.2}$$

where A_c = area of collector, (m^2).

I_c = rate of solar radiation incident on collector, (W/m^2).

τ, α = transmittance and absorpivity of collector surface respectively.

q = rate of heat transfer, (W). The suffixes *u* and *l* represent useful and lost energy respectively.

q_s = rate of increase in internal energy of the collector, (W).

The *collector efficiency* η_c is defined as the ratio of useful energy delivered in a certain period of time to the incident solar energy over the same period,

$$\eta_c = \left\{\int_0^\theta q_u \, d\theta\right\} \div \int_0^\theta A_c \, I_c \, d\theta \qquad (8.3)$$

where θ is the duration over which performance is being averaged.

8.6 TEMPERATURE DISTRIBUTION IN A FLAT-PLATE COLLECTOR

Figure 8.3 shows qualitatively the temperature distribution in a flat-plate collector. Radiation falls on the collector plate connecting two adjacent tubes in portion (a) of the figure. It is absorbed by the plate and conducted laterally towards the flow-tubes, where it is then transferred to the following fluid by convection. Thus the temperature at any cross-section between the tubes will be higher than the temperature in the vicinity of the tubes as shown in portion (b) of the figure.

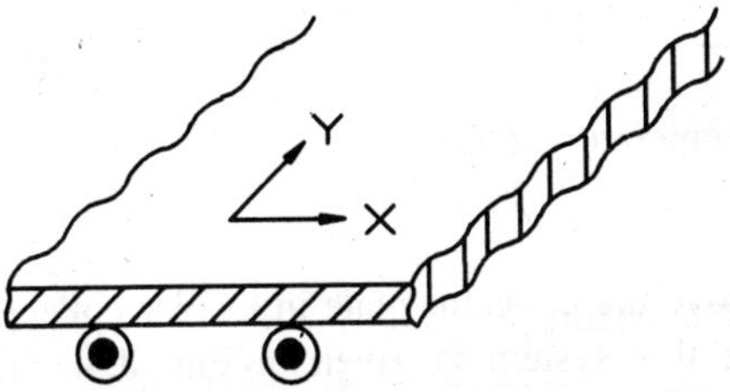

(a) Absorber-plate-tubes arrangement

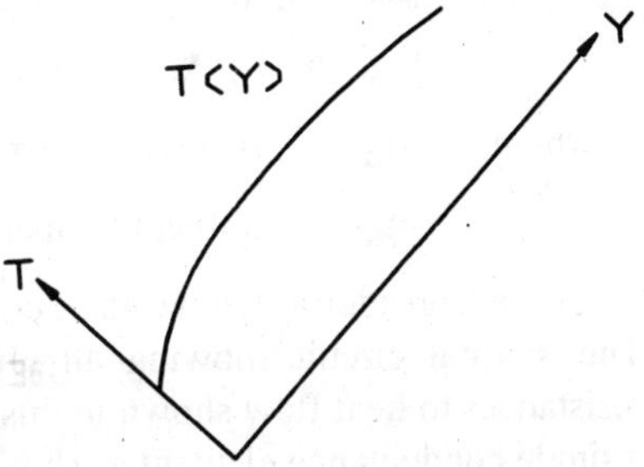

(b) Temperature variation at certain value of X

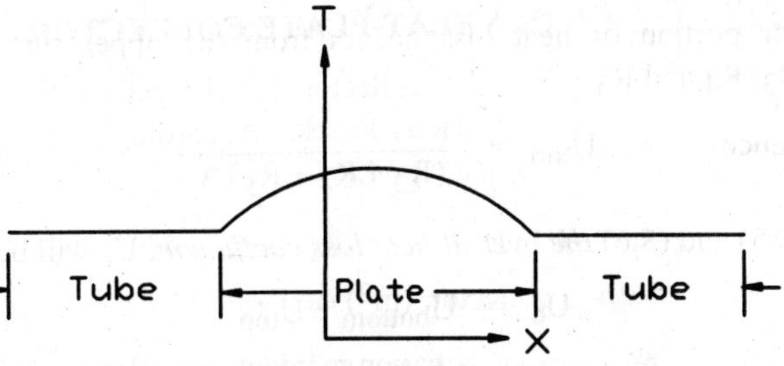

(c) Temperature variation in fluid flow direction

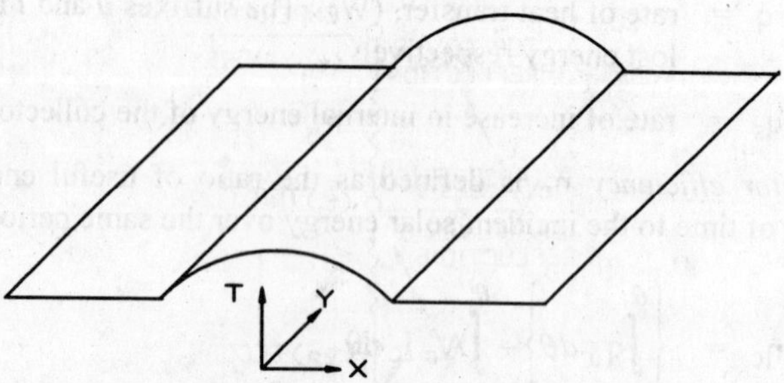

(d) 3-dim temperature distribution in absorber-plate of a flat-plate collector

Fig. 8.3 Temperature Distribution in Absorber-Plate of a Flat-Plate Collector

Further, the temperature of the working fluid and collector will increase in the direction of fluid-flow. The temperature distribution in y-direction at mid point between the tubes is shown in portion (c) of the figure. The three dimensional view of the temperature distribution in absorber plate is shown in Fig. 8.3 (d).

8.7 OVERALL LOSS-COEFFICIENT OF A COLLECTOR

To understand the parameters that determine the thermal efficiency of a solar collector, it is important to develop the concept of *over-all collector heat-loss coefficient*, U_c, which is given by ;

$$q_{loss} = U_c A_c (t_c - t_a) \tag{8.4}$$

where t_c = average collector-plate temperature, (°C)

t_a = ambient temperature, (°C)

Let us consider a double-glass covering above the collector plate in a solar collector. The thermal circuit showing all elements of this system is given in Fig. 8.4. The resistances to heat flow shown in this circuit will be analysed before combining them into a single conductance element as shown in Fig. 8.5.

If the lower surface of the collector is insulated (with resistance = R_1), it will offer resistance to heat flow (loss) towards the bottom. Let $R_2 = 0$ then ;

$$U_{bottom} = \frac{1}{R_1 A_c} \tag{8.5}$$

The major portion of heat loss occurs from the upper surface, through thermal resistances R_3, R_4 and R_5

hence $$U_{top} = \frac{1}{(R_3 + R_4 + R_5) A_c} \tag{8.6}$$

From Eqs. (8.5) and (8.6) *the overall heat loss coefficient*, U_c will be given by ;

$$U_c = U_{bottom} + U_{top}$$

$$U_c = \frac{1}{R_1 A_c} + \frac{1}{(R_3 + R_4 + R_5) A_c} \tag{8.7}$$

where the resistances R_3, R_4 and R_5 are evaluated as below :

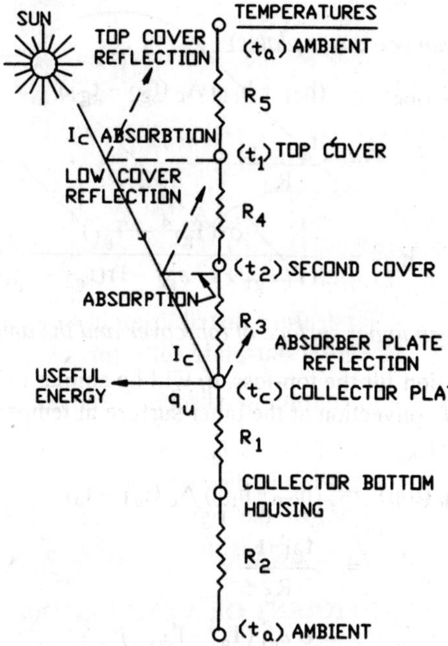

Fig. 8.4 Thermal Circuit for a Flat-plate Collector

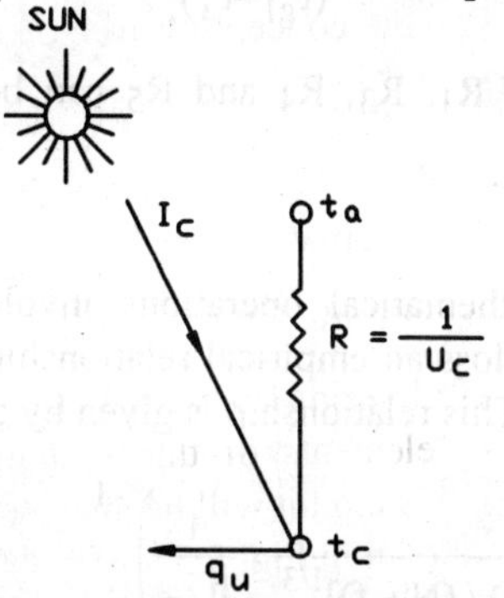

Fig. 8.5 Equivalent Circuit for that of Fig. 8.4

(a) R_3 (*resistance between absorber plate and second glass-cover*) :

Heat is transferred (loss) to the top by convection and radiation in parallel.

So $$q_l\,(\text{top}) = A_c\, h_{c2}\,(t_c - t_{g2}) + \frac{\sigma A_c\,(T_c^4 - T_{g2}^4)}{(1/\varepsilon_p)+(1/\varepsilon_{g2})-1} \tag{8.8}$$

[where suffix p and g refer to plate and glass and F_{12} is given by Eq. (6.82b)]

$$= (h_{c2} + h_{r2})\, A_c\,(t_c - t_{g2}) \tag{8.9}$$

$$= \frac{t_2 - t_{g2}}{R_3}$$

here $$h_{r2} = \frac{\sigma\,(T_c^4 - T_{g2}^4)}{(1/\varepsilon_p + 1/\varepsilon_{g2} - 1)(t_c - t_{g2})} \tag{1.11}$$

(b) R_4 *(resistance between two cover plates)*

similarly,
$$q_l\,(\text{top}) = (h_{c1} + h_{r1})\,A_c\,(t_{g2} - t_{g1}) \tag{8.10}$$
$$= \frac{t_{g2} - t_{g1}}{R_4}$$

where
$$h_{r1} = \frac{\sigma\,(T_{g2}{}^4 - T_{g1})^4}{(1/\varepsilon_{g1} + 1/\varepsilon_{g2} - 1)(t_{g2} - t_{g1})}$$

(c) R_5 *(resistance between upper surface of top-cover and the ambient air)*

Here also, the expression for the top-loss, q_l will be similar to (a) and (b) above but it will be a combination of convection at the outer surface at temperature t_{g1} plus radiation to the sky at temperature t_{sky}.

So
$$q_l\,(\text{top}) = (h_{c\infty} + h_{r\infty})\,A_c\,(t_{g1} - t_a) \tag{8.11}$$
$$= \frac{t_{g1} - t_a}{R_5}$$

where
$$h_{r\infty} = \frac{\sigma\,\varepsilon_{g1}\,(T_g{}^4 - T_{sky}{}^4)}{(t_{g1} - t_a)}$$

Accordingly, the values of R_1, R_3, R_4 and R_5 can be substituted in Eq. (8.7) to evaluate U_c.

(d) *Empirical Relation*

To avoid the tedious mathematical operations involved in the usage of above equations to evaluate top heat loss an empirical relationship is proposed by Klein which is often sufficiently accurate. This relationship is given by ;

$$q_l\,(\text{top}) = \left(\frac{N}{(366/T_c)\,[(t_c - t_a)/(N + f)]^{1/3}} + \frac{1}{h_{c\infty}}\right)^{-1} + \left(\frac{\sigma A_c\,(T_c{}^4 - T_a{}^4)}{[(\varepsilon_p + 0.05\,N(1 - \varepsilon_p)]^{-1} + [(2N + f - 1)/\varepsilon_g] - N}\right)$$

where N = number of glass covers

$f = (1 - 0.04\,h_{c\infty} + 0.0005\,h_{c\infty}^2)(1 + 0.091\,N)$

$h_{c\infty} = 5.7 + 3.8\,V_\infty\ (W/m^2K)$

V_∞ = Wind Velocity (m/s)

Example 8.1

Calculate the top heat loss coefficient and the heat loss for a flat-plate collector with a single-cover having the data as below :

Absorber plate temperature, t_c = 67° C; plate-glass cover spacing = 3 cm, ε_g = 0.93; ε_p = 0.2; α_p = 0.8, ambient air temperature, t_a = 10 C; k_{air} = 0.03 W/mK; wind velocity = 3 m/sec; back insulation = 2 cm of k = 0.04 W/mK and t_{sky} = – 10 C.

Solution :

Refer to Fig. 8.4 for thermal circuit of the collector. Here $R_4 = 0$. The collector arrangement is shown in Fig. 8.6.

The problem will be solved by iteration process by first assuming the temperature of glass cover, t_g. This assumed value will be set into the expressions of R_3 and R_5 whose values will then determine U_{top} and q_l (top). A new t'_g will then be obtained and compared with the assumed value, to ascertain if further iteration is needed.

(a)
$$R_1 A_c = \frac{l}{k} \tag{1.8}$$

$$= \frac{0.2}{0.04} = 5 \text{ m}^2\text{K/W}$$

$$U_{bottom} = \frac{1}{R_1 A_c} = 0.2 \text{ W/m}^2\text{K}$$

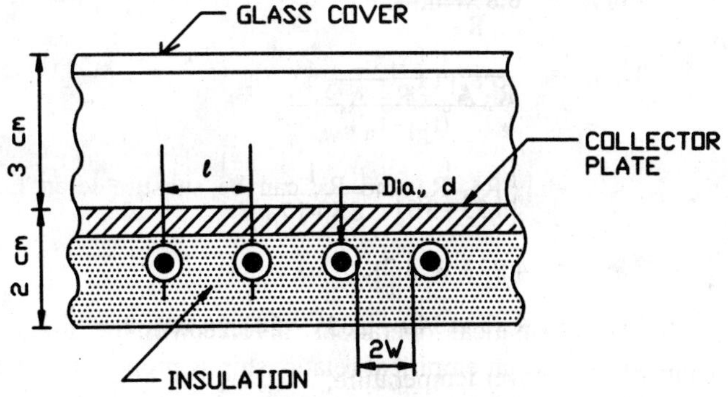

Fig. 8.6 Example 8.1

(b) Between plate and cover :

$$\frac{1}{R_3 A_c} = h_{c2} + h_{r2} \tag{8.9}$$

$$\text{where } h_{r2} = \frac{\sigma(T_c^4 - T_g^4)}{(1/\varepsilon_p + 1/\varepsilon_g - 1)(t_c - t_g)}$$

$$= \frac{56.69 \times 10^{-9}\,(340^4 - T_g^4)}{(1/0.2 + 1/0.93 - 1)(67 - t_g)} \tag{A}$$

In order to determine h_{c2}, let $Gr = 8 \times 10^4$ (for an assumed $\Delta t = 4$ C) and $Pr = 0.71$ at 63°C, therefore the value of Nu can be determined by Eq. (4.205) as;

$$Nu = 3.51$$

hence
$$h_{c2} = \frac{Nu\, l}{k} = \frac{3.51 \times 0.03}{0.03}$$

$$= 3.51 \text{ W/m}^2\text{K} \tag{B}$$

(c) Between cover and ambient :

$$\frac{1}{R_5 A_c} = h_{c\infty} + h_{r\infty} \tag{8.11}$$

where $h_{c\infty} = 5.7 + 3.8\, V_\infty$

$$= 5.7 + 3.8 \times 3 = 17.1 \text{ W/m}^2\text{K} \tag{C}$$

and $h_{r\infty} = \dfrac{\sigma\, \varepsilon_g (T_g{}^4 - T_{sky}{}^4)}{(t_g - t_a)}$

$$= \frac{(56.69 \times 10^{-9})(0.93)(T_g{}^4 - 263^4)}{(t_g - 10)} \tag{D}$$

(d) Now, assume $t_g = 37°C$ or $T_g = 310°K$ and substitute it in equations (A) and (D).

we get $h_{r2} = 1.55 \text{ W/m}^2\text{K}$

and $h_{r\infty} = 8.8 \text{ W/m}^2\text{K}$

Also, $$U_{top} = \frac{1}{R_3 A_c + R_5 A_c} \tag{8.7}$$

$$= \left\{\frac{1}{1.55 + 3.51} + \frac{1}{8.8 + 17.1}\right\}^{-1}$$

$$= 4.233 \text{ W/m}^2\text{K}$$

hence $q_l\,(\text{top}) = 4.233\,(340 - 283) = 241.28 \text{ W/m}^2$

(e) To evaluate the value of glass cover temperature,

$$q_l(\text{top}) = (h_{c2} + h_{r2})\,(t_c - t'_g) = 241.28$$

therefore

$$(3.51 + 1.55)\,(67 - t'_g) = 241.28$$

which gives the new value of temperature of glass cover as

$$t'_g = 19.3°C,\ T_g' = 292.3 \text{ K}$$

(f) Accordingly, second iteration is carried out with this value of $T_g = 292.3$ K which when substituted in Eqs. (A) and (D) gives ;

$$h_{r2} = 1.43 \text{ W/m}^2\text{K}$$

$$h_{r\infty} = 14.1 \text{ W/m}^2\text{K}$$

so $$U_{top} = \left\{\frac{1}{1.43 + 3.51} + \frac{1}{14.1 + 17.1}\right\}^{-1}$$

$$= 4.25 \text{ W/m}^2\text{K}$$ Ans.

Also, $q_l\,(\text{top}) = 4.25\,(340 - 283) = 242.25 \text{ W/m}^2$ Ans.

(g) In order to satisfy that no further iteration is required :

$$(3.51 + 1.55)(67 - t_g'') = 242.25$$

or $$t_g'' = 19.1°C$$

which is fairly in agreement with t_g = 19.3°C. Hence assumption O. K.

Aliter

The value of q_l (top) and U_{top} can be determined by the application of empirical relationship of Eq. (8.12) as below :

$$A_c = 1 \text{ m}^2$$

and $$f = (1 - 0.04 \times 17.1 + 0.0005 \times 17.1^2)(1.091) = 0.5$$

so $$q_{l(top)} = 1 \times (67 - 10)\left[\frac{1}{(366/340)[(67-10)/1.5]^{1/3}} + \frac{1}{17.1}\right]^{-1} + \left(\frac{56.69 \times 10^{-9} \times 1(340^4 - 283^4)}{(0.2 + 0.05 \times 1 \times 0.8)^{-1} + (1.5/0.93) - 1}\right)$$

$$= 170.1 + 82.43$$

$$= 252.5 \text{ W/m}^2 \qquad \text{Ans.}$$

Hence $$U_{top} = \frac{252.5}{67-10}$$

$$= 4.43 \text{ W/m}^2\text{K} \qquad \text{Ans.}$$

Now, $$U_c = U_{botom} + U_{top}$$

$$= 0.2 + 4.43 = 4.63 \text{ W/m}^2\text{K}$$

So, $$\text{total heat loss} = A_c U_c (t_c - t_a)$$

$$= 1 \times 4.63 \times 57$$

$$= 263.9 \text{ W/m}^2 \qquad \text{Ans.}$$

8.8 EFFICIENCY OF A SOLAR COLLECTOR

The collector efficiency over a certain period of time was defined by Eq. (8.3). In order to calculate the average efficiency of collector on a full day basis, the averaged hourly figures of useful energy q_u should be evaluated for respective solar incidence, I_c. These figures will then be totalled to get daily-averaged efficiency of the flat plate collector.

The rate of heat transfer to the fluid flowing in the collector depends on the collector surface temperature t_c, the temperature of the fluid and the heat transfer coefficient between the fluid and collector, h_i. Solar radiant energy impinging on the upper face of the collector plate is conducted in the transverse direction towards the flow channels, and the plate acts as a fin attached to the walls of the flow tubes as evidenced by the temperature distribution of Fig. 8.3(b). Incorporating the concept of heat-flow to the fluid due to fin-effect and avoiding detailed mathematical derivations the pertinent results can be obtained as below :

(a)
$$q_{total} = 2\,w\,\eta_f\,\{\alpha_c I_c - U_c (t_b - t_a)\} \tag{8.13}$$

where w = half of the distance between walls of two adjacent tubes, (m)

$= \dfrac{l-d}{2}$, (m) see Fig. 8.6,

$$\eta_f = \frac{\tanh mw}{mw}$$

$m^2 = U_c/kt$

k, t = thermal conductivity and thickness of plate, (W/mK, m)

t_b = temperature of the fin-base, (°C). At y = 0 (Fig. 8.3) it is t_c,

t_a = ambient air temperature, (°C)

U_c = heat loss coefficient, (W/m^2K)

(b) The collector efficiency factor, F′ is given by ;

$$F' = \frac{1/U_c}{l\left(\dfrac{1}{U_c\,(d+2w\eta_f)} + \dfrac{1}{h_i\,\pi d_i}\right)} \tag{8.14}$$

where h_i = inside film coefficient in pipe, (W/m^2K)

d_i = inner pipe diameter, (m)

(c) *The collector heat removal ratio*, F_R is defined as the ratio of the actual rate of heat transfer to the working fluid to the rate of heat transfer at the maximum temperature difference between absorber and the environment. It is given by ;

$$F_R = \frac{G_c\,C_p}{U_c}\left\{1 - \exp\left(\frac{U_c F'}{G_c C_p}\right)\right\} \tag{8.15}$$

where G_c = flow rate of fluid per unit area of collector, (kg/m^2 sec)

C_p = specific heat of fluid, (kJ/kg K)

(d) The rate of useful heat transfer, q_u is finally expressed in terms of fluid inlet temperature as ;

$$q_u = A_c F_R \{\alpha_c I_c - U_c (t_{in} - t_a)\} \tag{8.16}$$

where t_{in} = fluid temperature at inlet, (°C)

Example 8.2

A collector (1 m × 2 m) has an overall loss-conductance of 8.0 W/m^2K on the upper surface. It is made of copper tubes 1 cm ID 0.05 cm thick connected by 0.05 cm plate (k = 390 W/m^2K) at centre to centre distance of 15 cm. The heat transfer coefficient for water flow in tubes is 1500 W/m^2K. $\alpha_c = 0.9$, $\tau_c = 0.9$. Water enters the collector at 57°C @ 0.02 kg/sec.

Evaluate the daily efficiency on the basis of hourly averaged figures of solar radiation and heat loss. The hourly value of solar insolance, I_c and t_a are as given in columns 2 and 3 of the Table 8.1.

Solution :

Here $U_c = 8.0\ W/m^2K$, $k = 390\ W/mK$,

$d_i = 1\ cm$, $d_o = 1.1\ cm$, $l = 15\ cm$,

$t = 0.05\ cm$, $h_i = 1500\ W/m^2K$,

$\alpha_c = 0.9$, $\tau_c = 0.9$, $t_{in} = 57°C$

$G_c = 0.02/2 = 0.01\ kg/m^2\ sec$

and $w = (15 - 1.1)/2 = 6.95\ cm$

(a) from Eq. (8.13)

$$\eta_f = \frac{\tanh mw}{mw}$$

where

$$m = \left(\frac{U_c}{kt}\right)^{1/2}$$

or

$$m = \left(\frac{8}{390 \times 5 \times 10^{-4}}\right)^{1/2} = 6.4$$

hence

$$\eta_f = \frac{\tanh(6.4 \times 0.0695)}{6.4 \times 0.0695} = 0.938$$

(b) from Eq. (8.14), the efficiency factor, F′ is given by

$$F' = \frac{1/U_c}{l\left(\dfrac{1}{U_c\ (d + 2w\eta_f)} + \dfrac{1}{h_1\ \pi d_i}\right)}$$

$$= \frac{1/8.0}{0.15\left(\dfrac{1}{8.0(0.01 + 2 \times 0.0695 \times 0.938)} + \dfrac{1}{1500\pi \times 0.01}\right)} = 0.920$$

(c) The heat removal factor F_R is given by Eq. (8.15)

$$F_R = \frac{G_c C_p}{U_c}\ [1 - \exp(-U_c F'/G_c C_p)]$$

$$= \frac{0.01 \times 4184}{8.0}\ (1 - e^{-(8.0 \times 0.92/0.01 \times 4184)}) = 0.844$$

(d) Thus, by incorporating the transmittance τ_c, the Eq. (8.16) will take the form :

$$q_u = 2 \times 0.844\ \{0.9 \times 0.9 \times I_c - 8.0\ (57 - t_a)\}$$

$$= 1.688\ \{0.81\ I_c - 8.0\ (57 - t_a)\} \qquad \text{(A)}$$

also $\eta_c = q_u/0.81\ A_c\ I_c$ (for each hour) (B)

(e) The values of I_c and t_a as available in the data are plugged in Eq. (A) for each hour, and q_u is obtained. Further, η_c is obtained for each hour from Eq. (B). These values are shown in column 4 and 5 respectively of Table 8.1.

Note, that during early morning and evening the sun is not delivering useful energy which is evidenced by figures of zero in column 4 and 5, between 0700 – 1000 hrs. and 1500 – 1700 hrs.

$$\text{Daily average } \eta_c = \frac{\sum q_u}{0.81 A_c \sum I_c} = \frac{2099}{0.81 \times 2 \times 4837} = 0.265 \qquad \text{Ans.}$$

Table 8.1 Data and Calculated Values for Example 8.2

Time (hrs.)	I_c (W/m^2)	t_a (C)	*q_u (W)	η_c*
7-8	94	–3	0	0
8-9	208	7	0	0
9-10	387	10	0	0
10-11	583	13	221	0.190
11-12	804	17	584	0.363
12-13	828	17	619	0.374
13-14	720	15	438	0.304
14-15	579	15	237	0.205
15-16	387	11	0	0
16-17	267	7	0	0
	= 4,837 W/m²		= 2,099 W	

*As calculated from Eqs. (A) and (B)

Example 8.3

The following data were obtained from a single glass cover flat plate solar heat collector : Mean plate temperature = 70°C; Ambient and sky temperature = 10°C; Back insulation thickness = 5 cm; Insulation thermal conductivity = 0.05 W/mK; Coefficient of heat transfer by convection from plate to cover = 3 W/m^2K; Equivalent coefficient of heat transfer by radiation from plate to cover = 6 W/m^2K; Coefficient of convective heat transfer from cover to ambient air = 25 W/m^2K and equivalent coefficient of radiant heat transfer from cover to sky = 5 W/m^2K. Compute the total heat loss per m^2 collector area. Assume that resistance to heat flow by convection at the back surface of the insulation is negligible. (UPSC' 93).

Solution :

The analogous electrical circuit for the problem is shown in Fig. 8.7. Heat will flow from the plate to the top as well as the back of the collector. Let suffixes p, c, a, s and b denote plate, cover, ambient, sky and back respectively.

Now, consider 1 m² of plate area

$$R_{p-a} = R_{p-c} + R_{c-a}$$

$$\frac{1}{R_{p-c}} = \frac{1}{R_1} + \frac{1}{R_2} = 3 + 6 = 9$$

so $$R_{p-c} = \frac{1}{9} = 0.1111 \text{ m}^2 \text{ K/W}$$

and $$\frac{1}{R_{c\text{-}a}} = \frac{1}{R_3} + \frac{1}{R_4} = 25 + 5 = 30$$

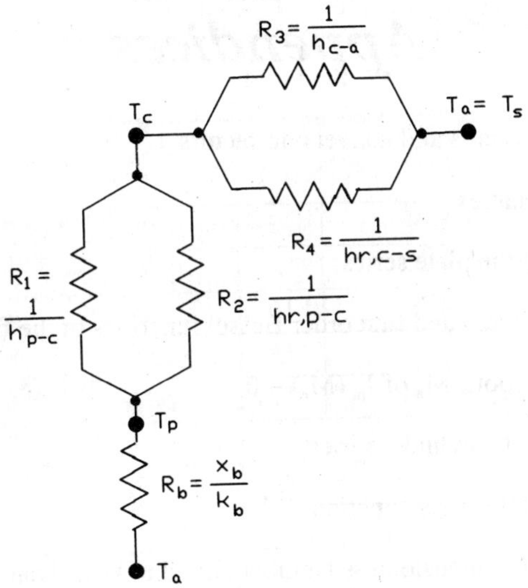

Fig. 8.7 Example 8.3

or $$R_{c-a} = \frac{1}{30} = 0.0333 \text{ m}^2 \text{ K/W}$$

Thus, $$R_{p-a} = 0.1111 + 0.0333 = 0.1444 \text{ m}^2 \text{ K/W}$$

Hence, rate of heat flow from top,

$$q_{top} = \frac{T_p - T_a}{R_{p-a}} = \frac{70-10}{0.1444} = 415.5 \text{ W/m}^2$$

Now $$R_b = \frac{x_b}{k_b} = \frac{5 \times 10^{-2}}{0.05} = 1 \text{ m}^2 \text{ K/W}$$

Heat flow rate from back,

$$q_{back} = \frac{T_p - T_a}{R_b} = \frac{70-10}{1} = 60 \text{ W/m}^2$$

Therefore, total heat flow rate,

$$q = 415.5 + 60 = 475.5 \text{ W/m}^2$$ Ans.

Appendices

Appendix A : System of units and conversion factors

Appendix B : Property tables

Appendix C : Values of the plate series

Appendix D : Values of zero and first order Bessel functions of the first kind

Appendix E : First five roots, M_n of $J_m (M_n) = 0$

Appendix F : Values of the cylinder series

Appendix G : Values of the error function

Appendix H : Important Dimensionless Numbers Used in Heat Transfer

Appendix A

System of Units and Conversion Factors

A.1 SYSTEMS OF UNITS

Numerical calculations in engineering require a consistent system of units. There are several systems existing and each is equally correct. Any of the systems may be chosen for use but confusion should be avoided.. The following table summarises the relationships between mass and force for various systems of units. ($F = ma/g_c$).

Table A-1 : Magnitude and Units of g_c

System	Mass (M)	Length (L)	Time (θ)	Force (F)	g_c
Absolute	gm	cm	sec	dyne	1.0 gm cm/dyne sec^2
S. I.	kg	m	sec	newton	1.0 m kg/newton sec^2
MKS	kg	m	sec	kg_f	9.806 m kg/kg_f sec^2
English	lb_m	ft	sec	lb_f	32.174 lb_m ft/lb_f sec^2
English (Engg.)	slug	ft	sec	lb_f	1.0 slug ft/lb_f sec^2
English (Absolute)	lb_m	ft	sec	poundal	1.0 lb_m ft/poundal sec^2

A.2 SI UNITS

A common system of units for international usage has been accepted by all countries of the world. This system is known as *System International* abbreviated as SI units. It is the purest form of metric units and is an extension and refinement of traditional metric system. There are seven *basic* units as given in Table A-1, which are arbitrarily defined (Ref. IS : 5-1969) internationally.

Table A-2 Basic Units in SI

S.No.	Physical Quantity	Unit	Symbol
1	length	meter	m
2	mass	kilogram	kg
3	time	second	g
4	electrical current	ampere	A
5	luminous intensity	candela	cd
6	thermodynamic temp.	kelvin	K
7	amount	mole	mole

Table A-3 shows some of the important *derived units* which are often used by mechanical engineers. In SI units, only decimal multiples and submultiples are used. The different factors may be abbreviated by use of *prefix* or *symbols* as shown in Table A-4.

Table A-3 Derived Units in SI

Physical Quantity	Unit	Symbol
force	newton	N = kg m/s^2
frequency	hertz	Hz = c/s
work, energy or heat	joule	J = Nm
power, heat flow rate	watt	W = J/s
torque, bending moment	newton	Nm = J
Pressure, stress or modulus of elasticity	newton per square meter	N/m^2
dynamic viscosity	newton second per square meter	Ns/m^2
thermal conductivity	watt per meter per degree kelvin	W/mK = J/smK
specific heat	joule per kilogram per degree kelvin	J/kgK

Table A-4 Multiples of Ten

Factor	Prefix	Symbol
10^{12}	tera	T
10^{9}	giga	G
10^{6}	mega	M
10^{3}	kilo	k
10^{-3}	mili	m
10^{-6}	micro	μ
10^{-9}	nano	n
10^{-12}	pico	p

A-3 WRITING OF UNITS

Some accepted recommendations about the representation of SI units and symbols are given below :

1. Full names of units or prefixes shall be written in upright roman, lower case letters. No capitals are used even if unit is named after a person, e.g. kelvin, newton, joule, watt, volt, etc.
2. Symbols of units are written (i) in capitals if named after a person e.g. K, N, J and W etc. (ii) otherwise in lower case letters e.g. m, s, kg etc. (see Table A-2)
3. Algebraic quantities are written in italics say E for e. m. f. and I for current.
4. No *Abbreviation* be used for units. Write either as a symbol or full name.
5. *Unit* remain unaltered in *plurals*. No *s* be used either in full name or in symbol.

6. No full *stop* (period) or a punctuation mark be used e.g. 3 m; SI units ; 100.25 kg.
7. If a prefix is used with a symbol no space be left in between e.g. MJ and not M J.
8. Not more than one prefix be used e.g. for 10^{-9} m use nm and not mμm.
9. Power to basic unit applies to whole and not to base alone e.g. $km^2 = (km)^2 = 10^6\ m^2 \neq k\ (m)^2$ or $10^3\ m^3$
10. Leave a space between numeral and its unit. 64 kW and not 64kW.
11. In a combined unit a space may or may not be left. ISI has recommended that even a full stop may be used e.g. Nm, N m, N.m are all correct but m N is not correct because it may imply milinewton. Hence it should be checked that it does not give some other meaning.
12. A solidus or oblique may be used or a negative power may be used. e.g. $\frac{m}{s}$, m/s or ms^{-1} are all correct but ms^{-1} may mean 'per milisecond'; but since no prefix should have been used in denominator, this complication may not occur.
13. For large numerals use groups of three without a coma (though in trade accounting it may continue). Grouping should start both sides from decimal. No grouping is done if figures are only 4.

Correct	Incorrect
234 869.739 45	23 486 9.7 394 5
40 000	40000
3456.72	3 456.72
0.123 457	0.123457

A.4 CONVERSION FACTORS

The following relations for conversion of units from one system to another (between British, Metric and SI) will be often used in calculations :

(*a*) *Length*

$$1\ \text{ft} = 0.3048\ \text{m} = 30.48\ \text{cm}$$

$$1\ \text{m} = 100\ \text{cm} = 1{,}000\ \text{mm} = 3.2808\ \text{ft}$$

$$= 39.37\ \text{in} = 1.0936\ \text{yd}$$

$$1\ \text{in} = 2.54\ \text{cm} = 0.0254\ \text{m}$$

$$1\ \text{km} = 0.62137\ \text{mile}$$

$$1\ \text{mm} = 1{,}000\ \text{micron}\ \ (\mu)$$

$$1\ \mu = 10^{-6}\ \text{m} = 10^{-4}\ \text{cm}$$

(*b*) *Mass*

$$1\ \text{kg} = 2.205\ \text{lb}_m$$

$$1\ \text{lb}_m = 453.592\ \text{gm} = 0.45359\ \text{kg}$$

$$1\ \text{gm} = 0.002205\ \text{lb}_m = 15.432\ \text{grain}$$

$$1 \text{ quintal} = 100 \text{ kg}$$

$$1 \text{ ton} = 10 \text{ q} = 1000 \text{ kg}$$

$$1 \text{ slug} = 32.1739 \text{ lb}_m$$

(*c*) *Area*

$$1 \text{ in}^2 = 6.4516 \text{ cm}^2$$

$$1 \text{ ft}^2 = 929.03 \text{ cm}^2 = 0.0929 \text{ m}^2$$

$$1 \text{ cm}^2 = 0.165 \text{ in}^2$$

$$1 \text{ m}^2 = 10.7693 \text{ ft}^2 = 1.196 \text{ yd}^2$$

$$1 \text{ hectare} = 10^4 \text{ m}^2 = 2.19 \text{ acre}$$

(*d*) *Volume*

$$1 \text{ in}^3 = 16.387 \text{ cc}$$

$$1 \text{ liter} = 1{,}000 \text{ cc} = 61.024 \text{ in}^3$$

$$1 \text{ gallon (English)} = 4.546 \text{ liters} = 0.1605 \text{ ft}^3$$

$$1 \text{ m}^3 = 35.3148 \text{ ft}^3$$

$$1 \text{ ft}^3 = 0.028 \text{ m}^3$$

(*e*) *Density*

$$1 \text{ gm/cc} = 62.33 \text{ lbm/ft}^3$$

$$1 \text{ lbm/ft}^3 = 0.016 \text{ gm/cc} = 16 \text{ kg/m}^3$$

$$1 \text{ cft (air)} = 0.0809 \text{ lb}_m = 36.5 \text{ gm at N.T.P.}$$

$$1 \text{ liter (air)} = 1.2982 \text{ gm at N.T.P.}$$

(f) *Temperature*

$$°\text{F} = (1.8 \times °\text{C}) + 32$$

$$°\text{C} = \frac{°\text{F} - 32}{1.8}$$

$$°\text{R} = °\text{F} + 459.62$$

$$°\text{K} = °\text{C} + 273.15$$

$$°\text{R} = 1.8 \times °\text{K}$$

(*g*) *Force*

$$1 \text{ newton (N)} = 10^5 \text{ g-cm-sec}^{-2} \text{ (dyne)}$$

$$= 1 \text{ kg-m-sec}^{-2} = 0.102 \text{ kg}_f$$

$$= 0.2481 \text{ lb}_f$$

$$1 \text{ g cm-sec}^{-2} \text{ (dyne)} = 10^{-5} \text{ kg-m-sec}^{-2} \text{ (newton)}$$

$$= 7.233 \times 10^{-5} \text{ lb}_m\text{-ft-sec}^{-2} \text{ (poundal)}$$

$$= 2.481 \times 10^{-6} \text{ lb}_f$$

$$1 \text{ kg-m-sec}^{-2} = 10^5 \text{ gm-cm-sec}^{-2} = 1 \text{ newton}$$

$= 7.233\ lb_m\text{-ft-sec}^{-2}$

$= 2.481 \times 10^{-1}\ lbf$

$1\ lb_m\text{-ft-sec}^{-2} = 1.3826 \times 10^4\ \text{gm-cm-sec}^{-2}$

$= 1.3826 \times 10^{-1}\ \text{kg-m-sec}^{-2} = 0.13826\ N$

$= 3.108 \times 10^{-2}\ lb_f$

$1\ lb_f = 4.4482 \times 10^5\ \text{gm-cm-sec}^{-2}$

$= 4.482\ \text{kg-m-sec}^{-2} = 4.482\ N$

$= 32.174\ lb_m\text{-ft-sec}^{-2}$

$1\ kg_f = 2.205\ lb_f$

$= 9.80665\ N$

(*h*) *Pressure*

$1\ \text{bar} = 750.06\ \text{mm Hg}$

$= 401.85\ \text{in}\ H_2O$

$= 29.53\ \text{in Hg}$

$= 0.98692\ \text{atm}$

$= 14.504\ lb_f/in^2$

$= 2.0886 \times 10^3\ lb_f/ft^2$

$= 10^5\ N/m^2$

$1\ \text{bar} = 10^5\ \text{kg/m-sec}^2$

$= 10^6\ \text{gm/cm-sec}^2$

$1\ N/m^2 = 1\ \text{pascal}$

$= 10^{-5}\ \text{bar}$

$= 1\ \text{kg/m-sec}^2$

$= 10\ \text{gm/cm-sec}^2$

$= 7.5006 \times 10^{-3}\ \text{mm Hg}$

$= 4.0185 \times 10^{-3}\ \text{in}\ H_2O$

$= 2.953 \times 10^{-4}\ \text{in Hg}$

$= 0.98692 \times 10^{-5}\ \text{atm}$

$= 1.4504 \times 10^{-4}\ lb_f/in^2$

$= 0.020886\ lb_f/ft^2$

$1\ \text{gm/cm-sec}^2 = 10^{-1}\ \text{kg/m-sec}^2$

$= 2.0886 \times 10^{-3}\ lb_f/ft^2$

$= 1.4504 \times 10^{-5}\ lb_f/in^2$

$= 9.8692 \times 10^{-7}\ \text{atm}$

$= 7.5006 \times 10^{-4}$ mm Hg

$= 10^{-6}$ bar = 0.1 N/m^2

1 lb$_f$/ft^2 $= 4.788 \times 10^2$ gm/cm-sec^2

$= 4.788 \times 10$ kg/m-sec^2

$= 6.9444 \times 10^{-3}$ lb$_f$/in^2

$= 4.7254 \times 10^{-4}$ atm

$= 3.5913 \times 10^{-1}$ mm Hg

$= 4.788 \times 10^{-4}$ bar

$= 47.88$ N/m^2

1 lbf/in^2 $= 6.8947\ 10^4$ gm/cm-sec^2

$= 6.8947 \times 10^3$ kg/m-sec^2

$= 144$ lb$_f$/ft^2

$= 6.8046 \times 10^{-2}$ atm

$= 0.06895$ bar

$= 6.895 \times 10^3$ N/m^2

$= 51.715$ mm Hg

$= 0.0703$ kg$_f$/cm^2

1 atm $= 1.0133 \times 10^6$ gm/cm-sec^2

$= 1.0133 \times 10^5$ kg/m-sec^2

$= 2.1162 \times 10^3$ lb$_f$/ft^2

$= 14.696$ lb$_f$/in^2

$= 760$ mm Hg

$= 1.03$ kg$_f$/cm^2 = 1.01325 bar

$= 1.01325 \times 10^5$ N/m^2

$= 1.3332 \times 10^5$ N/m^2

1 mm Hg $= 1.3332 \times 10^3$ gm/cm-sec^2

$= 1.3332 \times 10^2$ kg/m-sec^2

$= 2.7845$ lb$_f$/ft^2

$= 1.9337 \times 10^{-2}$ lb$_f$/in^2

$= 1.3158 \times 10^{-3}$ atm

$= 1.333 \times 10^{-3}$ bar

$= 1.333 \times 10^2$ N/m^2

(*i*) *Viscosity*, μ

1 gm/cm-sec (poise) $= 10^{-1}$ kg/m-sec

	=	1 dyne sec/cm^2 = (1/98.1) kg$_f$ sec/m^2
	=	0.067197 lb$_m$/ft-sec
	=	2.0886 × 10^{-3} lb$_f$-sec/ft^2
	=	100 cp (centipoise)
	=	2.04191 × 10^2 lb$_m$/ft-hr
	=	360 kg/m-hr
1 kg/m-sec	=	10 gm/cm-sec = 0.67197 lb$_m$/ft-sec
	=	2.0886 × 10^{-2} lb$_f$-sec/ft^2 = 10^3 cp
	=	2.4191 × 10^3 lb$_m$/ft-hr
	=	3600 kg/m-hr
1 lb$_m$/ft-sec	=	14.882 gm/cm-sec
	=	1.4882 kg/m-sec
	=	3.1081 × 10^{-2} lb$_f$-sec/ft^2
	=	148.82 cp = 3600 lb$_m$/ft-hr
1 lb$_f$-sec/ft^2	=	478.8 gm/cm-sec = 4788 kg/m-sec
	=	32.174 lb$_m$/ft-sec
1 lb$_f$-sec/ft^2	=	4.788 × 10^4 cp
	=	1.1583 × 10^4 lb$_m$/ft-hr
1 cp	=	10^{-2} gm/cm-sec = 10^{-3} kg/m-sec
	=	6.7197 × 10^{-4} ln$_m$/ft-sec
	=	2.0868 × 10^{-5} lb$_f$-sec/ft^2
	=	2.4191 lb$_m$/ft-hr
1 lb$_m$/ft-hr	=	4.1338 × 10^{-3} gm/cm-sec
	=	2.7778 × 10^{-4} lb$_m$/ft-sec
	=	0.41338 cp = 1.488 kg/m-hr
1 kg/m-hr	=	0.6719 lb$_m$/ft-hr

(j) *Diffusivity, α or Kinematic Viscosity, ν*

1 cm^2/sec (stoke)	=	10^{-4} m^2/sec = 3.875 ft^2/hr
	=	0.36 m^2/hr
1 m^2/sec	=	10^4 cm^2/sec = 3.875 × 10^4 ft^2/hr
	=	3600 m^2/hr
1 ft^2/hr	=	0.25807 cm^2/sec = 0.09291 m^2/hr
1 m^2/hr	=	2.7778 cm^2/sec = 10.764 ft^2/hr

(*k*) *Thermal Conductivity, k*

1 watt/m-K	=	2.3901 × 10^{-3} cal/sec-cm-°K

	=	0.5778 btu/hr-ft-°F
	=	0.8598 kcal/hr-m-°C
1 cal/sec-cm-°K	=	418.4 watt/m-°K
	=	241.75 btu/hr-ft-°F
	=	359.724 kcal/hr-m-°C
1`btu/hr-ft-°F	=	1.7307 watt/m-°K
	=	4.1365×10^{-3} cal/sec-cm-°K
	=	1.488 kcal/hr-m-°C
1 kcal/hr-m-°C	=	2.778×10^{-3} cal/sec-cm-°K
	=	0.6719 btu/hr-ft-°F
	=	1.1623 watt/m-°K

(*l*) *Heat Transfer Coefficient, h*

1 watt/m^2-°K	=	2.3901×10^{-5} cal/cm^2-sec-°K
	=	10^{-4} watt/cm^2-°K
	=	1.7611×10^{-1} btu/ft^2-hr-°F
	=	0.86 kcal/m^2-hr-°C
1 cal/cm^2-sec-°K	=	4.184×10^4 watt/m^2-°K
	=	4.184 watt/cm^2-°K
	=	7.3686×10^3 btu/ft^2-hr-°F
	=	36,000 kcal/m^2-hr-°C
1 watt/cm^2-°K	=	10^4 watt/m^2-°K
	=	2.3901×10^{-1} cal/cm^2-sec-°K
	=	8.6×10^3 kcal/m^2-hr-°C
1 btu/ft^2-hr-°F	=	5.6782 watt/m^2 -°K
	=	1.3571×10^{-4} cal/cm^2-sec-°K
	=	5.6782×10^{-4} watt/cm^2-K
	=	4.883 kcal/m^2-hr-°C
1 kcal/m^2-hr-°C	=	1.163 watt/m^2-°K
	=	2.78×10^{-5} cal/cm^2-sec-°K
	=	1.163×10^{-4} watt/cm^2-°K
	=	0.2048 btu/ft^2-hr-°F

(*m*) *Work or Energy*

1 erg	=	10^{-7} joule = 7.3756×10^{-8} ft-lb$_f$
	=	2.3901×10^{-8} cal

	=	9.4783×10^{-11} btu
	=	3.7251×10^{-14} hp-hr
	=	2.7778×10^{-14} kW-hr
	=	1 g cm^2 sec^{-2}
1 joule	=	1 newton meter
	=	1watt-sec
	=	10^7 erg = 0.73756 ft-lb$_f$
	=	0.23901 cal = 9.4783×10^{-4} btu
	=	1 kg m^2 sec^{-2}
1 joule	=	3.7251×10^{-7} hp-hr
	=	2.7778×10^{-7} kW-hr
1 ft-lb$_f$	=	1.3558×10^7 erg = 1.3558 joule
	=	0.32405 cal = 1.2851×10^{-3} btu
	=	5.0505×10^{-7} hp-hr
	=	3.7662×10^{-7} kW-hr
1 cal	=	4.184×10^7 erg = 4.184 joule
	=	3.086 ft-lb$_f$ = 3.9657×10^{-3} btu
	=	1.5586×10^{-6} hp-hr
	=	1.1622×10^{-6} kW-hr
1 btu	=	1.055×10^{10} erg
	=	1.055×10^3 joule
	=	778.16 ft-lb$_f$ = 251.16 cal
	=	3.9301×10^{-4} hp-hr
	=	2.9307×10^{-4} kW-hr
1 kW-hr	=	3.6×10^{13} erg = 3.6×10^6 joule
	=	2.6552×10^6 ft-lb$_f$
	=	8.6042×10^5 cal
	=	3.4122×10^3 btu = 1.341 hp-hr
	=	860.42 kcal
1 kcal	=	4.184×10^{10} erg
	=	4.184 kJ
	=	3086 ft-lb$_f$ = 3.9657 btu
	=	1.5586×10^{-3} hp-hr
	=	1.1622×10^{-3} kW-hr

(n) Power

1 watt	=	1 joule/sec = 0.80 kcal/hr
	=	978×10^{-4} btu/sec
1 hp (FPS)	=	550 ft-lb_f/sec = 42.41 btu/min
	=	746 watt = 596.8 kcal/hr
	=	1.015 hp (MKS)
1 hp (MKS)	=	75 m-kg_f/sec = 0.17569 kcal/sec
	=	735.3 watt

(o) Energy Flow Rate, q

1 watt	=	0.86 kcal/hr = 0.23901 cal/sec
	=	3.4143 btu/hr
1 kcal/hr	=	0.2778 cal/sec = 3.968 btu/hr
	=	1.163 watt
1 cal/sec	=	3600 cal/hr = 14.285 btu/hr
	=	4.184 watt
1 btu/hr	=	0.252 kcal/hr = 0.2930 watt
1 ton (refrig)	=	200 btu/min = 50 kcal/min
	=	3.517 kW

(p) Energy Flux, q/A

1 kcal/hr-m^2	=	0.3686 btu/hr-ft^2 = 1.163 watt/m^2
	=	27.78×10^{-6} cal/sec-cm^2
1 cal/sec-cm^2	=	36,000 kcal/hr-m^2 = 30,960 watt/m^2
	=	14.285 btu/hr-ft^2
1 btu/hr-ft^2	=	2.713 kcal/hr-m^2
	=	3.1552 watt/m^2
1 watt/m^2	=	0.86 kcal/hr-m^2
	=	0.23901×10^{-4} cal/sec-cm^2
	=	0.317 btu/hr-ft^2

(q) Enthalpy, h

1 btu/lb	=	0.5556 kcal/kg
	=	2.321 J/g
	=	2.321 kJ/kg
	=	0.5556 chu/kg

Appendix B

Property Tables

The following property tables are given in the succeeding pages :

B–1 Physical Properties of Air

B–2 Physical Properties of Water (liq.)

B–3 Physical Properties of Saturated Steam

B–4 Physical Properties of Some Solids

B–5 Physical Properties of Some Metals and Alloys

B–6 Physical Properties of Some Liquids

B–7 Physical Properties of Liquid Metals

Table B–1 : Physical Properties of Air

t (°C)	ρ (kg/m³)	C_p (kJ/kg K)	$k \times 10^2$ (watt/m K)	$\alpha \times 10^2$ (m²/hr)	$\mu \times 10^2$ (kg/hr m)	$\nu \times 10^6$ (m²/sec)	Pr –
– 50	1.584	1.013	2.034	4.57	5.264	9.23	0.728
– 40	1.515	1.013	2.115	4.96	5.475	10.04	0.728
– 30	1.453	1.913	2.197	5.37	5.645	10.80	0.723
– 20	1.395	1.009	2.278	5.38	5.822	12.09	0.716
–10	1.342	1.009	2.360	6.28	5.996	12.43	0.712
0	1.293	1.005	2.441	6.77	6.188	13.28	0.707
10	1.247	1.005	2.511	7.22	6.346	14.16	0.705
20	1.205	1.005	2.592	7.21	6.533	15.06	0.703
30	1.165	1.005	2.673	8.23	6.717	16.00	0.701
40	1.128	1.005	2.755	8.75	6.904	16.96	0.699
50	1.093	1.005	2.824	9.29	7.067	17.95	0.698
60	1.060	1.005	2.894	9.79	7.221	18.97	0.696
70	1.029	1.009	2.946	10.28	7.344	20.02	0.694
80	1.000	1.009	3.045	10.87	7.523	21.09	0.692
90	0.972	1.009	3.127	11.48	7.701	22.10	0.690
100	0.946	1.009	3.208	12.11	7.880	23.13	0.688
120	0.898	1.009	3.336	13.26	8.170	25.45	0.686
140	0.854	1.013	3.487	14.52	8.479	27.80	0.684
160	0.815	1.017	3.638	15.80	8.786	30.08	0.682
180	0.779	1.022	3.778	17.10	9.070	32.59	0.681
200	0.746	1.026	3.929	18.49	9.380	34.85	0.680
250	0.674	1.038	4.266	21.49	10.020	40.61	0.677

Table B–2 : Physical Properties of Water (Liquid)

t (°C)	ρ (kg/m³)	C_p (kJ/kg K)	k × 10^2 (watt/m K)	α × 10^4 (m^2/hr)	μ × 10^2 (kg/hr m)	ν × 10^6 (m^2/sec)	Pr –
0	999.9	4.212	55.093	4.71	644.093	1.789	13.67
10	999.7	4.191	57.418	4.94	469.818	1.306	9.52
20	998.2	4.183	59.859	5.16	361.892	1.006	7.02
30	995.7	4.174	61.718	5.35	288.668	0.805	5.42
40	992.2	4.174	63.345	5.51	235.602	0.659	4.31
50	998.1	4.174	64.740	5.65	197.771	0.556	3.54
60	983.3	4.178	65.902	5.78	169.305	0.478	2.98
70	977.8	4.187	66.716	5.87	146.370	0.415	2.55
80	971.8	4.195	67.413	5.96	127.924	0.365	2.21
90	965.3	4.208	67.995	6.03	113.507	0.326	1.95
100	958.4	4.220	68.227	6.09	101.910	0.295	1.75
110	951.0	4.233	68.460	6.13	93.215	0.272	1.60
120	943.1	4.250	68.576	6.16	85.448	0.252	1.47
130	934.8	4.266	68.576	6.19	78.744	0.233	1.36
140	926.1	4.287	68.460	6.2	72.475	0.217	1.26
150	917.0	4.312	68.343	6.22	66.792	0.203	1.17
160	907.4	4.346	68.227	6.23	62.206	0.191	1.10
170	897.3	4.379	67.878	6.22	58.623	0.181	2.05
180	886.9	4.417	67.413	6.20	54.976	0.173	1.00
190	876.0	4.459	66.949	6.17	51.921	0.165	0.96
200	863.0	4.505	66.251	6.14	49.266	0.158	0.93
220	840.3	4.614	64.508	5.99	44.823	0.148	0.89
240	813.6	4.756	62.764	5.84	41.356	0.141	0.87
260	784.3	4.949	60.440	5.61	38.274	0.135	0.87
280	750.7	5.229	57.418	5.27	35.596	0.131	0.90
300	712.5	5.736	53.931	4.75	32.835	0.129	0.97

Table B–3 : Physical Properties of Saturated Steam

t (°C)	ρ (kg/m³)	C_p (kJ/kg K)	$k \times 10^2$ (watt/m K)	$\alpha \times 10^3$ (m²/hr)	$\mu \times 10^2$ (kg/hr m)	$\nu \times 10^6$ (m²/sec)	Pr —
100	0.598	2.135	2.371	67.9	4.320	20.02	1.08
110	0.826	2.177	2.487	49.8	4.487	15.07	1.09
120	1.121	2.206	2.592	37.8	4.613	11.46	1.09
130	1.469	2.257	2.685	28.7	4.757	8.85	1.11
140	1.966	2.315	2.790	22.07	4.861	6.89	1.12
150	2.572	2.395	2.883	17.02	5.030	5.47	1.16
160	3.258	2.479	3.010	13.40	5.162	4.36	1.18
170	4.122	2.583	3.127	10.58	5.376	3.57	1.21
180	5.157	2.709	3.266	8.42	5.430	2.93	1.25
190	6.394	2.855	3.417	6.74	5.604	2.44	1.30
200	7.862	3.023	3.545	5.37	5.604	2.03	1.36
210	9.488	3.199	3.719	4.37	5.906	1.71	1.41
220	11.62	3.408	3.894	3.54	6.050	1.45	1.47
230	13.99	3.634	4.091	2.90	6.245	1.24	1.54
240	16.76	3.881	4.289	2.37	6.409	1.06	1.61
250	19.98	4.158	4.510	1.96	6.564	0.913	1.68
260	23.72	4.467	4.800	1.63	6.774	0.794	1.75
270	28.09	4.815	5.103	1.36	6.948	0.688	1.82
280	33.19	5.234	5.486	1.14	7.174	0.600	1.90
290	39.15	5.694	5.823	0.941	7.404	0.526	2.01
300	46.21	6.280	6.265	0.778	7.654	0.461	2.13
320	64.72	8.206	7.509	0.509	8.240	0.353	2.50
340	92.76	12.351	9.763	0.292	9.539	0.272	3.35
360	144.00	23.027	12.785	0.139	10.460	0.202	5.23

Table B-4 Physical properties of Some Solids

Material	t (°C)	ρ (kg/m^3)	k (watt/m K)	C_p (kJ/kg C)	$\alpha \times 10^3$ (m^2/hr)
Non-metals					
Aluminum Foil	50	20	0.047	-	-
Asbestos, Fibre	50	470	0.110	0.816	1.04
Brick, masonry	20	800 to 1500	0.23 to 0.29	-	-
Cellotex	20	215	0.047	-	-
Chalk	50	2000	0.093	0.879	1.91
Coal	20	1400	0.186	1.424	0.41
Concrete	20	2300	1.279	1.130	1.77
Cork, sheet	30	190	0.042	1.884	0.42
Firebrick	100	550	0.140	-	-
Glass	20	2500	0.744	0.670	1.60
Ice	0	920	2.249	2.261	3.89
Leather	30	1000	0.159	-	-
Marble	90	2700	1.302	0.419	4.15
Mica	-	290	0.581	0.879	8.20
Mineral wool	50	200	0.047	0.921	0.91
Plaster	20	1680	0.779	-	-
Rubber	0	1200	0.163	1.382	0.353
Sand, dry	20	1500	0.325	0.796	9.85
Sand, wet	20	1650	1.127	2.093	1.77
Slate	100	2800	1.488	-	-
Soil dry	-	1500	0.138	-	-
Soil wet	-	1700	0.657	2.010	0.693
Metals					
Aluminum	0	2670	203.403	0.921	328.0
Brass	0	8600	85.429	0.877	95.0
Copper	0	8800	383.259	0.381	412.9
Iron	20	7220	62.764	0.502	62.5
Lead	0	11400	34.869	0.1298	85.0

Table B-5 Physical Properties of Some Metals and Alloys at 20°C

Material	ρ (kg/m³)	C_p (kJ/kg C)	k (watt/m K)	$\alpha \times 10^6$ (m²/s)
Aluminum	2,702	0.896	236	97.5
Beryllium	1,850	1.750	205	63.3
Chromium	7,160	0.440	91.4	29.0
Copper	8,933	0.383	399	116.6
Gold	19,300	0.129	316	126.9
Iron	7,870	0.452	31.1	22.8
Lead	11,340	0.129	35.3	24.1
Magnesium	1,740	1.017	156	88.2
Manganese	17,290	0.486	7.78	2.2
Molybdenum	10,240	0.251	138	53.7
Nickel	8,900	0.446	91	22.9
Platinum	21,450	0.133	71.4	25.0
Potassium	860	0.741	103	161.6
Silicon	2,330	0.703	153	93.4
Silver	10,500	0.234	427	173.8
Tin	5,750	0.227	67.0	51.3
Titanium	4,500	0.611	22.0	8.0
Tungsten	19,300	0.134	179	69.2
Uranium	19,070	0.113	27.4	12.7
Vanadium	6,100	0.502	31.4	10.3
Zinc	7,140	0.385	121	44.0
Al-Bronze 5Al 95 Cu	8,666	0.410	83	23.3
Brass 70 Cu, 30 Zn	8,522	0.385	111	18.0
Bronze 75 Cu, 25 Sn	8,666	0.343	26	8.6
Iron, cast	7,272	0.420	52	17.0
Iron, wrought	7,849	0.460	59	16.3
Steel, 1C	7,801	0.473	43	11.7
Steel, 1.5C	7,753	0.486	36	9.7
Steel, stainless	7,817	0.461	14.4	4.17

Table B-6 : Physical Properties of Some Liquids

Material	*t* (°C)	ρ (kg/m³)	*k* (watt/m K)	C_P (kJ/kg C)	$\alpha \times 10^4$ (m²/hr)	$\mu \times 10^2$ (kg/hr m)
Ammonia	20	617	0.570	4.731	0.703	-
Ammonia	– 32	665	0.558	4.606	0.656	-
Benzene	0	900	0.151	1.675	0.30	235.205
Gasoline	0	900	1.145	1.800	0.24	-
Petroleum	200	-	0.090	-	-	-
Petroleum oil	200	890	0.080	2.42	0.173	-

Table B-7 : Properties of Liquid Metals

Metal	Melt Pt (°C)	Temp (°C)	k (watt/m K)	ρ (kg/m³)	C_P (kJ/kg K)	μ (kg/m-hr)	Pr -
Bismuth	271.1	300	16.505	10,020	0.144	5.95	0.0144
		500	15.575	9,770	0.152	4.27	0.0116
		700	15.575	9,530	0.162	3.3	0.0097
Lead	327.2	350	16.272	10,500	0.159	8.65	0.0235
		500	15.575	10,350	0.155	6.92	0.019
		700	14.006	10,100	0.155	4.93	0.015
Mercury	– 38.3	15	8.136	13,550	0.138	5.73	0.0274
		150	11.623	13,200	0.138	3.95	0.0124
		300	14.064	12,800	0.134	3.11	0.0084
Potassium	63.9	150	44.981	810	0.796	1.34	0.0066
		400	39.518	740	0.754	0.64	0.0035
		700	33.126	670	0.754	0.46	0.0031
Sodium	97.8	100	86.591	930	1.382	2.51	0.011
		350	72.295	858	1.298	1.04	0.0050
		700	59.626	780	1.256	0.64	0.0038

Appendix C

(Ref. Page 136)

Table C-1 : Values of the Plate Series, P $\left(\frac{\alpha\tau}{S^2}\right)$

$\frac{\alpha\tau}{S^2}$	$P\left(\frac{\alpha\tau}{S^2}\right)$	$\frac{\alpha\tau}{S^2}$	$P\left(\frac{\alpha\tau}{S^2}\right)$	$\frac{\alpha\tau}{S^2}$	$P\left(\frac{\alpha\tau}{S^2}\right)$	$\frac{\alpha\tau}{S^2}$	$P\left(\frac{\alpha\tau}{S^2}\right)$
0.020	1.00000	0.120	0.91755	0.220	0.73668	0.320	0.57776
0.024	0.99999	0.124	0.91072	0.224	0.72968	0.324	0.57211
0.028	0.99996	0.128	0.90379	0.228	0.72274	0.328	0.56651
0.032	0.99984	0.132	0.89675	0.232	0.71584	0.332	0.56097
0.036	0.99961	0.138	0.88963	0.236	0.70930	0.336	0.55548
0.040	0.99919	0.140	0.88244	0.240	0.70220	0.340	0.55004
0.044	0.99850	0.144	0.87519	0.244	0.69546	0.344	0.54466
0.048	0.99750	0.148	0.86788	0.248	0.68877	0.348	0.53932
0.052	0.99614	0.152	0.86065	0.252	0.68214	0.352	0.53404
0.056	0.99439	0.156	0.85318	0.256	0.67555	0.356	0.52881
0.060	0.99222	0.160	0.84580	0.260	0.66903	0.360	0.52363
0.064	0.98962	0.164	0.83840	0.264	0.66256	0.364	0.51850
0.068	0.98661	0.168	0.83100	0.268	0.65614	0.368	0.51342
0.072	0.98318	0.172	0.82360	0.272	0.65078	0.372	0.50838
0.076	0.97936	0.176	0.81622	0.276	0.64347	0.376	0.50340
0.080	0.97516	0.180	0.80884	0.280	0.63722	0.380	0.49846
0.084	0.97061	0.184	0.80848	0.284	0.63103	0.384	0.49357
0.088	0.96572	0.188	0.79414	0.288	0.62489	0.388	0.48873
0.092	0.96052	0.192	0.78684	0.292	0.61881	0.392	0.48394
0.096	0.95504	0.196	0.77956	0.296	0.61278	0.396	0.47919
0.100	0.94930	0.200	0.77231	0.300	0.60680	0.400	0.47449
0.104	0.94330	0.204	0.76510	0.304	0.60088	0.404	0.46983
0.108	0.93715	0.208	0.75793	0.308	0.59502	0.408	0.46522
0.112	0.93078	0.212	0.75080	0.312	0.58921	0.412	0.46066
0.116	0.92424	0.216	0.74372	0.316	0.58346	0.416	0.45614

Table C-1 (Contd.) : Values of the Plate Series, P $\left(\frac{\alpha\tau}{S^2}\right)$

$\frac{\alpha\tau}{S^2}$	$P\left(\frac{\alpha\tau}{S^2}\right)$	$\frac{\alpha\tau}{S^2}$	$P\left(\frac{\alpha\tau}{S^2}\right)$	$\frac{\alpha\tau}{S^2}$	$P\left(\frac{\alpha\tau}{S^2}\right)$	$\frac{\alpha\tau}{S^2}$	$P\left(\frac{\alpha\tau}{S^2}\right)$
0.420	0.45166	0.560	0.31976	0.700	0.22636	1.00	0.1080
0.424	0.44723	0.564	0.31662	0.704	0.22414	1.04	0.0978
0.428	0.44284	0.568	0.31351	0.708	0.22194	1.08	0.0886
0.432	0.43849	0.572	0.31043	0.712	0.21976	1.12	0.0803
0.436	0.43419	0.576	0.30738	0.716	0.21760	1.16	0.0728
0.440	0.42992	0.580	0.30436	0.720	0.21546	1.20	0.0659
0.444	0.42570	0.584	0.30138	0.724	0.21335	1.24	0.0597
0.448	0.42152	0.588	0.29841	0.728	0.21125	1.28	0.0541
0.452	0.41738	0.592	0.29548	0.732	0.20918	1.32	0.0490
0.456	0.41329	0.596	0.29258	0.736	0.20712	1.36	0.0444
0.460	0.40923	0.600	0.28971	0.740	0.20509	1.40	0.0402
0.464	0.40521	0.604	0.28686	0.744	0.20307	1.44	0.0365
0.468	0.40124	0.608	0.28405	0.748	0.20108	1.48	0.0330
0.472	0.39729	0.612	0.28126	0.752	0.19910	1.52	0.0299
0.476	0.39339	0.616	0.27850	0.756	0.19715	1.56	0.0271
0.480	0.38953	0.620	0.27576	0.760	0.19521	1.60	0.0246
0.484	0.38571	0.624	0.27305	0.764	0.19330	1.68	0.0202
0.488	0.38192	0.628	0.27037	0.768	0.19140	1.76	0.0166
0.492	0.37817	0.632	0.26772	0.772	0.18952	1.84	0.0136
0.496	0.37446	0.636	0.26509	0.776	0.18766	1.92	0.0112
0.500	0.37078	0.640	0.26248	0.780	0.18581	2.00	0.0092
0.504	0.36714	0.644	0.25990	0.784	0.18399	2.08	0.0075
0.508	0.36353	0.648	0.25735	0.788	0.18218	2.16	0.0062
0.512	0.35996	0.652	0.25482	0.792	0.18039	2.24	0.0051
0.516	0.35643	0.656	0.25232	0.796	0.17862	2.32	0.0042
0.520	0.35293	0.660	0.24984	0.800	0.17690	2.40	0.0034
0.524	0.24946	0.664	0.24739	0.820	0.16840	2.48	0.0028
0.528	0.34603	0.668	0.24496	0.840	0.16020	2.56	0.0023
0.532	0.34263	0.672	0.24256	0.860	0.15250	2.64	0.0019
0.536	0.33927	0.676	0.24017	0.880	0.14520	2.72	0.0016
0.540	0.33593	0.680	0.23781	0.900	0.13820	2.80	0.0013
0.544	0.33264	0.684	0.23548	0.920	0.13150	2.88	0.0010
0.548	0.32937	0.688	0.23357	0.940	0.12520	2.96	0.0009

Appendix D

(Ref. Page 140)

Table D-1 : Zero and First Order Bessel Functions of the First Kind; $J_0(u)$ and $J_1(u)$

u	$J_0(u)$	$J_1(u)$	u	$J_0(u)$	$J_1(u)$	u	$J_0(u)$	$J_1(u)$
0.0	1.0000	0.0000	3.0	– 0.2600	0.3391	6.0	0.1506	– 0.2767
0.1	0.9975	0.0499	3.1	– 0.2921	0.3009	6.1	0.1773	– 0.2559
0.2	0.9900	0.0995	3.2	– 0.3202	0.2613	6.2	0.2016	– 0.2329
0.3	0.9776	0.1483	3.3	– 0.3453	0.2207	6.3	0.2238	– 0.2081
0.4	0.9604	0.1960	3.4	– 0.3643	0.1792	6.4	0.2433	– 0.1816
0.5	0.9385	0.2423	3.5	– 0.3801	0.1374	6.5	0.2601	– 0.1538
0.6	0.9120	0.2867	3.6	– 0.3918	0.0955	6.6	0.2740	– 0.1250
0.7	0.8812	0.3290	3.7	– 0.3992	0.0538	6.7	0.2851	– 0.0953
0.8	0.8463	0.3688	3.8	– 0.4026	0.0128	6.8	0.2931	– 0.0652
0.9	0.8075	0.4059	3.9	– 0.4018	–0.0272	6.9	0.2981	– 0.0349
1.0	0.7652	0.4400	4.0	– 0.3971	– 0.0660	7.0	0.3001	– 0.0047
1.1	0.7196	0.4709	4.1	– 0.3887	– 0.1033	7.1	0.2991	0.0252
1.2	0.6711	0.4983	4.2	– 0.3766	– 0.1386	7.2	0.2951	0.0543
1.3	0.6201	0.5220	4.3	– 0.3610	– 0.1719	7.3	0.2882	0.0826
1.4	0.5669	0.5419	4.4	– 0.3423	– 0.2028	7.4	0.2786	0.1096
1.5	0.5118	0.5579	4.5	– 0.3205	– 0.2311	7.5	0.2663	0.1352
1.6	0.4554	0.5699	4.6	– 0.2961	– 0.2566	7.6	0.2516	0.1592
1.7	0.3980	0.5778	4.7	– 0.2693	– 0.2791	7.7	0.2346	0.1813
1.8	0.3400	0.5815	4.8	– 0.2404	– 0.2985	7.8	0.2154	0.2054
1.9	0.2818	0.5812	4.9	– 0.2097	– 0.3147	7.9	0.1944	0.2192
2.0	0.2239	0.5767	5.0	– 0.1776	– 0.3276	8.0	0.1716	0.2346
2.1	0.1666	0.5683	5.1	– 0.1443	– 0.3371	8.1	0.1475	0.2476
2.2	0.1104	0.5560	5.2	– 0.1103	– 0.3432	8.2	0.1222	0.2580
2.3	0.0555	0.5399	5.3	– 0.0758	– 0.3460	8.3	0.0960	0.2657
2.4	0.0025	0.5202	5.4	– 0.0412	– 0.3453	8.4	0.0692	0.2708
2.5	–0.0484	0.4971	5.5	– 0.0068	– 0.3414	8.5	0.0419	0.2731
2.6	–0.0968	0.4708	5.6	0.0270	– 0.3343	8.6	0.0146	0.2728
2.7	–0.1424	0.4416	5.7	0.0599	– 0.3241	8.7	– 0.0125	0.2697
2.8	–0.1850	0.4097	5.8	0.0179	– 0.3110	8.8	– 0.0392	0.2641
2.9	–0.2243	0.3754	5.9	0.1220	– 0.2951	8.9	– 0.0652	0.2559

Table D-1 (Contd.) **: Zero and First Order Bessel Functions of the First Kind; $J_0(u)$ and $J_1(u)$**

u	$J_0(u)$	$J_1(u)$	u	$J_0(u)$	$J_1(u)$	u	$J_0(u)$	$J_1(u)$
9.0	– 0.0903	0.2453	11.0	– 0.1712	– 0.1768	13.0	0.2060	– 0.0703
9.1	– 0.1142	0.2324	11.1	– 0.1528	– 0.1913	13.1	0.2129	– 0.0488
9.2	– 0.1368	0.2174	11.2	– 0.1330	– 0.2038	13.2	0.2167	– 0.0271
9.3	– 0.1577	0.2004	11.3	– 0.1121	– 0.2143	13.3	0.2183	– 0.0052
9.4	– 0.1768	0.1816	11.4	– 0.0920	– 0.2224	13.4	0.2177	0.0166
9.5	– 0.1939	0.1613	11.5	– 0.0677	– 0.2284	13.5	0.2150	0.0380
9.6	– 0.2090	0.1395	11.6	– 0.0446	– 0.2320	13.6	0.2101	0.0590
9.7	– 0.2218	0.1166	11.7	– 0.0213	– 0.2333	13.7	0.2032	0.0791
9.8	– 0.2323	0.0928	11.8	0.0020	– 0.2323	13.8	0.1943	0.0984
9.9	– 0.2403	0.0684	11.9	0.0250	– 0.2290	13.9	0.1836	0.1165
10.0	– 0.2459	0.0435	12.0	0.0477	– 0.2234	14.0	0.1711	0.1334
10.1	– 0.2490	0.0184	12.1	0.0697	– 0.2158	14.1	0.1570	0.1488
10.2	– 0.2496	– 0.0066	12.2	0.0908	– 0.2060	14.2	0.1414	0.1626
10.3	– 0.2477	– 0.0313	12.3	0.1108	– 0.1943	14.3	0.1245	0.1747
10.4	– 0.2434	– 0.0555	12.4	0.1296	– 0.1807	14.4	0.1066	0.1850
10.5	– 0.2366	– 0.0788	12.5	0.1469	– 0.1655	14.5	0.0875	0.1934
10.6	– 0.2276	– 0.1012	12.6	0.1626	– 0.1407	14.6	0.0679	0.1998
10.7	– 0.2164	– 0.1224	12.7	0.1766	– 0.1307	14.7	0.0476	0.2043
10.8	– 0.2032	– 0.1422	12.8	0.1887	– 0.1114	14.8	0.0271	0.2066
10.9	– 0.1881	– 0.1604	12.9	0.1988	– 0.0912	14.9	0.0064	0.2069
						15.0	– 0.0142	0.2051

Appendix E

(Ref : page 140)

Table E-1 : First Five Roots, M_n of $J_m(M_n) = 0$

m	M_1	M_2	M_3	M_4	M_5
0	2.4048	5.5201	8.6537	11.7915	14.9309
1	3.8317	7.0156	10.1735	13.3237	16.4706
2	5.1356	8.4172	11.6198	14.7960	17.9598
3	6.3802	9.7610	13.0152	16.2235	19.4094
4	7.5883	11.0647	14.3725	17.6160	20.8269

Appendix F

(Ref. : page 141)

Table F-1 : Values of the Cylinder Series, C $\left(\frac{\alpha\tau}{S^2}\right)$

$\frac{\alpha\tau}{S^2}$	$C\left(\frac{\alpha\tau}{S^2}\right)$	$\frac{\alpha\tau}{S^2}$	$C\left(\frac{\alpha\tau}{S^2}\right)$	$\frac{\alpha\tau}{S^2}$	$C\left(\frac{\alpha\tau}{S^2}\right)$	$\frac{\alpha\tau}{S^2}$	$C\left(\frac{\alpha\tau}{S^2}\right)$	$\frac{\alpha\tau}{S^2}$	$C\left(\frac{\alpha\tau}{S^2}\right)$
0.020	0.99999	0.140	0.69798	0.260	0.35577	0.380	0.17792	0.80	0.0157
0.024	0.99994	0.144	0.68339	0.264	0.34767	0.384	0.17385	0.85	0.0117
0.028	0.99974	0.148	0.66897	0.268	0.33975	0.388	0.16988	0.90	0.0088
0.032	0.99921	0.152	0.65475	0.272	0.33201	0.392	0.16599	0.95	0.0066
0.036	0.99813	0.156	0.64073	0.276	0.32444	0.396	0.16229	1.00	0.0049
0.040	0.99627	0.160	0.62692	0.280	0.31704	0.400	0.1585	1.05	0.0037
0.044	0.99345	0.164	0.61332	0.284	0.30981	0.410	0.1496	1.10	0.0028
0.048	0.98951	0.168	0.59996	0.288	0.30275	0.420	0.1412	1.15	0.0021
0.052	0.98439	0.172	0.58683	0.292	0.29584	0.430	0.1332	1.20	0.0016
0.056	0.97806	0.176	0.57392	0.296	0.28909	0.440	0.1258	1.25	0.0012
0.060	0.97054	0.180	0.56126	0.300	0.28249	0.45	0.1187	1.30	0.0009
0.064	0.96189	0.184	0.54883	0.304	0.26604	0.46	0.1120	1.35	0.0007
0.068	0.95219	0.188	0.53664	0.308	0.26973	0.47	0.1057	1.40	0.0005
0.072	0.94152	0.192	0.52468	0.312	0.26358	0.48	0.0998	1.50	0.0003
0.076	0.93000	0.196	0.51297	0.316	0.25756	0.49	0.0942	1.60	0.0002
0.080	0.91772	0.200	0.50149	0.320	0.25167	0.50	0.0887	1.70	0.0001
0.084	0.90478	0.204	0.49624	0.324	0.24592	0.52	0.0792		
0.088	0.89129	0.208	0.47922	0.328	0.24030	0.54	0.0704		
0.092	0.87733	0.212	0.46844	0.332	0.23481	0.56	0.0628		
0.096	0.86299	0.216	0.45788	0.336	0.22945	0.58	0.0560		
0.100	0.84836	0.220	0.44755	0.340	0.22420	0.60	0.0499		
0.104	0.83349	0.224	0.43743	0.344	0.21908	0.62	0.0444		
0.108	0.81846	0.228	0.42753	0.348	0.21407	0.64	0.0396		
0.112	0.80333	0.232	0.41785	0.352	0.20918	0.66	0.0352		
0.116	0.78813	0.236	0.40838	0.356	0.20440	0.68	0.0314		
0.120	0.77293	0.240	0.39912	0.360	0.19973	0.70	0.0280		
0.124	0.75776	0.244	0.39005	0.364	0.19516	0.72	0.0249		
0.128	0.74265	0.248	0.38119	0.368	0.19070	0.74	0.0222		
0.132	0.72763	0.252	0.37253	0.372	0.18634	0.76	0.0198		
0.136	0.71273	0.256	0.36405	0.376	0.18208	0.78	0.0176		

Appendix G

(Ref. : page 149)

Table G-1 : Values of the Error Function, F(u)

u	F(u)	u	F(u)	u	F(u)	u	F(u)
0.00	0.00000	0.60	0.60386	1.20	0.91031	1.80	0.98909
0.02	0.02256	0.62	0.61941	1.22	0.91553	1.82	0.98994
0.04	0.04411	0.64	0.63459	1.24	0.92059	1.84	0.99074
0.06	0.06762	0.66	0.64938	1.26	0.92524	1.86	0.99147
0.08	0.09008	0.68	0.66378	1.28	0.92973	1.88	0.99216
0.10	0.11246	0.70	0.67780	1.30	0.93401	1.90	0.99279
0.12	0.13476	0.72	0.69143	1.32	0.93806	1.92	0.99338
0.14	0.15695	0.74	0.70468	1.34	0.94191	1.94	0.99392
0.16	0.17901	0.76	0.71754	1.36	0.94556	1.96	0.99443
0.18	0.20094	0.78	0.73001	1.38	0.94902	1.98	0.99489
0.20	0.22270	0.80	0.74210	1.40	0.95228	2.00	0.99532
0.22	0.24430	0.82	0.75381	1.42	0.95538	2.04	0.99609
0.24	0.26570	0.84	0.76514	1.44	0.95830	2.08	0.99673
0.26	0.28690	0.86	0.77601	1.46	0.96105	2.12	0.99728
0.28	0.30788	0.88	0.78669	1.48	0.96365	2.16	0.99775
0.30	0.32863	0.90	0.79691	1.50	0.96610	2.20	0.99814
0.32	0.34913	0.92	0.80677	1.52	0.96841	2.24	0.99846
0.34	0.36936	0.94	0.81627	1.54	0.97059	2.28	0.99874
0.36	0.38933	0.96	0.82542	1.56	0.97263	2.32	0.99897
0.38	0.40901	0.98	0.83423	1.58	0.97455	2.36	0.99916
0.40	0.42839	1.00	0.84270	1.60	0.97635	2.40	0.99931
0.42	0.44747	1.02	0.85084	1.62	0.97804	2.50	0.99959
0.44	0.46622	1.04	0.85865	1.64	0.96962	2.60	0.99976
0.46	0.48466	1.06	0.86614	1.66	0.98110	2.70	0.99987
0.48	0.50275	1.08	0.87333	1.68	0.98249	2.80	0.99993
0.50	0.52050	1.10	0.88020	1.70	0.98379	2.90	0.99996
0.52	0.53790	1.12	0.88679	1.72	0.98500	3.00	0.99998
0.54	0.55494	1.14	0.89308	1.74	0.98613	3.20	0.999994
0.56	0.57162	1.16	0.89910	1.76	0.98719	3.40	0.999998
0.58	0.58792	1.18	0.90484	1.78	0.98817	3.60	0.999999

Appendix H

Table H-1 : Important Dimensionless Numbers Used in Heat Transfer

Name	Symbol	Definition	Significance
Biot Number	Bi	hL/k	Ratio of internal to external thermal resistance for a solid
Coefficient of Friction	C_f	$\tau_\omega / \frac{1}{2} \rho V^2$	Ratio of wall shear stress to kinetic energy storage rates
Fourier number	Fo	$\alpha\tau/L^2$	Ratio of heat conduction to thermal energy storage rates
Friction factor	f	$D\Delta p / \frac{1}{2} \rho V^2 L$	Ratio of pressure drop to kinetic energy of the flow
Grashof number	Gr	$\frac{L^3 g\beta(T_w - T_\infty)}{\nu^2}$	Ratio of buoyancy force to viscous force
Nusselt number	Nu	hL/k	Ratio of convective to conductive heat transfer rates in a fluid
Peclet number	Pe = (Re) (Pr)	VL/α	-
Prandtl number	Pr	$\mu C_p/k = \nu/\alpha$	Ratio of kinematic viscosity to thermal diffusivity of a fluid
Rayleigh number	Ra = (Gr) (Pr)	$\frac{L^3 g\beta(T_w - T_\infty)}{\nu\alpha}$	-
Reynolds number	Re	$VD\rho/\mu$	Ratio of inertia force to viscous force in fluid flow
Stanton number	$St = \frac{Nu}{(Re)(Pr)}$	$h/\rho V C_p$	Ratio of convective heat transfer per unit surface area per unit temperature difference to flow energy change per unit flow area per unit temperature change.

Index

D

I

J

K